# GEOMETRY MEETS SUPERSYMMETRY

# GEOMETRY MEETS SUPERSYMMETRY

Andrei Smilga
University of Nantes, France

World Scientific

NEW JERSEY • LONDON • SINGAPORE • BEIJING • SHANGHAI • HONG KONG • TAIPEI • CHENNAI • TOKYO

*Published by*

World Scientific Publishing Co. Pte. Ltd.
5 Toh Tuck Link, Singapore 596224
*USA office:* 27 Warren Street, Suite 401-402, Hackensack, NJ 07601
*UK office:* 57 Shelton Street, Covent Garden, London WC2H 9HE

Library of Congress Control Number: 2025942256

**British Library Cataloguing-in-Publication Data**
A catalogue record for this book is available from the British Library.

**GEOMETRY MEETS SUPERSYMMETRY**



ISBN 978-981-98-1766-5 (hardcover)
ISBN 978-981-98-1767-2 (ebook for institutions)
ISBN 978-981-98-1768-9 (ebook for individuals)

For any available supplementary material, please visit
https://www.worldscientific.com/worldscibooks/10.1142/14431#t=suppl

Desk Editor: Carmen Teo Bin Jie

Typeset by Stallion Press
Email: enquiries@stallionpress.com

# Preface

This book represents the second corrected and expanded edition of the book [1] published in 2020 by World Scientific.

I have corrected the inevitable inaccuracies and added new material. This primarily concerns the geometry of hyperkähler manifolds, which is now discussed in significantly greater detail. As a result, Chapter 10 has expanded so much that it had to be split into two. Hyperkähler geometry is now discussed in the first half of Chapter 11. New hyperkähler material has also been added to Chapters 3, 12 and 13, and I am indebted to Martin Roček and Vladinir Rubtsov for illuminating discussion of these issues. A discussion of gauge fields on $\mathbb{CP}^n$ manifolds has been added to Chapter 14 (Chapter 13 of the first edition), and the spectra of the corresponding Hamiltonians have been found.

# Contents

## Part 2. PHYSICS 81



## Part 3. SYNTHESIS 171

# Introduction

In 1982, Edward Witten published his seminal paper [2] where he showed that some classical problems of differential geometry and differential topology such as the de Rham complex and Morse theory can be described in a very simple and transparent way using the language of supersymmetric quantum mechanics. However, according to what I could judge from my conversations with mathematicians, the language of supersymmetry has not become yet a common working tool in this branch of mathematics. The experts working in this field prefer to use more traditional methods.

That is a pity because supersymmetry not only allows one to describe in a simple way the results known before, but it also allows one to derive many *new* mathematical results. And this is the *raison d'être* for this book. I try to explain the supersymmetry formalism in a way understandable both for physicists and mathematicians and use it to derive both old and new mathematical results for differential geometry of manifolds.

One can make here a general remark. The relationship of mathematics and theoretical physics is obvious. One can recall that geometry was first developed in ancient Babylon as an applied discipline to serve the needs of farmers and tax assessors. Only later people realized that geometry is not only useful but also beautiful and began to play with geometric problems for their own sake. And still later they were surprised to realize that purely abstract, as it seemed, multidimensional geometric constructions turned out to be useful for description of real physical world.

The dialog between pure maths and theoretical physics is mutually benefitial — there is no need to list here a vast number of examples justifying this point. Unfortunately, this dialog is now often hampered by a difference of scientific languages used by the two communities. It is difficult for a physicist to understand a paper written by a pure mathematician, even if it is written on a subject that they know fairly well, because it is expressed in a different form, to which they are not accustomed. On the other hand, it is difficult for a mathematician to understand papers written by physicists.

At this junction, I would like to make an important comment. I used above the words *physics* and *physicist*. But *what* do they precisely mean? Well, *physics* is understood in a traditional sense: it is the science studying the fundamental properties of the world we live in. And a *physicist* has been traditionally understood as a scientist doing physics and studying these laws. Physicists did so in the time of Galileo and Newton and still did a half-a-century ago. That concerned both theorists and experimentalists. A theorist could use complicated mathematical tools, but the essense of his work was to explain the experimental results and make predictions, which his colleage experimentalist could verify. The supreme judge was the experiment, and all the rest, including the logical rigour of the predictions, was not so important.

And now the situation is very much different. We practically reached the technological limits in building new accelerators, and high-energy physics faces a serious crisis: there are practically no new interesting experiments and the living link between theory and experiment is practically broken. Under these circumstances, many theorists ceased to study nature, hence ceased to be physicists in a traditional sense of the word and are now studying a multitude of many *imaginal* worlds using methods that were developed in the last century to study the physical universe. In the absense of experiment, the logical consistency of the derivations rests the only criterium of truth. In fact, a new branch of mathematics has been created, but it is being developed by people who got their education at the physical departments of universities and are often having an experience of working as "physical physicists". And the mutual Babylonian misunderstanding mentioned above is now, in a considerable extent, the misunderstanding between the scientists working in this new branch of science, one may call it *physical mathematics*, and the mathematicians working in traditional fields.

Such a rift did not exist 200 years ago, while 100 years ago, it was already there but not so deep. It deepened in the middle of the last century after Nicolas Bourbaki published an influential series of monographs in different branches of mathematics written in an extremely austere formal manner. Like it or not, but this rift is now a fact of life that one cannot ignore.

Going back to our book, we have written it in a slang representing a mixture of the two languages. We did so in a hope that this slang will be understandable to both communities.

More precisely:

(1) When studying geometry of a manifold, we are interested in its invariant properties that do not depend on the choice of coordinates. It is possible (and mathematicians do so) to formulate the statements of Riemannian geometry in an austere way without any use of the coordinates. We find it, however, convenient to use in our description *some* coordinate parameterization of a manifold (bearing in mind, of course, that many different such parameterizations are possible). In particular, a *vector field* on the manifold of dimension $D$ will be undestood simply as a set of $D$ functions $V^N$ of the coordinates $x^M$ that transform under general coordinate transformations in an appropriate way. By *affine connection* we understand the object $G^P_{MN}$ carrying three indices, which enters the definition of covariant derivative. $G^P_{MN}$ may be symmetric under the interchange $M \leftrightarrow N$ if there is no torsion or not symmetric if the torsion is present.

(2) Your author is a physicist and speaks maths with a heavy accent. Still, as the subject of this book is mathematical, I will try to organize it in a mathematical way, including theorems, lemmas, etc. though heuristic reasoning and clarifying examples also play an essential role. I do so (though not quite consistently) even when describing purely physical issues such as classical mechanics. All the notions (except the most elementary) are defined in this book so that both the physicists and mathematicians will be able to understand what am I talking about. A warning: these definitions will inevitably have a distinct physical flavour. And the proofs of the theorems that we give are not incorrect, but they will lack an austere mathematical rigour. First of all, being not a mathematician, I am simply not able to provide it and, second, our goal is to be understandable to the both communities.

The basic idea of Witten was the following. Consider a smooth manifold of dimension $D$ parameterized by the coordinates $x^M$. Consider the classical de Rham complex, i.e. the set of all $p$-forms,

$$\alpha^{(p)} = \alpha_{M_1 \ldots M_p}(x^N)\, dx^{M_1} \wedge \cdots \wedge dx^{M_p}, \tag{0.1}$$

$p = 0, \ldots, D$. One defines the operator of exterior derivative $d$:

$$d\alpha = \partial_N \alpha_{M_1 \ldots M_p}\, dx^N \wedge dx^{M_1} \wedge \cdots \wedge dx^{M_p} \tag{0.2}$$

and the conjugate operator $d^\dagger$ that acts on $p$-forms as $(-1)^{pD+D+1} \star d\star$, where $\star$ is the duality operator (see Chapter 1 for further details).

The operators $d$ and $d^\dagger$ are nilpotent, while the anticommutator $\{d, d^\dagger\}$ is a certain second-order differential operator. In fact, it is (up to a sign) the Laplace–Beltrami operator $\triangle$, generalized to operate on differential forms.

One can now observe that the algebra

$$d^2 = (d^\dagger)^2 = 0, \quad \{d, d^\dagger\} = -\triangle \tag{0.3}$$

is isomorphic to the simplest supersymmetry algebra

$$\boxed{\hat{Q}^2 = (\hat{Q}^\dagger)^2 = 0, \quad \{\hat{Q}, \hat{Q}^\dagger\} = \hat{H},} \tag{0.4}$$

where $\hat{Q}$ and $\hat{Q}^\dagger$ are Hermitially conjugate supercharges and $\hat{H}$ is the Hamiltonian. It follows that $[\hat{Q}, \hat{H}] = [\hat{Q}^\dagger, \hat{H}] = 0$ so that a system involves one complex or two real integrals of motion. In this language, a $p$-form (0.1) is interpreted as a *wave function* belonging to the Hilbert space where the supercharges and the Hamiltonian act. This wave function depends on the ordinary real dynamical variables $x^M$ and also on the *Grassmann* anticommuting variables $\psi^M$. Any wave function

$$\Psi(x^M, \psi^M) \tag{0.5}$$

can be expanded into a Taylor series in $\psi^M$. It is noteworthy that this series is finite, as the product of $D+1$ anticommuting factors $\prod_{i=1}^{D+1} \psi^{M_i}$ includes at least one factor entering twice and vanishes. Thus, a $p$-form (0.1) maps into a wave function (0.5) involving a product of $p$ Grassmanian factors.

The usefulness of this mapping stems from two remarks:

(1) It allows one to use the whole bunch of physical methods that were developed since Newton and Schrödinger to study dynamical systems.
(2) All dynamical systems relevant for geometric applications are supersymmetric. And one can use powerful methods (in particular, the *superspace* and *superfield* formalism) developed by physicists during half a century that passed since 1971 when supersymmetry was born [3].

One should note the following. Supersymmetry was discovered and mostly studied then by physicists as a new interesting symmetry of *field theory* systems. However, for geometrical applications, we need to deal only with supersymmetric quantum mechanical (SQM) systems where the dynamical variables depend only on time but not on spatial coordinates, as they also do in field theories.[1]

Equation (0.4) describes the simplest SQM algebra with one complex or two real supercharges. We see later that this algebra is relevant for the description of arbitrary real manifolds equipped with the de Rham differential and its conjugate and also of arbitrary complex manifolds equipped with the Dolbeault differential and its conjugate. But the supersymmetry algebra can be richer and involve more than two real supercharges. We will see that such *extended* supersymmetries are relevant for the description of certain special classes of the manifolds, in particular, of the *Kähler* and *hyperkähler* manifolds and also the so-called HKT manifolds.[2] All these and some other manifolds are discussed at length later.

The plan of this book is the following.

The part *Geometry* includes mathematical preliminaries — some standard facts of the theory of smooth manifolds and is mainly addressed to physicists. In particular, we briefly describe Riemannian geometry (though a physicist who studied general relativity should already be familiar with this subject) and the de Rham complex, then complex geometry and the Dolbeault complex, and finally the hyperkähler and HKT manifolds (the latter subject is less known and the corresponding section may also represent an interest to mathematicians).

*Physics* provides the reader with a necessary information concerning ordinary and supersymmetric dynamical systems. We start with a brief review of the ordinary analytical mechanics and of quantum mechanics. In principle, this is studied at not only physical but also mathematical departments in good universities, but my experience tells me that many my mathematical colleagues are not so well familiar with these issues. What is definitely not studied at the universities are dynamical systems with Grassmann variables. We explain what they are in the second part of Chapter 4.

And then we go over to supersymmetric systems. In Chapter 5, a supersymmetric quantum system is defined as a system admitting the algebra (0.4) (a mathematical definition) or as a system where all excited eigenstates in the spectrum of the Hamiltonian are double degenerate (a physical

[1] For this reason, such independent spatial coordinates will not or almost will not appear in our book. Still, to prevent a possible confusion, we note that they have nothing to do with the dynamical variables $x^M(t)$.

[2] HKT stands for "hyperkähler with torsion".

definition). We discuss the simplest physical example of such a system (it is the motion of an electron in a homogeneous magnetic field — the problem solved by Landau almost hundred years ago [4]), some other simple examples, and unravel in Chapter 7 their mathematical structure introducing the notion of *superspace* (for quantum-mechanical system, it is rather "supertime") and one-dimensional *superfields* (or "supervariables"). Finally, we introduce one-dimensional *harmonic* superspace. This notion is indispensable for the description of generic HKT manifolds. In Chapter 6, we review the path integral approach to ordinary and supersymmetric systems. The important notion of the *Witten index* is introduced.

The part *Synthesis* is central in this book. We go back to the geometric structures discussed in the first part but describe now the known and not so well known new results in supersymmetric language. Chapter 8 is devoted to the de Rham complex (we also touch upon Morse theory, which can be explained in very simple terms using supersymmetry). Chapter 9 is devoted to the Dolbeaux complex. Among other things, we give there the supersymmetric interpretation of the Newlander–Nirenberg theorem and show that the Dolbeaux complex can be well defined for not only complex manifolds but also certain noncomplex even-dimensional manifolds, such as $S^4$.

In Chapters 10 and 11, we discuss SQM systems enjoying *extended* supersymmetry with several pairs of Hermitially conjugated supercharges $(\hat{Q}_j, \hat{Q}_j^\dagger)$. These models describe Kähler, hyperkähler and HKT manifolds as well as some other types of manifolds (quasicomplex, spinorial, bi-spinorial, bi-Kähler and bi-HKT) that have not yet attracted much attention of mathematicians.

In Chapter 12, we establish a genetic relationship between the different kinds of models (and hence between the different kinds of manifolds). We show in particular that any supersymmetric model described in Chapters 8–11 can be derived from a trivial noninteracting flat space model by applying two operations: (i) similarity transformation of the supercharges and (ii) Hamiltonian reduction. At the end of this chapter, we discuss the procedures of Kähler and hyperkähler reduction, which are related but not identical to Hamiltonian reduction. The hyperkähler reduction allows one to construct many nontrivial hyperkähler geometries.

Chapter 13 is devoted to generic hyperkähler and HKT manifolds described using the harmonic superspace language. We show that, by the same token as any Kähler metric is derived from the Kähler potential, $h_{j\bar{k}} = \partial_j \partial_{\bar{k}} \mathcal{K}(z^n, \bar{z}^n)$, any hyperkähler metric can be derived from the harmonic prepotential $\mathcal{L}^{+4}$. Similarly, the data including the pair of prepotentials $(\mathcal{L}^{+3}, \mathcal{L})$ provide us at the end of a long day[3] with a particular HKT metric.

In Chapter 14, we give a detailed discussion of the gauge fields living on manifolds and the associated supersymmetric models. In mathematical description, gauge fields represent connections of principal fiber bundles. Such a bundle is characterized by an integer topological charge (the integral of an appropriate *Chern class*). On the physical side, it is possible to consider systems with fractional topological charges (in particular, a model living on $S^2$ with fractional magnetic flux), but such systems are not supersymmetric.

The last chapter is devoted to the Atiyah–Singer theorem that relates the index of certain elliptic operators to topological characteristics of the manifold where they are defined. We show that this index can be interpreted as the Witten index of an appropriate SQM system and calculate it in some nontrivial cases.

## Our inconsistent conventions

- To distinguish the ordinary functions from the operators, the latter will be in most cases equipped with hats: $H$ is a classical Hamiltonian and $\hat{H}$ is a quantum Hamiltonian. But sometimes we will not put a hat: $d$ and $d^\dagger$ in Eq. (0.3) are operators, even though they do not carry hats.
- We mostly use the bar for the complex conjugation of ordinary functions or Grassmann functions, $A \to \bar{A}$. In some cases, the star $A^*$ will also be used but only for the ordinary functions.
- The Hermitially conjugated operators will be denoted as $\hat{\bar{A}}$ but sometimes also by $\hat{A}^\dagger$.
- The real tensor world indices will be denoted by capital latin letters: $M, N$, etc.[4] For the corresponding tangent space indices, we take the letters from the beginning of the alphabet: $A, B$, etc. The world indices are written at two levels: this distingushes covariant and contravariant vectors and tensors. The position of the tangent space indices does not

[3]The procedure here is essentially more complicated than for the Kähler manifolds where the metric is obtained simply by differentiating the Kähler potential $\mathcal{K}$.

[4]But in the discussion of four-dimensional manifolds, we will usually (not always) endow the coordinates with Greek indices $\mu, \nu, \ldots$, rejoicing over the fact that the world we live in has four dimensions and following the conventions adopted in physical literature.

have a mathematical significance and will be chosen exclusively out of aesthetic considerations.

- In complex spaces, the holomorphic world tensor indices will be denoted by small letters: $m, n, \ldots$ The antiholomorphic indices will be denoted as $\bar{m}, \bar{n}, \ldots$ The corresponding tangent space indices are $a, b, \ldots; \bar{a}, \bar{b}, \ldots$ Their position carries some significance and is not quite arbitrary (see the footnote on p. 28).
- In some cases (but not always), one of the bars in the expressions like $\bar{V}^{\bar{n}}$ will be skipped and we will write $V^{\bar{n}}$ or $\bar{V}^n$ instead.

## Aknowledgements

I am deeply indebted to my Dubna collaborators: Evgeny Ivanov and Sergei Fedoruk, without whom I would not be able to derive many results reported in this book. Sergei Fedoruk, Zohar Komargodski, George Papadopoulos and Alexei Rosly took the pain to read the manuscript or its parts at the preliminary stage and made a lot of valuable comments. It is also a pleasure for me to appreciate the help of Boris Smilga in drawing the figures. Finally, I would like to thank the whole World Scientific team, and especially Lakshmi Narayanan, who was the editor of the first edition, and Carmen Teo Bin Jie, the editor of this edition, for encouragement and support.

# PART 1

# GEOMETRY

# Chapter 1

# Real Manifolds

This and the following two chapters of this book are addressed to a reader-physicist. We can also recommend to him two excellent reviews by T. Eguchi, P.B. Jilkey and A.J. Hanson [5] and P. Candelas [6].

## 1.1 Riemannian geometry

Consider a $D$-dimensional manifold $\mathcal{M}$ without boundaries. A neighbourhood[1] of each point on $\mathcal{M}$ is homeomorphic to $\mathbb{R}^D$.

Any such manifold can be represented as a union

$$\mathcal{M} = \bigcup_\alpha U_\alpha \tag{1.1}$$

of a finite number of open subsets $U_\alpha \subset \mathcal{M}$, each of these subsets being homeomorphic to $\mathbb{R}^D$. For example, the surface of the Earth can be represented as a union of two open disks: one of them has a center at the North Pole and extends down to the Tropic of Capricorn and the second is centered at the South Pole and extends up to the Tropic of Cancer.

Introduce the natural Cartesian coordinates on $\mathbb{R}^D$. Then, an invertible map $U_\alpha \leftrightarrow \mathbb{R}^D$ provides us with a certain *coordinate chart* on $U_\alpha$ — each point of $U_\alpha$ is characterized by a set of coordinates $\{x^M\}$. There are many different such maps and, correspondingly, many different charts $\{x^M\}$. The transitions from one chart into another are called *reparameterizations* or *general coordinate transfromations.*

The sets $U_\alpha$ may overlap. The intersection $U_\alpha \cap U_\beta$ belongs thus to two different charts (a map of Europe and a map of Russia both contain Yalta)

[1] We said "neighbourhood" and this means that our manifold is *metric*, i.e. a distance between any two of its points is defined. Mathematicians like to play with more general not necessarily metric manifolds, but we will not do so in our book.

and is described by two different sets of coordinates $\{x^M_{(\alpha)}\}$ and $\{x^M_{(\beta)}\}$. One set can be expressed into another,

$$x^N_{(\alpha)} = f^N_{\alpha\beta}(x^M_{(\beta)}). \tag{1.2}$$

The functions $f^N_{\alpha\beta}$ are called the *transition functions.* If three charts $U_\alpha$, $U_\beta$, $U_\gamma$ have a nonempty intersection, the transition functions should satisfy there an obvious condition $f_{\alpha\beta} \circ f_{\beta\gamma} = f_{\alpha\gamma}$. A set of all charts is called an *atlas* of the manifold.

**Definition 1.1.** A *smooth* manifold is a manifold where the transition functions have an infinite number of derivatives.

**Definition 1.2.** An *orientable* manifold is a manifold where one can choose the charts such that all the transition Jacobians $|\partial_M f^N_{\alpha\beta}|$ are positive.

A well-known example of a nonorientable manifold is the so-called Klein bottle. But we will be interested only in the smooth orientable manifolds in our book.

One can define scalar, vector and tensor fields living on $\mathcal{M}$. A scalar field depends only on a manifold point and is invariant under reparameterizations. There are two types of vectors.

**Definition 1.3.** A set of $D$ functions of coordinates that transform under reparameterizations in the same way as $dx^M$,

$$V'^M = \frac{\partial x'^M}{\partial x^N} V^N, \tag{1.3}$$

is called a *contravariant vector.*

**Definition 1.4.** A set of $D$ functions of coordinates that transform under reparameterizations in the same way as the gradient operator $\partial_M$,

$$V'_M = \frac{\partial x^N}{\partial x'^M} V_N, \tag{1.4}$$

is called a *covariant vector.*

Following the usual convention, we wrote the index for the contravariant vector upstairs and for the covariant vector downstairs. The summation over the repeated indices is assumed. If $\mathcal{M}$ is the flat Euclidean space $\mathbb{R}^D$ and we limit our consideration to the transformations representing rigid rotations and translations, there is no distinction between these two types of vectors.

We can further consider tensors with an arbitrary number of covariant or contravariant indices. For example, a tensor $T_N^M$ of rank $(1,1)$ is defined as a set of $D^2$ functions of the coordinates that transform under reparameterizations as

$$(T_N^M)' = \frac{\partial x'^M}{\partial x^P} \frac{\partial x^Q}{\partial x'^N} T_Q^P. \tag{1.5}$$

The generalization to an arbitrary case is obvious. The product of two tensors also represents a tensor. This product can keep all the indices of the factors or some of them may be contracted. We leave to the reader the proof of the fact that, if $T_N^M$ and $S_N^M$ are tensors [i.e. transform as in (1.5)], the object $R_N^M = T_P^M S_N^P$ is also a tensor, while $T_N^M S_M^N$ is a scalar.

Metric smooth manifolds are also called *Riemannian*. All the manifolds to be discussed in this book are Riemannian. If the distance between two points $X$ and $Y$ is small enough, one can assume that these points belong to the same chart $U_\alpha$ and their coordinates are close to each other. Let $ds$ be the distance between the points $X$ and $X + dX$ characterized by the coordinates $x^M$ and $x^M + dx^M$. Generically, $ds^2$ represents a quadratic form

$$ds^2 = g_{MN}(x^P) dx^M dx^N. \tag{1.6}$$

In our book, we consider only the manifolds with $ds^2 > 0$ for nonvanishing $dx^M$. But the reader understands that the latter condition is not fulfilled for flat Minkowski space or curved space with Minkowskian signature — the space we live in.

Bearing in mind that $ds^2$ is a scalar, we conclude that $g_{MN}$ is a symmetric covariant tensor of the second rank. It is called the *metric tensor.* One can define also the inverse metric tensor $g^{MN}$ satisfying $g_{MN} g^{NP} = \delta_M^P$. For any contravariant vector $V^M$, the vector $V_M = g_{MN} V^N$ is covariant. For any covariant vector $V_M$, the vector $V^M = g^{MN} V_N$ is contravariant.

**Theorem 1.1.** *The object*

$$E_{M_1 \dots M_D} = \sqrt{g}\, \varepsilon_{M_1 \dots M_D}, \tag{1.7}$$

*where* $g = \det(g_{MN})$ *and* $\varepsilon_{M_1 \dots M_D}$ *is totally antisymmetric with normalization* $\varepsilon_{1 \dots D} = 1$, *is an invariant tensor* (i.e. *it preserves its form under reparameterizations*).

**Proof.** Any antisymmetric tensor of rank $D$ has the form $T_{M_1 \dots M_D} = f(x) \varepsilon_{M_1 \dots M_D}$. The tensor nature of $T_{M_1 \dots M_D}$ dictates its transformation law:

$$T'_{M_1\ldots M_D} = \frac{\partial x^{N_1}}{\partial x'^{M_1}} \cdots \frac{\partial x^{N_D}}{\partial x'^{M_D}} T_{N_1\ldots N_D}, \tag{1.8}$$

which means that

$$f'(x') = Jf(x), \tag{1.9}$$

where

$$J = \frac{\partial(x^1,\ldots,x^D)}{\partial(x'^1,\ldots,x'^D)}$$

is the Jacobian of the transformation. Based on the invariance of (1.6), it is not difficult to check that (1.9) holds iff $f(x) = \sqrt{g(x)}$, and hence $T_{M_1\ldots M_D}$ coincides with (1.7). □

Our next task is to learn how to differentiate tensor fields. The gradient of a scalar $\partial_M \phi$ is a covariant vector, but the object $\partial_N V^M$ is not a tensor, it does not transform according to (1.5).

**Definition 1.5.** The *covariant derivative* of the contravariant vector field $V^M(x)$ is defined as

$$\boxed{\nabla_N V^M = \partial_N V^M + G^M_{NP} V^P} \tag{1.10}$$

with the condition that $\nabla_N V^M$ transforms as a tensor.

The object $G^M_{NP}$ will be called in this book the *connection*.[2] While $\nabla_N V^M$ is a tensor, the connection is not. Its transformation law is

$$(G^M_{NP})' = \frac{\partial x'^M}{\partial x^S}\left(\frac{\partial x^R}{\partial x'^N}\frac{\partial x^Q}{\partial x'^P}G^S_{RQ} + \frac{\partial^2 x^S}{\partial x'^N \partial x'^P}\right). \tag{1.11}$$

**Definition 1.6.** The covariant derivative of the covariant vector field $V_M(x)$ is defined as

$$\boxed{\nabla_N V_M = \partial_N V_M - G^P_{NM} V_P,} \tag{1.12}$$

with the same $G^P_{NM}$ as in (1.10).

It is also a tensor *(prove it)*.

[2]This definition is convenient for us. Mathematicians usually define it somewhat differently: for them the whole covariant derivative (1.10) is the affine connection.

The generalization of the above definitions for an arbitrary tensor is quite straightforward. For example,

$$\nabla_P T_N^M = \partial_P T_N^M + G_{PQ}^M T_N^Q - G_{PN}^Q T_Q^M. \qquad (1.13)$$

The "physical meaning" of the covariant derivatives (1.10) and (1.12) is the following. The ordinary partial derivatives give the difference of the values of the function in neighbouring points: $f(x + dx) - f(x) \approx (\partial_M f) dx^M$. The presence of the second term in (1.10) indicates that before being subtracted from $V^M(x + dx)$, the vector $V^M(x)$ should be modified a little bit:

$$V^M(x) \to \tilde{V}^M_{\Rightarrow x+dx}(x) = V^M(x) - G_{NP}^M V^P dx^N + o(dx). \qquad (1.14)$$

This modification can be interpreted as a *parallel transport* of the vector $V^M$ from the point $x$ to the point $x + dx$.

Similarly, the presence of the second term in (1.12) indicates a parallel transport of the covariant vector $V_M$:

$$V_M(x) \to \tilde{V}_{M,\Rightarrow x+dx}(x) = V_M(x) + G_{NM}^P V_P dx^N + o(dx). \qquad (1.15)$$

The origin of the term "connection" is now clarified. The connection $G_{NP}^M$ specifies a particular way that the different points of the manifold are connected by the operator of the parallel transport. A lot of different choices for $G_{NP}^M$ are possible. A distinguished role is played by the *Levi-Civita connection*[3] $\Gamma_{NP}^M$.

It satisfies two conditions:

(1) The covariant derivative of the metric tensor vanishes.
(2) The connection is symmetric, $\Gamma_{NP}^M = \Gamma_{PN}^M$.

The first condition (the *metricity condition*) actually means that, after a parallel transport to the point $x + dx$, the vectors $\tilde{V}^M$ and $\tilde{V}_M$ are still related to each other by the metric tensor at the point $x + dx$:

$$\tilde{V}_M(x + dx) = g_{MN}(x + dx)\tilde{V}^N(x + dx).$$

Or speaking differently, the contravariant vector $V^M$ is rotated a little bit after an infinitesimal parallel transport, but its length is not changed:

$$g_{MN}(x + dx)\tilde{V}^M(x + dx)\tilde{V}^N(x + dx) = g_{MN}(x)V^M(x)V^N(x). \qquad (1.16)$$

The same is true for scalar products.

[3]Another name for $\Gamma_{NP}^M$ is *Christoffel symbols.*

When the second condition is also imposed, the connection is determined uniquely and reads

$$\Gamma^M_{NP} = \frac{1}{2} g^{MQ} \left(\partial_N g_{QP} + \partial_P g_{NQ} - \partial_Q g_{NP}\right). \tag{1.17}$$

People say that this connection is *torsionless*.

**Definition 1.7.** The *torsion* is the antisymmetric part of the generic connection $G^M_{NP}$,

$$C^M_{NP} = G^M_{NP} - G^M_{PN}. \tag{1.18}$$

**Remark:** In contrast to $G^M_{NP}$, the torsion is a tensor. Indeed, the unhandy term with second derivatives in the transformation law (1.11) is symmetric in the indices $N, P$ and disappears after antisymmetrization.

An important local characteristics of a manifold is its *curvature*. We will be interested not with the extrinsic curvature that depends on embedding but with the invariant intrinsic curvature (for example, the intrinsic curvature vanishes for a cylinder but not for a sphere).

The ordinary Riemannian curvature is defined by inspecting the rotation of a vector after its parallel transport (defined with $G^M_{NP} = \Gamma^M_{NP}$) along a small closed contour. To understand that the vector *can* be rotated after this, the reader is welcome to meditate over Fig. 1.1.

For two-dimensional surfaces, it does not matter what was the original vector direction and how the contour in the manifold is oriented. But for higher dimensions, it does. Consider a parallel transport of $V^M$ along a

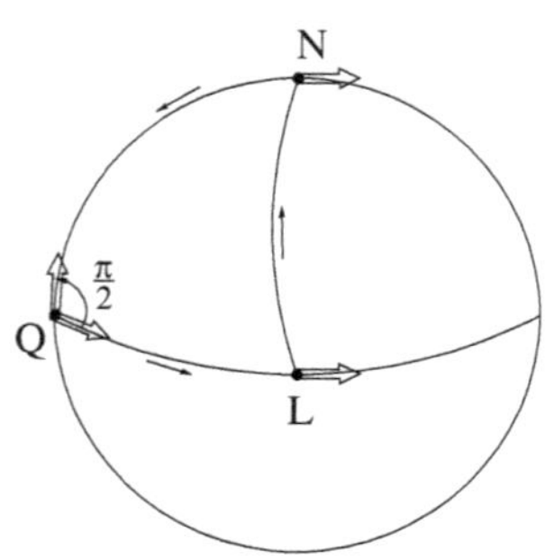


**Fig. 1.1.** Rotation of a tangent vector after going around a closed circuit on the Earth surface. $N$ is the North Pole and $Q$ and $L$ represent Quito in Ecuador and Libreville in Gabon — these cities lie almost at the equator and their longitudes differ by $\sim 90°$.

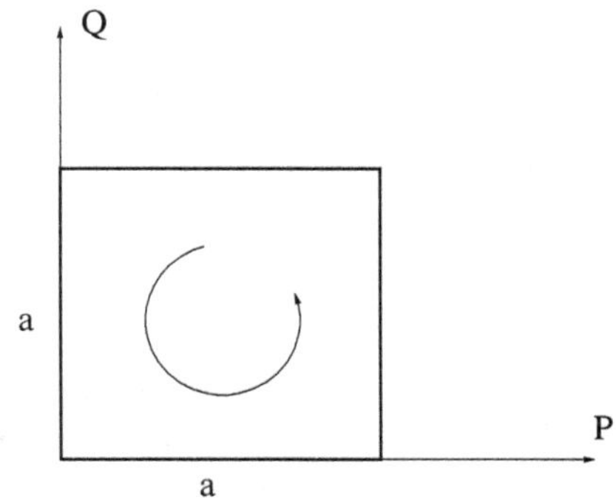


**Fig. 1.2.** Parallel transport along a small square.

small square lying in the plane $(PQ)$ (see Fig. 1.2). Bearing in mind (1.14) and its relationship to (1.10), it is easy to see that the vector is shifted by

$$\Delta V^M = -a^2 [\nabla_P, \nabla_Q] V^M, \tag{1.19}$$

where $a$ is the size of the square.[4] Now, $\nabla_P V^M$ is a tensor, hence $[\nabla_P, \nabla_Q] V^M$ is also a tensor. The latter can be presented as

$$[\nabla_P, \nabla_Q] V^M = R^M{}_{NPQ} V^N. \tag{1.20}$$

The object $R^M{}_{NQP}$ is called the *Riemann curvature tensor.* It follows from (1.20) and (1.10) that

$$\boxed{R^M{}_{NPQ} = \partial_P \Gamma^M_{QN} - \partial_Q \Gamma^M_{PN} + \Gamma^M_{PR} \Gamma^R_{QN} - \Gamma^M_{QR} \Gamma^R_{PN}.} \tag{1.21}$$

It is convenient to lower its first index and consider the totally covariant tensor of the 4th rank $R_{MNQP}$, which can be represented as

$$\begin{aligned} R_{MNQP} = {} & \frac{1}{2} \left( \partial_M \partial_P g_{NQ} + \partial_N \partial_Q g_{MP} - \partial_M \partial_Q g_{NP} - \partial_N \partial_P g_{MQ} \right) \\ & + g_{ST} \left( \Gamma^S_{MP} \Gamma^T_{NQ} - \Gamma^S_{MQ} \Gamma^T_{NP} \right). \end{aligned} \tag{1.22}$$

It has the following properties:

$$\begin{aligned} R_{MNQP} = -R_{NMQP} = -R_{MNPQ} = R_{QPMN}, \\ R_{MNQP} + R_{MQPN} + R_{MPNQ} = 0. \end{aligned} \tag{1.23}$$

---

[4]For a small flat contour of another form, $a^2$ is substituted by the area of soap film stretched over the contour.

For a $D$-dimensional manifold, the Riemann tensor has $D^2(D^2-1)/12$ independent components.

The second line in Eq. (1.23) is the first *Bianchi identity*. And the second Bianchi identity reads

$$\nabla_S R_{MPNQ} + \nabla_N R_{MPQS} + \nabla_Q R_{MPSN} = 0. \tag{1.24}$$

It immediately follows from the definition (1.20) and from the property $\nabla_S g_{MN} = 0$.

Besides the ordinary Riemann tensor, one can define a curvature tensor for any connection $G^M_{NP}$. It is given by the same formula (1.21), but with $G^M_{NP}$ substituted for $\Gamma^M_{NP}$. A situation when the ordinary Riemann tensor (1.21) vanishes (so that the manifold is flat in the standard meaning of this word), but some other curvature tensor does not vanish or vice versa, is quite possible and not uncommon.

The contractions of the Riemann tensor give the *Ricci tensor* $R_{MQ} = R_{MNQ}{}^N$ and the *Riemann scalar curvature* $R = R_M{}^M$. For two-dimensional surfaces, $R$ coincides with the *Gaussian curvature* $\mathbb{R}$ (the inverse product of the principal radii of the tangent ellipsoid), multiplied by 2.

## 1.2 Differential forms

Consider a special class of tensor fields, the totally antisymmetric covariant tensors $T_{[M_1 \cdots M_p]}$. These fields play a distinguished very important role in many mathematical and physical applications. A nice fact is that one can describe these fields using a convenient compact notation not explicitly displaying the indices.[5] For example, instead of a covariant vector $V_M$, one can consider the linear form $V = V_M dx^M$, where $dx^M$ is an infinitesimal displacement.[6]

For an antisymmetric covariant tensor of the second rank $T_{[MN]}$ or higher ranks, we meet some problem, however. The contraction $T_{[MN]} dx^M dx^N$ is simply zero. But one can define the so-called *wedge product* $dx^M \wedge dx^N$, which is antisymmetric under the permutation of the differentials, $dx^M \wedge dx^N = -dx^N \wedge dx^M$. Likewise, $dx^{M_1} \wedge \cdots \wedge dx^{M_p}$ is totally

[5]The plethora of indices is an inherent feature of the Riemann formalism described above. This makes explicit calculations (for example, the calculations in general relativity) difficult. Sometimes, one has to live with it, but for antisymmetric tensors, the life can be considerably simplified.

[6]Let $F_M$ be the field of forces that act on a particle moving along the manifold. Then, $F_M dx^M$ is the *work* done by the force $F_M$ as the particle moves from $x^M$ to $x^M + dx^M$.

antisymmetric in $M_1, \ldots, M_p$. With this definition, $T = T_{[MN]} dx^M \wedge dx^N$, etc. do not vanish. On a $D$-dimensional manifold, one can define such forms of degree $p$ varying from 0 (for the scalars) up to $D$.

The forms of degrees $p$ and $D - p$ are interrelated by the *Hodge duality transformation* $\star$.

**Definition 1.8.** If

$$\alpha = \alpha_{M_1 \cdots M_p} dx^{M_1} \wedge \cdots \wedge dx^{M_p}, \tag{1.25}$$

then

$$\boxed{\star\alpha = \frac{1}{(D-p)!} E^{M_1 \ldots M_p}{}_{M_{p+1} \ldots M_D} \alpha_{M_1 \cdots M_p} dx^{M_{p+1}} \wedge \cdots \wedge dx^{M_D},} \tag{1.26}$$

where $E_{M_1 \ldots M_D}$ is the covariant totally antisymmetric tensor defined in (1.7).

With this definition, the property

$$\star\star\, \alpha_p = (-1)^{p(D-p)} \alpha_p \tag{1.27}$$

holds.

Thus, in ordinary three-dimensional flat space, the magnetic field can be described either by the vector $B_i$ or by the dual antisymmetric tensor[7] $F_{ij} = \frac{1}{2}\varepsilon_{ijk} B_k$.

The dual of the *invariant volume* form[8]

$$\alpha_V = \sqrt{g} dx^1 \wedge \cdots \wedge dx^D \tag{1.28}$$

is the constant scalar $\alpha_0 = 1$.

The set of all forms is obviously a vector space. But besides adding, one can also multiply forms using the wedge product. Clearly, $\alpha_p \wedge \beta_q$ is a (p+q)-form. If $p + q > D$, it vanishes.

[7]In this physical example, we use the physical convention for the spatial indices.

[8]If the manifold is compact, the integral

$$\int_{\mathcal{M}} \alpha_V = \int_{\mathcal{M}} \sqrt{g} \prod_{N=1}^{D} dx^N$$

gives its volume.

In addition, with the definition (1.26) in hand, we introduce the *inner product*. It is usually defined only for the forms of the same degree $p$ and reads[9]

$$(\alpha_p, \beta_p) = \int_{\mathcal{M}} \alpha_p \wedge \star\bar{\beta}_p = p! \int_{\mathcal{M}} \alpha_{M_1\cdots M_p} \overline{\beta^{M_1\cdots M_p}} \sqrt{g} \prod_{N=1}^{D} dx^N. \tag{1.29}$$

Note that $(\alpha_p, \alpha_p) \geq 0$ and $(\star\alpha_p, \star\alpha_p) = (\alpha_p, \alpha_p)$.

One can also define the interior product $\langle V, \alpha_p \rangle$ when the first factor is a 1-form, $V = V_M dx^M$, and $p \geq 1$. In fact, we are rather dealing here not with a product but with an operator $i_V$ that acts in the space of forms and downgrades the degree of form by 1.

**Definition 1.9.** If

$$\alpha = \alpha_{M_1,\ldots,M_p} dx^{M_1} \wedge \ldots \wedge dx^{M_p},$$

then

$$\langle V, \alpha \rangle \equiv i_V \alpha = p V^N \alpha_{N M_2 \ldots M_p} dx^{M_2} \wedge \ldots \wedge dx^{M_p}. \tag{1.30}$$

The following property holds:

$$i_V(\alpha_p \wedge \beta) = (i_V \alpha_p) \wedge \beta + (-1)^p \alpha_p \wedge (i_V \beta). \tag{1.31}$$

The forms can also be differentiated. We define the exterior derivative operator $d$ as follows [it was already written in (0.2), but we repeat it here]:

**Definition 1.10.** If

$$\alpha = \alpha_{M_1 \ldots M_p} dx^{M_1} \wedge \cdots \wedge dx^{M_p},$$

then

$$d\alpha = \partial_N \alpha_{M_1 \ldots M_p} dx^N \wedge dx^{M_1} \wedge \cdots \wedge dx^{M_p}. \tag{1.32}$$

The exterior derivative of a $p$-form is a $(p+1)$-form. The operation (1.32) is a generalization of the ordinary curl. In flat three-dimensional space,

$$A_i dx^i \xrightarrow{d} \partial_i A_j dx^i \wedge dx^j \xrightarrow{\star} (\varepsilon_{ijk} \partial_i A_j) dx^k. \tag{1.33}$$

[9] In most mathematical textbooks, the coefficients $\alpha_{M_1\cdots M_p}$ of the differential forms on the real manifolds are assumed to be real. But they may also be assumed to be complex and that is what we will do, anticipating the mapping of the forms to the wave functions of an appropriate supersymmetric quantum mechanical model to be discussed in Chapter 8.

The operator $d$ is nilpotent, $d^2\alpha = 0$. For two forms $\alpha_p$ and $\beta_q$, the property

$$d(\alpha_p \wedge \beta_q) = d\alpha_p \wedge \beta_q + (-1)^p \alpha_p \wedge d\beta_q \tag{1.34}$$

holds.

**Remark:** Note that one could replace in the definition (1.32) the ordinary derivative $\partial_N$ by the covariant Levi-Civita derivative $\nabla_N$. The extra terms in $\nabla_N \alpha_{M_1 \ldots M_p}$ vanish after antisymmetrization. This means that the operator $d$ may be viewed of as an operator acting on the Hilbert space including all antisymmetric tensor fields. It makes a tensor out of a tensor.

Finally, we define the adjoint of $d$ with respect to the norm (1.29):

$$d^\dagger \alpha_p = (-1)^{Dp+D+1} \star d \star \alpha_p. \tag{1.35}$$

For any even $D$ and for all $p$, $d^\dagger = - \star d \star$. For an odd $D$, $d^\dagger = (-1)^p \star d \star$. The result of the action of $d^\dagger$ on a $p$-form is a $(p-1)$-form. For any couple of forms $\alpha_p$ and $\beta_{p-1}$, the property

$$(\alpha_p, d\beta_{p-1}) = (d^\dagger \alpha_p, \beta_{p-1}) \tag{1.36}$$

holds. $d^\dagger$ is a generalization of the divergence, as can be easily seen by inspecting the action of $d^\dagger$ on a 1-form in flat space: $d^\dagger(A_i dx^i) = -(\partial_i A^i)$. The operator $d^\dagger$ is nilpotent, as immediately follows from the definition (1.35), from the property (1.27) and from the nilpotency of $d$. Consider the operator

$$\triangle = -(dd^\dagger + d^\dagger d). \tag{1.37}$$

It is a differential operator of the second order which makes $p$-forms out of $p$-forms. It is called the *Laplace–de Rham operator*. When one restricts the action of $\triangle$ on ordinary functions ($p = 0$), we are dealing with the *Laplace–Beltrami operator*

$$\triangle f = \frac{1}{\sqrt{g}} \partial_M (\sqrt{g} g^{MN} \partial_N f), \tag{1.38}$$

a generalization of the ordinary Laplacian for curved manifolds.

The operator $-\triangle$ is positive definite:

$$\begin{aligned}(\alpha_p, -\triangle \alpha_p) &= (\alpha_p, d^\dagger d\alpha_p) + (\alpha_p, dd^\dagger \alpha_p) \\ &= (d\alpha_p, d\alpha_p) + (d^\dagger \alpha_p, d^\dagger \alpha_p) \geq 0. \end{aligned} \tag{1.39}$$

The properties

$$[\triangle, d] = [\triangle, d^\dagger] = 0 \tag{1.40}$$

hold.

The Hilbert space of the differential forms with the operators $d, d^\dagger$ and $\triangle$ acting on it is called the *de Rham complex*.[10]

Now we give three more definitions.

**Definition 1.11.** The form $\alpha_p$ is *closed* if $d\alpha_p = 0$. The form $\alpha_p$ is *co-closed* if $d^\dagger \alpha_p = 0$.

**Definition 1.12.** The form $\alpha_p$ is *exact* if it can be represented as $\alpha_p = d\beta_{p-1}$. The form $\alpha_p$ is *coexact* if it can be represented as $\alpha_p = d^\dagger \gamma_{p+1}$.

**Definition 1.13.** The form $\alpha_p$ is *harmonic* if $\triangle \alpha_p = 0$.

An exact form is always closed (this follows from the nilpotency of $d$), but a closed form is not always exact. The following theorem, however, holds[11]:

**Theorem 1.2.** *In a topologically trivial space homeomorphic to $\mathbb{R}^D$, a closed form is also exact.*

**Proof.** The proof is especially simple for 1-forms, in which case the theorem can be reformulated as follows: *In $\mathbb{R}^D$, a vector field $A_M(x)$ with zero curl, $\partial_{[M} A_{N]} = 0$, can be presented as a gradient of a scalar function $\phi(x)$.* The proof is simple. Given the value $\phi(0)$ at the origin, we determine $\phi(x)$ at any other point as

$$\phi(x) = \phi(0) + \int_0^x A_M(x) dx^M, \tag{1.41}$$

where the intergal is done along any line connecting 0 and $x$ (see Fig. 1.3a). The condition $\partial_{[M} A_{N]} = 0$ and the Stokes theorem ensures that the integral (1.41) does not depend on the path, and $\phi(x)$ is a single-valued function. Obviously, $\partial_M \phi(x) = A_M(x)$.

A similar proof can be given for higher $p$. □

---

[10] I learned from my conversations with mathematicians that they do not always realize that the de Rham complex has a structure of Hilbert space. But it has, and this fact is of paramount importance for us.

[11] We later use it on several occasions to prove the existence of a solution to an equation or equation system.

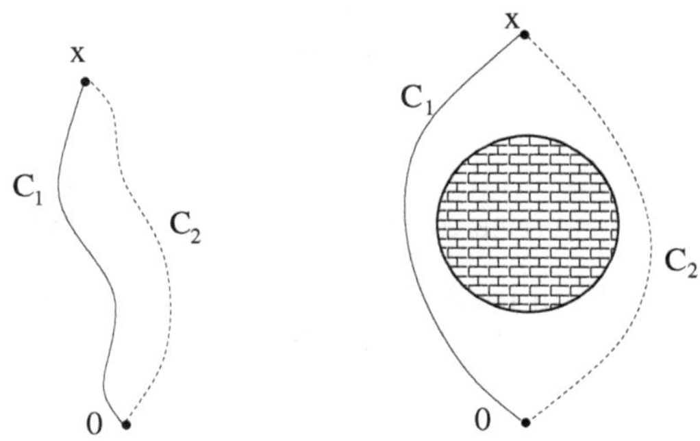


(a) No obstacles (b) An obstacle

**Fig. 1.3.** Exact and not exact.

**Remark:** Note that the requirement that our space has a trivial topology is essential. If the space has a "hole" (has a nontrivial fundamental group), not all paths can be continuously transformed to one another, and the velocity field of liquid that flows around a hole cannot be presented as a gradient of some potential (see Fig. 1.3b). For a nontrivial manifold $\mathcal{M}$ and a given $p$, the coset *closed/exact* is a linear space known as the *de Rham cohomology group* $H^p_{dR}(\mathcal{M})$. We discuss at length cohomology of manifolds in Chapter 8.

## 1.3 Principal fiber bundles and gauge fields

A fiber bundle $\mathcal{E}$ over the base $\mathcal{M}$ with fiber $\mathcal{F}$ is a manifold which looks *locally* as a direct product of $\mathcal{M}$ and $\mathcal{F}$. To be more exact: if we represent $\mathcal{M}$ as a union (1.1) of the open subsets $U_j$, the bundle represents a union of the direct products $U_j \times \mathcal{F}$.

Note that the bundle may be glued from the pieces $U_j \times \mathcal{F}$ in a nontrivial "twisted" way so that the bundle does not necessarily coincide globally with the product $\mathcal{M} \times \mathcal{F}$. The simplest nontrivial bundle is the Möbius strip. In this case, the base is the circle $S^1$ and the fiber is the interval $I$. But the Möbius strip does not coincide with $S^1 \times I$.

**Definition 1.14.** A *principal fiber bundle* is a bundle with the fiber representing a Lie group $\mathcal{G}$.

To avoid unnecessary (for us) complications, we assume in the following that this group is compact.

**Definition 1.15.** A *vector bundle* associated with a given principal bundle is a bundle with the fiber being a representation space $\Phi$ of the group $\mathcal{G}$.

It will be more convenient for us to work with this latter object. Note here that what mathematicians call "representation space" physicists simply call "representation", and we stick in the following to the physical terminology.

**Definition 1.16.** Take in each fiber that grows over the point $x \in \mathcal{M}$ a particular element $\phi(x) \in \Phi$. We obtain what is called a *section* of our vector bundle.

Given a section $\phi(x)$, we can act at each point on $\phi(x)$ by a group element $g(x)$ to obtain another section:

$$\phi'(x) = g(x)\phi(x). \tag{1.42}$$

The transition (1.42) will be called a *gauge transformation*. The word "gauge" means that the group transformation (1.42) is *local*: its parameters depend on $x \in \mathcal{M}$ so that we are dealing here not just with the group $\mathcal{G}$ but with an infinite-dimensional group $\prod_x \mathcal{G}(x)$.

We are interested only in the smooth functions $\phi(x), g(x)$ that can be differentiated. However, an ordinary derivative $\partial_M \phi(x)$ is an inconvenient object because is does not transform as in (1.42): $\partial_M \phi'(x)$ involves an extra term $\sim \partial_M g(x)$. Proceeding in the same way as we did in (1.10) and (1.12), we define a *covariant* derivative

$$\nabla_M \phi(x) = [\partial_M + iA_M(x)]\phi(x), \tag{1.43}$$

where the components of $A_M(x)$ belong to the Lie algebra[12] $\mathfrak{g}$.

$A_M(x)$ is section-dependent. We require that it transforms when going from $\phi(x)$ to $\phi'(x)$ in such a way that

$$\nabla'_M \phi'(x) = g(x)\nabla_M \phi(x). \tag{1.44}$$

It is easy to derive the law

$$\boxed{A'_M(x) = gA_M(x)g^{-1} + i(\partial_M g)g^{-1}.} \tag{1.45}$$

A kinship of $A_M(x)$ to the affine connection $G^M_{NP}$ with the transformation law $(1, 11)$, which was discussed in the previous section, is obvious. Mathematicians call $A_M(x)$ the *principal $\mathcal{G}$-connection*, but the physicists call it

[12] We use the physical convention: $A_M(x)$ are represented by Hermitian matrices without distinction between a Lie algebra and its envelopping algebra.

a *gauge field* living on the manifold. It is the latter name that we are going to use in our book.

We know that the manifold represents generically a union of several different charts. Up to now, we only defined $A_M(x)$ in a particular chart. The fields $A_M^{(\alpha)}(x_{(\alpha)})$ and $A_M^{(\beta)}(x_{(\beta)})$ in two different charts may coincide when $x_{(\beta)}$ is expressed via $x_{(\alpha)}$, but it is not a necessary condition. In many interesting cases,[13] $A_M^{(\alpha)}$ and $A_M^{(\beta)}$ do not coincide but are related by a nontrivial gauge transformation

$$A_M^{(\beta)} = g_{\alpha\beta} A_M(x) g_{\alpha\beta}^{-1} + i(\partial_M g_{\alpha\beta}) g_{\alpha\beta}^{-1}. \tag{1.46}$$

The only condition is that, if by any chance three different charts $U_\alpha$, $U_\beta$ and $U_\gamma$ overlap, the functions $g_{\alpha\beta}, g_{\beta\gamma}$ and $g_{\alpha\gamma}$ satisfy in the overlap region the consistency condition (the *cocycle condition* in the mathematical language) $g_{\alpha\beta} g_{\beta\gamma} = g_{\alpha\gamma}$.

Gauge fields play a tremendously important role in physics. An example is the electromagnetic Maxwell field. In that case, $\mathcal{G} = U(1)$, $g(x) = \exp\{i\chi(x)\}$ and the law (1.45) is reduced to $A'_\mu(x) = A_\mu(x) - \partial_\mu \chi(x)$, where $\mu = 0, 1, 2, 3$ is the Minkowski space index. Gauge fields associated with the non-Abelian groups $SU(3)$ and $SU(2)$ play the fundamental role in the physics of strong and weak interactions.

In the Riemann geometry section, we discussed the affine connection related to the infinitesimal parallel transport of vectors and tensors and also the curvature tensor describing how the vectors and tensors transform after going around a small closed contour. The notion of curvature can also be introduced for principal fiber bundles. To this end, one should consider, as in Eqs. (1.19) and (1.20) the commutator of two covariant derivatives (1.43),

$$-i[\nabla_M, \nabla_N] = \partial_M A_N - \partial_N A_M + i[A_M, A_N] \stackrel{\text{def}}{=} F_{MN}. \tag{1.47}$$

Clearly, $F_{MN} \in \mathfrak{g}$. It is an antisymmetric tensor.[14] Mathematicians call the form $F = (1/2) F_{MN} dx^M \wedge dx^N$ "the curvature form of a principal $\mathcal{G}$-connection". We will use the physical language and call the tensor $F_{MN}$ the *field density tensor* (and $A_M$ will be called the *gauge potential* or just

[13]They are discussed in Chapter 14.

[14]To prove the tensor nature of $F_{MN}$, note that

$$\partial_M A_N - \partial_N A_M = \nabla_M A_N - \nabla_N A_M + C_{MN}^P A_P, \tag{1.48}$$

where $\nabla_M$ is the *Levi-Civita* covariant derivative and $C_{MN}^P$ is the torsion tensor defined in (1.18).

the *gauge field*). In the Maxwell case, the components of the tensor $F_{\mu\nu} = \partial_\mu A_\nu - \partial_\nu A_\mu$ are the electric and magnetic fields.

The Abelian field density is invariant under the gauge transformation (1.45). In the non-Abelian case, it transforms as

$$F'_{MN} = gF_{MN}g^{-1}. \tag{1.49}$$

The notation of differential forms allows one to represent (1.47) in a somewhat more compact way. Introduce a matrix-valued 1-form $A = A_M dx^M$. Then, the definition (1.47) is equivalent to

$$\mathcal{F} = dA + iA \wedge A = \frac{1}{2} F_{MN} dx^M \wedge dx^N. \tag{1.50}$$

## 1.4 Vielbeins and tangent space

Suppose we define at each point $x$ of the manifold $\mathcal{M}$ a set of $D$ orthonormal vectors $\mathbf{e}_A$; $\mathbf{e}_A \cdot \mathbf{e}_B = \delta_{AB}$. In coordinates, the latter condition reads

$$g_{MN} e_A^M e_B^N = \delta_{AB}. \tag{1.51}$$

**Definition 1.17.** The object $e_A^M$ is called the *vielbein.*[15]

One can imagine (though it is not necessary) that $\mathcal{M}$ is embedded in a higher-dimensional Euclidean space and a *tangent space*[16] (denoted as $T_P\mathcal{M}$) is traced at each point P of the manifold (Fig. 1.4). The points in $T_P\mathcal{M}$ are related to the points in $\mathcal{M}$ at the vicinity of $P$ by the orthogonal projection. The set $\{\mathbf{e}_A\}$ represents then an orthonormal frame in $T_P\mathcal{M}$.

Consider the tensor $Q^{MN} = e_A^M e_A^N$. Multiplying it by $e_{MB}$ and using (1.51), we derive $Q^{MN} e_{BM} = e_B^N$ for all $B$. Bearing in mind the completeness of the basis $\{\mathbf{e}_A\}$ in $T_P\mathcal{M}$, this implies that

$$Q^{MN} = e_A^M e_A^N = g^{MN}. \tag{1.52}$$

Thus, the vielbein $e_A^M$ is in some sense a "square root" of the metric. The indices $M, N$, etc. will be called *world indices* and $A, B$, etc. will be called *tangent space indices.* Any tensor can be projected onto the tangent space by contracting it with an appropriate number of the vielbein factors $e_A^M$

[15] A "multipede".

[16] Mathematicians distinguish the notions of tangent space and cotangent space, and in this case, it is rather cotangent than tangent. This distinction is not, however, relevant for us in this book and we use a sloppy terminology and omit the prefix *co*.

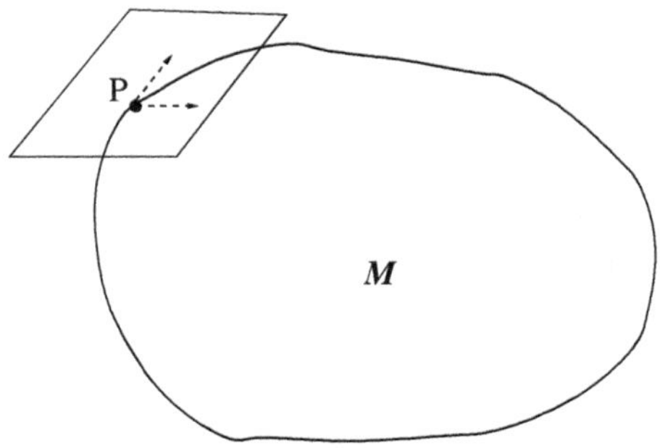


**Fig. 1.4.** A zweibein.

and $e_{AM}$. In particular, the tensor (1.7) can be alternatively represented as

$$E_{M_1 \ldots M_D} = e_{A_1 M_1} \cdots e_{A_D M_D} \varepsilon_{A_1 \ldots A_D} . \tag{1.53}$$

It is clear that there is a considerable freedom in the choice of the vielbein. One can multiply $e_A^M$ at each point by an orthogonal matrix $O_{AB}(x)$, and this does not affect the metric. This is nothing but *gauge freedom* discussed in the previous section. For a mathematician, the manifold $\mathcal{M}$ with the tangent space attached at each its point defines the *tangent bundle* of $\mathcal{M}$, but one can also define the so-called *orthonormal fiber bundle* with the fiber representing the set of all orthonormal frames in $T_P\mathcal{M}$, which is homeomorphic to $O(D)$. This is a particular case of the principal fiber bundle.

Define now the covariant derivative of the vielbein. We want it to transform as a tensor under general coordinate transformations and as a vector in the tangent space. The latter condition means that the rotation $e'^M_A = O_{AB} e_B^M$ implies

$$(\nabla_N e_A^M)' = O_{AB} \nabla_N e_B^M . \tag{1.54}$$

The only way to ensure it is to add to the covariant derivative defined in (1.10) an extra term and write

$$\nabla_N e_A^M = \partial_N e_A^M + G^M_{NP} e_A^P + \omega_{AB,N} e_B^M . \tag{1.55}$$

The object $\omega_{AB,M}$ is called the *spin connection.*[17] $\omega_{AB,M}$ is anti-Hermitian and belongs to the Lie algebra $so(D)$. It is nothing but the gauge field for

[17]This term is due to the fact that $\omega_{AB,M}$ is indispensible for describing *spinor* fields on the manifolds. It is used in general relativity, where $\mathcal{M}$ represents a physical curved spacetime and the spinor fields are physical fermion fields. In our book, $\mathcal{M}$ will never

the group $SO(D)$ (multiplied by $i$ in our convention): $-i\omega$ transforms under rotations in the same way as in (1.45), giving

$$\omega'_{AB,M} = O_{AC}(x)\omega_{CD,M}O_{BD}(x) - [\partial_M O_{AC}(x)]O_{BC}(x). \qquad (1.56)$$

Many different choices for $\omega_{AB,M}$ are in principle possible. But the cleverest (and the standardly used) choice consists in imposing the requirement

$$\nabla_N e_A^M = 0. \qquad (1.57)$$

Its meaning is the following. Define the vielbein field at all points of the manifold. Take the vielbein at point $x$ and perform its parallel transport to $x + dx$. It is rotated due to nonzero affine connections, as dictated in (1.14). But then we use the gauge freedom of local frame rotations associated with nonzero $\omega_{AB,M}$ to "neutralize" the effect due to nonzero $G^M_{NP}$. The condition (1.57) says that, after a parallel transport from $x$ to $x + dx$ and a subsequent frame rotation, the vielbein coincides with $e_A^M(x + dx)$!

In this case, the spin connection $\omega_{AB,M}$ is expressed via the vielbeins and affine connections: $\omega_{AB,M} = -e_{BN}(\partial_M e_A^N + G^N_{MP}e_A^P)$. One can prove (we leave it to the reader) that, if $G^N_{MP}$ are chosen so that the covariant derivative of the metric tensor vanishes, the spin connection is antisymmetric under $A \leftrightarrow B$ and hence belongs to $so(D)$, indeed! We arrive then at the standard textbook expression:

$$\omega_{AB,M} = e_{AN}\left(\partial_M e_B^N + G^N_{MP}e_B^P\right). \qquad (1.58)$$

This result is quite natural. As was mentioned, the metricity condition $\nabla_P g_{MN} = 0$ means that the vector lengths and scalar products do not change under parallel transport. Thus, a parallelly transported vielbein is still a vielbein — an orthonormal frame in the tangent space. And then the local rotations generated by the antisymmetric spin connection can bring, indeed, the vielbein to $e_A^M(x + dx)$.

There exist affine connections not respecting the metricity (one of them, the so-called *Obata connection*, is discussed in Chapters 3 and 13). Then, $\omega_{AB,M}$ as defined in (1.58) is not antisymmetric and does not belong to $so(D)$. But in this case, the very notions of vielbein and spin connection are probably not so useful.

---

be interpreted as physical space (the latter will involve only time) but rather as an external *target space* whose coordinates are time-dependent dynamical variables. Still, spin structures may also be considered for many such target spaces, and we do so in Chapters 14 and 15, where we discuss the Dirac operator and the index theorem.

The relationship (1.58) can be rewritten in a nice compact way using the language of differential forms. Introduce the 1-forms

$$e_A = e_{AM} dx^M, \quad \omega_{AB} = \omega_{AB,M} dx^M. \tag{1.59}$$

Equation (1.58) is equivalent to a *Cartan structure equation*

$$de_A + \omega_{AB} \wedge e_B = T_A, \tag{1.60}$$

where

$$T_A = \frac{1}{2} e_{AM} C^M_{NP} dx^N \wedge dx^P = \frac{1}{2} C_{A,BC} e_B \wedge e_C \tag{1.61}$$

is the torsion 2-form. When there is no torsion, we simply obtain $de_A + \omega_{AB} \wedge e_B = 0$.

Even nicer is the second Cartan's structure equation,[18]

$$d\omega_{AB} + \omega_{AC} \wedge \omega_{CB} = \frac{1}{2} R_{ABCD} e_C \wedge e_D \stackrel{\text{def}}{=} R_{AB}, \tag{1.62}$$

which relates the curvature tensor to the spin connection. More explicitly,

$$\boxed{R_{MN,AB} = \partial_M \omega_{AB,N} - \partial_N \omega_{AB,M} + \omega_{AC,M} \omega_{CB,N} - \omega_{AC,N} \omega_{CB,M}.} \tag{1.63}$$

Equation (1.62) has exactly the same meaning as (1.50). It is remarkable that the same curvature tensor may be expressed via the affine connections as in (1.21) or via the spin connections as in (1.63). But this is stipulated by the requirement (1.57) that relates parallel transports to local frame rotations.

Taking the exterior derivative of Eq. (1.60) and using (1.62), we derive the so-called *consistency condition*

$$dT_A + \omega_{AB} \wedge T_B = R_{AB} \wedge e_B. \tag{1.64}$$

When the torsion is absent, we derive the constraint

$$R_{AB} \wedge e_B = 0 \tag{1.65}$$

for the ordinary Riemann curvature. This condition is none other than the second line in (1.23).

---

[18] Do not mix up $R_{AB}$ thus defined with the Ricci tensor.

# Chapter 2

# Complex Manifolds

## 2.1 Complex description

**Definition 2.1.** An even-dimensional manifold representing a union (1.1) of open subsets $U_\alpha$ homeomorphic to $\mathbb{R}^{D=2d}$ is called a *complex manifold* if each subset can be parameterized by the complex coordinates $\{z_{(\alpha)}^{n=1,\ldots,d}\}$ in such a way that the transition functions are analytic:

$$z_{(\alpha)}^n = f_{\alpha\beta}^n(z_{(\beta)}^m). \tag{2.1}$$

The manifold may or may not be equipped with a metric, but we will always assume that it is.

**Theorem 2.1.** *For a complex manifold, one can always choose the metric in the Hermitian form:*

$$\boxed{ds^2 = 2h_{n\bar{m}}\, dz^n d\bar{z}^{\bar{m}},} \tag{2.2}$$

*with* $\overline{h_{m\bar{n}}} = h_{n\bar{m}}$.

The factor 2 is introduced for further convenience.

**Proof.** For a particular chart $U_\alpha$ homeomorphic to $\mathbb{C}^d$, it is obvious. The property $\overline{h_{m\bar{n}}} = h_{n\bar{m}}$ follows from the reality of $ds^2$. Then, the existence of a globally defined Hermitian metric can be quite rigourously proven using the so-called *partition of unity* technique. The crucial observation is that, in the intersection $U_\alpha \cap U_\beta$ of two overlapping charts, a Hermitian metric in $U_\alpha$ stays Hermitian in $U_\beta$ after a holomorphic change of variables.

Let a Hermitian metric $ds^2_{(\alpha)}$ be defined on each $U_\alpha$. We introduce a set of real nonnegative smooth functions $f_\alpha(z_{(\alpha)}, \bar{z}_{(\alpha)}) \in (0,1)$ such that

- $f_\alpha = 0$ at the border of $U_\alpha$.
- $\sum_\alpha f_\alpha = 1$ at every point of the manifold.

Then, the metric

$$ds^2 = \sum_\alpha f_\alpha \, ds^2_{(\alpha)} \tag{2.3}$$

is the metric on the union $\cup_\alpha U_\alpha$ that we are looking for. □

To understand it, imagine a manifold consisting of only two overlapping charts, like in the example that follows. Then, the metric (2.3) coincides with $ds^2_{(1)}$ at all points of $U_{(1)}$ not belonging to $U_{(2)}$, coincides with $ds^2_{(2)}$ at all points of $U_{(2)}$ not belonging to $U_{(1)}$ and represents a smooth interpolation between the two in the overlap region.

**Remark:** The theorem above does not claim that one can always choose complex coordinates such that any Riemannian metric $ds^2 = g_{MN} dx^M dx^N$ is expressed in a Hermitian form (2.2). That is true for two-dimensional real manifolds but not when $D = 4$ or higher: generically, the terms $\propto dz^m dz^n$ and $\propto d\bar{z}^m d\bar{z}^n$ might be present. The theorem only says that a Hermitian metric can always *be chosen*, and we always do so.

We express the inverse metric tensor as $h^{\bar{n}m}$ so that $h_{n\bar{m}} h^{\bar{m}p} = \delta_n^p$ and $h^{\bar{p}n} h_{n\bar{m}} = \delta_{\bar{m}}^{\bar{p}}$.

The simplest nontrivial example of a complex manifold is $S^2$ (Fig. 2.1). It can be represented as a union of two charts. One of them includes all

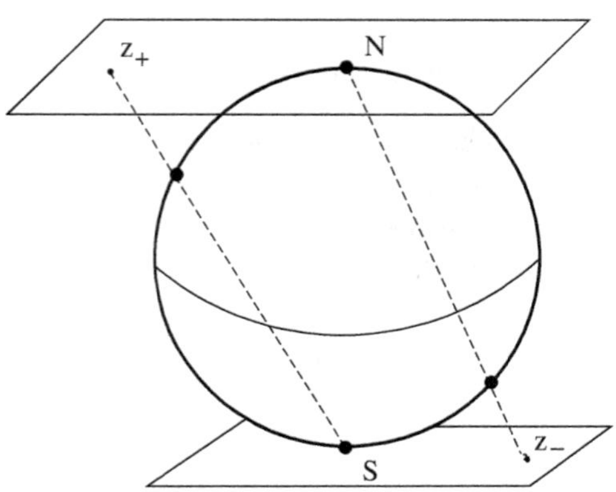


**Fig. 2.1.** Complex coordinates on $S^2$.

the points of the sphere except its north pole. It can be parameterized by the complex coordinate

$$z_- = \cot\frac{\theta}{2}e^{i\phi} \tag{2.4}$$

given by the stereographic projection on the plane tangent to the south pole. Another chart includes all the points of the sphere except its south pole. It can be parameterized by the complex coordinate

$$z_+ = \tan\frac{\theta}{2}e^{-i\phi} \tag{2.5}$$

given by the stereographic projection on the plane tangent to the north pole. In the overlap region, which includes all the points of the sphere except its poles, the transition function $z_+ = 1/z_-$ is analytic. The metric is convenient to choose in the form

$$ds^2 = 2h\,dzd\bar{z} = \frac{2\,dzd\bar{z}}{(1+z\bar{z})^2} \tag{2.6}$$

for the both charts. Then, $ds_+^2 = ds_-^2$ on the whole sphere except the poles, and we do not need to bother in this case about the partition of the unity.

We also pose $z = (x+iy)/\sqrt{2}$ so that the area of our sphere is equal to[1]

$$A = \int h\,dzd\bar{z} = \int dxdy\,\frac{1}{\left(1+\frac{x^2+y^2}{2}\right)^2} = 2\pi, \tag{2.7}$$

and its radius is equal to $1/\sqrt{2}$.

**Theorem 2.2.** *Any orientable two-dimensional manifold is complex.*

**Proof.** Consider two overlapping charts on $\mathcal{M}$. In the overlap region, the manifold can be described by the coordinates $\{x, y\}$ or $\{u, v\}$. It is convenient to choose the coordinates such that

$$ds^2 = \lambda(x,y)(dx^2+dy^2) = \mu(u,v)(du^2+dv^2). \tag{2.8}$$

Choose the complex coordinates in each chart:

$$z = \frac{x+iy}{\sqrt{2}}, \qquad w = \frac{u+iv}{\sqrt{2}}. \tag{2.9}$$

[1]See the footnote 8 on p. 31.

Then,

$$\lambda\, dz d\bar{z} = \mu\, dw d\bar{w}. \tag{2.10}$$

Generically,

$$dw = \frac{\partial w}{\partial z} dz + \frac{\partial w}{\partial \bar{z}} d\bar{z}. \tag{2.11}$$

Substituting it into (2.10), we obtain, however,

$$\frac{\partial w}{\partial z} \frac{\partial \bar{w}}{\partial z} = 0,$$

which means that $w$ is either holomorphic or antiholomorphic function of $z$. If it were antiholomorphic, the Cauchy–Riemann equations would read

$$\frac{\partial u}{\partial x} = -\frac{\partial v}{\partial y}, \quad \frac{\partial u}{\partial y} = \frac{\partial v}{\partial x}$$

and the Jacobian

$$\frac{\partial(u,v)}{\partial(x,y)} = -\left(\frac{\partial u}{\partial x}\right)^2 - \left(\frac{\partial u}{\partial y}\right)^2 \tag{2.12}$$

would be negative. But that would contradict the assumption that the manifold is orientable. □

Not all orientable four-dimensional manifolds are complex, however. Neither is it true in higher even dimensions. For example, $S^4$ is not. The reader can understand why, if they try to repeat the construction in Fig. 2.1 for $S^4$. It is easy to find a couple of complex coordinates $z_-^{n=1,2}$ parameterizing the "southern" hemisphere on $S^4$. It is also easy to find a couple of complex coordinates $z_+^{n=1,2}$ parameterizing the "northern" hemisphere. But it is impossible to perform it in a way that the transition functions are holomorphic. If $z_-^n$ and $z_+^n$ are obtained by stereographic projections, like in Fig. 2.1, they are related as

$$z_+^n = \frac{\bar{z}_-^n}{\bar{z}_-^m z_-^m} \quad \text{or} \quad z_+^n = \frac{z_-^n}{\bar{z}_-^m z_-^m}. \tag{2.13}$$

In neither case, the transition functions are holomorphic.

Of course, the fact that the simplest attempt to built up a complex description of $S^4$ does not work is not yet a proof that it is impossible. In the case of $S^4$, it can be proven, however. Surprisingly, no such proof exists for $S^6$. We are not able to construct complex charts with analytic transition functions, but we are not quite sure that it is impossible.

The simplest example of a complex four-dimensional manifold is[2] $\mathbb{CP}^2$. We write its metric at the end of this chapter.

A manifold of complex dimension $d$ can, of course, be also treated as a real manifold of dimension $D = 2d$. The translation from the real to the complex language and inverse is performed by trading the set of coordinates $x^M$ to the set $\{z^m, \bar{z}^{\bar{m}}\}$. Also,

$$\partial_M = (\partial_m, \partial_{\bar{m}}), \quad g_{m\bar{n}} = g_{\bar{m}n} = h_{m\bar{n}}, \quad g_{mn} = g_{\bar{m}\bar{n}} = 0. \tag{2.14}$$

With this choice, the convention (2.2) for the Hermitian metric exactly matches the definition (1.6).

For $d = 1$, these conventions mean that $z$ is expressed via $x, y$ as in (2.9). Also,

$$\frac{\partial}{\partial z} = \frac{1}{\sqrt{2}}\left(\frac{\partial}{\partial x} - i\frac{\partial}{\partial y}\right). \tag{2.15}$$

In a multidimensional case,

$$z^1 = \frac{x^1 + ix^2}{\sqrt{2}}, \quad z^2 = \frac{x^3 + ix^4}{\sqrt{2}}, \ldots \tag{2.16}$$

The tensor fields living on the manifold may now have the indices of four types: holomorphic or antiholomorphic and covariant or contravariant. When an index is raised, it changes its holomorphicity: $V^{\bar{n}} = h^{\bar{n}m}V_m$, etc. Holomorphicity is also changed, of course, after a complex conjugation: $\overline{V_n} = \bar{V}_{\bar{n}}$.

In these terms, the only nonvanishing components of the Levi-Civita connection (1.17) are

$$\begin{aligned}
&\Gamma_{\bar{m},np} = \frac{1}{2}(\partial_n h_{p\bar{m}} + \partial_p h_{n\bar{m}}), \quad \Gamma_{m,\bar{n}\bar{p}} = \frac{1}{2}(\partial_{\bar{n}} h_{m\bar{p}} + \partial_{\bar{p}} h_{m\bar{n}}), \\
&\Gamma_{m,n\bar{p}} = \Gamma_{m,\bar{p}n} = \frac{1}{2}C_{mn\bar{p}}, \quad \Gamma_{\bar{m},p\bar{n}} = \Gamma_{\bar{m},\bar{n}p} = \frac{1}{2}C_{\bar{m}\bar{n}p}
\end{aligned} \tag{2.17}$$

with

$$C_{mn\bar{p}} = \partial_n h_{m\bar{p}} - \partial_m h_{n\bar{p}}, \quad C_{\bar{m}\bar{n}p} = \overline{C_{mn\bar{p}}} = \partial_{\bar{n}} h_{p\bar{m}} - \partial_{\bar{m}} h_{p\bar{n}}. \tag{2.18}$$

Note that $\partial_{[l}C_{mn]\bar{p}} = \partial_{[\bar{l}}C_{\bar{m}\bar{n}]p} = 0$.

---

[2]A complex projective space $\mathbb{CP}^n$ is defined as a set of all complex $(n+1)$-tuples $\{w^0, w^1, \ldots, w^n\}$ with at least one nonzero $w^j$, factorized over identification $\{w^0, \ldots, w^n\} \equiv \{\lambda w^0, \ldots, \lambda w^n\}$ with any nonzero complex $\lambda$.

We now introduce a very important notion of the *Bismut connection* [7,8].

**Definition 2.2.** The Bismut connection is a torsionful connection

$$\boxed{G^{(B)}_{M,NP} = \Gamma_{M,NP} + \frac{1}{2}C_{MNP},} \tag{2.19}$$

with the torsion tensor $C_{MNP} \equiv C_{M,NP}$ completely antisymmetric under exchange of *any* pair of indices, whose only nonzero components are those given in (2.18) and those restored from them by antisymmetry.

The nonzero components of $G^{(B)}_{m,n\bar{p}}$ are[3]

$$\begin{aligned} G^{(B)}_{\bar{m},np} &= \partial_p h_{n\bar{m}}, \quad G^{(B)}_{m,\bar{n}\bar{p}} = \partial_{\bar{p}} h_{m\bar{n}}; \\ G^{(B)}_{m,n\bar{p}} &= C_{mn\bar{p}}, \quad G^{(B)}_{\bar{m},\bar{n}p} = C_{\bar{m}\bar{n}p}, \end{aligned} \tag{2.20}$$

while

$$G^{(B)}_{m,\bar{n}p} = G^{(B)}_{\bar{m},n\bar{p}} = G^{(B)}_{m,np} = G^{(B)}_{\bar{m},\bar{n}\bar{p}} = 0. \tag{2.21}$$

We define next the complex vielbeins $e^a_m$ and the complex conjugated $e^{\bar{a}}_{\bar{m}}$ with the properties[4]

$$\begin{aligned} e^a_m e^{\bar{a}}_{\bar{n}} &= h_{m\bar{n}}, \quad e^{\bar{n}}_{\bar{a}} e^m_a = h^{\bar{n}m}, \\ e^a_m e^n_a = \delta^n_m, \quad e^{\bar{a}}_{\bar{m}} e^{\bar{n}}_{\bar{a}} &= \delta^{\bar{n}}_{\bar{m}}, \quad e^m_a e^b_m = \delta^b_a, \quad e^{\bar{m}}_{\bar{a}} e^{\bar{b}}_{\bar{m}} = \delta^{\bar{b}}_{\bar{a}}. \end{aligned} \tag{2.23}$$

The metric is invariant under unitary tangent space rotations

$$e^a_m \to U^a_b\, e^b_m, \quad e^{\bar{a}}_{\bar{m}} \to e^{\bar{b}}_{\bar{m}} (U^\dagger)^{\bar{a}}_{\bar{b}}. \tag{2.24}$$

The spin connections (1.58) can also be expressed in this complex notation. The Levi-Civita spin connections read

$$\begin{aligned} \omega_{\bar{b}a,m} &= -\omega_{a\bar{b},m} = e^b_p(\partial_m e^p_a + \Gamma^p_{mk} e^k_a), \\ \omega_{ab,m} &= 0, \quad \omega_{\bar{a}\bar{b},m} = \frac{1}{2} e^{\bar{p}}_{\bar{a}} e^{\bar{k}}_{\bar{b}}\, C_{\bar{p}\bar{k}m}. \end{aligned} \tag{2.25}$$

and other components are related to (2.25) by complex conjugation:

$$\omega_{\bar{b}a,\bar{m}} = \overline{\omega_{b\bar{a},m}}, \quad \omega_{ab,\bar{m}} = \overline{\omega_{\bar{a}\bar{b},m}}. \tag{2.26}$$

---

[3]See e.g. Eq. (2.7) in Ref. [9].

[4]In contrast to the real case, we write here the tangent space indices at two levels, bearing in mind that they are complex conjugated when lowered or lifted. The scalar product $X_A Y_A$ of two tangent space vectors acquires in the complex notation the form

$$X_A Y_A = X_a Y^a + X_{\bar{a}} Y^{\bar{a}} = X_a Y_{\bar{a}} + X_{\bar{a}} Y_a. \tag{2.22}$$

For the spin connections associated with the Bismut connection (2.19), the components $\omega^{(B)}_{ab,M}$ and $\omega^{(B)}_{\bar{a}\bar{b},M}$ vanish and only the components $\omega^{(B)}_{a\bar{b},M}$ and $\omega^{(B)}_{\bar{a}b,M}$ satisfying the relations (2.26) are left.

Consider now the form

$$\omega^{(B)}_{AB} = \omega^{(B)}_{AB,M} dx^M. \tag{2.27}$$

It follows from the analysis above that the only nonzero components of this form are $\omega^{(B)}_{a\bar{b}} \equiv [\omega^{(B)}]_a{}^b$ and $\omega^{(B)}_{\bar{a}b} \equiv [\omega^{(B)}]_{\bar{a}}{}^{\bar{b}}$. In view of (2.26), $\overline{[\omega^{(B)}]_a{}^b} = -[\omega^{(B)}]_b{}^a$, i.e. $\omega^{(B)} \in u(d)$.

The same concerns the curvature form $R^{(B)}_{AB}$ related to $\omega^{(B)}_{AB}$ by the Cartan structure equation (1.62): its only nonzero components are $[R^{(B)}]_a{}^b$ and[5] $[R^{(B)}]_{\bar{a}}{}^{\bar{b}}$. This implies that the curvature tensor of the Bismut connection has the only nonzero components $R^{(B)}_{m\bar{n},p\bar{q}}$ and those obtained by permutations $m \leftrightarrow \bar{n}$, $p \leftrightarrow \bar{q}$, which is not true for the ordinary Riemann tensor.

Bearing in mind (1.19) and (1.20), this means that a holomorphic vector stays holomorphic after a parallel transport with Bismut connection along a closed contour. The group describing the rotations along closed contours is called *holonomy group*. Thus, in the case considered, the holonomy group is $U(d) \subset SO(2d)$. The same concerns the group describing local rotations of vielbeins in tangent space. This follows from the condition (1.57) that we impose on the spin connections.

Let us emphasize again that this is all true for the frame rotations and parallel transports induced by the *Bismut* spin and affine connections. For a generic complex manifold, the holonomy group corresponding to the ordinary Levi-Civita connection is still $SO(D)$, like it is for a generic $D$-dimensional real manifold.

A very important class of complex manifolds are *Kähler manifolds.*

**Definition 2.3.** A complex manifold endowed with a Hermitian metric (2.2) is Kähler if

$$C_{mn\bar{p}} = \partial_n h_{m\bar{p}} - \partial_m h_{n\bar{p}} = 0. \tag{2.28}$$

One can make the following remarks:

(1) We have seen that all two-dimensional orientable manifolds are complex. They carry complex dimension $d = 1$ and it is clear from (2.28) that they are all Kähler.

[5]Do not forget the prescription (2.22) when contemplating the term $\omega_{AC} \wedge \omega_{CB}$ in the LHS.

(2) A metric satisfying the condition (2.28) can be presented as

$$h_{m\bar{n}} = \partial_m \partial_{\bar{n}} K(z, \bar{z}). \tag{2.29}$$

The function $K(z, \bar{z})$ is called the *Kähler potential.*[6]

(3) If $C_{mn\bar{p}} = 0$, the Bismut connection coincides with the torsionless Levi-Civita connection. Hence, the ordinary holonomy group of a Kähler manifold is $U(d)$ (and if the holonomy group is $U(d)$, the manifold is Kähler).

(4) In this case, the only nonzero components of the Christoffel symbols (1.17) are holomorphic and antiholomorphic:

$$\Gamma^p_{mn} = h^{\bar{q}p} \partial_m h_{n\bar{q}}, \quad \Gamma^{\bar{p}}_{\bar{m}\bar{n}} = h^{\bar{p}q} \partial_{\bar{m}} h_{q\bar{n}}. \tag{2.30}$$

Inversely, if the Christoffel components of mixed holomorphicity like $\Gamma^p_{m\bar{n}}$ vanish, the metric is Kähler.

(5) The only nonzero components of the Kähler spin connections are

$$\omega_{\bar{b}a,m} = -\omega_{a\bar{b},m} = e^{\bar{n}}_{\bar{b}} \partial_m e^{\bar{a}}_{\bar{n}}, \quad \omega_{b\bar{a},\bar{m}} = -\omega_{\bar{a}b,\bar{m}} = e^n_b \partial_{\bar{m}} e^a_n. \tag{2.31}$$

The only nonzero components of the Kähler Riemann tensor are

$$\begin{aligned} R_{m\bar{p}n\bar{q}} &= -R_{\bar{p}mn\bar{q}} = -R_{m\bar{p}\bar{q}n} = R_{\bar{p}m\bar{q}n} \\ &= \partial_m \partial_{\bar{p}} h_{n\bar{q}} - h^{\bar{s}t}(\partial_m h_{n\bar{s}})(\partial_{\bar{p}} h_{t\bar{q}}). \end{aligned} \tag{2.32}$$

The relation

$$R_{m\bar{p}n\bar{q}} = e^a_n e^{\bar{b}}_{\bar{q}} (\partial_m \omega_{a\bar{b},\bar{p}} - \partial_{\bar{p}} \omega_{a\bar{b},m} + \omega_{a\bar{c},m} \omega_{c\bar{b},\bar{p}} - \omega_{a\bar{c},\bar{p}} \omega_{c\bar{b},m}) \tag{2.33}$$

[it follows from (1.63)] also holds.

### 2.1.1 *Complex differential forms*

In complex terms, a generic differential form (1.25) is expressed as

$$\alpha_{p,q} = \alpha_{m_1 \cdots m_p \, \bar{n}_1 \cdots \bar{n}_q} dz^{m_1} \wedge \cdots \wedge dz^{m_p} \wedge d\bar{z}^{\bar{n}_1} \wedge \cdots \wedge d\bar{z}^{\bar{n}_q}. \tag{2.34}$$

---

[6]Note that Eq. (2.29) provides the full classification of the Kähler metrics in a particular chart defining the manifold: invent any real function $K$ of the complex coordinates and the metric will follow. But it says nothing about the topological structure of the manifold — how different charts are glued together and what are the holomorphic transition functions.

The expression (2.34) is called a $(p,q)$-form. $p$ and $q$ do not exceed the complex dimension of the manifold $d$, otherwise the wedge product in the right-hand side vanishes.

We can now adjust all the constructions and the definitions of Section 1.2 to the complex case. The Hodge duality operator is convenient to define as

$$*\alpha_{p,q} = \frac{1}{(d-p)!(d-q)!} E^{\bar{m}_1\cdots\bar{m}_p}{}_{m_{p+1}\cdots m_d}{}^{n_1\cdots n_q}{}_{\bar{n}_{q+1}\cdots\bar{n}_d}$$

$$\overline{\alpha_{m_1\cdots m_p\,\bar{n}_1\cdots\bar{n}_q}}\, dz^{m_{p+1}} \wedge \cdots \wedge dz^{m_d} \wedge d\bar{z}^{\bar{n}_{q+1}} \wedge \cdots \wedge d\bar{z}^{\bar{n}_d}, \qquad (2.35)$$

where

$$E_{m_1\cdots m_d\,\bar{n}_1\cdots\bar{n}_d} = \det(h_{m\bar{n}})\,\varepsilon_{m_1\cdots m_d}\,\varepsilon_{\bar{n}_1\cdots\bar{n}_d} \qquad (2.36)$$

is the invariant tensor (1.7) in the complex notation and the indices are raised with $h^{\bar{m}n}$. The main distinction of (2.35) compared with the real duality operator (1.26) is the presence of complex conjugated coefficients $\overline{\alpha_{m_1\cdots m_p\,\bar{n}_1\cdots\bar{n}_q}}$ in the RHS.[7] The operator (2.35) transforms a $(p,q)$-form to a $(d-p, d-q)$-form. The property

$$**\,\alpha_{p,q} = (-1)^{p(d-p)+q(d-q)}\alpha_{p,q} \qquad (2.37)$$

holds.

The inner product is defined for the forms of the same type $p,q$ in analogy with[8] (1.29):

$$(\alpha_{p,q}, \beta_{p,q}) = \int_{\mathcal{M}} \alpha_{p,q} \wedge *\beta_{p,q} = p!q! \int_{\mathcal{M}} \alpha_{r_1\cdots r_p\bar{s}_1\cdots\bar{s}_q}\,\overline{\beta_{m_1\cdots m_p\bar{n}_1\cdots\bar{n}_q}}$$
$$\times\, h^{\bar{m}_1 r_1}\cdots h^{\bar{m}_p r_p} h^{\bar{s}_1 n_1}\cdots h^{\bar{s}_q n_q}\,\det(h_{n\bar{m}}) \prod_t dz^t d\bar{z}^t. \qquad (2.39)$$

The dual form $*\alpha_{p,q}$ has the same norm as $\alpha_{p,q}$.

---

[7]Actually, we could also include complex conjugation in the definition of duality for the de Rham complex, but we preferred to keep the standard definition there.

[8]The convention

$$\prod_t dz^t d\bar{z}^t = \prod_M dx^M \qquad (2.38)$$

(without the factors $i$ that are sometimes inserted) is consistently chosen throughout this book.

Consider now a subset of (2.34) including only the $(p, 0)$-forms (or else holomorphic $p$-forms):

$$\alpha_{p,0} = \alpha_{m_1 \ldots m_p} \, dz^{m_1} \wedge \cdots \wedge dz^{m_p} . \tag{2.40}$$

Define the operator of holomorphic exterior derivative $\partial$ in analogy with (1.32):

$$\partial\alpha = \partial_n \alpha_{m_1 \ldots m_p} \, dz^n \wedge dz^{m_1} \wedge \cdots \wedge dz^{m_p} . \tag{2.41}$$

We introduce also the conjugate [with respect to the inner product (2.39)] operator $\partial^\dagger$. In the full analogy with (1.35), it is expressed as

$$\partial^\dagger = (-1)^{dp+d+1} * \partial * . \tag{2.42}$$

The operators $\partial$ and $\partial^\dagger$ are nilpotent. The anticommutator $-\{\partial, \partial^\dagger\}$ is a Hermitian operator called the *Dolbeault laplacian.*

**Definition 2.4.** The set of all holomorphic forms (2.40) equiped by the operators $\partial$ and $\partial^\dagger$ is called the *Dolbeault complex.*

One can also define the *anti-Dolbeault complex* involving the antiholomorphic forms

$$\bar\alpha = \alpha_{\bar m_1 \ldots \bar m_p} \, d\bar z^{\bar m_1} \wedge \cdots \wedge d\bar z^{\bar m_p} \tag{2.43}$$

and the nilpotent operators $\bar\partial$ and $\bar\partial^\dagger$. Generically, the anti-Dolbeault laplacian equal to $-\{\bar\partial, \bar\partial^\dagger\}$ does not coincide with the Dolbeault one.

There is, however, a very important class of the manifolds where it does — these are Kähler manifolds.

Before discussing this, note that the action of the operators $\partial$, $\partial^\dagger$, $\bar\partial$ and $\bar\partial^\dagger$ can also be defined on the space of *all* $(p, q)$-forms. For $\partial$ and $\bar\partial$, this action represents a rather obvious generalization of (2.41), and $\partial^\dagger$ and $\bar\partial^\dagger$ are their Hermitian conjugates. It is rather obvious that

$$\begin{aligned} \partial\alpha_{p,q} &= \alpha_{p+1,q}, \quad & \bar\partial\alpha_{p,q} &= \alpha_{p,q+1}, \\ \partial^\dagger\alpha_{p,q} &= \alpha_{p-1,q}, \quad & \bar\partial^\dagger\alpha_{p,q} &= \alpha_{p,q-1}. \end{aligned} \tag{2.44}$$

The sums $\partial + \bar\partial$ and $\partial^\dagger + \bar\partial^\dagger$ are our old friends, the operators $d$ and $d^\dagger$ of the de Rham complex.

**Definition 2.5.** For a Kähler manifold, the set of all forms (2.34) and the operators $\partial$, $\partial^\dagger$, $\bar\partial$ and $\bar\partial^\dagger$ is called the *Kähler–de Rham complex.*

One may ask why we impose in this definition the condition for the manifold to be Kähler. Have we not just seen that the action of the holomorphic exterior derivatives and their conjugates can be defined for any Hermitian complex manifold? The matter is that Kähler manifolds have a following special property:

**Theorem 2.3.** *For a Kähler manifold, the Dolbeault and anti-Dolbeault laplacians coincide,* $\{\partial, \partial^\dagger\} = \{\bar\partial, \bar\partial^\dagger\}$. *In addition, the anticommutators* $\{\partial, \bar\partial^\dagger\}$ *and* $\{\partial^\dagger, \bar\partial\}$ *vanish.*

**Proof.** We do not prove this theorem here. We do so in Chapter 10 using supersymmetric methods. □

It follows that for Kähler manifolds, the Dolbeault laplacian coincides up to the factor 1/2 with the Laplace–de Rham operator $\triangle = -\{d, d^+\}$. The latter commutes with both exterior derivatives and their conjugates. This means that the operators $\partial$, $\bar\partial$, $\partial^\dagger$ and $\bar\partial^\dagger$ form a closed *superalgebra* (while for a generic complex manifold, it is not the case).

This is the first time that the word "superalgebra" appeared in the main body of our book. Superalgebra is a generalization of Lie algebra including not only commutators but also anticommutators. The simplest superalgebra (0.3) describes the de Rham complex. And for the Kähler–de Rham complex, we are dealing with an *extended* superalgebra. Rich mathematics associated with the notion of superalgebra is displayed and discussed later in this book. For the time being, we simply illustrate the construction that we just described by the picture in Fig. 2.2 — an adaptation of what is called the *Hodge diamond.*

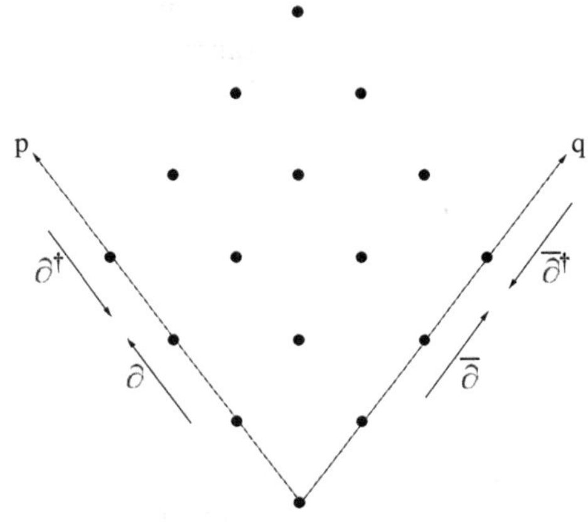


**Fig. 2.2.** Kähler–de Rham complex. Each dot corresponds to a set of forms with definite values of $p$ and $q$.

For any complex manifold, one can define a real (1, 1)-form $\mathcal{K} = ih_{n\bar{m}}\, dz^n \wedge d\bar{z}^{\bar{m}}$. It follows from the definition (2.28) that, if the manifold is Kähler, this form is closed:

$$d\mathcal{K} = (\partial + \bar{\partial})\mathcal{K} = 0. \tag{2.45}$$

In this case, the form $\mathcal{K}$ is called the *Kähler form.* The closedness of $\mathcal{K}$ is also a sufficient condition for the manifold to be Kähler.

## 2.2 Real description

An interesting and important fact is that one can describe complex manifolds without explicitly introducing complex charts but working exclusively in the real terms.

**Definition 2.6.** An *almost complex manifold* is a manifold equipped with a tensor field $I_{MN}$ satisfying the properties (i) $I_{MN} = -I_{NM}$ and (ii) ${I_M}^N {I_N}^P = -\delta_M^P$. The tensor ${I_M}^N$ is called the *almost complex structure.*

It is clear that the tangent space projection $I_{AB}$ belongs to $SO(D)$.

Putting down the index $P$ in the condition (ii) and using the antisymmetry (i), we derive

$$g_{MN} = {I_M}^P {I_N}^Q g_{PQ}. \tag{2.46}$$

It is easy to prove that an almost complex manifold is necessarily even-dimensional. Indeed, the determinant of an odd-dimensional skew-symmetric matrix $I_{AB}$ vanishes, which contradicts the condition $I^2 = -\mathbb{1}$. But not any even-dimensional manifold is almost complex. One can prove (though it is not so easy) that $S^4$ is not. On the other hand, $S^6$ *is* almost complex, as we will see soon.

A reader who meets for the first time this definition might be somewhat perplexed. It might not be clear for him why an antisymmetric *real* tensor is called *complex* structure and the word *almost* in the definition looks enigmatic.

To understand this, consider first the simplest possible example — flat two-dimensional Euclidean space. It can be parametrized by the real Cartesian coordinates $x^1, x^2$ or by the complex coordinate $z = (x^1 + ix^2)/\sqrt{2}$. An obvious relation $\partial z/\partial x^2 = i\partial z/\partial x^1$ holds, which can also be presented in the form[9]

$$\boxed{\frac{\partial z}{\partial x^M} + i\varepsilon_{MN}\frac{\partial z}{\partial x^N} = 0} \tag{2.47}$$

(with $\varepsilon_{12} = 1$). The tensor $\varepsilon_{MN}$ satisfies both conditions in the definition above and *is* the complex structure in this case. (Up to the sign. It will be convenient for us to choose $I_{MN} = -\varepsilon_{MN}$.)

For a curved two-dimensional surface, the complex structure looks a little more complicated. For example, for $S^2$ with the metric $ds^2 = d\theta^2 + \sin^2\theta\, d\phi^2$, it may be chosen as

$$\begin{aligned} &I_{\theta\theta} = I_{\phi\phi} = 0, \quad I_{\theta\phi} = -I_{\phi\theta} = \sin\theta, \\ \Rightarrow\ &I_\theta{}^\theta = I_\phi{}^\phi = 0, \quad I_\phi{}^\theta = -\sin\theta \quad I_\theta{}^\phi = \frac{1}{\sin\theta}. \end{aligned} \tag{2.48}$$

However, when we project it on a tangent plane, we obtain

$$I_{AB} = I_{MN} e_A^M e_B^N = \pm\varepsilon_{AB}, \tag{2.49}$$

with the sign depending on the choice of zweibeins. $I_{AB}$ may be interpreted as the orthogonal matrix rotating the vielbein by $\pm\pi/2$.

Note also that one can describe $I_{AB}$ as a tensor in $\mathbb{R}^3$,

$$I_{ij} = \varepsilon_{ijk} n^k \tag{2.50}$$

($n^k$ being a unit vector orthogonal to the surface of $S^2$), projected on the tangent plane.

Now, consider a general multidimensional case.

**Theorem 2.4.** *Suppose that a tensor $I_M{}^N$ satisfying the conditions above exists. With a proper vielbein choice, its tangent space projection can be brought to the canonical form*[10]

$$I_{AB} = \operatorname{diag}(-\varepsilon, \dots, -\varepsilon) \quad \text{with} \quad \varepsilon = \begin{pmatrix} 0 & 1 \\ -1 & 0 \end{pmatrix}. \tag{2.51}$$

---

[9]The property holds for not only $z$, but also any holomorphic function $f(z)$. In the latter case, the real and imaginary parts of (2.47) are none other than the Cauchy–Riemann conditions.

[10]Again, $I_{AB}$ can be interpreted as an orthogonal matrix rotating the vielbeins and satisfying the condition $I^T = -I$.

**Proof.** To construct an orthonormal base in the tangent space $E$ where the complex structure acquires the form (2.51), we start with choosing in $E$ an arbitrary unit vector $e_0$. It follows from $I^T = -I$ and $I^2 = -\mathbb{1}$ that the vector $e_1 = Ie_0$ has also unit length and is orthogonal to $e_0$. Obviously, $Ie_1 = I^2 e_0 = -e_0$. Consider the subspace $E^* \subset E$ that is orthogonal to $e_0$ and $e_1$. If it is not empty, choose there an arbitrary unit vector $f_0$ and consider $f_1 = If_0$. One can easily see that $f_1$ also belongs to $E^*$, is orthogonal to $f_0$ and has unit length. Now, consider the subspace $E^{**} \subset E^* \subset E$ that is orthogonal to $e_0, e_1, f_0, f_1$ and, if $E^{**}$ is not empty, repeat the procedure. We arrive at the matrix (2.51). □

We further suppose that the equation system

$$\frac{\partial z^n}{\partial x^M} - iI_M{}^N \frac{\partial z^n}{\partial x^N} = 0 \tag{2.52}$$

has $d$ independent solutions. We may trade now $x^M$ for $\{z^n, \bar{z}^{\bar{n}}\}$ and express $I_M{}^N$ in this frame. The expression [it follows from (2.52)] is very simple:

$$I_m{}^n = \frac{\partial x^M}{\partial z^m} I_M{}^N \frac{\partial z^n}{\partial x^N} = \frac{\partial x^M}{\partial z^m} I_M{}^N \left( iI_N{}^P \frac{\partial z^n}{\partial x^P} \right) = -i \frac{\partial z^n}{\partial x^M} \frac{\partial x^M}{\partial z^m} = -i\delta^n_m,$$

$$I_{\bar{m}}{}^{\bar{n}} = i\delta^{\bar{n}}_{\bar{m}}, \quad I_m{}^{\bar{n}} = I_{\bar{m}}{}^n = 0. \tag{2.53}$$

One can formulate and prove the following theorem:

**Theorem 2.5.** *If the equation system* (2.52) *has $d$ independent solutions, the manifold is complex and its metric has a Hermitian form* (2.2).

**Proof.** Consider first one of the charts and express there the metric tensor $g_{MN}$ in the frame with the coordinates $\{z^n, \bar{z}^{\bar{n}}\}$. Using (2.52) and (2.46), we derive

$$h^{mn} = \frac{\partial z^m}{\partial x^M} \frac{\partial z^n}{\partial x^N} g^{MN} = -I_M{}^P I_N{}^Q \frac{\partial z^m}{\partial x^P} \frac{\partial z^n}{\partial x^Q} g^{PQ} = -h^{mn} = 0.$$

In the same way, we can derive $h^{\bar{m}\bar{n}} = 0$. It follows that $h_{mn} = h_{\bar{m}\bar{n}} = 0$ and the metric has the Hermitian form.[11]

To prove that the manifold is complex globally, we have to show that the transition functions in the overlap region of two charts parameterized by the coordinates $\{z^n, \bar{z}^{\bar{n}}\}$ and $\{w^p, \bar{w}^{\bar{p}}\}$ are holomorphic, $z^n = f^n(w^p)$, so that $\partial z^n / \partial \bar{w}^{\bar{p}} = 0$.

[11]Incidentally, the conditions (i) and (ii) in the Definition 2.6 above, which imply (2.46) were crucial in deriving this property. Sometimes, people consider the pairs $(g, I)$ that do not satisfy (2.46), but we do not do so.

As we have just shown, if the condition (2.52) is fulfilled, the tensor ${I_M}^N$ acquires the form[12] (2.53), and this is true for both charts. Assume a generic dependence $z^n = f^n(w^p, \bar{w}^{\bar{p}})$ and consider the transformations of the component of ${I_M}^N$ when going from $z$ to $w$. We derive

$$\frac{\partial w^q}{\partial z^n}\frac{\partial z^n}{\partial w^p} - \frac{\partial w^q}{\partial \bar{z}^{\bar{n}}}\frac{\partial \bar{z}^{\bar{n}}}{\partial w^p} = \delta^q_p,$$
$$\frac{\partial w^q}{\partial z^n}\frac{\partial z^n}{\partial \bar{w}^{\bar{p}}} - \frac{\partial w^q}{\partial \bar{z}^{\bar{n}}}\frac{\partial \bar{z}^{\bar{n}}}{\partial \bar{w}^{\bar{p}}} = 0. \tag{2.54}$$

On the other hand, we have the obvious relations

$$\frac{\partial w^q}{\partial z^n}\frac{\partial z^n}{\partial w^p} + \frac{\partial w^q}{\partial \bar{z}^{\bar{n}}}\frac{\partial \bar{z}^{\bar{n}}}{\partial w^p} = \delta^q_p,$$
$$\frac{\partial w^q}{\partial z^n}\frac{\partial z^n}{\partial \bar{w}^{\bar{p}}} + \frac{\partial w^q}{\partial \bar{z}^{\bar{n}}}\frac{\partial \bar{z}^{\bar{n}}}{\partial \bar{w}^{\bar{p}}} = 0. \tag{2.55}$$

It follows that

$$\frac{\partial w^q}{\partial z^n}\frac{\partial z^n}{\partial w^p} = \delta^q_p \quad \text{and} \quad \frac{\partial w^q}{\partial z^n}\frac{\partial z^n}{\partial \bar{w}^{\bar{p}}} = 0. \tag{2.56}$$

The first identity says that the matrix $\partial w^q/\partial z^n$ is not degenerate. But then the second identity dictates that $\partial z^n/\partial \bar{w}^{\bar{p}} = 0$. □

It is useful for some applications to introduce the tensors

$${Q_M}^N = \frac{\partial x^N}{\partial z^q}\frac{\partial z^q}{\partial x^M}, \quad {P_M}^N = \frac{\partial x^N}{\partial \bar{z}^q}\frac{\partial \bar{z}^q}{\partial x^M}. \tag{2.57}$$

Using (2.52) and its corollary (2.53), we derive

$${Q_M}^N = \frac{1}{2}\left(\delta^N_M + i{I_M}^N\right), \quad {P_M}^N = \frac{1}{2}\left(\delta^N_M - i{I_M}^N\right). \tag{2.58}$$

The tensor ${P_M}^N$ also enters the condition (2.52). It can be interpreted as the projector from the space of all vectors to the space of antiholomorphic vectors. Likewise, ${Q_M}^N$ is the projector onto the space of holomorphic vectors.

---

[12]It is called in this case the *complex structure*; not "almost" anymore.

The properties

$$Q_M{}^N Q_N{}^S = Q_M{}^S, \quad P_M{}^N P_N{}^S = P_M{}^S, \quad Q_M{}^N P_N{}^S = 0 \qquad (2.59)$$

hold.

It is now natural to ask in what case the system (2.52) has a solution and in what case it has not. The answer to this question is provided by the *Newlander–Nirenberg theorem* [10]:

**Theorem 2.6.** *The complex coordinates satisfying the condition* (2.52) *can be introduced and the manifold is complex iff the condition*

$$\boxed{\mathcal{N}_{MN}{}^K = \partial_{[M} I_{N]}{}^K - I_M{}^P I_N{}^Q \partial_{[P} I_{Q]}{}^K = 0} \qquad (2.60)$$

*holds.*[13]

**Proof. Necessity.** Represent the system (2.52) as $\mathcal{D}_M z^n = 0$ with

$$\mathcal{D}_M = \partial_M - i I_M{}^N \partial_N. \qquad (2.61)$$

For self-consistency, the conditions $[\mathcal{D}_M, \mathcal{D}_N] z^n = 0$ should also hold. We derive

$$\begin{aligned}
[\mathcal{D}_M, \mathcal{D}_N] z^n &= \left[-i\partial_{[M} I_{N]}{}^Q - I_{[M}{}^P (\partial_P I_{N]}{}^Q)\right] \partial_Q z^n \\
&= \left[-i\partial_{[M} I_{N]}{}^K - i I_{[M}{}^P (\partial_P I_{N]}{}^Q) I_Q{}^K\right] \partial_K z^n \\
&\quad - I_{[M}{}^P (\partial_P I_{N]}{}^Q) \mathcal{D}_Q z^n \\
&= -i \mathcal{N}_{MN}{}^K \partial_K z^n
\end{aligned} \qquad (2.62)$$

(we used $\mathcal{D}_Q z^n = 0$ and $I_Q{}^K \partial_P I_N{}^Q = -I_N{}^Q \partial_P I_Q{}^K$ that follows from $I^2 = -\mathbb{1}$). For this to vanish, the tensor $\mathcal{N}_{MN}{}^K$ should also vanish (to see that, choose the real coordinates $x^M$ as the real and imaginary parts of $z^n$).

[13] This combination is a tensor, in spite of the presence of the ordinary rather than covariant derivatives. This is so because the terms in the covariant derivatives involving the Christoffel symbols cancel out in this case. Using a sloppy language, we call the LHS of Eq. (2.60) the *Nijenhuis tensor.* A conventional definition of the Nijenhuis tensor is a little bit different:

$$\mathcal{N}_{MN}{}^{K(\text{conventional})} = I_M{}^P \mathcal{N}_{PN}{}^{K(\text{this book})} = I_M{}^P \partial_{[P} I_{N]}{}^K + I_N{}^P \partial_{[M} I_{P]}{}^K.$$

We will do so because the object (2.60) has a more transparent structure, and it is this combination that directly appears in (2.62).

**Sufficiency.** This part of the theorem [the proof of existence of the solution to the system (2.52) under the condition (2.60)] is more diffucult. Well, it might be not so difficult for mathematicians in the case when the complex structures ${I_M}^N$ represent analytic functions of the coordinates (it is more complicated when only a limited number of derivatives of ${I_M}^N$ over the coordinates exist, but we are definitely not interested in such fine mathematical issues). Then, the sufficiency of the conditions $[\mathcal{D}_M, \mathcal{D}_N] = {\mathcal{K}_{MN}}^Q \mathcal{D}_Q$ with some ${\mathcal{K}_{MN}}^Q$ (in our case, ${\mathcal{K}_{MN}}^Q = -{I_{[M}}^P(\partial_P {I_{N]}}^Q)$ for the equation system $\mathcal{D}_M z^n = 0$ to have a solution is a corollary of the classical Frobenius theorem [11]. We give here another "physical" proof of this fact [12] which will elucidate the meaning of the constraint (2.60). Its linearized version is similar in spirit to multidimensional Cauchy–Riemann conditions:

- Let the complex structure ${I_M}^N$ have a canonic form[14] (2.51). Then, the solutions to (2.52) obviously exist. One of the solutions is

$$z^1_{(0)} = \frac{x^1 + ix^2}{\sqrt{2}}, \quad z^2_{(0)} = \frac{x^3 + ix^4}{\sqrt{2}}, \ldots, \tag{2.63}$$

  as in (2.16), but any analytic function of $z^n_{(0)}$ is also a solution.

  Suppose now that the complex structure does not coincide with ${(I_0)_M}^N = \mathrm{diag}(-\varepsilon, \ldots, -\varepsilon)$ but is close to it: $I = I_0 + \Delta$, $\Delta \ll 1$. As a first step in the proof, we show that, after such an infinitesimal deformation, solutions to (2.52) still exist.
- Let us first do it in the simplest case $D = 2$. Then, the condition (2.60) is in fact not a constraint: it is fulfilled identically. For our purposes, it is sufficient to check it in the first order in $\Delta$. The LHS of (2.60) is reduced in this case to

$$\begin{aligned} &\partial_1 {\Delta_2}^K - \partial_2 {\Delta_1}^K - \varepsilon_{1P}\varepsilon_{2Q}(\partial_P {\Delta_Q}^K - \partial_Q {\Delta_P}^K) \\ &= \partial_1 {\Delta_2}^K - \partial_2 {\Delta_1}^K + (\partial_2 {\Delta_1}^K - \partial_1 {\Delta_2}^K) = 0. \end{aligned}$$

  The condition $I^2 = -\mathbb{1}$ means that $\{\Delta, I_0\} = 0$. This implies that $\Delta = \alpha\sigma^1 + \beta\sigma^3$ or

$$\Delta^1_1 = -\Delta^2_2, \quad \Delta^2_1 = \Delta^1_2. \tag{2.64}$$

---

[14] We know that, for any almost complex structure, its tangent space projection $I_{AB}$ can always be chosen in this form. But the requirement that the same is true for ${I_M}^N$ is not trivial.

Look now at the system (2.52). We set $z = z_{(0)} + \delta z$. The equations acquire the form

$$\frac{\partial}{\partial x^1}(\delta z) + i\frac{\partial}{\partial x^2}(\delta z) = \frac{i\Delta_1{}^1 - \Delta_1{}^2}{\sqrt{2}},$$
$$\frac{\partial}{\partial x^2}(\delta z) - i\frac{\partial}{\partial x^1}(\delta z) = \frac{i\Delta_2{}^1 - \Delta_2{}^2}{\sqrt{2}}. \tag{2.65}$$

Bearing in mind (2.64), these two equations coincide. They can be expressed as[15]

$$\frac{\partial(\delta z)}{\partial \bar{z}_{(0)}} = \frac{i}{2}\Delta_1{}^{1+i2}, \tag{2.66}$$

which can be easily integrated on complex plane.

- The simplest nontrivial case is $D = 4$. The condition $\{\Delta, I_0\} = 0$ implies

$$\Delta_1{}^1 = -\Delta_2{}^2, \quad \Delta_1{}^2 = \Delta_2{}^1,$$
$$\Delta_1{}^3 = -\Delta_2{}^4, \quad \Delta_1{}^4 = \Delta_2{}^3,$$
$$\Delta_3{}^1 = -\Delta_4{}^2, \quad \Delta_3{}^2 = \Delta_4{}^1,$$
$$\Delta_3{}^3 = -\Delta_4{}^4, \quad \Delta_3{}^4 = \Delta_4{}^3. \tag{2.67}$$

We pose $z^1 \to z, z^2 \to w$. A short calculation shows that, bearing the relations (2.67) in mind, Eqs. (2.52) are reduced to

$$\frac{\partial(\delta z)}{\partial \bar{z}_{(0)}} = \frac{i}{2}\Delta_1{}^{1+i2},$$
$$\frac{\partial(\delta z)}{\partial \bar{w}_{(0)}} = \frac{i}{2}\Delta_3{}^{1+i2},$$
$$\frac{\partial(\delta w)}{\partial \bar{z}_{(0)}} = \frac{i}{2}\Delta_1{}^{3+i4},$$
$$\frac{\partial(\delta w)}{\partial \bar{w}_{(0)}} = \frac{i}{2}\Delta_3{}^{3+i4}. \tag{2.68}$$

If $D > 2$, the conditions (2.60) provide nontrivial constraints. Their linearized version is

$$\partial_P \Delta_N{}^M - \partial_N \Delta_P{}^M = (I_0)_P{}^Q (I_0)_N{}^S \left[\partial_Q \Delta_S{}^M - \partial_S \Delta_Q{}^M\right]. \tag{2.69}$$

[15]The notation $\Delta_M^{1+i2} = \Delta_M^1 + i\Delta_M^2$ is introduced.

Again, bearing in mind (2.67), one can show that, for $D = 4$, out of 24 real conditions in (2.69), only 4 independent real or 2 independent complex constraints are left. The latter have a simple form

$$\frac{\partial}{\partial \bar{z}_{(0)}} {\Delta_3}^{1+i2} - \frac{\partial}{\partial \bar{w}_{(0)}} {\Delta_1}^{1+i2} = 0,$$
$$\frac{\partial}{\partial \bar{z}_{(0)}} {\Delta_3}^{3+i4} - \frac{\partial}{\partial \bar{w}_{(0)}} {\Delta_1}^{3+i4} = 0. \qquad (2.70)$$

The first equation in (2.70) is the integrability condition for the system of the first two equations in (2.68). It is *sufficient* for the solution of this system to exist. Indeed, it implies that the (0,1)-form

$$\omega = {\Delta_1}^{1+i2}\, d\bar{z}_{(0)} + {\Delta_3}^{1+i2}\, d\bar{w}_{(0)}$$

is closed, $\bar{\partial}_0\, \omega = 0$. Bearing in mind the trivial topology of a chart of our complex manifold that we are discussing, $\omega$ is also exact (see Theorem 1.2), which is tantamount to saying that the solution exists. The second relation in (2.70) is the necessary and sufficient integrability condition for the system of the third and fourth equations in (2.68).

- This reasoning can be translated to the case of higher dimensions. For an arbitrary $D = 2d$, Eqs. (2.52) are reduced, bearing in mind $I^2 = -\mathbb{1} \Rightarrow [\Delta, I_0] = 0$, to $d^2$ conditions similar to (2.68) involving differentiation over each antiholomorphic variable $\bar{z}^{\bar{n}}_{(0)}$ for each complex function $\delta z^n$. The conditions (2.60) lead to $d^2(d-1)/2$ complex constraints which represent integrability conditions of the type (2.70). They imply that the forms

$$\omega_1 = {\Delta_1}^{1+i2}\, d\bar{z}^1_{(0)} + {\Delta_3}^{1+i2}\, d\bar{z}^2_{(0)} + \cdots,$$
$$\omega_2 = {\Delta_1}^{3+i4}\, d\bar{z}^1_{(0)} + {\Delta_3}^{3+i4}\, d\bar{z}^2_{(0)} + \cdots, \qquad (2.71)$$

etc. are all closed. Due to the trivial topology of the chart, it also means that they are exact.

- Once the complex coordinates $z = z_{(0)} + \delta z$ satisfying Eqs. (2.52) are found, the complex structure acquires in these new coordinates the canonical form (2.53) and (2.51). Thus, we have actually proven that, after a small deformation, ${I_M}^N$ can be brought to the form (2.51) by an infitesimal diffeomorphism, provided the condition (2.60) is satisfied.
- Let now the complex structure be arbitrary, not necessarily close to $I_0$ of Eq. (2.51). If the tensor ${I_M}^N(x)$ is analytic in a given chart (and we

*assume* that), we can expand it into a formal series in a small parameter $\alpha$:

$$I(x) = I_0 + \alpha I_1(x) + \alpha^2 I_2(x) + \cdots \tag{2.72}$$

Do the same for the solutions $z^n(x)$ we are looking for:

$$z^n(x) = z^n_{(0)} + \alpha z^n_{(1)}(x) + \alpha^2 z^n_{(2)}(x) + \cdots \tag{2.73}$$

The correction $\alpha z^n_{(1)}(x)$ was determined before. Let $\tilde{z}^n(x) = z^n_{(0)} + \alpha z^n_{(1)}(x)$. As was just mentioned, the complex structure in these new coordinates has the canonical form (2.53) up to the terms $\propto \alpha^2$. Introducing the real and imaginary parts of $\tilde{z}^n(x)$ and calling them $\tilde{x}^M$, we may bring it in the form (2.51).

- Taking also into account the term $\alpha^2 I_2(x)$ in (2.72), we may express the complex structure in the new coordinates $\tilde{x}$ as

$$I(\tilde{x}) = I_0 + \alpha^2 \tilde{I}_2(\tilde{x}) + \text{ higher-order terms.} \tag{2.74}$$

Repeating the same procedure that we used to determine $z^n_{(1)}(x)$, we can now determine $\tilde{z}^n_{(2)}(\tilde{x})$, determine from that $z^n_{(2)}(x)$, and likewise all the terms in the series (2.73).

- With the only reservation that we did not address a difficult question of the convergence of the series (2.73), the theorem is proven.

□

In Chapter 9, we give a supersymmetric interpretation of this theorem.

The next important observation is as follows:

**Theorem 2.7.** *The complex manifold is Kähler iff the ordinary covariant derivatives of its complex structure vanish.*

**Proof.** Choose the frame where the complex structure has the form (2.53). One can then derive $\nabla_P I_m{}^n = \Gamma^n_{Pq} I_m{}^q - \Gamma^q_{Pm} I_q{}^n = 0$ for both $P = p$ and $P = \bar{p}$. The covariant derivative $\nabla_P I_{\bar{m}}{}^{\bar{n}}$ also vanishes. On the other hand,

$$\begin{aligned}
\nabla_P I_m{}^{\bar{n}} &= \Gamma^{\bar{n}}_{Pq} I_m{}^q - \Gamma^{\bar{q}}_{Pm} I_{\bar{q}}{}^{\bar{n}} = -2i\Gamma^{\bar{n}}_{Pm}, \\
\nabla_P I_{\bar{m}}{}^n &= \Gamma^n_{P\bar{q}} I_{\bar{m}}{}^{\bar{q}} - \Gamma^q_{P\bar{m}} I_q{}^n = 2i\Gamma^n_{P\bar{m}}.
\end{aligned} \tag{2.75}$$

These derivatives vanish iff the Christoffels of mixed holomorphicity vanish, i.e. iff the manifold is Kähler. □

Thus, the condition $\nabla_P I_{MN} = 0$ can serve as an alternative definition of the Kähler manifolds.

For a generic complex manifold, the Levi-Civita covariant derivatives (2.75) do not vanish. On the other hand, the complex structure *is* covariantly constant with respect to a class of affine connections with vanishing components $G^{\bar{n}}_{Pq}$ and $G^{n}_{P\bar{q}}$, as a glance at (2.75) reveals. Another glance at Eq. (2.21) reveals that the *Bismut* connection belongs to this class. We arrive at the following:

**Theorem 2.8.** *The complex structure of a generic complex manifold is covariantly constant with respect to its Bismut connection.*

There is a beautiful formula that expresses the torsion tensor of the Bismut connection in the real terms. It reads [13, 14]

$$\boxed{C_{MNP} = I_M{}^Q I_N{}^S I_P{}^R \left(\nabla_Q I_{SR} + \nabla_S I_{RQ} + \nabla_R I_{QS}\right),} \tag{2.76}$$

where $\nabla_M$ is the ordinary Levi-Civita covariant derivative.[16] Its validity is easy to check going into the frame $M \equiv \{m, \bar{m}\}$, using (2.53) and comparing (2.76) with (2.18).

## 2.3 Examples

Let us now see how all these general constructions work in some particular cases:

(1) $S^2 \equiv \mathbb{CP}^1$. We already discussed it earlier. Within the standard latitude-longitude parametrization, its metric is $ds^2 = d\theta^2 + \sin^2\theta\, d\phi^2$. The complex structure was written in (2.48). The reader can be easily convinced that the Nijenhuis tensor (2.60) vanishes, as it should.

The complex coordinates for the south and north hemispheres can be chosen as in (2.4) and (2.5) and the metric as in (2.6). This metric is derived from the Kähler potential:

$$h = \frac{1}{(1+z\bar{z})^2} = \partial\bar{\partial}\ln(1+z\bar{z}). \tag{2.77}$$

The Kähler form $K = ih\, dz \wedge d\bar{z}$ is closed.

[16] Actually, one can replace here the covariant derivatives by the ordinary ones because the contribution of the symmetric Christoffel symbols cancels out in the sum. But we have written $\nabla I$ to emphasize the tensor nature of $C_{MNP}$.

(2) Not all four-dimensional manifolds are complex ($S^4$ is not) and not all complex $D = 4$ manifolds are Kähler. As an example of a complex but not Kähler manifold, consider the *Hopf manifold* [15] $\mathcal{H}^2$. Topologically, it is $S^3 \times S^1$. It can be described by a couple of complex coordinates $z^{m=1,2}$ lying in the spherical band $1 \leq |z| < 2$, with the points $z^m$ on the border $|z| = 1$ and the points $2z^m$ being identified. We choose the conformally invariant metric

$$ds^2 = \frac{dz^m d\bar{z}^m}{z^k \bar{z}^k}, \tag{2.78}$$

which is consistent with this identification. Obviously, the metric (2.78) does not satisfy (2.28) and is not Kähler.

(3) An example of a Kähler manifold of complex dimension 2 is the complex projective space $\mathbb{CP}^2$. It is a set of complex triples $(w^0, w^1, w^2)$ identified under the multiplication by a nonzero complex number $\lambda$:

$$(w^0, w^1, w^2) \equiv (\lambda w^0, \lambda w^1, \lambda w^2).$$

A natural symmetrically looking "round" metric on this space reads

$$ds^2 = 2\frac{\bar{w}^\alpha w^\alpha \, d\bar{w}^\beta dw^\beta - \bar{w}^\alpha w^\beta \, d\bar{w}^\beta dw^\alpha}{(\bar{w}^\alpha w^\alpha)^2} \tag{2.79}$$

($\alpha = 0, 1, 2$). Now, $\mathbb{CP}^2$ is topologically nontrivial, but it can be covered by three disks with the topology $\mathbb{C}^2$. One of these charts excludes the points with $w^0 = 0$ (the set of all such points has the topology $\mathbb{CP}^1$). The complex coordinates uniquely describing the points on this chart can be chosen as $z^{1,2} = w^{1,2}/w^0$. Then, the metric (2.79) reduces to the *Fubini–Study* form:

$$\boxed{h_{m\bar{n}} = \partial_m \partial_{\bar{n}} \, \ln(1 + z\bar{z}) = \frac{1}{1 + z\bar{z}} \left( \delta_{m\bar{n}} - \frac{z^n \bar{z}^m}{1 + z\bar{z}} \right)}, \tag{2.80}$$

$z\bar{z} \equiv z^k \bar{z}^{\bar{k}}$, $k = 1, 2$. Evidently, this metric is Kähler.

The second chart excludes the points with $w^1 = 0$ and is conveniently described by the coordinates $u^1 = w^0/w^1, u^2 = w^2/w^1$. The third chart excludes the points with $w^2 = 0$ and is conveniently described by the coordinates $v^1 = w^1/w^2, v^2 = w^0/w^2$. The metrics in the second and the third charts expressed in these coordinates have exactly the same functional form as the metric (2.80).

A generalization for $\mathbb{CP}^{n>2}$ is rather obvious. This manifold represents a set of complex $(n+1)$-tuples, $(w^0, \ldots, w^n)$, identified under multiplication by $\lambda \neq 0$. It has complex dimension $n$. The minimal number of the charts that cover $\mathbb{CP}^n$ is $n+1$. In each chart, the metric can be chosen in the form (2.80).

(4) $S^6$. As was mentioned, we do not know today whether this manifold admits a complex structure or not. But an almost complex structure exists. It represents a generalization of the complex structure for $S^2$ expressed as the three-dimensional antisymmetric tensor (2.50) projected onto the surface $|\mathbf{x}| = 1$.

Similarly, we choose an almost complex structure on $S^6$ as

$$I_{MN}(S^6) = f_{MNP}\, x^P, \tag{2.81}$$

where $M, N, P = 1, \ldots, 7$, the condition $|\mathbf{x}| = 1$ is imposed, and $f_{MNP}$ is a totally antisymmetric tensor describing the algebra of octonions. One of the choices for $f_{MNP}$ is

$$f_{123} = f_{176} = f_{257} = f_{365} = f_{246} = f_{145} = f_{347} = 1, \tag{2.82}$$

with all other nonzero components being restored by antisymmetry.

Obviously, $I_{NM} = -I_{MN}$. The validity of $I_M{}^P I_P{}^N = -\delta_M^N$ is best seen if projecting (2.81) on the tangent plane. For example, for the "north pole" $x^P = \delta^{P7}$, we obtain

$$I_{AB} = \begin{pmatrix} 0 & 0 & 0 & 0 & 0 & -1 \\ 0 & 0 & 0 & 0 & 1 & 0 \\ 0 & 0 & 0 & 1 & 0 & 0 \\ 0 & 0 & -1 & 0 & 0 & 0 \\ 0 & -1 & 0 & 0 & 0 & 0 \\ 1 & 0 & 0 & 0 & 0 & 0 \end{pmatrix}, \tag{2.83}$$

and the property $I^2 = -\mathbb{1}$ is obvious. Thus, it is an almost complex structure, indeed. But it is not a complex structure: the Nijenhuis tensor (2.60) calculated with (2.81) does not vanish.

# Chapter 3

# Hyperkähler and HKT Manifolds

## 3.1 Hyperkähler manifolds

This notion was introduced in mathematics relatively late [16]. We begin with the following:

**Definition 3.1.** A *hypercomplex* manifold is a manifold admitting three different integrable complex structures $I^p$ that satisfy the quaternion algebra:

$$\boxed{(I^p)_M{}^R (I^q)_R{}^N = -\delta^{pq}\delta_M^N + \varepsilon^{pqr}(I^r)_M{}^N.} \tag{3.1}$$

The complex structures $(I^p)_{MN}$, as well as $(I^p)_{AB} = e_A^M e_B^N (I^p)_{MN}$, are skew-symmetric matrices. The following theorem holds:

**Theorem 3.1.** *The dimension of a hypercomplex manifold is a multiple integer of* 4. *Any triple* $(I^p)_{AB}$ *of skew-symmetric matrices satisfying the algebra* (3.1) *can be brought by an appropriate choice of basis to the following canonical form:*

$$\begin{aligned} &I^1 \equiv I = \mathrm{diag}(\mathfrak{I}, \ldots, \mathfrak{I}), \quad I^2 \equiv J = \mathrm{diag}(\mathfrak{J}, \ldots, \mathfrak{J}), \\ &I^3 \equiv K = \mathrm{diag}(\mathcal{K}, \ldots, \mathcal{K}), \end{aligned} \tag{3.2}$$

*where*

$$\mathcal{I} = \begin{pmatrix} 0 & -1 & 0 & 0 \\ 1 & 0 & 0 & 0 \\ 0 & 0 & 0 & -1 \\ 0 & 0 & 1 & 0 \end{pmatrix}, \quad \mathcal{J} = \begin{pmatrix} 0 & 0 & -1 & 0 \\ 0 & 0 & 0 & 1 \\ 1 & 0 & 0 & 0 \\ 0 & -1 & 0 & 0 \end{pmatrix},$$

$$\mathcal{K} = \begin{pmatrix} 0 & 0 & 0 & -1 \\ 0 & 0 & -1 & 0 \\ 0 & 1 & 0 & 0 \\ 1 & 0 & 0 & 0 \end{pmatrix}. \tag{3.3}$$

**Proof.** The proof follows the same pattern as the proof of Theorem 2.4 for the possibility to choose the canonical form (2.51) of a single complex structure. Choose a unit vector $e_0$ in the tangent space $E$. The algebra (3.1) and the skew symmetry of $I^p_{AB}$ dictate that the vectors $e_0$ and $e_p = I^p e_0$ have also unit length and are mutually orthogonal. Thus, the dimension of the tangent space is at least 4. In the subspace spanned by the vectors $e_0, e_1, e_2, e_3$, the matrices $I^p$ have the form (3.3) (e.g. $I^1 e_2 = I^1 I^2 e_0 = I^3 e_0 = e_3$.). Now, consider the space $E^* \subset E$ orthogonal to $e_0$ and $e_p$. If it is not empty, choose there a unit vector $f_0$ and add this vector, as well as the vectors $f_p = I^p f_0 \in E^*$, to the basis. Thus, the dimension of $E^*$ is either 0 or at least 4. Consider now the subspace $E^{**} \subset E^* \subset E$ orthogonal to $(e_0, e_p, f_0, f_p)$. If it is not empty, repeat the procedure. We arrive at (3.2). □

For sure, the vector $e_0$ is not special, and one could start with e.g. $e_1$ and restore the vectors $e_{0,2,3}$ as $e_0 = -I^1 e_1, e_2 = I^3 e_1, e_3 = -I^2 e_1$. In the following, we restore the "democracy" between the basis vectors and use the notation $(e_0, e_p) \to (e_1, e_2, e_3, e_4)$, $(f_0, f_p) \to (e_5, e_6, e_7, e_8)$, etc. Then, $\mathcal{I}_{12} = -\mathcal{I}_{21} = -1, \mathcal{I}_{34} = -\mathcal{I}_{43} = -1$, $\mathcal{J}_{13} = -\mathcal{J}_{31} = -1$, etc.

**Remark:** For the complex manifolds with a single complex structure, one can always choose the coordinates such that the tensor $I_M{}^N$ acquires the canonical form (2.53). For a hypercomplex manifold, this is always possible to do for one of the complex structures but not for all of them. Generically, $J_M{}^N$ and $K_M{}^N$ would have in this case a nontrivial complicated form.[1]

[1]They obey, however, certain restrictions spelled out in Theorems 3.5 and 3.8.

The matrices (3.3) are skew-symmetric and belong to the algebra $so(4)$. But an important observation is that they are *self-dual*:

$$(\mathcal{I}, \mathcal{J}, \mathcal{K})_{AB} = \frac{1}{2}\,\varepsilon_{ABCD}(\mathcal{I}, \mathcal{J}, \mathcal{K})_{CD}. \tag{3.4}$$

This means that $\mathcal{I}, \mathcal{J}, \mathcal{K}$ belong to one of the $su(2)$-components of $so(4) = su(2) \oplus su(2)$.

A reader-physicist may recognize in these matrices the so-called *'t Hooft symbols* $\eta^p_{AB}$ [17]. The latter are defined as follows[2]:

$$\boxed{\eta^p_{44} = 0, \quad \eta^p_{a4} = -\eta^p_{4a} = \delta_{ap}, \quad \eta^p_{ab} = \varepsilon_{abp}} \qquad (a, b = 1, 2, 3). \tag{3.6}$$

The identity

$$\eta^p_{AB}\eta^q_{BC} = -\delta^{pq}\delta_{AC} - \varepsilon^{pqr}\eta^r_{AC} \tag{3.7}$$

holds. In these terms,

$$\mathcal{I}_{AB} = -\eta^3_{AB}, \quad \mathcal{J}_{AB} = \eta^2_{AB}, \quad \mathcal{K}_{AB} = -\eta^1_{AB}. \tag{3.8}$$

We now introduce the convenient symplectic notation which plays the same role for hypercomplex manifolds as the complex notation for complex manifolds. Represent a vector tangent-space index $A$ by the pair of indices $(ia)$, where $i = 1, 2$ and $a = 1, \ldots 2n$. We raise and lower the indices according to $X^i = \varepsilon^{ij}X_j, Y^a = \Omega^{ab}Y_b$, where $\varepsilon^{jk} = -\varepsilon_{jk}$ with the convention $\varepsilon_{12} = 1$ and $\Omega^{ab}$ is the symplectic matrix, which we choose in the form $\Omega^{ab} = \text{diag}(-\varepsilon, \ldots, -\varepsilon)$ [like in Eq. (2.51), though that formula had, of course, a completely different meaning]; $\Omega_{ab} = -\Omega^{ab} = \text{diag}(\varepsilon, \ldots, \varepsilon)$. In this notation, the subgroup $SU(2) \times Sp(n)$ of the tangent space rotation group plays a special role. These restricted rotations leave the matrix $\varepsilon^{ij}\Omega^{ab}$ invariant.

---

[2]In the following, we also need the dual 't Hooft symbols $\bar\eta^p_{AB}$, which differ from $\eta^p_{AB}$ by the signs for the components with $B = 4$ or $A = 4$:

$$\bar\eta^p_{a4} = -\bar\eta^p_{4a} = -\delta_{ap}, \quad \bar\eta^p_{ab} = \varepsilon_{abp}. \tag{3.5}$$

The symbols $\bar\eta^p_{AB}$ are anti-self-dual.

To establish the precise relation between the vector and symplectic notations, we introduce $4n$ rectangular matrices $\Sigma_A$:

$$(\Sigma_{1,2,3,4})_{aj} = (\sigma_\mu, 0, \dots)_{aj}, \quad (\Sigma_{5,6,7,8})_{aj} = (0, \sigma_\mu, 0, \dots)_{aj}, \quad \text{etc.}$$
$$(\Sigma^\dagger_{1,2,3,4})^{ja} = (\sigma^\dagger_\mu, 0, \dots)^{ja}, \quad (\Sigma^\dagger_{5,6,7,8})^{ja} = (0, \sigma^\dagger_\mu, 0, \dots)^{ja}, \quad \text{etc.} \tag{3.9}$$

with[3]

$$(\sigma_\mu)_{aj} = (\boldsymbol{\sigma}, i\mathbb{1})_{aj} \quad (\sigma^\dagger_\mu)^{ja} = (\boldsymbol{\sigma}, -i\mathbb{1})^{ja}, \tag{3.11}$$

where $\boldsymbol{\sigma}$ are the Hermitian *Pauli matrices*:

$$\sigma^1 = \begin{pmatrix} 0 & 1 \\ 1 & 0 \end{pmatrix}, \quad \sigma^2 = \begin{pmatrix} 0 & -i \\ i & 0 \end{pmatrix}, \quad \sigma^3 = \begin{pmatrix} 1 & 0 \\ 0 & -1 \end{pmatrix}. \tag{3.12}$$

Then, for any vector (in the flat Euclidean tangent space), we have the correspondence

$$V_{aj} = -\frac{i}{\sqrt{2}} (\Sigma_A)_{aj} V_A, \quad V^{ja} = \frac{i}{\sqrt{2}} (\Sigma^\dagger_A)^{ja} V_A,$$
$$V_A = \frac{i}{\sqrt{2}} V_{aj} (\Sigma^\dagger_A)^{ja} = -\frac{i}{\sqrt{2}} V^{ja} (\Sigma_A)_{aj} \tag{3.13}$$

and similarly for tensors. In these terms, the flat metric is expressed as

$$g_{aj,\,bk} = -\frac{1}{2} (\Sigma_A)_{aj} (\Sigma_A)_{bk} = \varepsilon_{jk} \Omega_{ab}, \quad g^{ja,\,kb} = \varepsilon^{jk} \Omega^{ab}. \tag{3.14}$$

It follows:

$$X_A Y_A = X_{ai} Y^{ia} = Y_{ai} X^{ia}. \tag{3.15}$$

Note that, for a real vector $V^A$, the symplectic components obey the *pseudoreality* condition

$$\overline{V^{ja}} = V_{aj} = \epsilon_{jk} \Omega_{ab} V^{kb}. \tag{3.16}$$

In the symplectic notation, the canonical flat complex structures (3.2), (3.3) can be expressed as

$$(I^p)_{aj,bk} = -\frac{1}{2} (\Sigma_A)_{aj} (\Sigma_B)_{bk} (I^p)_{AB} = -i(\sigma^3, \sigma^2, \sigma^1)_{jk}\, \Omega_{ab}, \tag{3.17}$$

---

[3] Note the identities

$$\sigma^\dagger_\mu \sigma_\nu = \delta_{\mu\nu} + i\eta^p_{\mu\nu} \sigma^p, \quad \sigma_\mu \sigma^\dagger_\nu = \delta_{\mu\nu} + i\bar\eta^p_{\mu\nu} \sigma^p,$$
$$\varepsilon_{jk} \varepsilon_{ab} (\sigma^\dagger_\mu)^{kb} = -(\sigma_\mu)_{aj}. \tag{3.10}$$

where

$$(\sigma^p)_{jk} = \varepsilon_{jl}(\sigma^p)_k{}^l, \quad (\sigma^p)^{jk} = \varepsilon^{jl}(\sigma^p)_l{}^k \tag{3.18}$$

and $(\sigma^p)_k{}^l$ are the Pauli matrices (3.12). The matrices (3.18) are symmetric.[4] In the following, when we need to express the complex structures in the symplectic notation, we use another, more convenient representation:

$$(I^p)_{aj,bk} = i(\sigma^p)_{jk}\,\Omega_{ab} \quad \text{and} \quad (I^p)^{ja,kb} = i(\sigma^p)^{jk}\Omega^{ab}. \tag{3.19}$$

This corresponds to another choice of the complex structures: $(I, J, K) \to (-K, -J, -I)$. The triple $(-K, -J, -I)$ is also quaternionic.

**Definition 3.2.** A *hyperkähler* manifold is a hypercomplex manifold with all three complex structures covariantly constant with respect to the Levi-Civita connection.

**Remark:** In fact, it is not necessary to stipulate the hypercomplex nature of the manifold in the definition. It is sufficient to require the presence of two different *anticommuting* covariantly constant complex structures $I, J$. Indeed, it is not difficult to prove that the matrix $K = IJ$ is also in this case a covariantly constant complex structure.

A hyperkähler manifold is Kähler with respect to any complex structure (see Theorem 2.7). As was discussed in the previous chapter, the holonomy group of a generic Kähler manifold of complex dimension $2n$ is $U(2n) \subset SO(4n)$. But the presence of two extra complex structures brings about additional constraints on the holonomy group. We now prove an important theorem:

**Theorem 3.2.** *A manifold of real dimension $4n$ is hyperkähler iff its holonomy group is $Sp(n) \subset U(2n) \subset SO(4n)$.*

**Proof.** This statement is very well known to mathematicians. We give here its detailed proof in explicit "physical" terms.

---

[4] Note that the matrices $(\sigma^p)^{ja}$ in (3.11) (with both upper or both lower indices) and $(\sigma^p)_l{}^k$ in (3.18) (with the indices placed at the different levels) coincide. The difference in conventions is justified by the fact that the indices in (3.18) refer to one and the same subgroup $SU(2) \subset SO(4n)$, whereas the indices $j, a$ in (3.11) have a different nature.

**Direct theorem**

We prove first the direct theorem: the holonomy group of a hyperkähler manifold is $Sp(n)$. After passing to the tangent space, the covariant constancy condition $\nabla_M I^p_{NS} = 0$ for the triplet of the complex structures $I^p_{NS}$ takes the form

$$\partial_M I^p_{AB} + (\omega_{AC,M}\, I^p_{CB} - I^p_{AC}\, \omega_{CB,M}) = 0, \tag{3.20}$$

where $\omega_{AB,P}$ is the spin connection (1.58) with $G^N_{MP} = \Gamma^N_{MP}$. Substituting the constant expressions (3.19) for $I^p_{AB}$ in (3.20), we observe that this condition reduces to

$$\boxed{\omega_{AC,M}\, I^p_{CB} - I^p_{AC}\, \omega_{CB,M} = 0,} \tag{3.21}$$

which tells us that the spin connection, understood as a matrix $\omega_{AB}$ in tangent space whose entries are 1-forms, commutes with all complex structures.

The four-dimensional case ($n = 1$) is especially simple. As was noted, the complex structures (3.3) are the elements of $so(4)$ forming a basis in one of its $su(2)$ components. A skew-symmetric matrix $\omega_{AB}$ that commutes with them belongs to another $su(2)$ components. If $\omega_{AB} \in su(2)$, the same concerns the curvature form $R_{AB}$. And[5] $su(2) \equiv sp(1)$.

If $n > 1$, we use the symplectic notation introduced above. Bearing in mind (3.19) and (3.15), the condition (3.21) takes the form

$$\omega_{ai\,bj}\, (\sigma^p)^j{}_k + (\sigma^p)_i{}^j\, \omega_{aj\,bk} = 0. \tag{3.22}$$

A generic antisymmetric connection $\omega_{ai\,bj} = -\omega_{bj\,ai}$ can be parameterized as

$$\omega_{ai\,bj} = \varepsilon_{ij} T_{(ab)} + B_{[ab]\,(ij)} = \varepsilon_{ij} T_{(ab)} + B^q_{[ab]} (\sigma^q)_{ij} \tag{3.23}$$

with arbitrary $T_{(ab)}$ and $B^q_{[ab]}$. When one substitutes this into (3.22), the first term $\propto \varepsilon_{ij}$ does not contribute and we are led to

$$B^q_{[ab]}\, [\sigma^q, \sigma^p]_{ik} = 0. \tag{3.24}$$

This holds for any $p$, which implies

$$B^q_{[ab]} = 0. \tag{3.25}$$

[5] It is instructive to rediscuss from this angle the holonomy group of a Kähler four-dimensional manifold equiped with only one complex structure $\mathcal{J}$. A spin connection that commutes with $\mathcal{J}$ belongs to the subalgebra $su(2) \times u(1) = u(2)$, and the corresponding holonomy group is $U(2)$.

We thus derived

$$\omega_{AB} = \omega_{ai\,bj} = \varepsilon_{ij}\, T_{(ab)}. \tag{3.26}$$

But any symmetric matrix of dimension $2n$ can be presented as

$$T_{(ab)} = T_a{}^c\, \Omega_{bc}, \tag{3.27}$$

where[6] $T_a{}^c \in sp(n)$. Thus, $\omega_{ai\,bj}$ belongs to $sp(n)$. But then $R_{ai\,bj} = d\omega_{ai\,bj} + \omega_{ai\,ck} \wedge \omega^{kc}{}_{bj}$ also belongs to $sp(n)$, and the theorem is proven.

**Inverse theorem**

The proof of the inverse theorem — that, if the holonomy group is $Sp(n)$, the existence of three quaternionic covariantly constant complex structures follows — is based on the following lemma:

**Lemma.** Let $\mathfrak{g}$ be a Lie algebra and $\mathfrak{h}$ be its subalgebra. Let $A_M$ and $F_{MN}$ be the gauge potential and the field density for the algebra $\mathfrak{g}$. Let $F_{MN} \in \mathfrak{h}$. Then, one can always choose a gauge where also $A_M \in \mathfrak{h}$.

**Proof.** Such a gauge is well known; it is the Fock–Schwinger gauge $x^M A_M \doteq 0$ [18]. In this gauge, the potential is expressed via the field density as

$$\boxed{A_M = \int_0^1 d\alpha\, \alpha x^N F_{NM}(\alpha x).} \tag{3.28}$$

Indeed, in the gauge $x^M A_M = 0$, we may write

$$\begin{aligned} A_M(y) &= \frac{\partial}{\partial y^M}[y^P A_P(y)] - y^P \frac{\partial A_P(y)}{\partial y^M} \\ &= y^P \left[ F_{PM}(y) - \frac{\partial A_M(y)}{\partial y^P} \right]. \end{aligned} \tag{3.29}$$

Substitute $\alpha x$ for $y$ and rewrite it in the form

$$\alpha x^P F_{PM}(\alpha x) = A_M(\alpha x) + x^P \frac{\partial}{\partial x^P} A_M(\alpha x) = \frac{\partial}{\partial \alpha}[\alpha A_M(\alpha x)]. \tag{3.30}$$

Integrating this over $\alpha$ between 0 and 1, we arrive at (3.28). □

[6] Indeed, an element $h$ of $sp(n)$ is a Hermitian $2n$-dimensional matrix satisfying $h^T\Omega + \Omega h = 0$.

In the case of interest, $\mathfrak{g} = so(4n)$, $\mathfrak{h} = sp(n)$, $A \equiv \omega$ and $F \equiv R$. If $R \in sp(n)$, one can choose the coordinates and vielbeins with $\omega \in sp(n)$. And once $\omega \in sp(n)$, it commutes with the quaternionic complex structures (3.19). Bearing in mind (3.20), it follows that the convolutions of these flat structures with the vielbeins are covariantly constant. □

The proven theorem means that one can *define* a hyperkähler manifold as a manifold where the Riemann curvature form $R_{AB}$ lies in the $sp(n)$ algebra. This definition and Definition 3.2 are equivalent.

Other important theorems stipulate the conditions under which a Kähler manifold with complex structure $I$ admits two extra covariantly constant complex structures $J, K$ forming together with $I$ a quaternionic triple. It is convenient for us to present and discuss them not here but in Chapter 11 together with the discussion of a supersymmetric method to construct hyperkähler metric (see Theorems 11.3 and 11.4).

### 3.1.1 *Examples*

Evidently, flat space $\mathbb{R}^{4n}$ is a hyperkähler manifold. The flat complex structures are the constant matrices (3.2). $\mathbb{R}^{4n}$ is not compact, but one can easily impose periodic boundary conditions for all Cartesian coordinates to obtain a compact flat torus.

There are many nontrivial compact hyperkähler manifolds, but in four dimensions, there is only one such example — the so-called $K3$ manifold. It is, however, rather complicated and is not discussed here.

We describe instead two most popular and most discussed in the literature examples of the *noncompact* hyperkähler four-dimensional manifolds: the Taub–NUT manifold [19] and the Eguchi–Hanson manifold [20]. According to the theorem just proven, the curvature form $R_{AB}$ of a four-dimensional hyperkähler manifold belongs to $sp(1) \equiv su(2)$. This means that $R_{AB}$ belongs to one of the $su(2)$ components of $so(4)$ and is either self-dual or anti-self-dual. If the complex structures are chosen to be self-dual, as in (3.3), the form $R_{AB}$, which commutes with them, must be anti-self-dual, and vice versa.

Another (physical) name for four-dimensional hyperkähler manifolds is *gravitational instantons*. The point is that the Ricci tensor $(Ricci)_{MN}$ for these manifolds vanish and they thus represent solutions to the Einstein equations in empty Euclidean space, $(Ricci)_{MN} = 0$.

Indeed, the (anti)-self-duality of the form $R_{AB}$ implies that the Riemann tensor satisfies the condition

$$R_{ABCD} = \pm\frac{1}{2}\varepsilon_{ABEF}R_{EFCD}. \tag{3.31}$$

Multiplying this by $\delta_{AC}$, we derive

$$(Ricci)_{BD} = R_{ABAD} = \pm\frac{1}{2}\varepsilon_{ABEF}R_{EFAD} = 0 \tag{3.32}$$

[the last relation is a corollary of the second line in (1.23)].

This explains the word "gravitational". And "instantons" is a common name by which the physicists call Euclidean solutions of the classical equations of motion. In Chapter 14, we write such solutions for non-Abelian gauge fields, representing a particular interest for physics. We say few more words about the physical interpretation of the gravitational instantons at the end of this section.

### Taub–NUT manifold and its multidimensional generalizations

Consider $\mathbb{R}^4$ parametrized by the ordinary Cartesian coordinates $x^M$ and impose the constraint $r^2 = (x^1)^2 + (x^2)^2 + (x^3)^2 + (x^4)^2 > a^2$ (the topology is thus $S^3 \times \mathbb{R}$). Choose the metric

$$ds^2 = \frac{1}{4}\frac{r+a}{r-a}\,dr^2 + (r^2 - a^2)(\sigma_1^2 + \sigma_2^2) + 4a^2\frac{r-a}{r+a}\,\sigma_3^2, \tag{3.33}$$

where $\sigma_{1,2,3}$ are the *Maurer–Cartan forms*,[7]

$$\sigma_p = \frac{\eta^p_{MN}x^M\,dx^N}{r^2} \tag{3.34}$$

with $\eta^p_{MN}$ defined in (3.6). The property

$$d\sigma_p = \varepsilon_{pqr}\sigma_q \wedge \sigma_r \tag{3.35}$$

holds.

[7]Do not confuse them with the Pauli matrices.

A natural choice of the vierbein 1-forms is

$$e_{1,2} = \sqrt{r^2 - a^2}\,\sigma_{1,2}, \quad e_3 = 2a\sqrt{\frac{r-a}{r+a}}\,\sigma_3, \quad e_4 = \frac{1}{2}\sqrt{\frac{r+a}{r-a}}\,dr. \tag{3.36}$$

The connections $\omega_{AB}$ can be found as a solution to the Cartan equation $de_A + \omega_{AB} \wedge e_B = 0$. They are[8]

$$\omega_{12} = \left[2 - \frac{4a^2}{(r+a)^2}\right]\sigma_3, \quad \omega_{23} = \frac{2a}{r+a}\sigma_1, \quad \omega_{31} = \frac{2a}{r+a}\sigma_2,$$
$$\omega_{14} = \frac{2r}{r+a}\sigma_1, \quad \omega_{24} = \frac{2r}{r+a}\sigma_2, \quad \omega_{34} = \frac{4a^2}{(r+a)^2}\sigma_3. \tag{3.37}$$

These connections are neither self-dual nor anti-self-dual, but the curvatures (1.62) *are* anti-self-dual:

$$\begin{aligned}
R_{14} = -R_{23} &= \frac{4a}{(r+a)^3}\,(e_4 \wedge e_1 + e_2 \wedge e_3), \\
R_{24} = R_{13} &= \frac{4a}{(r+a)^3}\,(e_4 \wedge e_2 - e_1 \wedge e_3), \\
R_{34} = -R_{12} &= -\frac{8a}{(r+a)^3}\,(e_4 \wedge e_3 + e_1 \wedge e_2).
\end{aligned} \tag{3.38}$$

Hence, the holonomy group of the manifold is $SU(2)$, and this ensures that the manifold is hyperkähler.

The fact that the connections (3.37) do not have a definite duality is due to our choice of vierbeins in (3.36). However, as was mentioned above, one can always bring $\omega_{AB}$ into the subalgebra $su(2)$ where $R_{AB}$ belongs by using the gauge freedom $O(4)$ to modify the vielbein choice.

The form (3.33) of the metric is convenient for calculating the connections and curvatures and unravelling thereby its hyperkähler nature, but for field-theory applications, another form, related to (3.33) by a variable change,[9] is more convenient and, probably, more familiar for a reader-physicist [22]. It reads

$$\boxed{ds^2 = V^{-1}(r)\,(d\Psi + \mathbf{A}\cdot d\mathbf{x})^2 + V(r)\,d\mathbf{x}d\mathbf{x},} \tag{3.39}$$

---

[8]Attention! In the review [5], this and the next formula are given with wrong coefficients.

[9]See e.g. Eq. (3.3) in Ref. [21].

where[10] $\mathbf{x}$ is a 3-vector, $\Psi \in (0, 4\pi)$,

$$V(r) = \frac{1}{r} + \lambda \tag{3.40}$$

($r = |\mathbf{x}|$) is a kind of "Coulomb potential", and $\boldsymbol{A}(\boldsymbol{x})$ is the vector potential of a *Dirac magnetic monopole* [23] satisfying

$$\boldsymbol{\nabla} \times \boldsymbol{A} = \boldsymbol{\nabla} \frac{1}{r} = -\frac{\boldsymbol{x}}{r^3}.$$

$\boldsymbol{A}$ depends on the gauge. It can be chosen in the form

$$A_x = \frac{y}{r\,(r+z)}, \quad A_y = -\frac{x}{r\,(r+z)}, \quad A_z = 0, \tag{3.41}$$

but it is sometimes more convenient to impose the additional requirement $\boldsymbol{A}(\boldsymbol{x}) = \boldsymbol{A}(-\boldsymbol{x})$ and choose

$$A_x = -\frac{yz}{r(x^2+y^2)}, \quad A_y = \frac{xz}{r(x^2+y^2)}, \quad A_z = 0. \tag{3.42}$$

We observe a singularity along the $z$-axis.

The metric (3.39) has axial symmetry $U(1)$ related to the rotations around the $z$-axis. But this distinguished direction where the potential is singular[11] can be arbitrarily chosen. There is an infinite family of such metrics parameterized by a point in the quotient $SU(2)/U(1) \simeq S^2$. The corresponding potentials are interrelated by gauge transformations (1.45). One may say that the Taub–NUT manifold "knows" not only about $U(1)$, where the gauge field belongs, but also about $SU(2)$.

The singularity of the metric (3.38) on the Dirac string does not lead to the singularity of the curvature. For negative $\lambda$, there also is a less cosher singularity at $r = -1/\lambda$. If we want to avoid it, we must keep $\lambda$ positive. It can be rescaled to unity.

But one can as well consider the metric (3.39), (3.40) with negative $\lambda$ away of singularity. Or rather with

$$V(r) = \mu - \frac{1}{r}, \quad \mu > 0. \tag{3.43}$$

[10]The reader might be perplexed by an unusual range for the angular variable, but it is not a mistake. As we explain in Chapter 11, the metric (3.39) is reduced to the flat $\mathbb{R}^4$ metric in the limit $\lambda = 0$. And the corresponding variable change (11.54) involves $\Psi/2$ rather than $\Psi$. The range $\Psi \in (0, 2\pi)$ would correspond to $\mathbb{R}^4/Z_2$.

[11]A Dirac string in the physical terminology. The physical requirement for this singularity to be unobservable leads to the quantization of magnetic charge. We discuss it in detail in Chapter 14.

Such a metric has several interesting physical interpretations. In particular, the Lagrangian describing the motion of a particle along such manifold also describes the dynamics of two interacting magnetic monopoles of a special type (the so-called $BPS$ monopoles [24]) at large distances $r \gg \mu^{-1}$ [25].[12]

The dynamics of the system of an arbitrary number $N$ of monopoles is described by a hyperkähler manifold of dimension $D = 4(N-1)$. The latter enjoys $SU(N)$ isometry. Its metric represent a multidimensional generalization of (3.39). One can represent it in the form [27]

$$\boxed{ds^2 = V_{ml} d\boldsymbol{r}_m d\boldsymbol{r}_l + V^{-1}_{ml} \Lambda_m \Lambda_l,} \tag{3.44}$$

where $V$ is a $N \times N$ matrix with the components

$$\begin{aligned} V_{mm} &= \mu - \sum_{l \neq m} \frac{1}{|\boldsymbol{r}_m - \boldsymbol{r}_l|} \quad \text{(no summation over } m \text{ in the LHS)}, \\ V_{m \neq l} &= \frac{1}{|\boldsymbol{r}_m - \boldsymbol{r}_l|}, \end{aligned} \tag{3.45}$$

and

$$\Lambda_m = d\Psi_m + \sum_{l \neq m} \boldsymbol{A}(\boldsymbol{r}_m - \boldsymbol{r}_l) \cdot (d\boldsymbol{r}_m - d\boldsymbol{r}_l) \tag{3.46}$$

with the same $\boldsymbol{A}(\boldsymbol{x})$ as in (3.42).

The metric (3.44) with

$$V_{mm} = \lambda + \sum_{l \neq m} \frac{1}{|\boldsymbol{r}_m - \boldsymbol{r}_l|}, \quad V_{m \neq l} = -\frac{1}{|\boldsymbol{r}_m - \boldsymbol{r}_l|} \tag{3.47}$$

has no singularities. It describes a generalized Taub–NUT hyperkähler manifold.

A mathematician would probably prefer the choice (3.47), but, being a physicist who likes the monopole interpretation, I will stick to the choice (3.45) in the following discussion, assuming that the singularities at the small intermonopole distances can be patched preserving hyperkählerity in a similar way to what was explicitly done in Ref. [26] for the two-monopole system.

[12]The dynamics of the monopoles at arbitrary distances is also described by a somewhat more complicated hyperkähler metric — the Atiyah–Hitchin metric [26]. It is expressed in elliptic functions and differs from (3.39), (3.43) by a series of exponentially suppressed at large distances terms.

The expression (3.44) involves $4N$ coordinates: the positions of the monopoles $\boldsymbol{r}_m$ and their phases[13] $\Psi_m$. But the metric depends only on the differences $\boldsymbol{r}_m - \boldsymbol{r}_l, \Psi_m - \Psi_l$. The variables describing the position of the center of inertia of the monopoles and their mean phase are splitted apart and the manifold described by the metric (3.44) represents a direct product of $\mathbb{R}^4$ and a nontrivial hyperkähler manifold we are interested in.

This manifold "knows" about $SU(N)$ in the same sense as the ordinary Taub–NUT knows about $SU(2)$. Indeed, the differences of the monopole positions of which the metric depends have the same structure as the roots of $su(N)$.

Using *Hamiltonian reduction* or else *hyperkähler reduction* [29,30] (these methods are related; we discuss in detail both of them in Chapter 12), one can derive from the generalized Taub–NUT $SU(N)$ metrics written above the metrics associated with many other simple compact Lie groups [31]:

- The simplest nonunitary simple group is $SO(5)$. The metric of the corresponding eight-dimensional hyperkähler manifold may be derived from the metric (3.44) for four monopoles after imposing the constraints[14]

$$\boldsymbol{r}_1 + \boldsymbol{r}_3 = \boldsymbol{r}_2 + \boldsymbol{r}_4 = \Psi_1 + \Psi_3 = \Psi_2 + \Psi_4 = 0. \tag{3.48}$$

Setting $\mu = 1$, we obtain [31]

$$\frac{1}{2} ds^2_{SO(5)} = d\boldsymbol{r}_1^2 \left(1 - \frac{1}{r_1}\right) + d\boldsymbol{r}_2^2 \left(1 - \frac{1}{r_2}\right) - \sum_{\pm} \frac{(d\boldsymbol{r}_1 \pm d\boldsymbol{r}_2)^2}{|\boldsymbol{r}_1 \pm \boldsymbol{r}_2|} + \text{phase part}. \tag{3.49}$$

We see two short ($\boldsymbol{r}_1$ and $\boldsymbol{r}_2$) and two long ($\boldsymbol{r}_1 \pm \boldsymbol{r}_2$) roots of $so(5)$.
- To derive the twelve-dimensional hyperkähler metric associated with the group $SO(7)$, one should start with the metric for $SU(6)$ describing the dynamics of six monopoles, put a half of them behind the mirror and

[13]A BPS monopole, being a particular case of the 't Hooft–Polyakov monopole [28], has an additional phase parameter related to the global $U(1)$ symmetry of the solution. But we do not plunge further here in physical details referring the reader to Ref. [25].

[14]We put a mirror in the configuration space and require for the monopoles 3 and 4 to represent mirror images of the monopoles 1 and 2. String theorists call the manifolds thus obtained *orientifolds* [32].

impose the constraints

$$\boldsymbol{r}_1 + \boldsymbol{r}_4 = \boldsymbol{r}_2 + \boldsymbol{r}_5 = \boldsymbol{r}_3 + \boldsymbol{r}_6 = 0,$$
$$\Psi_1 + \Psi_4 = \Psi_2 + \Psi_5 = \Psi_3 + \Psi_6 = 0. \tag{3.50}$$

A similar procedure may be applied to derive the metrics corresponding to the groups $SO(2r+1)$ of higher rank. We obtain

$$\frac{1}{2} ds^2_{SO(2r+1)} = \sum_{m=1}^{r} d\boldsymbol{r}_m^2 \left(1 - \frac{1}{r_m}\right) - \sum_{m>l}^{r} \sum_{\pm} \frac{(d\boldsymbol{r}_m \pm d\boldsymbol{r}_l)^2}{|\boldsymbol{r}_m \pm \boldsymbol{r}_l|}$$
$$+ \text{phase part.} \tag{3.51}$$

- For $Sp(r)$, we start with $2r+1$ monopoles and put $r$ of them before the mirror, $r$ more behind the mirror and the last one *right on* the mirror. The constraints are

$$\boldsymbol{r}_1 + \boldsymbol{r}_{r+1} = \ldots = \boldsymbol{r}_r + \boldsymbol{r}_{2r} = \boldsymbol{r}_{2r+1} = 0,$$
$$\Psi_1 + \Psi_{r+1} = \ldots = \Psi_r + \Psi_{2r} = \Psi_{2r+1} = 0. \tag{3.52}$$

The metric has the same structure as (3.51), but the roles of the short and long roots are now interchanged.
- The hyperkähler reduction method also allows one to treat the case of $G_2$. One has to take seven monopoles and impose the constraints

$$\boldsymbol{r}_1 + \boldsymbol{r}_4 = \boldsymbol{r}_2 + \boldsymbol{r}_5 = \boldsymbol{r}_3 + \boldsymbol{r}_6 = \boldsymbol{r}_7 = 0,$$
$$\Psi_1 + \Psi_4 = \Psi_2 + \Psi_5 = \Psi_3 + \Psi_6 = \Psi_7 = 0, \tag{3.53}$$

as for $Sp(3)$, but on top of that, one should require $\boldsymbol{r}_1 + \boldsymbol{r}_2 + \boldsymbol{r}_3 = \Psi_1 + \Psi_2 + \Psi_3 = 0$. The metric can be expressed as

$$\frac{1}{2} ds^2_{G_2} = d\boldsymbol{r}_1^2 \left(1 - \frac{3}{r_1}\right) + d\boldsymbol{r}_2^2 \left(1 - \frac{3}{r_2}\right) + d\boldsymbol{r}_3^2 \left(1 - \frac{3}{r_3}\right)$$
$$- \frac{(d\boldsymbol{r}_1 - d\boldsymbol{r}_2)^2}{|\boldsymbol{r}_1 - \boldsymbol{r}_2|} - \frac{(d\boldsymbol{r}_1 - d\boldsymbol{r}_3)^2}{|\boldsymbol{r}_1 - \boldsymbol{r}_3|} - \frac{(d\boldsymbol{r}_2 - d\boldsymbol{r}_3)^2}{|\boldsymbol{r}_2 - \boldsymbol{r}_3|}$$
$$+ \text{phase part.} \tag{3.54}$$

We see three long roots, $\boldsymbol{r}_1 - \boldsymbol{r}_2$, $\boldsymbol{r}_2 - \boldsymbol{r}_3$ and $\boldsymbol{r}_1 - \boldsymbol{r}_3$, and three short roots, $\boldsymbol{r}_{1,2,3}$.

For the groups $SO(2r)$ and higher exceptional groups, hyperkähler reduction does not work so well. There is, however, a universal method based on the analysis of a certain three-dimensional supersymmetric gauge field theory [31]. Supersymmetric field theories are beyond the scope of our book, and we do not explain how it is derived but only present the result.

**Theorem 3.3.** *The metric of the 4r-dimensional hyperkähler manifold associated with a simple Lie groups $\mathcal{G}$ and belonging to the class described above has the form*

$$ds^2_{\mathcal{G}} \propto Q^{ab} d\boldsymbol{x}^a \cdot d\boldsymbol{x}^b + (Q^{-1})^{ab} \Lambda^a \Lambda^b, \tag{3.55}$$

*where*[15] $a, b = 1, \ldots, r$,

$$Q^{ab} = \delta^{ab} - \sum_j \frac{\alpha_j^a \alpha_j^b}{|\alpha_j(\boldsymbol{x})|} \tag{3.56}$$

*and*

$$\Lambda^a = d\Psi^a + \sum_j \alpha_j^a \, \boldsymbol{A}[\alpha_j(\boldsymbol{x})] \cdot \alpha_j(d\boldsymbol{x}) \tag{3.57}$$

*with* $\boldsymbol{A}(\boldsymbol{y})$ *written in* (3.42).

*The sums in* (3.56) *and* (3.57) *run over all positive roots of* $\mathfrak{g}$, *and* $\alpha_j(\boldsymbol{y}) = \alpha_j^a \boldsymbol{y}^a$. *The length of the long roots is normalized to 1.*

A dedicated reader may check that the metric (3.55) coincides after an appropriate coordinate change and rescaling with (3.44) for $SU(3)$ that has the roots

$$\alpha_1 = (1, 0), \quad \alpha_2 = \left(\frac{1}{2}, \frac{\sqrt{3}}{2}\right), \quad \alpha_3 = \left(-\frac{1}{2}, \frac{\sqrt{3}}{2}\right), \tag{3.58}$$

with (3.49) for $SO(5)$ that has the roots

$$\alpha_1 = (1, 0), \quad \alpha_2 = (0, 1), \quad \alpha_3 = \left(\frac{1}{2}, \frac{1}{2}\right), \quad \alpha_4 = \left(-\frac{1}{2}, \frac{1}{2}\right) \tag{3.59}$$

and with (3.54) for $G_2$ that has the roots

$$\alpha_1 = (1, 0), \quad \alpha_2 = \left(\frac{1}{2}, \frac{\sqrt{3}}{2}\right), \quad \alpha_3 = \left(-\frac{1}{2}, \frac{\sqrt{3}}{2}\right),$$
$$\alpha_4 = \left(0, \frac{1}{\sqrt{3}}\right), \quad \alpha_5 = \left(\frac{1}{2}, \frac{1}{2\sqrt{3}}\right), \quad \alpha_6 = \left(-\frac{1}{2}, \frac{1}{2\sqrt{3}}\right). \tag{3.60}$$

---

[15] We put upstairs the indices $a, b$ that refer to an orthogonal basis in the Cartan subalgebra not to confuse them with the "monopole" indices $m, l$ in (3.44), which run from 1 to $N = r + 1$.

Finally, we mention that we derive in Chapter 13, using the harmonic superspace technique, an expression for a certain eight-dimensional hyperkähler metric with the same $O(5)$ symmetry as (3.49). One can suppose that this metric can be reduced to (3.49) [its nonsingular counterpart derived by hyperkähler reduction from a variant of (3.44) with $V_{ml}$ given by (3.47) with positive $\lambda$] by an appropriate coordinate change.

### Eguchi–Hanson manifold

Topologically, this manifold represents a tangent bundle (i.e. a bundle with the fiber representing a tangent space at each point of the manifold)[16] over $S^2$. Its metric can be chosen in the form [20]

$$\boxed{ds^2 = \frac{dr^2}{1-(a/r)^4} + r^2\left\{\sigma_1^2 + \sigma_2^2 + \left[1 - \left(\frac{a}{r}\right)^4\right]\sigma_3^2\right\},} \tag{3.61}$$

where $\sigma_{1,2,3}$ are the Maurer–Cartan forms (3.34) and $r \geq a$. (We included the value $r = a$ in the allowed range because the apparent singularity of the metric at $r = a$ is in this case removable.)

The vierbeins may be chosen as

$$e_{1,2} = r\,\sigma_{1,2}, \quad e_3 = r\sqrt{1-(a/r)^4}\,\sigma_3, \quad e_4 = \frac{dr}{\sqrt{1-(a/r)^4}}. \tag{3.62}$$

The connections are self-dual:

$$\begin{aligned}
\omega_{14} = \omega_{23} &= \sqrt{1-(a/r)^4}\,\sigma_1, \\
\omega_{24} = \omega_{31} &= \sqrt{1-(a/r)^4}\,\sigma_2, \\
\omega_{34} = \omega_{12} &= \left[1+(a/r)^4\right]\sigma_3.
\end{aligned} \tag{3.63}$$

The curvature forms are also, of course, self-dual:

$$\begin{aligned}
R_{14} = R_{23} &= \frac{2a^4}{r^6}(e^4 \wedge e^1 + e^3 \wedge e^2), \\
R_{24} = R_{31} &= \frac{2a^4}{r^6}(e^4 \wedge e^2 + e^1 \wedge e^3), \\
R_{34} = R_{12} &= -\frac{4a^4}{r^6}(e^4 \wedge e^3 + e^2 \wedge e^1).
\end{aligned} \tag{3.64}$$

[16] Actually, "cotangent bundle" and "cotangent space", but, as you remember, we have chosen to use sloppy language at this point.

A nice feature of the Eguchi–Hanson metric is its *asymptotic flatness*. Indeed, it is clear from (3.61) that the metric becomes flat in the limit $r \to \infty$. However, the boundary of the Eguchi–Hanson manifold at infinity is not $S^3$ (as for the ordinary four-dimensional space), but rather $RP^3 = S^3/\mathbb{Z}_2$.

Asymptotic flatness allows one to suggest a nice physical interpretation of the Eguchi–Hanson solution [33]. As was mentioned, it represents a "gravitational instanton". In physics, instantons (Euclidean solutions) can often be interpreted as the quasiclassical trajectories for the *tunnelling transitions* from one vacuum to another. And this is also the case for the Eguchi–Hanson solution. It can be interpreted as a tunnelling trajectory between two flat $\mathbb{R}^3$ spaces of opposite orientations. The 3-space turns inside out during such a transition.

Another name for $S^2$ is $\mathbb{CP}^1$. One can also consider tangent bundles on $\mathbb{CP}^{n>1}$. Their dimensions are $D = 4n$ — twice as much as for $\mathbb{CP}^{n>1}$. One can show [16] that these manifolds are hyperkähler.

## 3.2 HKT manifolds

Up to now we discussed the geometrical issues worked out first by mathematicians and then described by physicists in the language of supersymmetry. The reader will learn about the supersymmetric description of generic Riemannian, complex, Kähler and hyperkähler geometries in a proper place in this book. But HKT manifolds is a different story: they were discovered *first* by physicists doing supersymmetry [34, 35] and *then* these geometries were grasped and assimilated by mathematicians [36, 37]. In this geometrical chapter, we do not talk about supersymmetry, however, and present the HKT geometry using the same language as in the preceding discussions.

**Definition 3.3.** An HKT manifold is a hypercomplex manifold where the complex structures are covariantly constant with respect to one and the same Bismut connection.

The following remarks are in order:

(1) As was the case for the hyperkähler manifolds, it is not necessary to stipulate the hypercomplex nature of the manifold in the definition. It is sufficient to require the presence of two anticommuting complex structures $I, J$ that are covariantly constant with respect to one and the same Bismut connection. It follows then that $K = IJ$ is also

a complex structure that is covariantly constant with respect to the same connection.

(2) "HKT" stands for "hyperkähler with torsion". This term is in fact misleading because HKT manifolds are not hyperkähler and not even Kähler. But it is now firmly established in the literature, nothing better has been suggested, and we follow the crowd.

(3) The words "one and the same" in the definition are crucial. Indeed, according to Theorem 2.8, each complex structure of a hypercomplex manifold is covariantly constant with respect to its Bismut connection. Generically, these three Bismut connections are different. But they all coincide for an HKT manifold, and this is not trivial.

(4) Not *any* hypercomplex manifold is HKT. But any four-dimensional hypercomplex manifold *is* (see Theorem 3.7). Examples of hypercomplex non-HKT manifolds of dimension 8 and higher were found in Ref. [38]. The bi-HKT models to be described in Section 10.3.1 are also hypercomplex but not HKT.

The following important theorem holds.

**Theorem 3.4.** *The holonomy group of an HKT manifold with respect to its universal Bismut connection is $Sp(n)$. Inversely, if there exists a torsionful metric-preserving connection whose curvature form lies in $sp(n)$, one can find three quaternionic complex structures that are covariantly constant with respect to this connection, and we are dealing with an HKT manifold.*

**Proof.** We can use literally the same reasoning as in the proof of Theorem 3.2. The presence or absence of torsion in the universal connection with respect to which three quaternionic complex structures are covariantly constant is irrelevant. □

There is a 2-form associated with each complex structure:

$$\omega_I = I_{MN}\, dx^M \wedge dx^N, \quad \omega_J = J_{MN}\, dx^M \wedge dx^N,$$
$$\omega_K = K_{MN}\, dx^M \wedge dx^N. \tag{3.65}$$

For a hyperkähler manifold, all these forms are closed [see Eq. (2.45)], but for a generic hypercomplex or HKT manifold, they are not. These forms possess, however, some special properties. We prove two theorems. One of them is rather simple and another one is a little bit more complicated.

**Theorem 3.5.** *Consider a hypercomplex manifold with the quaternionic complex structures $I, J, K = IJ$ and the corresponding forms* (3.65). *Then,* *the form $\omega_J + i\omega_K$ is purely holomorphic and the form $\omega_J - i\omega_K$ is purely antiholomorphic with respect to the complex structure $I$.*

**Proof.** Consider the tensor $\mathcal{I}^+ = (J + iK)/2$. Bearing in mind the quaternion algebra for $I, J, K$, we obtain $I\mathcal{I}^+ = -i\mathcal{I}^+$ and $\mathcal{I}^+ I = i\mathcal{I}^+$. In the frame where $I$ has the form (2.53), we derive that the only nonvanishing components of $(\mathcal{I}^+)_M{}^N$ are $(\mathcal{I}^+)_m{}^{\bar{n}}$. And this means that $(\mathcal{I}^+)_{MN}$ has only the holomorphic components and the form $\omega_J + i\omega_K$ is holomorphic. The antiholomorphicity of the form $\omega_J - i\omega_K$ is proven in exactly the same way. □

**Corollary:** The only nonzero components of the tensor $J_M{}^N$ are $J_m{}^{\bar{n}}$ and $J_{\bar{m}}{}^n$, and the same is true for $K_M{}^N$. It follows in addition that

$$\boxed{K_m^{\ \bar{n}} = -iJ_m^{\ \bar{n}}, \quad K_{\bar{m}}^{\ n} = iJ_{\bar{m}}^{\ n}.} \tag{3.66}$$

**Remark:** The tensor $(\mathcal{I}^+)_{mn}$ plays for a hypercomplex manifold roughly the same role as the complex structure $I_{MN}$ for a complex manifold. It is thus natural to call $(\mathcal{I}^+)_{mn}$ a *hypercomplex structure.*[17] In the flat case, it has a simple form

$$(\mathcal{I}^+)_m{}^{\bar{n}} = \text{diag}(-\varepsilon, \ldots, -\varepsilon), \tag{3.67}$$

which looks like the canonical complex structure (2.51) but, of course, the dimension of the latter is two times larger.

One can also define the tensor $\mathcal{I}^- = (J - iK)/2$. Its only nonzero components are $(\mathcal{I}^-)_{\bar{m}}{}^n$. In the flat case, it has exactly the same form (3.67) as $\mathcal{I}^+$ modulo the interchange $n \leftrightarrow \bar{n}, m \leftrightarrow \bar{m}$. The products $-\mathcal{I}^+\mathcal{I}^-$ and $-\mathcal{I}^-\mathcal{I}^+$ give the holomorphic and antiholomorphic projectors (2.58).

**Theorem 3.6.** *A hypercomplex manifold is HKT iff the $(2,0)$-form $\omega_J + i\omega_K$ is annihilated by the action of the operator of the holomorphic (with respect to the complex structure $I$) exterior derivative $\partial_I$.*

**Proof.** We follow Ref. [36].

---

[17]People usually understand by hypercomplex structure simply the presence of a quaternionic triple of complex structures. But in our book, this term will always refer to the tensors $\mathcal{I}^\pm$.

### Direct theorem

We prove first that $\partial_I(\omega_J + i\omega_K) = 0$ for an HKT manifold.

Let $d$ be the ordinary exterior derivative operator. It can be represented as the sum $d = \partial_I + \bar{\partial}_I$. Introduce the operator $d_I$ representing (with the factor 2) the "imaginary part" of $\partial_I$:

$$d_I = i(\bar{\partial}_I - \partial_I). \tag{3.68}$$

Working in the frame where $I$ has the form (2.53), one derives

$$d_I\alpha = I_M{}^S\, \partial_S \alpha_{N_1 \ldots N_P}\, dx^M \wedge dx^{N_1} \wedge \cdots \wedge dx^{N_P}. \tag{3.69}$$

Another operator $\alpha \to \hat{I}\alpha$ is defined as

$$\hat{I}\alpha = I_{M_1}{}^{N_1} \ldots I_{M_P}{}^{N_P}\, \omega_{N_1 \ldots N_P}\, dx^{M_1} \wedge \cdots \wedge dx^{M_P}. \tag{3.70}$$

When acting on the form of the type $(p, q)$, it multiplies it by the factor $i^{q-p}$. The operators $\hat{I}, \hat{J}, \hat{K}$ satisfy the same quaternionic algebra as the complex structures.

We prove now some simple lemmas.

**Lemma 1.**

$$d_I\alpha = (-1)^P \hat{I} d(\hat{I}\alpha), \tag{3.71}$$

*where $P = p + q$ is the real degree of the form.*

**Proof.** Choose the complex coordinates. Consider the RHS of (3.71) and use the complex expression (2.53) for $I$. The components $I_M{}^N$ are thus constant and the partial derivatives do not act upon them. The form $d(\hat{I}\alpha)$ has the order $P + 1$ and, according to (3.70), the expression $\hat{I}d(\hat{I}\alpha)$ has altogether $(P + 1) + P = 2P + 1$ factors of $I$. This involves $P$ pairs giving $I^2 = -\mathbb{1}$ [this compensates the factor $(-1)^P$] and we are left with just one unpaired factor. We obtain the expression (3.69). □

**Lemma 2.** *For any complex manifold,*

$$d_I\omega_I = \frac{1}{3} C_{MRS}\, dx^M \wedge dx^R \wedge dx^S. \tag{3.72}$$

**Proof.** Choosing complex coordinates so that $\omega_I = -2ih_{m\bar{n}} dz^m \wedge d\bar{z}^{\bar{n}}$, and bearing in mind (2.53), (3.69) and (2.18), we derive

$$d_I\omega_I = C_{mr\bar{s}}\, dz^m \wedge dz^r \wedge d\bar{z}^{\bar{s}} + C_{\bar{m}\bar{r}s} d\bar{z}^{\bar{m}} \wedge d\bar{z}^{\bar{r}} \wedge dz^s, \tag{3.73}$$

which coincides with (3.72). □

**Corollary:** For the HKT manifolds where the Bismut torsions for $I, J, K$ coincide,

$$d_I \omega_I = d_J \omega_J = d_K \omega_K. \tag{3.74}$$

**Lemma 3.** *Let $I, J, K$ be quaternion complex structures. Then,*

$$\begin{gathered} \hat{I}\omega_I = \omega_I, \quad \hat{J}\omega_J = \omega_J, \quad \hat{K}\omega_K = \omega_K, \\ \hat{J}\omega_I = \hat{K}\omega_I = -\omega_I, \quad \hat{I}\omega_J = \hat{K}\omega_J = -\omega_J, \quad \hat{I}\omega_K = \hat{J}\omega_K = -\omega_K. \end{gathered} \tag{3.75}$$

**Proof.** Let us prove the relation $\hat{J}\omega_I = -\omega_I$. By definition,

$$\hat{J}\omega_I = J_M{}^R J_N{}^S I_{RS}\, dx^M \wedge dx^N.$$

On the other hand,

$$J_M{}^R J_N{}^S I_{RS} = J_M{}^R K_{NR} = -I_{MN}.$$

Other relations are proved similarly. □

**Remark:** The condition (3.74) can be rewritten, bearing in mind (3.71) and the first line in (3.75), as

$$\hat{I} d\omega_I = \hat{J} d\omega_J = \hat{K} d\omega_K. \tag{3.76}$$

We are ready now to prove the main theorem and show that

$$\partial_I(\omega_J + i\omega_K) = 0. \tag{3.77}$$

The real and imaginary parts of (3.77) give a kind of Cauchy–Riemann conditions[18]

$$d\omega_J - d_I \omega_K = 0, \quad d\omega_K + d_I \omega_J = 0. \tag{3.78}$$

Consider the first relation. We obtain

$$d_I \omega_K \stackrel{1}{=} \hat{I} d(\hat{I}\omega_K) \stackrel{3}{=} -\hat{I} d\omega_K = -\hat{J}\hat{K} d\omega_K \stackrel{remark}{=} -\hat{J}^2 d\omega_J = d\omega_J. \tag{3.79}$$

The number "1" above the equality sign means *in virtue of Lemma 1*, etc.

The relation $d\omega_K + d_I\omega_J = 0$ is proved in a similar way.

[18] Cf. Eq.(14) in Ref. [34].

### Inverse theorem

All the steps of this reasoning can be retraced. Indeed, (3.77) implies (3.78). Take e.g. the relation $d\omega_J = d_I\omega_K$ and act by $\hat{J}$ upon it. We see that

$$\hat{J}d\omega_J = \hat{J}d_I\omega_K \stackrel{1}{=} \hat{J}\hat{I}d(\hat{I}\omega_K) \stackrel{3}{=} \hat{K}d\omega_K. \tag{3.80}$$

From this and Lemma 1, one derives that $d_J\omega_J = d_K\omega_K$. Bearing in mind (3.72), this means that the Bismut connections for the complex structures $J$ and $K$ coincide. But then $I = JK$ is also covariantly constant with respect to the same connection and the manifold is HKT. □

It follows that the condition (3.77) together with the quaternionic algebra of the three complex structures can be taken as an alternative definition of HKT manifolds. The conditions $\partial_J(\omega_K + i\omega_I) = 0$ and $\partial_K(\omega_I + i\omega_J) = 0$ follow in virtue of the direct theorem just proven.

With this definition in hand, we can prove the theorem:

**Theorem 3.7.** *Any four-dimensional hypercomplex manifold is HKT.*

**Proof.** According to Theorem 3.4, the form $\omega_J + i\omega_K$ has type $(2,0)$ with respect to the complex structure $I$. But then, for a manifold of complex dimension 2, $\partial_I(\omega_J + i\omega_K)$ vanishes identically. □

One can now observe that HKT manifolds relate to generic hypercomplex manifolds in roughly the same way as Kähler manifolds relate to generic complex manifolds. Indeed, (i) Kähler manifolds are complex manifolds with the closed Kähler form and (ii) any complex manifold of real dimension 2 is Kähler. On the other hand, (i) HKT manifolds are hypercomplex manifolds for which the form $\omega_J + i\omega_K$ is closed with respect to $\partial_I$ and (ii) any hypercomplex manifold of complex dimension 2 is HKT.

We have discussed up to now the ordinary Levi-Civita connection and the torsionful Bismut connection. Another very useful notion that is taylor-made for studying the hypercomplex and HKT manifolds is the *Obata connection* [39].

**Definition 3.4.** The Obata connection is a torsionless connection with respect to which all three quaternionic complex structures of a hypercomplex manifold are covariantly constant.

For a hyperkähler manifold, the Obata connection coincides with the Levi-Civita connection, but it is not so in a generic HKT or still more generic hypercomplex case. The essential difference is that the Obata covariant derivative of the metric tensor does not vanish if the manifold is not hyperkähler! This means in particular that vectors may not only rotate under parallel transports along closed contours but also change their length; the holonomy group is not compact and complicated (see e.g. [40]).

**Theorem 3.8.** [41] *Let $I, J, K$ be three integrable quaternionic complex structures. Choose the complex coordinates associated with $I$. Then, the Obata connection is given by the formula*

$$(G^O)^k_{mn} = J_n{}^{\bar{l}}\partial_m J_{\bar{l}}{}^k = K_n{}^{\bar{l}}\partial_m K_{\bar{l}}{}^k,$$
$$(G^O)^{\bar{k}}_{\bar{m}\bar{n}} = J_{\bar{m}}{}^l\partial_{\bar{n}} J_l{}^{\bar{k}} = K_{\bar{m}}{}^l\partial_{\bar{n}} K_l{}^{\bar{k}}, \tag{3.81}$$

*and all other components of $(G^O)^K_{MN}$ vanish.*

**Proof.** It consists of several steps.

**Lemma 1.** *The expressions for $(G^O)^k_{mn}$ are symmetric under permutation $m \leftrightarrow n$, including no torsion. The same holds for $(G^O)^{\bar{k}}_{\bar{m}\bar{n}}$.*

**Proof.** This follows from integrability. Indeed, the condition (2.60) for the structure $J$ implies

$$J_S{}^M(\partial_M J_N{}^K - \partial_N J_M{}^K) = J_N^Q(\partial_Q J_S{}^K - \partial_S J_Q{}^K).$$

Choose $S = s, N = n, K = k$. Then, bearing in mind Theorem 3.5 and its corollary, we obtain

$$J_s{}^{\bar{m}}\partial_n J_{\bar{m}}{}^k = J_n{}^{\bar{m}}\partial_s J_{\bar{m}}{}^k.$$

We can repeat this reasoning for the expressions $K_n{}^{\bar{l}}\partial_m K_{\bar{l}}{}^k$, $J_{\bar{m}}{}^l\partial_{\bar{n}} J_l{}^{\bar{k}}$ and $K_{\bar{m}}{}^l\partial_{\bar{n}} K_l{}^{\bar{k}}$. □

**Lemma 2.** *The covariant derivatives of all complex structures, including the connection (3.81), vanish.*

**Proof.** Consider first the holomorphic covariant derivative $\nabla_p$.

- $I_M{}^N$ is given by (2.53) so that its ordinary derivative vanishes. When $M = \bar{m}$ or $N = \bar{n}$, the terms in $\nabla_p I_M{}^N$ involving the connections also vanish due to the absence of the components $(G^O)^N_{p\bar{m}}$ and $(G^O)^{\bar{n}}_{pM}$. And $\nabla_p I_m{}^n = (G^O)^n_{pq} I_m{}^q - (G^O)^q_{pm} I_q{}^n$ vanishes due to the cancellation of the two terms.

- For the derivative $\nabla_p J_M{}^N$, we use the expression of $(G^O)^k_{mn}$ via the complex structure $J$ and capitalize on its invariance under interchange $m \leftrightarrow n$ — the fact that we proved above. We derive

$$\begin{aligned}
\nabla_p J_m{}^{\bar n} &= \partial_p J_m{}^{\bar n} - (G^O)^k_{pm} J_k{}^{\bar n} = \partial_p J_m{}^{\bar n} - J_m{}^{\bar l}(\partial_p J_{\bar l}{}^k) J_k{}^{\bar n} \\
&= \partial_p J_m{}^{\bar n} + J_k{}^{\bar n} J_{\bar l}{}^k (\partial_p J_m{}^{\bar l}) = \partial_p J_m{}^{\bar n} - \partial_p J_m{}^{\bar n} = 0; \\
\nabla_p J_{\bar m}{}^n &= \partial_p J_{\bar m}{}^n + (G^O)^n_{pq} J_{\bar m}{}^q = \partial_p J_m{}^{\bar n} + J_q{}^{\bar l}(\partial_p J_{\bar l}{}^n) J_{\bar m}{}^q \\
&= \partial_p J_{\bar m}{}^n - \partial_p J_{\bar m}{}^n = 0. \qquad (3.82)
\end{aligned}$$

- To prove that $\nabla_p K_M{}^N = 0$, we express the Obata connection via the complex structure $K$ and proceed similarly.

The proof of the relations

$$\nabla_{\bar p} I = \nabla_{\bar p} J = \nabla_{\bar p} K = 0$$

can be built along the same lines. □

The assertion that the expressions via $J$ and $K$ displayed in Eq.(3.81) give the *same* Obata connection follows from the following fact:

**Lemma 3.** *The Obata connection is unique.*

**Proof.** Suppose there are two different Obata connections. Let their difference be $\Delta^N_{PM}$. The identities

$$\Delta^Q_{PM} I_Q^N - \Delta^N_{PQ} I_M^Q = 0, \qquad (3.83)$$

$$\Delta^Q_{PM} (J \pm iK)_Q^N - \Delta^N_{PQ} (J \pm iK)_M^Q = 0 \qquad (3.84)$$

should hold.

It follows from (3.83), (2.53) and from the symmetry $\Delta^N_{PM} = \Delta^N_{MP}$ that all the components except $\Delta^n_{pm}$ and $\Delta^{\bar n}_{\bar p \bar m}$ vanish.

Consider Eq. (3.84) for the structure $J + iK$. According to Theorem 3.5, the only nonzero components of $(J + iK)_Q^N$ are $(J + iK)_q{}^{\bar n}$ [resp. $(J + iK)_{\bar m}^{\bar q}$ for $(J + iK)_M^Q$]. Choose $N = \bar n, P = p, M = m$. Bearing in mind that $\Delta^{\bar n}_{p\bar q} = 0$, as we have just proved, we obtain

$$\Delta^q_{pm} \, (J + iK)_q{}^{\bar n} = 0.$$

Multiplying that by $(J - iK)_{\bar n}{}^s$ and using $JK = I$ and $I_q{}^s = -i\delta^s_q$, we derive $\Delta^q_{pm} = 0$.

Considering (3.84) for $J - iK$ and choosing there $N = n, P = \bar p, M = \bar m$, we prove that $\Delta^{\bar q}_{\bar p \bar m} = 0$. □
□

### 3.2.1 *Examples*

#### Four dimensions

**Theorem 3.9.** *Let $\mathcal{M}$ be a four-dimensional manifold with the hyperkähler metric $g_{\mu\nu}(x)$. Then, the manifold $\tilde{\mathcal{M}}$ with the metric $\tilde{g}_{\mu\nu}(x) = \lambda(x)\, g_{\mu\nu}(x)$ is HKT.*

**Proof.** $\mathcal{M}$ is Kähler with respect to any of the complex structures $I, J, K$ which we choose in the form (3.3). In particular, it is Kähler with respect to $I$. Choose the corresponding complex coordinates and write the metric on $\mathcal{M}$ as $ds^2 = 2h_{n\bar{m}}\, dz^m d\bar{z}^{\bar{n}}$. Then, the metric on $\tilde{\mathcal{M}}$ reads

$$d\tilde{s}^2 = 2\lambda\, h_{m\bar{n}}\, dz^m d\bar{z}^{\bar{n}}. \tag{3.85}$$

The corresponding Bismut connection with respect to which the complex structure $I$ is covariantly constant has the torsion (2.18). Bearing in mind that $h_{n\bar{m}}$ is Kähler, we derive

$$C_{mn\bar{p}} = (\partial_n \lambda)\, h_{m\bar{p}} - (\partial_m \lambda)\, h_{n\bar{p}}. \tag{3.86}$$

The corresponding expressiion for the torsion tensor $C_{\mu\nu\rho}$ in a generic frame reads

$$C_{\mu\nu\rho} = \tilde{E}_{\mu\nu\rho\sigma}\, \tilde{g}^{\sigma\tau}\, \partial_\tau(\ln\lambda) = \sqrt{\det(g)}\, \varepsilon_{\mu\nu\rho\sigma}\, g^{\sigma\tau}\, \partial_\tau \lambda, \tag{3.87}$$

where $\tilde{E}_{\mu\nu\rho\sigma}$ is the invariant antisymmetric tensor (1.7) in $\tilde{\mathcal{M}}$.

Repeat this reasoning for the complex structures $J$ and $K$. We arrive at the same result (3.87) for the corresponding Bismut torsions. Thus, the Bismut connections for the three complex structures coincide, and the manifold $\tilde{\mathcal{M}}$ is HKT. □

As a simplest example, consider a conformally flat four-dimensional metric

$$\boxed{ds^2 = \lambda(x)\, dx^\mu dx^\mu.} \tag{3.88}$$

This manifold can represent a deformation of $\mathbb{R}^4$ or be topologically nontrivial. The choice $\lambda(x) = 1/(x^\mu x^\mu)$ together with the restriction $1 \le |x^\mu| \le 2$ and identification of the points with $|x^\mu| = 1$ and $|x^\mu| = 2$ gives the Hopf manifold $\mathcal{H}^2 = S^3 \times S^1$ [see Eq. (2.78)].

Note that the complex structures $(I^p)_\mu{}^\nu$ for all the manifolds (3.88) coincide with their constant flat expressions (3.3). This means that the Obata connection (3.81) vanishes in this case.

We will see in Chapter 13 that the theorem just proven represents a very particular case of a much more general theorem. Using powerful supersymmetric methods, we will show there that all HKT manifolds are grouped in multiparametric families stemming from a hyperkähler manifold (we will call such HKT manifolds *reducible*) or *irreducible* families not involving a HK manifold as a member. For all the members of one family, the Obata connections (or, better to say, the Obata curvatures — connections depend on the frame choice) coincide.

In a certain sense, the Hopf manifold is irreducible because its topology differs from that of $R^4$. However, as we have seen, its metric can be *locally* reduced to the flat hyperkähler metric by a conformal transformation. An example of a four-dimensional HKT metric which cannot be reduced by a conformal transformation to any HK metric even locally was constructed in Ref. [42]. We discuss this Delduc–Valent metric in Chapter 13.

### Higher dimensions: group manifolds

The Hopf manifold $\mathcal{H}^2$ is a group manifold, $\mathcal{H}^2 = S^3 \times S^1 \equiv SU(2) \times U(1)$. Its metric can be alternatively presented as

$$ds^2 = 2d\phi^M d\phi^N \mathrm{Tr}\{\partial_M \omega \, \partial_N \omega^{-1}\} + d\chi^2, \tag{3.89}$$

where $\phi^M$ parametrize $SU(2)$, $\omega \in SU(2) = \exp\{i\sigma_M \phi^M/2\}$, and $\chi$ is the angle on $U(1)$.

Consider now the next in complexity example, which is $SU(3)$. It is a manifold of dimension $8 = 4 \cdot 2$. It was shown in Refs. [43, 44] that this manifold (as well as many other group manifolds whose dimension represents a multiple integer of 4) is HKT.

We start with proving the *Samelson theorem* [45].

**Theorem 3.10.** *Any group manifold $\mathcal{G}$ of even dimension is complex.*

**Proof.** We write the metric on $\mathcal{G}$ as

$$g_{MN} = 2\mathrm{Tr}\{\partial_M \omega \, \partial_N \omega^{-1}\}, \tag{3.90}$$

where $\omega = \exp\{it_M x^M\}$ is a matrix describing an element of a Lie group $\mathcal{G}$ with the Hermitian generators $t_M$ normalized as $\mathrm{Tr}\{t_M t_N\} = \frac{1}{2}\delta_{MN}$. Obviously, this metric is invariant under the left and right group multiplications, $\omega \to U\omega V$ $(U, V \in \mathcal{G})$.

Consider a neighbourhood of any point $\omega$ on $\mathcal{G}$. Bearing in mind the invariance above, this point can be brought to unity by group rotations, and we can assume $\omega \approx \mathbb{1}$.

We can write

$$g_{MN}(x) = e_{MA}(x)e_{NA}(x), \tag{3.91}$$

where the vielbeins are convenient to choose in the form $e_{MA}t_A = -i\omega^{-1}\partial_M\omega$, so that

$$e_{MA} = \delta_{MA} + \frac{1}{2}f_{MAP}\,x^P + o(x), \tag{3.92}$$

$f_{MAP}$ being the structure constants of $\mathcal{G}$.

To prove the theorem, we have to define the almost complex structure $I_{MN}(x)$ and show that the corresponding Nijenhuis tensor vanishes. The complex structure is expressed via the vielbeins as

$$I_{MN}(x) = e_{MA}(x)e_{NB}(x)I_{AB},$$

where $I_{AB} \equiv I_{MN}(0)$ is its tangent projection, which is one and the same in all the points. For small $x$, this gives

$$I_{MN}(x) = I_{MN}(0) + \frac{1}{2}I_M{}^Q(0)f_{NQP}\,x^P - \frac{1}{2}I_N{}^Q(0)f_{MQP}\,x^P + o(x). \tag{3.93}$$

The complex structure (3.93) is covariantly constant with respect to the Bismut connection with the torsion tensor

$$C_{MNP} = f_{MNP}. \tag{3.94}$$

Indeed, it is straightforward to see that at the origin $x = 0$,

$$\nabla_P^{(B)}I_{MN} = \partial_P I_{MN} - \frac{1}{2}f_{QPM}I^Q{}_N - \frac{1}{2}f_{QPN}I_M{}^Q = 0, \tag{3.95}$$

where we neglected the contribution of the ordinary Christoffel symbols $\Gamma^Q_{PM}$, which are of order $O(x)$ [the expansion of the metric (3.91) starts with the term $O(x^2)$]. Note that the torsion tensor (3.94) is invariant under group rotations, like the metric is, and does not depend on $x$.

Let us define now the matrix $I_{AB}$ as an operator acting on the Lie algebra $\mathfrak{g}$. Subdivide the set of all generators $t_A$ into the set of all positive root vectors $E_{\alpha_j}$, the set of all negative root vectors $E_{-\alpha_j} = (E_{\alpha_j})^\dagger$ and the elements of the Cartan subalgebra $H$. For a semi-simple group, the latter are spanned by the *coroots*[19] $\alpha_j^\vee = [E_{\alpha_j}, E_{-\alpha_j}]$. If the group includes a certain number of $U(1)$ factors, their generators also belong to $H$. Define the action of $\hat{I}$ on the root vectors as

$$\hat{I}E_{\alpha_j} = -iE_{\alpha_j}, \quad \hat{I}E_{-\alpha_j} = iE_{-\alpha_j} \tag{3.96}$$

---

[19] This choice defines the *Chevalley basis*.

and assume that $\hat{I}H = H$. For example, for $SU(3)$ with the standard nomenclature of the generators

$$t_{1+i2} = \begin{pmatrix} 0 & 1 & 0 \\ 0 & 0 & 0 \\ 0 & 0 & 0 \end{pmatrix} = E_\alpha, \quad t_{4+i5} = \begin{pmatrix} 0 & 0 & 1 \\ 0 & 0 & 0 \\ 0 & 0 & 0 \end{pmatrix} = E_{\alpha+\beta},$$

$$t_{6+i7} = \begin{pmatrix} 0 & 0 & 0 \\ 0 & 0 & 1 \\ 0 & 0 & 0 \end{pmatrix} = E_\beta,$$

$$t_3 = \frac{1}{2}\mathrm{diag}(1,-1,0), \quad t_8 = \frac{1}{2\sqrt{3}}\mathrm{diag}(1,1,-2) \tag{3.97}$$

[$\alpha$ and $\beta$ are the simple roots of $SU(3)$], this means that

$$\begin{aligned} &I_{21} = I_{54} = I_{76} = -I_{12} = -I_{45} = -I_{67} = 1, \\ &I_{83} = -I_{38} = 1, \end{aligned} \tag{3.98}$$

where the first line is equivalent to (3.96) and the second line defines one of the possible actions of $\hat{I}$ on $H$ compatible with the properties $I_{AB} = -I_{BA}$ and $I^2 = -\mathbb{1}$. Generically, we subdivide the generators of the Cartan subalgebra into pairs in an arbitrary way and proceed analogously. Naturally, the dimension of the Cartan subalgebra and hence of the whole manifold should be even.

We now substitute the complex structure (3.93) in the Nijenhuis tensor (2.60) and require it to vanish. We arrive at the *algebraic* condition

$$\boxed{f_{ABC} - I_{AD}I_{BE}f_{DEC} - I_{BD}I_{CE}f_{ADE} - I_{AD}I_{CE}f_{DBE} = 0.} \tag{3.99}$$

For $SU(3)$, one can check explicitly that this identity is satisfied, but we will now show how to prove it for any group. Represent each positive root vector as

$$E_{\alpha_j} = t_{A_j} + it_{A_j^*}, \tag{3.100}$$

with Hermitian $t_{A_j}$ and $t_{A_j^*}$. The corresponding components of $I$ are

$$I_{A_j^*A_j} = -I_{A_jA_j^*} = 1. \tag{3.101}$$

For example, for $SU(3)$, $A_j = 1,4,6$ and $A_j^* = 2,5,7$.

Note that the commutator $[t_A, t_B]$ can only give a "starred" generator $t_{C^*}$ and the commutator $[t_A, t_{B^*}]$ can only give $t_C$. In other words,

$$f_{ABC} = f_{AB^*C^*} = 0, \tag{3.102}$$

and only the structure constants of the type $f_{ABC^*}$ or $f_{A^*B^*C^*}$ do not vanish.

Now consider several cases:

(1) Consider the LHS of (3.99) with $A, B$ associated with the same root vector $E_\alpha$ so that $B = A^*$. Bearing in mind (3.101), it is easy to see that the first term in (3.99) cancels the second one, while the third and the fourth terms vanish.

(2) Let now $A$ and $B$ be associated with different root vectors $E_\alpha$ and $E_\beta$, with $t_C \equiv h$ belonging to the Cartan subalgebra. The commutator $[h, E_\alpha]$ is proportional to $E_\alpha$ and hence $f_{ABC} = 0$ in this case. Also, the other terms in (3.99) vanish.

(3) A somewhat less trivial case is when $A, B, C$ are associated with three different root vectors. Note that, for any triple $\alpha, \beta, \gamma$, one can find a couple $(\alpha, \beta)$ such the commutator $[E_\alpha, E_\beta]$ has no projection on $E_\gamma$. Indeed, if by any chance $[E_\alpha, E_\beta]$ *is* proportional to $E_\gamma$, then $\gamma = \alpha + \beta$. But then, for the couple $(\alpha, \gamma)$, the commutator $[E_\alpha, E_\gamma]$ may only be proportional to $E_{2\alpha+\beta}$ (if such a root exists) but not to $E_\beta$ or $E_\alpha$.

Thus, we choose a couple $\alpha, \beta$ for which $[t_A + it_{A^*}, t_B + it_{B^*}]$ has no projection on $t_C$ or $t_{C^*}$. It follows that

$$f_{ABC^*} - f_{A^*B^*C^*} = f_{A^*BC} + f_{AB^*C} = 0. \tag{3.103}$$

Bearing this and (3.102) in mind, it is easy to see that the relation (3.99) holds for all star attributions. For example, if $A$ and $B$ do not carry a star, while $C \to C^*$ does, we may use (3.101) to represent the LHS of (3.99) as

$$f_{ABC^*} - f_{A^*B^*C^*} + f_{AB^*C} + f_{A^*BC}, \tag{3.104}$$

which vanishes due to (3.103). □

We concentrate now on $SU(3)$.

**Theorem 3.11.** *The $SU(3)$ manifold is HKT.*

**Proof.** We prove first the lemma:

**Lemma 4.** *Let* $\Omega$ *be an automorphism of the algebra* $\mathfrak{g}$ *such that* $\Omega$ *is an orthogonal matrix satisfying*

$$\Omega_{AD}\Omega_{BE}\Omega_{CF}f_{DEF} = f_{ABC}. \tag{3.105}$$

*Let* $I$ *be an integrable complex structure. Then,*

$$J_{AB} = (\Omega I \Omega^T)_{AB} \tag{3.106}$$

*also gives, after multiplying by* $e_{MA}e_{NB}$*, an integrable complex structure, which is covariantly constant with the same Bismut connection.*

**Proof.** It is straightforward to see that $J$ is antisymmetric and squares to $-\mathbb{1}$. Bearing in mind the invariance of $f_{ABC}$, the matrix $J_{AB}$ satisfies the same condition (3.99) as $I_{AB}$ and hence $J_{MN}$ is integrable. Also, the covariant derivative (3.95) with the torsion tensor (3.94) vanishes: when deriving this fact for the matrix $I$, we used the expression (3.92) for the vielbeins but not a specific form of $I_{AB}$. □

To prove the theorem, we have to find a quaternion triple of the complex structures with coinciding Bismut connections. We take $I$ from Eq. (3.98) and search for $J$ in the form (3.106) such that $IJ + JI = 0$. The third structure is then $K = IJ$.

The automorphism $\Omega$ can be found by adjusting first its action on a subalgebra $su(2) \oplus u(1) \subset su(3)$. This subalgebra can be chosen in many different ways, and the theorem can be proven with any such choice [though a modification of the definition (3.98) for $I$ might be necessary in the generic case], but the most convenient and universal choice, which works well also for other groups, involves the $su(2)$ algebra associated with the *highest root* $\alpha + \beta$. It has three Hermitian generators: $t_4$, $t_5$ and

$$t_{\tilde{3}} = \frac{1}{2}\text{diag}(1, 0, -1) = \frac{t_3 + \sqrt{3}\, t_8}{2} \tag{3.107}$$

(a half of the coroot $\alpha^\vee + \beta^\vee$). The generator of $u(1)$ should then be chosen to be orthogonal to $t_{\tilde{3}}$:

$$t_{\tilde{8}} = \frac{1}{2\sqrt{3}}\text{diag}(-1, 2, -1) = \frac{-\sqrt{3}\, t_3 + t_8}{2}. \tag{3.108}$$

We choose the matrix $I$ in such a way that its elements in the rotated basis $(1, 2, \tilde{3}, 4, 5, 6, 7, \tilde{8})$ would have the same form as in (3.98). We see that the matrix $I$ splits in two blocks:

(1) the block in the subspace $(4, 5, \tilde{3}, \tilde{8})$ that corresponds to the subalgebra $su(2) \oplus u(1)$ of $su(3)$ described above,
(2) the block in the subspace (1,2,6,7) acting on the root vectors $E_{\pm\alpha}, E_{\pm\beta}$.

We note that each block has the form $\mathfrak{I}$ defined in (3.3).

To find the second complex structure, we calculate $\Omega I \Omega^T$ with the automorphism $\Omega : t_a \to U^\dagger t_a U$, where

$$U = \exp\left\{\frac{i\pi}{4}(E_{\alpha+\beta} + E_{-\alpha-\beta})\right\} = \exp\left\{\frac{i\pi}{2}t_4\right\}$$
$$= \frac{1}{\sqrt{2}}\begin{pmatrix} 1 & 0 & i \\ 0 & \sqrt{2} & 0 \\ i & 0 & 1 \end{pmatrix}. \tag{3.109}$$

The transformations of the generators $t_{4,5,\tilde{3},\tilde{8}}$ of the first block and of the generators $t_{1,2,6,7}$ in the second block can be found either directly by matrix multiplication or using the Hadamard formula:

$$e^R X e^{-R} = X + [R, X] + \frac{1}{2}[R, [R, X]] + \frac{1}{6}[R, [R, [R, X]]] + \dots \tag{3.110}$$

with $R = -i\frac{\pi}{4}(E_{\alpha+\beta} + E_{-\alpha-\beta})$.

We observe that the sectors $(45\tilde{3}\tilde{8})$ and $(1267)$ are not mixed. The generators in the first sector transform as

$$t_{4,\tilde{8}} \overset{\Omega}{\to} t_{4,\tilde{8}}, \quad t_5 \overset{\Omega}{\to} t_{\tilde{3}} \overset{\Omega}{\to} -t_5. \tag{3.111}$$

The nonzero matrix elements of $J = \Omega I \Omega^T$ in this sector are $J_{5\tilde{8}} = -J_{\tilde{8}5} = J_{\tilde{3}4} = -J_{4\tilde{3}} = 1$, which gives in the basis $(45\tilde{3}\tilde{8})$ the matrix $\mathfrak{J}$ in (3.3). In the sector (1267), the automorphism acts in a little more complicated way:

$$t_1 \overset{\Omega}{\to} \frac{t_1 - t_7}{\sqrt{2}}, \quad t_2 \overset{\Omega}{\to} \frac{t_2 - t_6}{\sqrt{2}}, \quad t_6 \overset{\Omega}{\to} \frac{t_2 + t_6}{\sqrt{2}}, \quad t_7 \overset{\Omega}{\to} \frac{t_1 + t_7}{\sqrt{2}}. \tag{3.112}$$

Still, the action of this automorphism on $I$ gives in this block the structure $-\mathfrak{J}$, the same up to a sign as in the first one!

The two matrices $I = \mathrm{diag}(\mathfrak{I}, \mathfrak{I})$ and $J = \mathrm{diag}(\mathfrak{J}, -\mathfrak{J})$ anticommute. The matrix $K = IJ$ gives the third member of the triple. The matrix $K$ can also be obtained directly from $I$ by applying the automorphism $t_a \to \tilde{U}^\dagger t_a \tilde{U}$ with

$$\tilde{U} = \exp\left\{\frac{\pi}{4}(E_{\alpha+\beta} - E_{-\alpha-\beta})\right\}. \tag{3.113}$$

□

As was mentioned, there are many other group manifolds admitting the HKT structure. The manifolds based on the group with only one non-Abelian factor are presented the following list [43]:

$$\boxed{\begin{gathered} SU(2l+1),\ SU(2l) \times U(1),\ Sp(l) \times [U(1)]^l, \\ SO(2l+1) \times [U(1)]^l,\ SO(4l) \times [U(1)]^{2l}, \\ SO(4l+2) \times [U(1)]^{2l-1}, \\ G_2 \times [U(1)]^2,\ F_4 \times [U(1)]^4,\ E_6 \times [U(1)]^2,\ E_7 \times [U(1)]^7, \\ E_8 \times [U(1)]^8. \end{gathered}} \tag{3.114}$$

We say here only few words concerning a general proof valid for all the manifolds from the list above, addressing the reader to Refs. [43, 44, 46] for details.

Consider the manifold $SU(4) \times U(1)$. $SU(4)$ has three simple roots. The positive roots of $SU(4)$ are schematically shown as follows:

$$\begin{pmatrix} * & \alpha & \alpha+\beta & \alpha+\beta+\gamma \\ * & * & \beta & \beta+\gamma \\ * & * & * & \gamma \\ * & * & * & * \end{pmatrix}. \tag{3.115}$$

We pick up the highest root $\theta = \alpha + \beta + \gamma$ there and define the action of the matrix $I$ in the Cartan subalgebra so that the coroot $\alpha^\vee + \beta^\vee + \gamma^\vee$ is mixed with the generator $t_0$ of $u(1)$ and two remaining basic elements of the Cartan subalgebra are mixed with one another. The action of $\hat{I}$ on the root vectors is still defined by (3.96). The matrix $I$ thus defined has a block-diagonal form. There are four distinct blocks: (I) the outer block including the generators $E_{\pm\theta}, \theta^\vee$ and $t_0$; (II) the inner block including the generators $E_{\pm\beta}$ and two remaining generators of the Cartan subalgebra; (III) the block including $E_{\pm\alpha}$ and $E_{\pm(\beta+\gamma)}$; (IV) the block including $E_{\pm(\alpha+\beta)}$ and $E_{\pm\gamma}$. All these blocks have the form $\mathfrak{I}$.

Next, we apply to $I$ the automorphism with

$$U_0 = \exp\left\{\frac{i\pi}{4}(E_\theta + E_{-\theta})\right\}. \tag{3.116}$$

This automorphism gives some matrix $\tilde{J}$, but in contrast to what we had for $su(3)$, it is not yet what we want: the conversion $\mathfrak{I} \to \pm\mathfrak{J}$ has been effectuated only in the three out of the four blocks. The inner block associated with the nontrivial *centralizer* of the root vectors $E_{\pm\theta}$ in $su(4)$ [which is $\mathfrak{g}^{(1)} = su(2) \oplus u(1)$] remains unconverted and $\tilde{J}$ does not anticommute with $I$. To convert the fourth block, we have to pick up the highest root $\theta_1$ in $\mathfrak{g}^{(1)}$ (in the considered case, $\mathfrak{g}^{(1)}$ has only *one* root $\beta \equiv \theta_1$) and act on $\tilde{J}$ by the *additional* automorphism with

$$U_1 = \exp\left\{\frac{i\pi}{4}(E_{\theta_1} + E_{-\theta_1})\right\}. \tag{3.117}$$

The consequent application of these two automorphisms give the matrix $J$ with the structure $\mathfrak{J}$ or $-\mathfrak{J}$ in all the blocks.

This procedure can be generalized to any group. One picks up the highest root in $\mathfrak{g}$, converts the "outer layer" in $I$ with the automorphism (3.116), finds the centralizer $\mathfrak{g}^{(1)}$ of the highest root, finds the highest root (or highest *roots* if $\mathfrak{g}^{(1)}$ is not simple, which is quite often the case) in this centralizer and then repeats this procedure until it is necessary.

Note that not *all* group manifolds of dimension $4n$ are HKT. For example, $SO(8)$ is not. The reason is that the Cartan subalgebra of $so(8)$ is too small to "serve" the four relevant basic roots — the highest root in $so(8)$ and three highest roots in its centralizer $\mathfrak{g}^{(1)} = su(2) \oplus su(2) \oplus su(2)$. One should bring about four $U(1)$ factors to make the construction possible. The manifold $SO(8) \times [U(1)]^4$ is HKT.

The fact that $SU(3)$ and other group manifolds in the list (3.2.1) are HKT was proven by traditional mathematical methods. For $SU(3)$, an alternative supersymmetric "physical" proof based on the formalism of harmonic superspace (see Chapters 7 and 13) also exists [47].

There are also many example of the HKT manifolds which are not group manifolds but *homogeneous spaces* $G/H$. This was discovered in Refs. [13, 44]. The simplest way to see that is to use the method based on automorphisms, which was outlined above [46]. One of the simplest examples is the manifold $SU(4)/SU(2)$. Its tangent space does not include the generators of the "internal" $SU(2)$ group, and the complex structures are the matrices $12 \times 12$ including only three blocks of the "outer layer" of the complex structures of $SU(4) \times U(1)$. The structure $J$ is obtained from the structure $I$ by the action of the outer automorphism (3.116). The automorphism (3.117) is pointless.

# PART 2
# PHYSICS

## Chapter 4

# Dynamical Systems with and without Grassmann Variables

The main idea of this book is that, to study the geometric properties of manifolds, it is very useful to consider the associated *dynamics*. We will be interested in particular with the dynamics of particles moving along the geodesic trajectories in the manifold. These trajectories realize the shortest way between two points. A geodesic satisfies the equation

$$\frac{d^2 x^M}{ds^2} = -\Gamma^M_{NP} \frac{dx^N}{ds} \frac{dx^P}{ds}. \tag{4.1}$$

In geometry, $s$ is just a parameter distinguishing different points on the line. But one can attribute to $s$ the meaning of *time*. Then, Eq. (4.1) is a dynamical equation describing the motion of a particle along the geodesic. Various dynamical systems have been intensely studied in physics since Galileo and Newton, a lot of experience and different know-hows have been accumulated, and we can capitalize on that.

Geodesics are, however, not enough. It turned out that, to obtain really strong and interesting results, one has to consider the motion along a *supermanifold* associated with a given manifold. This supermanifold is characterized by the ordinary coordinates but on top of that also *Grassmann* anticommuting coordinates. Certain general remarks concerning classical and quantum dynamics of the ordinary systems and of the systems involving Grassmann variables are the subject of this chapter.

### 4.1 Ordinary classical mechanics

In this and in the following section, we briefly recall some facts very well known from the university course in classical and quantum mechanics. We do so for a benefit of those readers who, being pure mathematicians, probably know the geometry of manifolds that we talked about in the first part much better than the author but might need to refresh their memory of certain basic physics notions.

By *dynamical system* we understand a system described by a set of ordinary differential equations

$$\frac{dx_i}{dt} = f_i(x_j). \tag{4.2}$$

The independent variable $t$ is interpreted as time.[1] The functions $x_i(t)$ are called *dynamical variables*. For the equation system (4.2), a Cauchy problem can be posed: if we know the initial conditions at $t = 0$, we can deduce the values of $x_i(t)$ at later times.

We restrict our consideration to a special class of dynamical systems, the *Hamiltonian systems*.

**Definition 4.1.** A Hamiltonian system is a dynamical system involving an even number $2n$ of dynamical variables which can be separated in to two classes — the *canonical coordinates* $q_{i=1,\ldots,n}$ and the *canonical momenta* $p_{i=1,\ldots,n}$ — in such a way that the dynamical equations (4.2) acquire the form

$$\boxed{\begin{aligned} \frac{dq_i}{dt} &\equiv \dot{q}_i = \frac{\partial H(p_j, q_j)}{\partial p_i}, \\ \frac{dp_i}{dt} &\equiv \dot{p}_i = -\frac{\partial H(p_j, q_j)}{\partial q_i}, \end{aligned}} \tag{4.3}$$

where the function $H(q_j, p_j)$ is called the Hamilton function or the *Hamiltonian*.

The set of all variables $\{q_i, p_i\}$ is called the *phase space*. It is obvious that[2]

$$\frac{dH}{dt} = \frac{\partial H}{\partial q_i}\dot{q}_i + \frac{\partial H}{\partial p_i}\dot{p}_i = \frac{\partial H}{\partial q_i}\frac{\partial H}{\partial p_i} - \frac{\partial H}{\partial p_i}\frac{\partial H}{\partial q_i} = 0. \tag{4.4}$$

---

[1]Mathematicians often discuss nowadays another kind of dynamical system where time is *discrete*. In this case, one writes instead of (4.2)

$$x_i^{(n+1)} = f_i[x_j^{(n)}].$$

*Cellular automata* including the famous Conway's Game of Life belong to this class. But discrete systems are beyond the scope of our book.

[2]We still assume the summation over the repeated indices, although neither $q_i$ nor $p_i$ represent vectors in the sense of Eqs. (1.3) and (1.4).

In other words, $H(q_j, p_j)$ is an *integral of motion* — a quantity preserving its value during the time evolution described by Eqs. (4.3). It has the physical meaning of energy.

Besides the energy, there might be other integrals of motion.

**Definition 4.2.** The structure[3]

$$\boxed{\{A, B\}_P = \frac{\partial A}{\partial p_i}\frac{\partial B}{\partial q_i} - \frac{\partial A}{\partial q_i}\frac{\partial B}{\partial p_i}} \tag{4.5}$$

is called the *Poisson bracket* of two phase space functions $A(q_i, p_i)$ and $B(q_i, p_i)$.

**Theorem 4.1.** *A quantity $f(q_i, p_i)$ is an integral of motion iff the Poisson bracket $\{H, f\}_P$ vanishes.*

**Proof.** Just substitute the Hamilton equations of motion (4.3) in

$$\frac{df}{dt} = \frac{\partial f}{\partial q_i}\dot{q}_i + \frac{\partial f}{\partial p_i}\dot{p}_i$$

to derive

$$\frac{df}{dt} = \{H, f\}_P. \tag{4.6}$$

□

Equation (4.6) describes the time evolution of an arbitrary phase-space function.

The Poisson bracket (4.5) has the following nice algebraic properties:

$$\{A, B\}_P = -\{B, A\}_P, \quad \{AB, C\}_P = A\{B, C\}_P + B\{A, C\}_P,$$
$$\{A, \{B, C\}_P\}_P + \{B, \{C, A\}_P\}_P + \{C, \{A, B\}_P\}_P = 0. \tag{4.7}$$

Substituting the Hamiltonian $H$ for $C$ in the last (Jacobi) identity, we derive an important theorem:

**Theorem 4.2.** *If $A(p_i, q_i)$ and $B(p_i, q_i)$ are two integrals of motion, their Poisson bracket either vanishes identically or, if not, it is also an integral of motion.*

[3] In different textbooks, one can find different sign conventions for the Poisson bracket. We stick to the conventions of Ref. [48].

Now, consider the function

$$L(q_i, \dot{q}_i) = p_i \dot{q}_i - H(p_i, q_i) \tag{4.8}$$

with the momenta $p_i$ in the RHS to be replaced by their solution of the first of the Hamilton equations,[4] $\dot{q}_i = \partial H/\partial p_i$. The function $L$ is called the *Lagrangian.* And the transformation (4.8) is called the *Legendre transformation.* The latter obviously works both ways: one can start from the Lagrangian and derive the Hamiltonian.

For the simplest one-dimensional Hamiltonian

$$H = p^2/(2m) + U(q), \tag{4.9}$$

we obtain

$$L = m\dot{q}^2/2 - U(q). \tag{4.10}$$

**Theorem 4.3.** *The set* (4.3) *of the Hamilton equations is equivalent to the following set of the equations of Lagrange:*

$$\boxed{\frac{d}{dt}\left(\frac{\partial L}{\partial \dot{q}_i}\right) - \frac{\partial L}{\partial q_i} = 0.} \tag{4.11}$$

**Proof.** Consider the differential

$$dL = \dot{q}_i dp_i + p_i d\dot{q}_i - dH. \tag{4.12}$$

Bearing in mind (4.3), we derive

$$dL = p_i d\dot{q}_i - \frac{\partial H}{\partial q_i} dq_i. \tag{4.13}$$

It follows that[5]

$$\frac{\partial L}{\partial \dot{q}_i} = p_i, \qquad \frac{\partial L}{\partial q_i} = -\frac{\partial H}{\partial q_i}. \tag{4.14}$$

Differentiating over time the first relation in (4.14) that expresses the canonical momenta via the Lagrangian, substituting for $\dot{p}_i$ the second Hamilton equation and taking also into account the second relation in (4.14), we derive (4.11). □

[4]We assume that the equation system $\dot{q}_i = \partial H/\partial p_i$ is not degenerate and has a unique solution for $p_i(q_j, \dot{q}_j)$.

[5]The partial derivatives in the LHS of Eq. (4.14) are done, of course, in the assumption of fixed $q_i$ or $\dot{q}_i$, whereas the partial derivatives $\partial H/\partial q_i$ in the RHS are done in the assumption of fixed $p_i$.

Introduce now the *action functional*

$$S[q(t), t_1 - t_0] = \int_{t_0}^{t_1} L[q_i(t), \dot{q}_i(t)]\, dt. \tag{4.15}$$

The fact that it depends, besides $q_i(t)$, only on the difference $t_1 - t_0$, but not on $t_0$ and $t_1$ separately, follows from the absence of an *explicit* time dependence in the Lagrangian. Let us fix the boundary conditions:

$$q_i(t_0) = q_i^{(0)}, \quad q_i(t_1) = q_i^{(1)}, \tag{4.16}$$

where $q_i^{(0,1)}$ is a set of constants. The following very important theorem can be proven:

**Theorem 4.4.** *A set of functions $q_i(t)$ realizing an extremum of the functional* (4.15) *with the boundary conditions* (4.16) *satisfies also the Lagrange equations* (4.11) *and represents the classical trajectory of the system.*

**Proof.** The variation of the action functional is

$$\delta S = \int_{t_0}^{t_1} \left[ \frac{\partial L}{\partial \dot{q}_i(t)} \delta \dot{q}_i(t) + \frac{\partial L}{\partial q_i(t)} \delta q_i(t) \right] dt. \tag{4.17}$$

Integrate the first term by parts, bearing in mind that $\delta q_i(t_0) = \delta q_i(t_1) = 0$. We obtain

$$\delta S = -\int_{t_0}^{t_1} \left[ \frac{d}{dt} \left( \frac{\partial L}{\partial \dot{q}_i(t)} \right) - \frac{\partial L}{\partial q_i(t)} \right] \delta q_i(t)\, dt. \tag{4.18}$$

At an extremum, this variation should vanish, and it should be so for any variation $\delta q_i(t)$. And that is only possible if the expression in the square brackets vanishes. □

**Remark:** For not too wild non-degenerate function $L$, the equation system (4.11) with the boundary conditions (4.16) has a unique solution. This means that the action functional has only one extremum. It is habitual to choose the sign of $L$ in such a way that the action takes a *minimal* value. The *least action principle* is one of the major guiding principles in physics.

Let us illustrate this general result with a simple example (Fig. 4.1). Take the Lagrangian

$$L = \frac{m\dot{z}^2}{2} - mgz \equiv T - U.$$

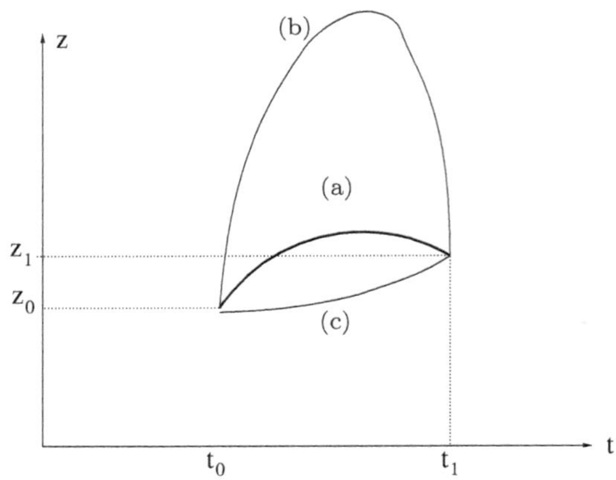


**Fig. 4.1.** Different trajectories of the stone: (a) the trajectory with minimal action; (b) too high kinetic energy; (c) too low potential energy.

It describes the vertical motion of a stone of mass $m$ under the action of gravity. Let us ask the following question: what should our stone do to minimize the action functional (4.15)? Well, to minimize the integral $\int(T-U)dt$, the stone would like to increase its potential energy $U(z) = mgz$ and to climb higher. But, bearing in mind that it has an appointment at $z_1$ at fixed time $t_1$, going too high would entail too high a velocity and a large positive contribution to the integral (4.15) coming from the first kinetic energy term, $T = m\dot{z}^2/2$. To minimize the integral of the kinetic energy, the stone should keep low profile and move from $z_0$ to $z_1$ along a straight line. The real trajectory is a negotiated compromise between these two conflicting requirements.

The last issue to be discussed in this section is the famous *Noether's theorem* [49]. It asserts that any continuous symmetry of the Lagrangian entails the existence of an integral of motion. It is convenient to prove first its restricted version and then generalize.

**Theorem 4.5.** *Suppose that the Lagrangian $L(q_i, \dot{q}_i)$ is invariant under an infinitesimal transformation of variables:*

$$q_i \to q_i + \delta q_i = q_i + \epsilon f_i(q_j, \dot{q}_j), \quad \epsilon \ll 1. \tag{4.19}$$

*Then, the Noether charge $Q = (\partial L/\partial \dot{q}_i)\, f_i$ is an integral of motion.*

**Proof.** The variation of the Lagrangian $\delta L$ vanishes and so does the variation of the action $\delta S$. The latter is expressed as in Eq. (4.17). Integrate it by parts, keeping now the boundary term. We obtain

$$\delta S = \epsilon \int_{t_0}^{t_1} \left[ \frac{\partial L}{\partial q_i} - \frac{d}{dt}\left(\frac{\partial L}{\partial \dot{q}_i}\right)\right] f_i \, dt + \epsilon \frac{\partial L}{\partial \dot{q}_i} f_i \Big|_{t_0}^{t_1} . \tag{4.20}$$

This should vanish for any $q_i(t)$. In particular, it is true for the classical trajectories satisfying the classical equations of motion. For such trajectories, the vanishing of $\delta S$ implies the vanishing of the boundary term in (4.20). This gives $Q(t_1) = Q(t_0)$. Thus, $Q(t)$ is an integral of motion, indeed. □

**Theorem 4.6.** *Now, suppose that the Lagrangian is not quite invariant under the transformation* (4.19) *but is shifted by a total time derivative,* $\delta L = \epsilon(d/dt)\Lambda(\dot{q}_i, q_i)$. *In this case, the integral of motion still exists, given by the expression*

$$Q(t) = \frac{\partial L}{\partial \dot{q}_i} f_i - \Lambda = p_i f_i - \Lambda. \tag{4.21}$$

**Proof.** The variation of the action (4.20) does not vanish any more. For a classical trajectory, we derive

$$\Lambda(t_1) - \Lambda(t_0) = \frac{\partial L}{\partial \dot{q}_i} f_i \Big|_{t_0}^{t_1},$$

from which the conservation of (4.21) follows. □

**Examples.**

(1) Consider the variation $\delta q_i = \epsilon \dot{q}_i$ generated by a infinitesimal time shift, $t \to t + \epsilon$. Under this variation, $\delta L = \epsilon \dot{L}$, which is a total derivative so that the conditions of Theorem 4.6 are fulfilled. The corresponding conserved Noether charge is simply the energy:

$$E = \frac{\partial L}{\partial \dot{q}_i} \dot{q}_i - L.$$

(2) Consider the Lagrangian

$$\boxed{L = m \frac{\dot{x}^2 + \dot{y}^2}{2} - U(x^2 + y^2).} \tag{4.22}$$

It is invariant under rotations. Infinitesimally,

$$\delta x = \epsilon y, \quad \delta y = -\epsilon x. \tag{4.23}$$

The conserved Noether charge,

$$Q = m(\dot{x}y - \dot{y}x) = p_x y - p_y x, \tag{4.24}$$

is none other than the anglular momentum $J_z$ (taken with the opposite sign).

To avoid a possible confusion of a reader-physicist, one should make the following remark.

In most textbooks on classical mechanics (see e.g. Ref. [50]) one understands by Noether's theorem a related, but somewhat different statement. One requires an *exact* invariance of the Lagrangian under the transformations $\delta q_i = \epsilon f_i(q_j, t)$ (no dependence on the general velocities $\dot{q}_i$ but a time dependence is allowed) supplemented by a possible transformation of time, $\delta t = \epsilon f_0(q_i, t)$. This is enough for conventional mechanical applications (for example, the energy conservation follows in that case from the invariance of the Lagrangian under constant time shifts *not* associated with shifts of $q_i$).

But to apply Noether's theorem for supersymmetric systems, as we do in further chapters, it is Theorem 4.6 that we need: the supersymmetry transformations do involve velocity dependence and the supersymmetric Lagrangian is not exactly invariant under these transformations but is shifted by a total time derivative.

In supersymmetric field theories, supersymmetry transformations shift the Lagrangian density by a gradient $\partial_\mu \Lambda$ of some function. People often say that such a transformation leaves the *action* invariant, but this assumes that the action integral is done over the whole space time and that the fields vanish at $t = \pm\infty$ and at spatial infinity. The action of a mechanical system, calculated as $\int_{t_0}^{t_1} L\, dt$ with finite limits, may be modified.

Note, however, that, though the action might be modified if a total derivative is added to the Lagrangian, the dynamical equations of motion are *not*. This is heuristically clear: a possible modification is due to the boundary terms, while the equations of motion determine the behavior of the system *inside* the time interval $(t_0, t_1)$. One may prove a simple theorem:

**Theorem 4.7.** *Let*

$$L' = L + \dot{\Lambda}(q_i) = L + \frac{\partial \Lambda}{\partial q_i}\dot{q}_i.$$

*Then, the Lagrange equations of motion following from $L'$ and $L$ coincide.*

**Proof.** We write

$$\frac{\partial L'}{\partial q_i} = \frac{\partial L}{\partial q_i} + \frac{\partial \dot{\Lambda}}{\partial q_i}$$

and

$$\frac{d}{dt}\frac{\partial L'}{\partial \dot{q}_i} = \frac{d}{dt}\left[\frac{\partial L}{\partial \dot{q}_i} + \frac{\partial \Lambda}{\partial q_i}\right].$$

It is clear that the shifts due to $\Lambda$ in the Lagrange equations (4.11) cancel out. □

The same is true if $\Lambda$ also depends on the velocities $\dot{q}_i$ or on still higher time derivatives of $q_i(t)$. One should only have in mind that, for the Lagrangians involving higher derivatives (the shifted Lagrangian $L'$ does in this case), the form of the equations of motion is modified. For example, for $L'(\ddot{q}_i, \dot{q}_i, q_i)$, the equations of motion read

$$\frac{d^2}{dt^2}\left(\frac{\partial L'}{\partial \ddot{q}_i}\right) - \frac{d}{dt}\left(\frac{\partial L'}{\partial \dot{q}_i}\right) + \frac{\partial L'}{\partial q_i} = 0. \tag{4.25}$$

It is *these* equations that coincide in this case with (4.11).

### 4.1.1 *Algebra of symmetries*

Suppose now that the action has two different symmetries: the symmetries under the infinitesimal transformations

$$\delta_f q_i = \epsilon_1 f_i(q_j, \dot{q}_j) \quad \text{and} \quad \delta_g q_i = \epsilon_2 g_i(q_j, \dot{q}_j), \tag{4.26}$$

which can, of course, be upgraded to finite ones.

**Definition 4.3.** The variation

$$\delta_{[f,g]} q_i = (\delta_f \delta_g - \delta_g \delta_f) q_i \tag{4.27}$$

is called the *Lie bracket* of two transformations[6] (4.26).

Consider the case when $f_i$ and $g_i$ do not depend on the velocities. Then,

$$\delta_{[f,g]} q_i = \epsilon_1 \epsilon_2 \left[f_j \frac{\partial g_i}{\partial q_j} - g_j \frac{\partial f_i}{\partial q_j}\right] \equiv \epsilon_1 \epsilon_2 h_i(q_k). \tag{4.28}$$

If the action is invariant under (4.26), it is also invariant under (4.28).

[6] We defined here the "physical" Lie bracket. The "mathematical" Lie bracket is defined as a commutator of abstract vector fields.

The Lagrangian may be shifted under the transformations $f$ and $g$ by total derivatives. Suppose that these are the derivatives of the functions depending only on $q_i$ but not on $\dot{q}_i$:

$$\delta_f L = \epsilon_1 \frac{d}{dt}\Lambda_f(q_i), \quad \delta_g L = \epsilon_2 \frac{d}{dt}\Lambda_g(q_i). \tag{4.29}$$

Then,

$$\delta_{h=[f,g]} L = (\delta_f \delta_g - \delta_g \delta_f) L = \epsilon_1 \epsilon_2 \frac{d}{dt}\Lambda_h(q_i), \tag{4.30}$$

where

$$\Lambda_h = \delta_f \Lambda_g - \delta_g \Lambda_f = f_i \frac{\partial \Lambda_g}{\partial q_i} - g_i \frac{\partial \Lambda_f}{\partial q_i}. \tag{4.31}$$

Consider now the Noether integrals of motion corresponding to the transformations $f, g, h$:

$$Q_f = p_i f_i - \Lambda_f, \quad Q_g = p_i g_i - \Lambda_g, \quad Q_{h=[f,g]} = p_i h_i - \Lambda_h. \tag{4.32}$$

**Theorem 4.8.** *The identity $Q_h = \{Q_f, Q_g\}_P$ holds.*

This can be verified quite straightforwardly.

The Lie bracket of the transformations $f$ and $h$ (and, correspondingly, the Poisson bracket $\{Q_f, Q_h\}_P$) may give still new symmetry of the system, and that is until the algebra of Lie brackets (resp. Poisson brackets) is closed.

The simplest nontrivial algebra includes only three symmetries. Consider the Lagrangian depending on three coordinates $x, y, z$ that is invariant under the infinitesimal rotations around the first and the second axes:

$$R_1 : \begin{cases} \delta_1 y = \epsilon_1 z \\ \delta_1 z = -\epsilon_1 y \end{cases}, \quad R_2 : \begin{cases} \delta_2 z = \epsilon_2 x \\ \delta_2 x = -\epsilon_2 z \end{cases}. \tag{4.33}$$

The Lie bracket $[R_1, R_2]$ gives obviously the rotation $R_3$ around the third axis. The Noether integrals are

$$Q_1 = p_y z - p_z y, \quad Q_2 = p_z x - p_x z, \quad Q_3 = p_x y - p_y x, \tag{4.34}$$

and the $so(3)$ algebra $\{Q_i, Q_j\}_P = \varepsilon_{ijk} Q_k$ holds.

A natural question is what happens if $f_i, g_i, \Lambda_f, \Lambda_g$ depend not only on $q_j$ but also on $\dot{q}_j$. As was mentioned, this is the case for the supersymmetry transformations — the main point of interest in our book. A *conjecture* is that Theorem 4.8 also holds in this general case.

The argument is the following. Noether's theorem guarantees the existence of the integrals of motion $Q_f$ and $Q_g$. Then, Theorem 4.1 tells us that $\{H, Q_f\}_P = \{H, Q_g\}_P = 0$. But then $\{H, \{Q_f, Q_g\}_P\}_P$ also vanishes — this is the assertion of Theorem 4.2, a corollary of the Jacobi identity. Thus, $Q_h = \{Q_f, Q_g\}_P$ is also an integral of motion. The only assertion that we cannot prove is that $Q_h$ *is* the Noether charge corresponding to the Lie bracket $[f, g]$.

Well, being confronted with such a difficulty, the author can recall that he is not a mathematician caring about rigorous proofs, but a physicist. In this capacity, I could well pose a rhetorical question "*What else can it be?*" and close the discussion. Still, I would not mind to learn such a rigorous proof (which I was not able to either find in the textbooks or cook up myself). One of the difficulties of proving this assertion is that the Lie bracket of generic $f$ and $g$ depends on not only the velocities but also accelerations. One should probably impose a restriction on $f, g$ that such dependence in $[f, g]$ is absent.

Finally, I would like to point out that, as far as the supersymmetry transformations are concerned, the property $Q_{[f,g]} = \{Q_f, Q_g\}_P$ holds in all the cases considered in this book. This can be verified explicitly.

## 4.2 Ordinary quantum mechanics

It is not proper to discuss here at length physical aspects of quantum mechanics. We concentrate on its mathematical structure.

In quantum world, classical trajectories have no sense and the main object of interest is *wave function* $\Psi(q_i)$. To find the wave functions $\Psi_n(q_i)$ of the stationary states of a physical system, one has to solve the spectral problem

$$\boxed{\hat{H}\, \Psi_n(q_i) = E_n\, \Psi_n(q_i)} \tag{4.35}$$

with appropriate boundary conditions. The quantum Hamiltonian $\hat{H}$ is an elliptic differential operator acting in the $\mathcal{L}^2$ Hilbert space containing the square integrable complex functions $\Psi(q_i)$. We require the existence of a well-defined inner product

$$\langle \Psi | \Phi \rangle = \int_Q d\mu\, \overline{\Psi(q_i)} \Phi(q_i), \tag{4.36}$$

where $d\mu$ is the measure on the space $Q$ where the dynamical variables $q_i$ belong. We have used here Dirac's notation $|\Phi\rangle$ for an element of our Hilbert space (*ket-vector*) and the notation $\langle\Phi|$ for the dual *bra-vector*.

**Definition 4.4.** The integral

$$\int_Q d\mu\, \overline{\Psi(q_i)} \hat{A} \Phi(q_i) \tag{4.37}$$

is called the *matrix element* of the operator $\hat{A}$ between the states $\Psi$ and $\Phi$ and is denoted as $\langle\Psi|\hat{A}|\Phi\rangle$.

The basic equation of quantum mechanics is not (4.35) but the time-dependent Schrödinger equation

$$i\hbar \frac{\partial\, \Psi(q_i, t)}{\partial t} = \hat{H}\, \Psi(q_i, t) \tag{4.38}$$

($\hbar$ is the Planck constant). Equation (4.35) is derived from (4.38) by plugging there a stationary ansatz

$$\Psi_n(q_i, t) = \Psi_n(q_i) \exp\left\{-\frac{iE_n t}{\hbar}\right\}. \tag{4.39}$$

The quantum Hamiltonian is obtained from the classical Hamiltonian $H(q_i, p_i)$ by substituting the operators $\hat{p}_i = -i\hbar\, \partial/\partial q_i$ for the classical momenta $p_i$. Thus, the quantum counterpart of the classical Hamiltonian (4.9) is

$$\boxed{\hat{H}(\hat{p}, q) = -\frac{\hbar^2}{2m}\frac{\partial^2}{\partial q^2} + U(q).} \tag{4.40}$$

For more complicated Hamiltonians that we meet later in this book, the transition from the classical to quantum Hamiltonian is not so trivial. It is not immediately clear, for example, what the quantum counterpart for the classical Hamiltonian $H^{\mathrm{cl}} = p^2 q^2$ is. Is it

$$\hat{H} = \hat{p}^2 q^2, \quad \hat{H} = q^2 \hat{p}^2 \tag{4.41}$$

or something else? (The two expressions above do not coincide because $\hat{p}$ and $q$ do not commute, $[\hat{p}, q] = -i\hbar$.)

The problem of choice between different quantum expressions that correspond to the same classical one is the *ordering ambiguity* problem. To resolve this ambiguity, one needs extra assumptions, extra guiding principles. One such assumption is rather natural: the quantum Hamiltonian should be Hermitian, otherwise the energies of the eigenstates representing the solutions to the Schrödinger equation (4.35) would become complex. It is unacceptable in physics and also leads to ugly nuisances in mathematics. Thus, the options (4.41) are excluded. On the other hand, the options $\hat{H} = \hat{p}q^2\hat{p}$ or $\hat{H} = q\hat{p}^2 q$ are not.

We discuss it at length later, in Chapters 5, 6, 8 and 9, but we would like to tell right now that among all ordering recipes, there is a distinguished one — the *Weyl ordering procedure*. We give here the definition, while its relevance and niceness are revealed later.

**Definition 4.5.** Let $A(p_i, q_i)$ be a function in phase space.[7] Consider its Fourier decomposition

$$A(p_i, q_i) = \int \prod_i d\alpha_i d\beta_i \, h(\alpha_i, \beta_i) e^{i(\alpha_i p_i + \beta_i q_i)}. \tag{4.42}$$

Then, the operator

$$\hat{A}(\hat{p}_i, q_i) = \int \prod_i d\alpha_i d\beta_i \, h(\alpha_i, \beta_i) e^{i(\alpha_i \hat{p}_i + \beta_i q_i)}. \tag{4.43}$$

is called the *Weyl-ordered* operator corresponding to the classical function $A(p_i, q_i)$. And the function $A(p_i, q_i)$ is called the *Weyl symbol* of the operator $\hat{A}$.

For monomials like $p^2q^2$, this amounts to the symmetric ordering:

$$p^2q^2 \to \frac{1}{6}(\hat{p}^2q^2 + q^2\hat{p}^2 + \hat{p}q^2\hat{p} + q\hat{p}^2q + \hat{p}q\hat{p}q + q\hat{p}q\hat{p}).$$

This follows from the simple fact that each term in the expansion of the exponential in (4.43) includes such symmetrized products.

To find the Weyl symbol of an arbitrary polynomial operator, one should first represent it as a sum of symmetric combinations using $[\hat{p}, q] = -i\hbar$. For example, the Weyl symbol of $\hat{p}q$ is $pq - i\hbar/2$ and the Weyl symbol of $q\hat{p}$ is $pq + i\hbar/2$.

Note that the Weyl symbol of the product of two operators does not coincide with the product of their Weyl symbols. Instead of that, one has

$$(\hat{A}\hat{B})_W = \exp\left[-\frac{i\hbar}{2}\left(\frac{\partial^2}{\partial p_i \partial Q_i} - \frac{\partial^2}{\partial P_i \partial q_i}\right)\right] A_W(p_i, q_i) B_W(P_i, Q_i)\bigg|_{p=P, q=Q}. \tag{4.44}$$

(the so-called *Grönewold–Moyal product* [51]).

[7]Physicists call such functions classical *observables*.

It follows that the Weyl symbol of the commutator of two operators is

$$\boxed{\begin{aligned}[\hat{A},\hat{B}]_W &= -2i\sin\left[\frac{\hbar}{2}\left(\frac{\partial^2}{\partial p_i \partial Q_i} - \frac{\partial^2}{\partial P_i \partial q_i}\right)\right] \\ &\quad \times A_W(p_i,q_i)B_W(P_i,Q_i)\Big|_{p=P,q=Q} \\ &\stackrel{\text{def}}{=} -i\hbar\{A_W,B_W\}_{GM}.\end{aligned}} \tag{4.45}$$

In the classical limit $\hbar \to 0$, the Grönewold–Moyal bracket $\{A,B\}_{GM}$ reduces to the Poisson bracket (4.5). For simple operators relevant for simple physical applications, the terms of order $\sim \hbar^3$, etc. in (4.45) vanish, and it is the Poisson bracket that goes over to the commutator under quantization.

An important corollary of the last statement and of Theorem 4.1 is that the quantum operator $\hat{A}$ corresponding to a classical integral of motion $A(p_i,q_i)$ commutes with the quantum Hamiltonian[8]: $[\hat{H},\hat{A}] = 0$.

The final (in this section) remark is the following. When a mathematician writes (4.35), they assume that the spectrum is pure point (*discrete* in physical language) and all the eigenstates $\Psi_n$ are normalizable. In physics, one often discusses the problems with continuous spectrum. For example, one can take the Hamiltonian (4.40) with vanishing $U(q)$ that describes free motion. In this case, the eigenfunctions represent plane waves $\sim e^{ipq}$ that are not normalizable. To make a mathematical sense of such a problem, one has to perform a *regularization* — introduce a parameter that slightly modifies the problem in such a way that the continuous spectrum becomes discrete and then explore the limit when the regularization is lifted. For the Hamiltonian like $-\partial^2/\partial q^2$, one has to introduce a *finite box* — allow for the particle to move only in a finite range $|q| < L$ and send $L \to \infty$ afterwards.

In our book, we mostly study compact manifolds. In the associate dynamical systems, the motion is finite and the spectrum is discrete. And if the manifold is not compact (for example, the Taub–NUT manifold and the Eguchi–Hanson manifold studied in Section 3.1.1 are not compact), such regularization procedure will always be had in mind.

[8]The exact statement is the following: if $A(p_i,q_i)$ is an integral of motion so that $\{H,A\}_P = 0$, one can always find an ordering such that the quantum counterparts $\hat{H}$ and $\hat{A}$ of $H$ and $A$ commute, $[\hat{H},\hat{A}] = 0$. I cannot prove it as a rigourous theorem, but it is so in all the cases that I know. For applications to supersymmetry, see Ref. [52] and discussion on p. 119.

## 4.3 Grassmann variables

The ordinary phase space discussed in the previous section was parameterized by coordinates and momenta representing ordinary commuting real numbers. One can, however, generalize this discussion and assume that the phase space of the system includes, besides ordinary commuting coordinates and momenta, also anticommuting *Grassmann* variables. Grassmann variables were introduced in physics half-a-century ago by Felix Berezin [53]. They are indispensible for a correct description of fermion fields in quantum field theory. In this book, we do not touch upon field theory issues, but Grassmann dynamical variables can also be introduced for classical and quantum *mechanical* systems. And that is what the rest of this chapter is about.

To begin with, we introduce the mathematical notions of Grassmann numbers and Grassmann algebra. The main definitions are the following:

- Let $\{a_i\}$ be a set of $n$ basic anticommuting variables: $a_i a_j + a_j a_i = 0$. The elements of the Grassmann algebra can be written as the functions $f(a_i) = c_0 + c_i a_i + c_{ij} a_i a_j + \ldots$ where the coefficents $c_0, c_i, \ldots$ are ordinary (real or complex) numbers. The series terminates at the $n$th term: the $(n+1)$th term of this series would involve a product of two identical anticommuting variables (say, $a_1^2$), which is zero. Note that even though a Grassmann number like $a_1 a_2$ commutes with any other number, it still cannot be treated as an ordinary number but represents instead an *even element* of the Grassmann algebra. There are, of course, also odd anticommuting elements. The variables $\{a_i\}$ are called the *generators* of the algebra.
- One can add Grassmann numbers, $f(a_i) + g(a_i) = c_0 + d_0 + (c_i + d_i)a_i + \ldots$, as well as multiply them. For example, $(1 + a_1 + a_2)(1 + a_1 - a_2) = 1 + 2a_1 - 2a_1 a_2$ (the anticommutation property of $a_i$ was used).
- One can also differentiate the functions $f(a_i)$ with respect to Grassmann variables: $\partial/\partial a_i\,(1) \stackrel{\text{def}}{=} 0$, $\partial/\partial a_i\,(a_j) \stackrel{\text{def}}{=} \delta_{ij}$, the derivative of a sum is the sum of derivatives, and the derivative of a product of two functions satisfies the Leibniz rule, except that the operator $\partial/\partial a_i$ should be thought of as a Grassmann variable, which sometimes leads to a sign change when $\partial/\partial a_i$ is pulled through to annihilate $a_i$ in the product. For example, $\partial/\partial a_1(a_2 a_3 a_1) = a_2 a_3$, but $\partial/\partial a_1(a_2 a_1 a_3) = -a_2 a_3$.
- One can also integrate over Grassmann variables. In contrast to the ordinary case, the integral cannot be obtained here as a limit of integral sums,

cannot be calculated numerically through "finite limits" (this makes no sense for Grassmann numbers) with Simpson's method and so on. What one can do, however, is to integrate over a Grassmann variable in the "whole range" ("from $-\infty$ to $\infty$" if you will, though this is, again, meaningless). The definition (due to Berezin) is

$$\boxed{\int da_j\, f(a) \stackrel{\text{def}}{=} \frac{\partial}{\partial a_j} f(a).} \tag{4.46}$$

We have in particular

$$\int da_i\, a_j = \delta_{ij}. \tag{4.47}$$

An obvious (and important for the following) corollary of (4.46) is

$$\int da_j\, \frac{\partial f}{\partial a_j} = 0 \tag{4.48}$$

(no summation over $j$).

- If the Grassmann algebra involves an even number of generators $2n$, one can divide them into two equal parts, $\{a_{j=1,\dots,2n}\} \to \{a_{j=1,\dots,n}, \bar{a}_{j=1,\dots,n}\}$, and introduce an involution, $a_j \leftrightarrow \bar{a}_j$, which we associate with complex conjugation. We assume in particular that, simultaneously with the involution of the generators, the ordinary numbers $c_0$, etc. are replaced by their complex conjugates:

$$f(a) = c_0 + c_i a_i + d_i \bar{a}_i + \dots \longrightarrow \overline{f(a)} = \overline{c_0} + \overline{c_i}\bar{a}_i + \overline{d_i} a_i + \dots. \tag{4.49}$$

It is also convenient to assume that, for any two elements $f, g$ of the Grassmann algebra, $\overline{(fg)} = \bar{g}\bar{f}$, as for the Hermitian conjugation (the order of the factors in (4.49) is, of course, not relevant).

The following important relation holds:

$$\boxed{\int \prod_{i=1}^{n} da_i d\bar{a}_i\, \exp\{M_{jk}\bar{a}_j a_k\} = \det(M).} \tag{4.50}$$

## 4.4 Grassmann dynamics

The fact that the Grassmann variables can be treated in the same way as the ordinary ones, that one can describe their classical dynamics using the methods of analytical mechanics outlined in Section 5.1 and quantum dynamics by solving an approriate Schrödinger equation, was dicovered comparatively recently [54, 55], *after* the discovery of supersymmetry. But in our book, the course of history is reversed. We first discuss non supersymmetric Grassmann systems, and supersymmetry is introduced in the following chapter.

As the simplest example, consider the classical Lagrangian

$$L = -i\dot{\psi}\bar{\psi} + \omega\bar{\psi}\psi, \tag{4.51}$$

where $\psi$ and $\bar{\psi}$ are anticommuting variables,

$$(\psi)^2 = (\bar{\psi})^2 = 0, \quad \psi\bar{\psi} + \bar{\psi}\psi = 0, \tag{4.52}$$

and $\omega$ is a real constant. The corresponding action $S = \int L dt$ is real (invariant under the involution defined above). For the second term in (4.51), it is seen immediately, and the conjugation of the first term gives $i\psi\dot{\bar{\psi}} = id(\psi\bar{\psi})/dt - i\dot{\psi}\bar{\psi}$. The total time derivative in $\bar{L}$ does not contribute in the action.

The Lagrange equations have the same form as for the ordinary commuting variables:

$$\frac{d}{dt}\left(\frac{\partial L}{\partial \dot{\psi}}\right) - \frac{\partial L}{\partial \psi} = 0 \tag{4.53}$$

and the complex conjugate. This gives

$$i\dot{\psi} + \omega\psi = 0, \quad i\dot{\bar{\psi}} - \omega\bar{\psi} = 0. \tag{4.54}$$

The variables $\psi$ and $\bar{\psi}$ "parameterize" phase space (whatever the word "parameterize" means for a Grassmann coordinate).

If the Lagrangian is written as in (4.51), involving time derivative of $\psi$ but not of $\bar{\psi}$, it is natural to call $\psi$ rather than $\bar{\psi}$ a Grassmann coordinate and $\Pi_\psi = \partial L/\partial\dot{\psi} = -i\bar{\psi}$ is then the canonical Grassmann momentum. The classical Hamiltonian is obtained from the (4.51) by the Legendre transformation[9]:

$$H = \dot{\psi}\Pi_\psi - L = \omega\psi\bar{\psi}. \tag{4.55}$$

---

[9]A reader-physicist has noted a similarity between the Hamiltonian (4.55) and the expression $H = \omega a\bar{a}$ for the Hamiltonian of the harmonic oscillator expressed via the holomorphic variables. It is by no means a coincidence. In fact, the system that we are now studying is called the *Grassmann oscillator*.

The Hamilton equations of motion are

$$\frac{\partial H}{\partial \psi} = -\dot{\Pi}_\psi \tag{4.56}$$

and the complex conjugate. They coincide with (4.54).

The Poisson bracket of two functions $f(\psi, \bar{\psi})$, $g(\psi, \bar{\psi})$ is now defined as

$$\{f, g\}_P = -i \left( \frac{\partial^2}{\partial \psi \partial \bar{\Psi}} + \frac{\partial^2}{\partial \bar{\psi} \partial \Psi} \right) f(\psi, \bar{\psi}) g(\Psi, \bar{\Psi}) \bigg|_{\psi=\Psi,\ \bar{\psi}=\bar{\Psi}} . \tag{4.57}$$

We derive in particular

$$\begin{aligned} \{\psi, \bar{\psi}\}_P &= \{\bar{\psi}, \psi\}_P = i, \\ \{\psi\bar{\psi}, \psi\}_P &= -\{\psi, \psi\bar{\psi}\}_P = i\psi, \\ \{\psi\bar{\psi}, \bar{\psi}\}_P &= -\{\bar{\psi}, \psi\bar{\psi}\}_P = -i\bar{\psi}. \end{aligned} \tag{4.58}$$

A generic expression for the Poisson bracket in a system involving several usual real commuting phase-space variables $(q_j, p_j)$ and several complex anticommuting phase-space variables $(\psi_\alpha, \bar{\psi}_\alpha)$ reads

$$\boxed{\begin{aligned} &\{f, g\}_P = \left[ \frac{\partial^2}{\partial p_j \partial Q_j} - \frac{\partial^2}{\partial q_j \partial P_j} - i \left( \frac{\partial^2}{\partial \psi_\alpha \partial \bar{\Psi}_\alpha} + \frac{\partial^2}{\partial \bar{\psi}_\alpha \partial \Psi_\alpha} \right) \right] \\ &f(p, q;\ \psi, \bar{\psi}) g(P, Q;\ \Psi, \bar{\Psi}) \big|_{p=P, q=Q,\ \psi=\Psi, \bar{\psi}=\bar{\Psi}} . \end{aligned}} \tag{4.59}$$

With this particular definition, the time evolution of an arbitrary phase-space function $A(p_j, q_j, \psi_\alpha, \bar{\psi}_\alpha)$ is still given by the relation (4.6). When the Poisson bracket $\{H, A\}_P$ vanishes, $A(p_j, q_j, \psi_\alpha, \bar{\psi}_\alpha)$ is an integral of motion.

We can now observe that the following:

- The Poisson bracket (4.59) satisfies the generalized Jacobi identity similar to that in (4.7) but with the signs depending on whether $A$, $B$ and $C$ are even or odd. If only one of these functions is odd or all of them are odd, the signs are the same as in (4.7). If $A$ and $B$ are odd, while $C$ is even, the middle term (where $C$ sits in the middle) acquires the negative sign and the identity reads

$$\{A, \{B, C\}_P\}_P - \{B, \{C, A\}_P\}_P + \{C, \{A, B\}_P\}_P = 0. \tag{4.60}$$

  If $A$ and $C$ are odd, while $B$ is even, the first term acquires the negative sign, and if $B$ and $C$ are odd, while $A$ is even, this concerns the last term. A corollary of the Jacobi identity, Theorem 4.2 is still valid.

- If at least one of the functions $f, g$ represents an even element of the Grassmann algebra generated by $\psi_\alpha, \bar{\psi}_\alpha$, the expression (4.59) is antisymmetric under exchange $f \leftrightarrow g$. When going from classical to quantum mechanics, the Poisson bracket [more exactly, the Grassmann generalization (5.54) of the Grönewold–Moyal bracket (4.45), but this elaboration is relevant only for complicated enough functions $f, g$] goes in this case to the commutator of two operators[10]:

$$\{f, g\}_P \to i[\hat{f}, \hat{g}]. \tag{4.61}$$

- When both $f$ and $g$ are odd anticommuting elements of the Grassmann algebra, the expression (4.59) is symmetric under exchange $f \leftrightarrow g$ and goes over to the *anticommutator* under quantization[11]:

$$\{f, g\}_P \to i\{\hat{f}, \hat{g}\}. \tag{4.62}$$

  We derive in particular that

$$\{\hat{\psi}_\alpha, \hat{\bar{\psi}}_\beta\} = \delta_{\alpha\beta} \quad \{\hat{\psi}_\alpha, \hat{\psi}_\beta\} = 0, \quad \{\hat{\bar{\psi}}_\alpha, \hat{\bar{\psi}}_\beta\} = 0. \tag{4.63}$$

A nontrivial anticommutator $\{\hat{\psi}_\alpha, \hat{\bar{\psi}}_\beta\}$ tells us that the quantum operators corresponding to classical Grassmann variables satisfy the *Clifford algebra*: "Grassmann" becomes "Clifford" after quantization.

For the toy model discussed above, the only nontrivial anticommutator is[12] $\{\hat{\bar{\psi}}, \hat{\psi}\} = 1$. A natural representation of this quantum algebra is $\hat{\psi} = \psi$ and $\hat{\bar{\psi}} = \partial/\partial\psi$.

To write the quantum counterpart for the classical Hamiltonian (4.55), we need to resolve the ordering ambiguity. A convenient choice is

$$\boxed{\hat{H} = \omega\psi\frac{\partial}{\partial\psi}.} \tag{4.64}$$

The Hilbert space where this Hamiltonian acts is very simple. It involves the wave functions $\Psi(\psi)$ which can be written as

$$\Psi(\psi) = a\psi + b. \tag{4.65}$$

[10] We have set now $\hbar = 1$ and will mostly stick to this choice throughout the rest of this book.

[11] That is why Grassmann variables are so interesting for physicists. It has been known since the old works of Fierz and Pauli [56] that the fermion field operators anticommute.

[12] Under this convention, the classical canonical momentum $\Pi_\psi = -i\bar{\psi}$ derived from the Lagrangian (4.51) goes over to $\hat{\Pi}_\psi = -i\partial/\partial\psi$.

All the higher-order terms in the Taylor expansion of $\Psi(\psi)$ vanish due to the property $\psi^2 = 0$! The Hamiltonian (4.64) has only two eigenfunctions:

$$\begin{aligned} \Psi(\psi) &= 1 \quad \text{with the eigenvalue} \quad E = 0, \\ \Psi(\psi) &= \psi \quad \text{with the eigenvalue} \quad E = \omega. \end{aligned} \tag{4.66}$$

Thus, our Grassmann oscillator is simpler than the ordinary harmonic oscillator: instead of the infinite tower of equidistant states, it has only two states.

Note that the Hamiltonian (4.64) also admits a matrix representation. One can trade the wave functions (4.65) of a holomorphic Grassmann variable $\psi$ for the two-component vectors[13]

$$\Psi = \begin{pmatrix} a \\ b \end{pmatrix}. \tag{4.67}$$

Then, the operators $\hat{\psi}$ and $\hat{\bar{\psi}}$ and the Hamiltonian $\hat{H}$ are represented by the matrices

$$\begin{aligned} \hat{\psi} &= \begin{pmatrix} 0 & 1 \\ 0 & 0 \end{pmatrix} = \sigma_+, \quad \hat{\bar{\psi}} = \begin{pmatrix} 0 & 0 \\ 1 & 0 \end{pmatrix} = \sigma_- \\ \hat{H} &= \omega \begin{pmatrix} 1 & 0 \\ 0 & 0 \end{pmatrix} = \omega \frac{1+\sigma_3}{2}. \end{aligned} \tag{4.68}$$

The matrix representation has been habitually used in quantum mechanics to describe e.g. the electron motion in a magnetic field (see the following chapter), but it is not convenient for the systems involving many ordinary and Grassmann degrees of freedom. For such systems, the Hilbert space includes the wave functions

$$\Psi(q_j, \psi_\alpha) = \Psi^{(0)}(q_j) + \Psi^{(1)}_\alpha(q_j)\psi_\alpha + \cdots . \tag{4.69}$$

The highest term in this expansion includes the product of all Grassmann variables available.

An alternative (a less convenient and rarely used) description would involve wave functions representing tensor products of a large number of spinors.

[13]Physicists prefer to call the objects (4.67) *spinors*. There are certain reasons for that, but in fact, as far as the quantum system (4.64) is concerned, they are neither spinors, nor vectors — they do not transform under rotations in any space.

# Chapter 5

# Supersymmetry

## 5.1 Basic definitions

**Definition 5.1.** A quantum system is called supersymmetric if its Hamiltonian can be represented as the anticommutator of two nilpotent mutually conjugated operators called *supercharges*:

$$\boxed{\hat{Q}^2 = (\hat{Q}^\dagger)^2 = 0, \quad \{\hat{Q}, \hat{Q}^\dagger\} = \hat{H}.} \tag{5.1}$$

In the Introduction, we already displayed this algebra [see Eq. (0.4)] and talked about it. As was mentioned there, it implies that $[\hat{Q}, \hat{H}] = [\hat{Q}^\dagger, \hat{H}] = 0$, i.e. the supercharges $\hat{Q}, \hat{Q}^\dagger$ are the integrals of motion.

Representing the complex supercharge $\hat{Q}$ as the sum of its real and imaginary part,

$$\hat{Q} = \frac{\hat{\mathcal{Q}}_1 + i\hat{\mathcal{Q}}_2}{2} \tag{5.2}$$

with Hermitian $\hat{\mathcal{Q}}_{1,2}$, one can rewrite (5.1) as

$$\hat{\mathcal{Q}}_1^2 = \hat{\mathcal{Q}}_2^2 = \hat{H}, \quad \{\hat{\mathcal{Q}}_1, \hat{\mathcal{Q}}_2\} = 0. \tag{5.3}$$

In other words, the Hamiltonian admits in this case *two different* anticommuting Hermitian square roots. Note that the existence of a single square root is a trivial property of any Hamiltonian whose spectrum is bounded from below. To be more precise, one has first to bring the ground state energy to a positive or zero value by adding, if necessary, a constant to the Hamiltonian and to extract the square root afterwards. But the presence of *two* different square roots imposes nontrivial constraints on the spectrum.

**Theorem 5.1.** *All eigenstates of a supersymmetric Hamiltonian have non-negative energies. If a state $\Psi_0$ with zero energy exists, it is annihilated by the action of the supercharges:*

$$\boxed{\hat{Q}\,\Psi_0 = \hat{Q}^\dagger\,\Psi_0 = 0.} \tag{5.4}$$

**Proof.** Consider an eigenstate $\Psi$ of the Hamiltonian such that either $\hat{Q}\Psi \neq 0$ or $\hat{Q}^\dagger\Psi \neq 0$. Then, $\hat{\mathcal{Q}}_1\Psi \neq 0$ or $\hat{\mathcal{Q}}_2\Psi \neq 0$. Let $\hat{\mathcal{Q}}_1\Psi \neq 0$. The Hermitian operator $\hat{\mathcal{Q}}_1$ commutes with the Hamiltonian and hence these two operators can be simultaneously diagonalized. Hence, $\Psi$ can be chosen to be an eigenstate of $\hat{\mathcal{Q}}_1$. Then, $\hat{\mathcal{Q}}_1\Psi = \lambda\Psi$ with nonzero $\lambda$. And then it follows from the first relation in (5.3) that $\hat{H}\Psi = \lambda^2\Psi$ and the energy is strictly positive. And the states annihilated by both $\hat{\mathcal{Q}}_1$ and $\hat{\mathcal{Q}}_2$ (if such states exist) have zero energy. The inverse is also true.[1] □

**Theorem 5.2.** *All eigenstates of $\hat{H}$ with $E \neq 0$ are double degenerate.*

**Proof.** Choose the basis in our Hilbert space including the states that are eigenstates of both $\hat{H}$ and $\hat{\mathcal{Q}}_1$. Pick up one of such state with positive energy $E$. The eigenvalue of $\hat{\mathcal{Q}}_1$ may in this case be equal to $\lambda = \sqrt{E}$ or $\lambda = -\sqrt{E}$. Let for definiteness $\lambda = \sqrt{E}$. Then, the state $\Psi' = \hat{\mathcal{Q}}_2\Psi/\sqrt{E}$ is also an eigenstate of $\hat{\mathcal{Q}}_1$ with the eigenvalue $\lambda' = -\sqrt{E}$. Indeed, bearing in mind that $\hat{\mathcal{Q}}_1$ and $\hat{\mathcal{Q}}_2$ anticommute, we derive

$$\hat{\mathcal{Q}}_1\Psi' = -\frac{\hat{\mathcal{Q}}_2\hat{\mathcal{Q}}_1\Psi}{\sqrt{E}} = -\hat{\mathcal{Q}}_2\Psi = -\sqrt{E}\,\Psi'. \tag{5.5}$$

The states $\Psi$ and $\Psi'$ have different eigenvalues of the operator $\hat{\mathcal{Q}}_1$ (and hence they are orthogonal), but the same energies and the same norm.

[1] One can make here a following side remark. It was essential for the proof that the operators $\hat{Q}$ and $\hat{Q}^\dagger$ entering (5.1) are Hermitially conjugate to one another so that $\hat{\mathcal{Q}}_1$ and $\hat{\mathcal{Q}}_2$ are Hermitian. If this condition is not fulfilled, the spectrum may include negative energies. It seems at the first sight that in that case one would obtain a non-Hermitian Hamiltonian that can have *any* eigenvalues, not necessarily real. Unexpectedly, there exist a certain *special* generalized supersymmetric system (and maybe other such systems) whose algebra includes the supercharges $\hat{Q}$ and $\hat{\hat{Q}} \neq \hat{Q}^\dagger$, but the spectrum is real. Seemingly, its Hamiltonian is not Hermitian, but actually it belongs to the class of so-called *pseudo-Hermitian* or, better to say, *crypto-Hermitian* Hamiltonians [57]. The spectrum is real, but it includes both positive and negative energies. This system represents an example of the system with *benign ghosts*: a ground state is absent, but still the evolution operator is unitary [58].

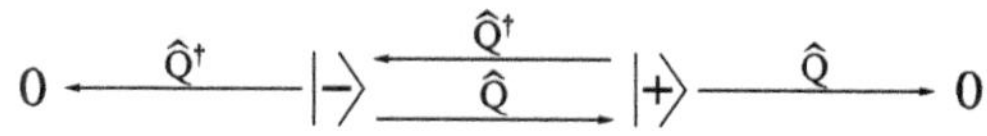


**Fig. 5.1.** A supersymmetric doublet.

They are obtained from one another by the action of $\hat{\mathcal{Q}}_2$ (note that $\hat{\mathcal{Q}}_2\Psi' = \sqrt{E}\,\Psi$) and represent a double degenerate pair in the spectrum of the Hamiltonian. Had we picked up in the beginning another eigenstate of $\hat{\mathcal{Q}}_1$, we would obtain another such degenerate pair (or maybe exactly the same if the state $\Psi'$ had been chosen). □

Consider the states $\Psi_\pm = \Psi \pm i\Psi'$. It is easy to see that

$$\hat{Q}\,\Psi_- = \sqrt{E}\,\Psi_+, \quad \hat{Q}^\dagger\,\Psi_- = 0,$$
$$\hat{Q}\,\Psi_+ = 0, \quad \hat{Q}^\dagger\,\Psi_+ = \sqrt{E}\,\Psi_-. \tag{5.6}$$

In other words, all the excited states can be divided into two sectors. The states $\Psi_-$ are annihilated by the operator $\hat{Q}^\dagger$ and go over to $\Psi_+$ under the action of $\hat{Q}$. The states $\Psi_+$ are annihilated by the operator $\hat{Q}$ and go over to $\Psi_-$ under the action of $\hat{Q}^\dagger$ (Fig. 5.1).

In quantum supersymmetric field theories that represent a physical interest (or, better to say, the theories that *will* represent a physical interest if supersymmetry is discovered in experiment), the double degeneracy of the levels of supersymmetric Hamiltonian is the degeneracy between the boson and fermion states. A supersymmetric version of electrodynamics includes on top of the massless photon a massless fermion called *photino.* In addition to electrons which are fermions, it includes *electrinos* — scalar charged bosons with the same mass. In our book, we practically do not deal with field theories, but we use the words *bosonic* and *fermionic* to distinguish the states from two sectors of the Hilbert space that we described above.

Besides the states belonging to supersymmetric doublets, the spectrum may include zero-energy *vacuum* states annihilated by both $\hat{Q}$ and $\hat{Q}^\dagger$. As we will see soon, it often makes sense to attribute some such states to the sector $|+\rangle$ and call them bosonic and some other states to the sector $|-\rangle$ and call them fermionic. But to do so, one has to concretize our discussion, to write down supersymmetric Hamiltonians for several particular systems and mark out their peculiarities and common features.

Before doing that, we prove a simple general theorem (the inverse of Theorem 5.2):

**Theorem 5.3.** *If all the excited eigenstates of the Hamiltonian are double degenerate, the system is supersymmetric.*

**Proof.** Divide all the eigenstates into two subsets such that each subset includes one member of the degenerate pair. The operators $\hat{Q}$ and $\hat{Q}^\dagger$ can be defined via their matrix elements as in (5.6). It is easy then to check that the algebra (5.1) is fulfilled. □

## 5.2 Supersymmetric oscillator

The simplest ordinary quantum system is the harmonic oscillator. And the simplest supersymmetric quantum system is the supersymmetric oscillator [59].[2] It is convenient to use the holomorphic representation where the wave functions of the ordinary oscillator depend on the holomorphic variable

$$\bar{a} = \frac{\omega q - ip}{\sqrt{2\omega}} \tag{5.7}$$

and the Hamiltonian reads[3]

$$\hat{H}^{\text{osc}} = \frac{\omega}{2}\left(\bar{a}\frac{\partial}{\partial \bar{a}} + \frac{\partial}{\partial \bar{a}}\bar{a}\right). \tag{5.8}$$

Its normalized eigenstates are

$$\Psi_n(\bar{a}) = \frac{(\bar{a})^n}{\sqrt{n!}}. \tag{5.9}$$

Its spectrum

$$E_n = \left(n + \frac{1}{2}\right)\omega \tag{5.10}$$

is very well known.

[2]In fact, one can supersymmetrize *any* ordinary system — supersymmetry is like a universal robe that can cover any body.

[3]Here, $\bar{a}$ and $\partial/\partial\bar{a}$ represent the creation and annihilation operators $\hat{a}^\dagger$ and $\hat{a}$.

To write down the Hamiltonian of the supersymmetric oscillator, one simply has to add to (5.8) the Hamiltonian (4.64) for the Grassmann oscillator, shifted by an appropriately chosen constant:

$$\boxed{\hat{H}^{\text{sup-osc}} = \omega\left(\bar{a}\frac{\partial}{\partial\bar{a}} + \psi\frac{\partial}{\partial\psi}\right).} \tag{5.11}$$

The wave functions depend now on $\bar{a}$ and the Grassmann variable $\psi$. The eigenfunctions and the eigenvalues read

$$\begin{aligned} \Psi_n(\bar{a},\psi) &= \frac{(\bar{a})^n}{\sqrt{n!}}, \quad E = \omega n, \\ \tilde{\Psi}_n(\bar{a},\psi) &= \frac{(\bar{a})^n}{\sqrt{n!}}\psi, \quad E = \omega(n+1). \end{aligned} \tag{5.12}$$

We see two towers of states. There are the states that do not involve the Grassmann factor $\psi$ and there are the states that do. For a physicist, it is natural to call the former *bosonic* and the latter *fermionic*. The ground state with zero energy is bosonic.[4] All excited states come in degenerate pairs including a bosonic and a fermionic state.

The system is thus supersymmetric, and it is easy to write down explicit expressions for the supercharges:

$$Q = \sqrt{\omega}\,\psi\frac{\partial}{\partial\bar{a}} \quad Q^\dagger = \sqrt{\omega}\,\bar{a}\frac{\partial}{\partial\psi}. \tag{5.13}$$

The supercharges (5.13) are nilpotent, and their anticommutator gives the Hamiltonian (5.11).

## 5.3 Electrons in a magnetic field

Historically, it was the first supersymmetric system constructed and studied. This was done by Landau back in 1930 [4].[5] This section is written in

[4]It is also possible to change signs in the appropriate places and consider the system where the ground state is fermionic. This is the convention that we used in Ref. [60] to illustrate physical Dirac's picture where the vacuum state involves an infinite number of fermions occupying the negative-energy levels in the Dirac sea. But in this book, we have preferred the convention that illustrates better the mathematical structure of the theory.

[5]Of course, Landau did not call it supersymmetry — the word did not exist yet. Landau can be compared in this respect to Mr. Jourdain, the *bourgeois gentilhomme*, who spoke prose without knowing what prose is.

a different style, compared to the rest of this book. It is mainly addressed to a reader-physicist who knows this system from kindergarten and might be surprised to find that it is in fact supersymmetric. A mathematician may skip it [maybe after having a brief look at Fig. 5.2 and Eq. (5.25)] and go directly to Section 5.4.

The generic Hamiltonian describing the motion in a magnetic field (the Pauli Hamiltonian) reads

$$\hat{H} = \frac{\left[\boldsymbol{\sigma}\cdot\left(\hat{\boldsymbol{P}} - \frac{e}{c}\boldsymbol{A}\right)\right]^2}{2m}, \tag{5.14}$$

where $e$ is the electron charge, $c$ is the speed of light, $\boldsymbol{A}$ is the vector potential and $\boldsymbol{\sigma}$ are the Pauli matrices (3.12). The latter satisfy the algebra

$$\sigma_j \sigma_k = \delta_{jk}\mathbb{1} + i\epsilon_{jkl}\sigma_l. \tag{5.15}$$

Bearing this and the definition $\hat{P}_j = -i\hbar\,\partial/\partial x^j$ in mind, we may rewrite (5.14) as

$$\boxed{\hat{H} = \frac{\left(\hat{\boldsymbol{P}} - \frac{e}{c}\boldsymbol{A}\right)^2}{2m} - \frac{e\hbar}{2mc}\boldsymbol{\sigma}\cdot\boldsymbol{B},} \tag{5.16}$$

where $\boldsymbol{B} = \boldsymbol{\nabla}\times\boldsymbol{A}$ is the magnetic field. The Hamiltonians (5.14) and (5.16) act on spinor wave functions

$$\Psi(\boldsymbol{x}) = \begin{pmatrix} a_+(\boldsymbol{x}) \\ a_-(\boldsymbol{x}) \end{pmatrix}. \tag{5.17}$$

The second term in (5.16) describes the interaction of the Dirac magnetic moment of an electron with an external magnetic field.

The first term in (5.16) is the quantum Hamiltonian of a spinless particle in a magnetic field. Its classical counterpart [the classical counterpart of the full Hamiltonian (5.16) also exists, we will present it at the end of this chapter] is $(\boldsymbol{P} - \frac{e}{c}\boldsymbol{A})^2/2m$, where $\boldsymbol{P}$ is the *canonical* momentum, to distinguish from the *kinetic* momentum $\boldsymbol{p} = m\dot{\boldsymbol{x}} = m\,\partial H/\partial\boldsymbol{P}$.

Generically, the Hamiltonian (5.16) is not supersymmetric. Let us assume, however, that the *direction* of the magnetic field is fixed. We may choose it to lie along the third axis:

$$\boldsymbol{B} = [0, 0, B(x, y)]. \tag{5.18}$$

Landau assumed that the field is constant. But that is not necessary for supersymmetry; one can allow it to depend on[6] $x$ and $y$. If the magnetic field has the form (5.18), one can choose the vector potential $\boldsymbol{A}$ to have only $x$ and $y$ components. In that case, the Hamiltonian represents a sum of two terms. The first term describes a nontrivial motion in the $(x, y)$ plane, while the second term has the form $\hat{P}_3^2/2m$ and describes free motion along the direction of the field (indeed, when $\boldsymbol{v}\|\boldsymbol{B}$, the Lorentz force is zero). Consider only the first nontrivial part. It coincides with (5.14), only now we assume that the vectors $\hat{\boldsymbol{P}}$ and $\boldsymbol{A}$ lie in the $(x, y)$ plane.

Consider the operators

$$\hat{Q} = \frac{1}{\sqrt{2m}} \boldsymbol{\sigma} \cdot \left( \hat{\boldsymbol{P}} - \frac{e}{c} \boldsymbol{A} \right) \frac{1 - \sigma_3}{2} \tag{5.19}$$

and

$$\hat{Q}^\dagger = \frac{1}{\sqrt{2m}} \boldsymbol{\sigma} \cdot \left( \hat{\boldsymbol{P}} - \frac{e}{c} \boldsymbol{A} \right) \frac{1 + \sigma_3}{2}, \tag{5.20}$$

with two-dimensional $\hat{\boldsymbol{P}}, \boldsymbol{A}$ and $\boldsymbol{\sigma}$. Capitalizing on the fact that $\sigma_{1,2}$ anticommute with $\sigma_3$, it is easy to check that these supercharges satisfy the algebra (5.1).

The problem is supersymmetric for any field (5.18). But for the homogeneous magnetic field, one can solve the Schrödinger problem exactly. In the following calculations, we make a simplification, getting rid of all dimensional constants $\hbar, c, m, |e|$ (so that the electron charge $e$ goes over to $-1$, cf. the footnote on p. 368). We wrote them before to make it clear that the problem under consideration is not just a mathematical construction but a real physical problem. Its solution gives an explanation of a real physical phenomenon, the so-called *Landau diamagnetism*.[7] But it is the mathematical structure of the problem that we are now mostly interested in, and keeping $\hbar$, etc. would obscure it.

---

[6]But not on the third coordinate, otherwise the divergence $\boldsymbol{\nabla} \cdot \boldsymbol{B}$ would be different from zero, which is not allowed by Maxwell's equations.

[7]Free electrons in metal rotating in an external magnetic field $\boldsymbol{B}$ induce their own small magnetic field in the direction opposite to $\boldsymbol{B}$.

Now, the vector potential can be chosen as $\boldsymbol{A} = (-By/2, Bx/2, 0)$. The Hamiltonian and the supercharges acquire the form

$$\boxed{\hat{H} = \frac{1}{2}\left[\left(-i\frac{\partial}{\partial x} - \frac{By}{2}\right)^2 + \left(-i\frac{\partial}{\partial y} + \frac{Bx}{2}\right)^2\right] + \frac{B}{2}\sigma_3,} \tag{5.21}$$

$$\hat{Q} = -i\left(\frac{\partial}{\partial z} + \frac{B\bar{z}}{2}\right)\sigma_+, \quad \hat{Q}^\dagger = -i\left(\frac{\partial}{\partial \bar{z}} - \frac{Bz}{2}\right)\sigma_-, \tag{5.22}$$

where[8] $z = (x+iy)/\sqrt{2}$ and $\sigma_\pm$ were defined in (4.68).

In agreement with general theorems, the whole Hilbert space (5.17) splits into two subspaces: subspace $|+\rangle$ with positive electron spin projection along the third axis and subspace $|-\rangle$ with negative spin projection. The spectrum of positive-energy states splits into degenerate doublets and the action of the supercharges follows the pattern displayed in Fig. 5.1.

In both sectors, the Hamiltonian (5.21) reduces to the Hamiltonian of the harmonic oscillator.[9] The spectrum is thus equally spaced, the spacing between the levels being equal to B (we assume, for definiteness, that $B > 0$). All the levels are infinitely degenerate due to the symmetry under translations in the $(x, y)$ plane. The difference between the two sectors is that the spin-down states are shifted down with respect to the spin-up states by $B$, which exactly coincides with the spacing between the oscillator levels (Fig. 5.2). This is the *physical* reason for the double degeneracy of all excited levels. [And we remember that the mathematical reason unsuspected by Landau is the supersymmetric algebra (5.1)!]

The left tower in Fig. 5.2 involves the vacuum zero-energy states. To find their wave functions is simpler than to find the whole spectrum. To this end, one only has to impose the condition $\hat{Q}\Psi = 0$ ($\hat{Q}^\dagger\Psi = 0$ is satisfied automatically), which gives the first-order equation

$$\left(\frac{\partial}{\partial z} + \frac{B\bar{z}}{2}\right) a_- = 0 \tag{5.23}$$

with the solutions

$$a_-(z, \bar{z}) = \exp\left\{-\frac{B\bar{z}(z+c)}{2}\right\}. \tag{5.24}$$

[8] We hope that the reader will not confuse this complex coordinate $z$ with the third real spatial coordinate that we left unnamed!

[9] And do not forget that for the real physical problem, one should also add to the Hamiltonian (5.21) the term $\hat{P}_3^2/2m$.

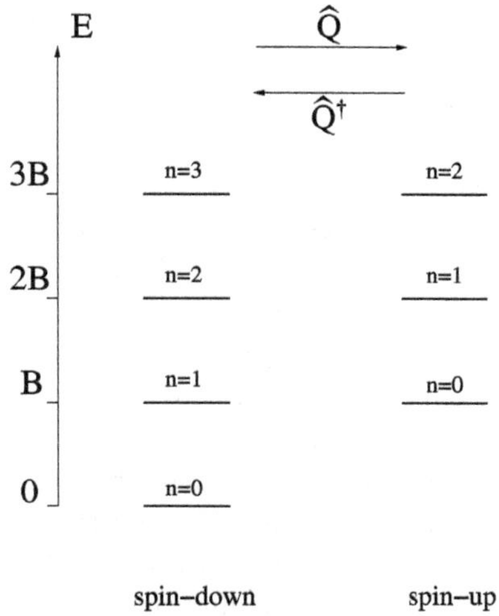


**Fig. 5.2.** Spectrum of the Landau Hamiltonian ($B > 0$). $n = 0, 1, \ldots$ mark the oscillator levels.

Different complexes $c$ correspond in the classical limit to different positions of the center of the circular electron orbits in the $(x, y)$ plane.

If $B$ is negative, the picture is opposite. There are no zero-energy states in the spin-down sector and infinitely many of them in the spin-up sector.

If the magnetic field in (5.18) is not homogeneous but depends nontrivially on $x, y$, the problem is still supersymmetric, and even though one cannot solve it analytically anymore, the generic pattern of the spectrum is the same as for the homogeneous field — some number of supersymmetric zero-energy states and a set of degenerate supersymmetric doublets.

In the following chapter, we prove the theorem:

**Theorem 5.4.** *The number of the normalized zero energy states (alias zero modes) in the generalized Landau problem with nontrivial $B(x, y)$ is proportional to the magnetic flux:*

$$\boxed{n_{E=0} = \frac{|\Phi|}{2\pi} = \frac{1}{2\pi}\left|\int B(x,y)\,dxdy\right|.} \tag{5.25}$$

The flux $\Phi$ must be an integer multiple of $2\pi$ — otherwise supersymmetry is lost (wait for Section 14.2.1 for a detailed explanation of this point).

For a homogeneous field, the flux is infinite, and the number of the zero modes is infinite too. If the flux is finite, so is the number of the zero modes.

The latter are spin-down states if the flux is positive and spin-up states if the flux is negative.

"And what if the flux is zero", the reader may ask. Well, in that case, there are no normalized supersymmetric ground states annihilated by both supercharges whatsoever, and all the states are paired into supersymmetry doublets. The absence of supersymmetric ground states means that the supersymmetry is *spontaneously broken* — the vacuum states are not invariant under supersymmetry transformations.[10] We talk more about spontaneous supersymmetry breaking in the following section and then in the following chapter where a general theory of this phenomenon is discussed.

When we discussed the Grassmann oscillator in Section 4.4, we mentioned that the same dynamics can be described in the matrix language. Obviously, this applies also to the supersymmetric oscillator treated in the preceding section. The dictionary of translations from one language into another was given in Eq. (4.68). This dictionary works both ways, and we can easily write the quantum supercharges of the Landau problem in the Grassmann form, simply replacing the matrices $\sigma_+$ and $\sigma_-$ in (5.22) by $\psi$ and $\hat{\bar{\psi}} = \partial/\partial\psi$, respectively. We obtain the expressions

$$\boxed{Q = -i\left(\frac{\partial}{\partial z} + \frac{B\bar{z}}{2}\right)\psi, \quad Q^\dagger = -i\left(\frac{\partial}{\partial \bar{z}} - \frac{Bz}{2}\right)\frac{\partial}{\partial\psi},} \tag{5.26}$$

and the Hamiltonian is given by the expression (5.21) where the matrix $\sigma^3$ is traded for

$$\sigma^3 = [\sigma^+, \sigma^-] \longrightarrow \psi\frac{\partial}{\partial\psi} - \frac{\partial}{\partial\psi}\psi.$$

The wave functions depend on $z, \bar{z}$ and on the holomorphic variable $\psi$. There are the states of the form $\Psi(z, \bar{z}; \psi) = a_-(z, \bar{z})$ (they are what we called before spin-down states) and the states of the form $\Psi(z, \bar{z}; \psi) = \psi a_+(z, \bar{z})$ (the spin-up states).

## 5.4 Witten's model

When Enrico Fermi met a nontrivial physical problem, he used to ask himself and his students and colleagues the following question: what is the

[10]The reader has the right to be confused at this point because we have not explained yet what the supersymmetry transformations *are*. But we do so soon.

*hydrogen atom* in this case? He meant to ask whether there exists a simple model bearing all the essential features of the complicated phenomenon he was set to understand.

The model of the supersymmetric oscillator treated in Section 5.2 is simple, indeed, but it is probably too simple and too rigid. The Landau problem discussed in the preceding section is both simple and complicated enough to play the role of such hydrogen atom, but there is an even better candidate for this role: the model suggested by E. Witten [61]. The wave functions in this model depend on a single ordinary variable $x$ and a holomorphic Grassmann variable $\psi$. The supercharges and Hamiltonian read

$$\hat{Q} = \frac{1}{\sqrt{2}}\psi\,[\hat{P} - iW'(x)], \quad \hat{Q}^\dagger = \frac{1}{\sqrt{2}}\hat{\bar{\psi}}\,[\hat{P} + iW'(x)]; \tag{5.27}$$

$$\boxed{\hat{H} = \frac{\hat{P}^2 + [W'(x)]^2}{2} + \frac{1}{2}W''(x)(\psi\hat{\bar{\psi}} - \hat{\bar{\psi}}\psi).} \tag{5.28}$$

where $\hat{P}$ and $\hat{\bar{\psi}}$ are the ordinary and Grassmann momentum operators, $\hat{P} = -i\partial/\partial x$ and $\hat{\bar{\psi}} = \partial/\partial\psi$, and $W(x)$ is an arbitrary real function called the *superpotential.* In the matrix formulation, the Hamiltonian (5.28) describes the one-dimensional motion of a spin $\frac{1}{2}$ particle in the external potential $V = \frac{1}{2}[W'(x)]^2$ and the magnetic field $W''(x)$.

In the particular case $W(x) = \omega x^2/2$, the Hamiltonian acquires the form

$$\hat{H} = \frac{\hat{P}^2 + \omega^2 x^2}{2} + \frac{\omega}{2}(\psi\hat{\bar{\psi}} - \hat{\bar{\psi}}\psi). \tag{5.29}$$

It is nothing but the Hamiltonian (5.11) of the supersymmetric oscillator discussed before but expressed in the ordinary representation: the operator (5.29) acts on the wave functions $\Psi(x,\psi)$ rather than on $\Psi(\bar{a},\psi)$.

For a generic $W(x)$, one cannot solve the complete spectral problem analytically, but one can always find explicit expressions for the wave function of the zero-energy state if such a state exists. As was the case also for two preceding problems, the Hilbert space is divided into two sectors: the fermionic states involving the factor $\psi$ in $\Psi(x,\psi)$ and the bosonic states without such factor.

**Theorem 5.5.** *A normalized bosonic zero-energy solution to the Schrödinger equation with the Hamiltonian* (5.28) *exists iff*

$$\lim_{x\to\pm\infty} W(x) = \infty. \tag{5.30}$$

*A normalized fermionic zero-energy solution to the Schrödinger equation with the Hamiltonian* (5.28) *exists iff*

$$\lim_{x \to \pm\infty} W(x) = -\infty. \tag{5.31}$$

**Proof.** All the bosonic states are clearly annihilated by $\hat{Q}^\dagger$ and all the fermionic states are annihilated by $\hat{Q}$. A zero-energy state in the bosonic sector should in addition be annihilated by $\hat{Q}$. This gives the equation

$$\left[\frac{\partial}{\partial x} + W'(x)\right] \Psi_0^B(x) = 0. \tag{5.32}$$

Its formal solution reads

$$\Psi_0^B(x) = \exp\left\{-W(x)\right\}. \tag{5.33}$$

The condition (5.30) is the condition for its normalizability.

Likewise, the fermion zero modes $\Psi_0^F(\psi, x) = \psi \Psi_0^F(x)$ should satisfy the equation

$$\left[\frac{\partial}{\partial x} - W'(x)\right] \Psi_0^F(x) = 0 \tag{5.34}$$

with the formal solution

$$\Psi_0^F(x) = \exp\left\{W(x)\right\}. \tag{5.35}$$

It is normalizable iff the condition (5.31) is fulfilled. □

If $W(\infty) = \pm\infty$, but $W(-\infty) = \mp\infty$, the zero-energy states are absent.[11] The simplest example is given by the cubic superpotential $W(x) = \lambda x^3$.

The ground states of the Hamiltonian have in this case nonzero energy. They are paired in supersymmetric doublets, like in Fig. 5.1. It is the same phenomenon of spontaneous supersymmetry breaking that we met in the previous section for the motion in a magnetic field with zero net flux (5.25).

## 5.5 Extended supersymmetry

The algebra of the systems considered in three preceding sections involved two different Hermitian supercharges. By the convention adopted now in

[11] A special case is when $W(\infty)$ or $W(-\infty)$ is a constant. In this case, the spectrum is continuous, the spectral problem should be regularized (see the discussion at the end of Section 4.2) and zero modes can appear as a result of this regularization.

the literature, such systems are called $\mathcal{N} = 2$ supersymmetric quantum mechanical (SQM) systems. This terminology is somewhat misleading. As we noted before, one can always extract a Hermitian square root of a Hermitian Hamiltonian with positive-definite spectrum and the algebra (5.1) is the *minimal* nontrivial supersymmetry algebra. Thus, it is better to count not real but complex supercharges, in which case the algebra (5.1) would be called $\mathcal{N} = 1$. But a universally adopted convention is a universally adopted convention, and we have to follow it, as we did earlier when adopting the term HKT, hyperkähler with torsion, for the manifolds that were not at all hyperkähler and not even Kähler.

Anyway, the symmetry of the system can be richer than (5.1). A supersymmetric system may involve several pairs of supercharges. The corresponding algebra is[12]

$$\boxed{\begin{aligned} &\{\hat{Q}_i, \hat{Q}_j\} = \{\hat{\bar{Q}}^i, \hat{\bar{Q}}^j\} = 0, \\ &\{\hat{Q}_i, \hat{\bar{Q}}^j\} = \delta_i^j \hat{H}, \end{aligned}} \qquad i, j = 1, \ldots \mathcal{N}/2, \tag{5.36}$$

with $\hat{\bar{Q}}^i \stackrel{\text{def}}{=} (\hat{Q}_i)^\dagger$.

The indices $i, j$ distinguishing different complex supercharges and the associated Grassmann variables are written at the different levels to highlight the fact that $\hat{Q}_i$ and $\hat{\bar{Q}}^i$ belong to the different — fundamental and antifundamental — representations of the $U(\mathcal{N}/2)$ group, a subgroup of the full automorphism group $SO(\mathcal{N})$ of the algebra (5.36) which rotates the complex supercharges and Grassmann variables. In all the extended models considered in this book, this symmetry (called *R-symmetry*) will play an important role. In many (though not in all) cases, it is a symmetry of the Hamiltonian.

$\mathcal{N} \geq 2$ is an even number. The minimal extention is $\mathcal{N} = 4$. In that case, the $R$-symmetry group is $U(2)$, its fundamental and the antifundamental representations coincide and the indices may be raised and lowered by the invariant $SU(2)$ tensors $\varepsilon^{ij} = -\varepsilon_{ij}$:

$$X_i = \varepsilon_{ij} X^j \quad \text{and} \quad X^i = \varepsilon^{ij} X_j. \tag{5.37}$$

The convention $\varepsilon_{12} = 1$ is chosen.

We give here one example of an $\mathcal{N} = 4$ SQM system. It describes the motion on the complex plane but, in contrast to the Landau problem,

[12] There are also SQM systems with more intricate extended algebras [62], but they do not play a role in geometrical applications and we do not talk about them in this book.

involves two Grassmann variables $\psi_{1,2}$ and two corresponding momenta $\Pi_{1,2} = -i\partial/\partial\psi_{1,2} \equiv -i\hat{\bar{\psi}}^{1,2}$.

The supercharges are[13]

$$\hat{Q}_1 = \psi_1 \hat{\pi} + i(\bar{z})^2 \hat{\bar{\psi}}^2, \quad \hat{Q}_2 = \psi_2 \hat{\pi} - i(\bar{z})^2 \hat{\bar{\psi}}^1,$$

$$\hat{\bar{Q}}^1 = \hat{\bar{\psi}}^1 \hat{\pi}^\dagger - i(z)^2 \psi_2, \quad \hat{\bar{Q}}^2 = \hat{\bar{\psi}}^2 \hat{\pi}^\dagger + i(z)^2 \psi_1, \tag{5.38}$$

with $\hat{\pi} = -i\partial/\partial z$, $\hat{\pi}^\dagger = -i\partial/\partial\bar{z}$. The Hamiltonian

$$\hat{H} = \hat{\pi}^\dagger \hat{\pi} + (z\bar{z})^2 - 2z\psi_1\psi_2 - 2\bar{z}\hat{\bar{\psi}}^2 \hat{\bar{\psi}}^1 \tag{5.39}$$

is not quadratic and describes nontrivial interactions.

More examples follow later. For geometrical applications, certain $\mathcal{N} = 4$ and $\mathcal{N} = 8$ systems are relevant. But the $\mathcal{N} = 6$ systems and more complicated, up to $\mathcal{N} = 32$, exist and have been studied.

## 5.6 Classical supersymmetry

At the beginning of this chapter, we defined what *quantum* supersymmetric system is. We considered then several particular quantum supersymmetric Hamiltonians but have not posed yet the following question: "What are their classical counterparts?" If one is interested only in quantum mechanics, using the Grassmann variables is convenient (especially in complicated systems) but not necessary. As we noted, one can well work with the Hamiltonians expressed in the matrix form. But Grassmann variables are undispensible if one wishes to write down a classical Hamiltonian or a classical Lagrangian. The classical Grassmann dynamics was already treated in Section 4.4, one only has to apply it to supersymmetric systems.

In all the examples considered heretofore, supersymmetric classical Hamiltonians can be written rather easily. Let us do so first for the supersymmetric oscillator. Take the quantum Hamiltonian (5.11) and replace $\partial/\partial\bar{a}$ by $i\Pi_{\bar{a}} \equiv a$ and $\partial/\partial\psi$ by $i\Pi_\psi \equiv \bar{\psi}$. We obtain

$$H^{\text{sup-osc}} = \omega(\bar{a}a + \psi\bar{\psi}). \tag{5.40}$$

The Hamilton equations of motion are

$$\dot{a} = -i\frac{\partial H}{\partial \bar{a}} = -i\omega a, \quad \dot{\psi} = -i\frac{\partial H}{\partial \bar{\psi}} = i\omega\psi \tag{5.41}$$

and complex conjugates.

[13]Do not confuse these complex $\mathcal{N} = 4$ supercharges with $\mathcal{Q}_{1,2}$ — the real and imaginary parts of the $\mathcal{N} = 2$ supercharge.

The classical supercharges are derived in a similar way from (5.13):

$$Q = \sqrt{\omega}\, a\psi \quad \bar{Q} = \sqrt{\omega}\, \bar{a}\bar{\psi}. \tag{5.42}$$

The quantum supersymmetry algebra (5.1) transforms to

$$\{Q, Q\}_P = \{\bar{Q}, \bar{Q}\}_P = 0, \quad \{\bar{Q}, Q\}_P = iH \tag{5.43}$$

or else[14]

$$\{\mathcal{Q}_1, \mathcal{Q}_1\}_P = \{\mathcal{Q}_2, \mathcal{Q}_2\}_P = 2iH, \quad \{\mathcal{Q}_1, \mathcal{Q}_2\}_P = 0, \tag{5.44}$$

where $\{A, B\}_P$ is the Poisson bracket defined in (4.59). By Jacobi identities, it follows that the Poisson brackets $\{Q, H\}_P$ and $\{\bar{Q}, H\}_P$ vanish, which means that $Q$ and $\bar{Q}$ are classical integrals of motion.

The classical Lagrangian is obtained from $H$ by the Legendre transformation:

$$L = \dot{\bar{a}}\Pi_{\bar{a}} + \dot{\psi}\Pi_\psi - H. \tag{5.45}$$

(This particular order of the Grassmann factors follows from the definition $\Pi_\psi = \partial L/\partial\dot{\psi}$.) Up to a total time derivative,[15] we obtain

$$L = i(\dot{a}\bar{a} - \dot{\psi}\bar{\psi}) - \omega(\bar{a}a + \psi\bar{\psi}). \tag{5.46}$$

By Noether's theorem, the existence of classical integrals of motion implies the invariance (up to a total derivative) of the Lagrangian under certain associated transformations. In our case, these are *supertransformations*. They read

$$\begin{aligned} \delta a &= i\epsilon\bar{\psi}, \quad \delta\bar{a} = i\bar{\epsilon}\psi, \\ \delta\psi &= -i\bar{a}\epsilon, \quad \delta\bar{\psi} = ia\bar{\epsilon}, \end{aligned} \tag{5.47}$$

where $\epsilon$ is a Grassmann transformation parameter.

For Witten's model, the classical expressions for the supercharges and the Hamiltonian can be written immediately. They simply coincide with (5.27) and (5.28), where now $P$ and $\bar{\psi}$ should be understood as the phase

[14] One may note here that, while for the ordinary phase space the relation $\{A, A\}_P = 0$ holds for any function $A(p_j, q_j)$, this is not so when the phase space includes also Grassmann variables.

[15] We have seen (see p. 90) that the Lagrangians that differ by a total derivative give the same equations of motion and are equivalent.

space variables rather than differential operators. The corresponding classical Lagrangian is

$$\boxed{L = \frac{\dot{x}^2 - [W'(x)]^2}{2} - i\dot{\psi}\bar{\psi} - W''(x)\psi\bar{\psi}.} \tag{5.48}$$

This Lagrangian is invariant up to a total derivative under

$$\begin{aligned} \delta x &= \epsilon\psi + \bar{\psi}\bar{\epsilon}, \\ \delta\psi &= \bar{\epsilon}[-i\dot{x} + W'(x)], \\ \delta\bar{\psi} &= \epsilon[i\dot{x} + W'(x)]. \end{aligned} \tag{5.49}$$

For the Landau problem, the classical supercharges are derived from the quantum expressions (5.26) by substituting the classical momenta $\pi, \bar{\pi}$ and $\Pi_\psi = -i\bar{\psi}$ for $-i\partial/\partial z$, $-i\partial/\partial\bar{z}$ and $-i\partial/\partial\psi$:

$$\begin{aligned} Q &= \left(\pi - \frac{iB\bar{z}}{2}\right)\psi, \\ \bar{Q} &= \left(\bar{\pi} + \frac{iBz}{2}\right)\bar{\psi}. \end{aligned} \tag{5.50}$$

Their Poisson bracket multiplied by $-i$ gives the classical Hamiltonian:

$$H = \left(\pi - \frac{iB\bar{z}}{2}\right)\left(\bar{\pi} + \frac{iBz}{2}\right) + B\psi\bar{\psi}. \tag{5.51}$$

The corresponding classical Lagrangian reads

$$\boxed{L = \dot{z}\dot{\bar{z}} + \frac{iB}{2}(\dot{z}\bar{z} - z\dot{\bar{z}}) - i\dot{\psi}\bar{\psi} - B\psi\bar{\psi}.} \tag{5.52}$$

It is invariant up to a total derivative under

$$\begin{aligned} \delta z &= \epsilon\psi, \quad \delta\psi = -i\bar{\epsilon}\dot{z} \\ \delta\bar{z} &= -\bar{\epsilon}\bar{\psi}, \quad \delta\bar{\psi} = i\epsilon\dot{\bar{z}}. \end{aligned} \tag{5.53}$$

In the systems considered so far, there were no ordering ambiguities and natural expressions for the classical supercharges could be easily written. But if the ordering ambiguities are present, the questions

(1) What particular expressions for the classical supercharges should be written?

(2) Is it always possible to do so in such a way that the quantum supersymmetry algebra would be preserved at the classical level, with the Poisson brackets taking over the role of commutators and anticommutators?

become nontrivial. The answers are also non-trivial.

The answer to the second question is "Not always!" The first example of a system where the quantum supersymmetry algebra cannot be preserved at the classical level was constructed in Ref. [63]. A simpler example was given in Ref. [64]. The latter concerns a supersymmetric system describing the motion of a spinorial particle in Euclidean space of even dimension $D$ equipped by an external gauge field. In this system, the Dirac operator $\not{D}$ and also the operator $\not{D}\gamma^{D+1}$ (with $\gamma^{D+1}$ being the multidimensional counterpart of the familiar to the physicists matrix $\gamma^5$) play the role of the quantum supercharges. The operator $\not{D}$ has a classical counterpart, but the operator $\not{D}\gamma^{D+1}$ has not (see Chapter 14 for detailed discussion).

In quantum field theories, we are used to have quantum anomalies when a classical symmetry cannot be preserved at the quantum level (the chiral anomaly, the conformal anomaly...). And here we are dealing with the opposite situation: supersymmetry that is realized at the quantum but not at the classical level.

The first question in the list above was studied in Ref. [52]. We analyzed there different supersymmetric systems where the supercharges and the Hamiltonians include products of canonical coordinates and momenta, and the ordering problem exists. Based on observations performed for various supersymmetric systems involving ordering ambiguities, we formulated the following **conjecture**:

(1) Consider a system enjoying an ordinary or extended classical supersymmetry algebra. Take the classical supercharges and write down the corresponding quantum operators using the *Weyl ordering prescription* (4.43). Then, these operators satisfy the quantum supersymmetry algebra.[16]

[16] The observation that the classical supersymmetry algebra holds at the quantum level under a particular ordering prescription was also made in Ref. [65]. But the universal recipe, *Take the classical supercharges and order the operators in their quantum version according to* (4.43), was not formulated there.

Note that Weyl ordering should be applied to the *supercharges* rather than to the Hamiltonian. We see later that, for complicated enough systems, the quantum Hamiltonian defined as the anticommutator of quantum supercharges *does* not coincide with the Weyl-ordered classical Hamiltonian.

(2) Inversely, *if* the quantum supersymmetry algebra can be preserved at the classical level, this can always be achieved by choosing the classical supercharges as the Weyl symbols of the quantum ones, the latter being expressed via the canonical fermion creation and annihilation operators $\psi_\alpha$ and $\hat{\bar{\psi}}^\alpha = \partial/\partial\psi_\alpha$.

This conjecture looks natural. After all, we know [see Eq. (4.45)] that the Weyl symbol of the commutator of two ordinary operators coincides with the Grönewold–Moyal bracket of their Weyl symbols. This result can be generalized on the case when the phase space involves Grassmann variables: the Weyl symbol of the commutator of two operators or their anticommutator if both operators are Grassmann-odd coincides with the GM bracket of their Weyl symbols, defined as

$$\boxed{\begin{aligned}\{A, B\}_{GM} = 2\sin\Bigg[&\frac{1}{2}\left(\frac{\partial^2}{\partial p_i \partial Q_i} - \frac{\partial^2}{\partial P_i \partial q_i}\right) \\ &- \frac{i}{2}\left(\frac{\partial^2}{\partial \psi_\alpha \partial \bar{\Psi}_\alpha} + \frac{\partial^2}{\partial \bar{\psi}_\alpha \partial \Psi_\alpha}\right)\Bigg] \\ &\times A(p, q;\, \psi, \bar{\psi}) B(P, Q;\, \Psi, \bar{\Psi})\Big|_{p=P, q=Q;\, \psi=\Psi, \bar{\psi}=\bar{\Psi}}\end{aligned}} \tag{5.54}$$

($\hbar = 1$).

This means that the quantum algebra is *always* preserved at the classical level if the GM bracket is chosen as the binary operation. It is not counterintuitive to expect that whatever holds for the GM brackets should also hold in most cases for the Poisson brackets. And if this rule of thumb fails, nothing helps.

But, of course, it would be interesting to rigorously prove (or disprove) it.

# Chapter 6

# Path Integrals and the Witten Index

This chapter provides the reader with the necessary prerequisits to understand the second half of Chapter 8, some parts of Chapters 9 and 14, and also Chapter 15, where we prove the famous *Atiyah–Singer index theorem* by supersymmetric methods. It was originally proven by Atiyah and Singer using traditional mathematical tools [66]. But the physical way of deriving this result by analyzing the spectrum of certain SQM systems appears to be both more simple and more beautiful [67].

We have to emphasize that the quantum systems to be considered in this chapter have all a *discrete* spectrum. Otherwise, the notion of index is ill-defined.

## 6.1 Path integrals in quantum mechanics

Classical mechanics has two basic ways of description — the Lagrange formalism and the Hamilton formalism. The same concerns quantum mechanics. Historically, it was first formulated in the framework of the quantum Hamiltonian formalism (see Section 4.2). The quantum counterpart of the Lagrange formalism is the path integral approach developed by Feynman. We give here its basics addressing the reader to the book [68] for details.

Consider the time-dependent Schrödinger equation (4.38). To focus on its mathematical structure, we set $\hbar = 1$. To make things still simpler, we assume the presence of only one dynamical real variable $q \in (-\infty, \infty)$ and the natural measure $d\mu = dq$ on this line. But all the results that we derive can be generalized to systems with several dynamical degrees of freedom and more complicated measures.

To solve this equation, one should fix the initial conditions — the values of $\Psi(q,t)$ at some $t = t_{\rm in}$. Then, the formal solution of Eq. (4.38) reads

$$\Psi(q,t) = \hat{U}(t - t_{\rm in})\Psi(q, t_{\rm in}), \tag{6.1}$$

where

$$\hat{U}(t - t_{\rm in}) = \exp\left\{-i(t - t_{\rm in})\hat{H}\right\} \tag{6.2}$$

is the *evolution operator*. We now consider the *kernel* of the evolution operator, that is the matrix element

$$\mathcal{K}(q_{\rm f}, q_{\rm in}; t_{\rm f} - t_{\rm in}) = \langle q_{\rm f}|\hat{U}(t_{\rm f} - t_{\rm in})|q_{\rm in}\rangle \equiv \langle q_{\rm f}, t_{\rm f}|q_{\rm in}, t_{\rm in}\rangle \tag{6.3}$$

($t_{\rm f} - t_{\rm in} \equiv \Delta t > 0$). It describes the probability amplitude that the system will find itself at the position $q_{\rm f}$ at $t = t_{\rm f}$ provided it was located[1] at $q = q_{\rm in}$ at $t = t_{\rm in}$.

In terms of (6.3), the solution (6.1) is expressed as

$$\Psi(q_{\rm f}, t_{\rm f}) = \int dq_{\rm in}\, \mathcal{K}(q_{\rm f}, q_{\rm in}; t_{\rm f} - t_{\rm in})\, \Psi(q_{\rm in}, t_{\rm in}). \tag{6.4}$$

The quantity $\mathcal{K}$ plays a fundamental role in the whole approach.

**Theorem 6.1.** *The kernel $\mathcal{K}$ is given by the sum (spectral decomposition)*

$$\boxed{\mathcal{K}(q_{\rm f}, q_{\rm in};\, \Delta t) = \sum_k \Psi_k(q_{\rm f})\overline{\Psi_k(q_{\rm in})}e^{-iE_k\Delta t},} \tag{6.5}$$

*where $\Psi_k(q)$ are the eigenstates of $\hat{H}$ with eigenvalues $E_k$.*

**Proof.** This follows from (6.4), from orthogonality,

$$\langle k|l\rangle = \int \bar{\Psi}_k(q)\Psi_l(q)\, dq = \delta_{kl}, \tag{6.6}$$

and from the fact that the time evolution of an eigenfunction $\Psi_k(q,t)$ amounts to the multiplication by the phase factor $e^{-iE_k\Delta t}$. □

[1] Note that, in the definition above, the bra-vector on the left describes the *final* state of the particle, while the ket-vector on the right describes its *initial* state. This unnatural for an air traveller convention follows from the definition of a matrix element in (4.37). It conforms with (6.4) and (6.6).

The reader can recognize in Eq. (6.5) the fundamental solution of the hyperbolic equation (4.38). Indeed,

$$\left(i\frac{\partial}{\partial t} - \hat{H}\right) \mathcal{K}(q_{\rm f}, q_{\rm in};\, \Delta t) = 0, \tag{6.7}$$

where the operator $\hat{H}$ acts on the first argument in $\mathcal{K}$. And in the limit when $\Delta t$ vanishes, the kernel acquires the form

$$\mathcal{K}(q_{\rm f}, q_{\rm in};\, 0) = \sum_k \Psi_k(q_{\rm f})\overline{\Psi_k(q_{\rm in})} = \delta(q_{\rm f} - q_{\rm in}). \tag{6.8}$$

In simple cases, the kernel can be found explicitly. For a free particle of unit mass $m = 1$ moving along a line,

$$\boxed{\mathcal{K}^{\rm free}(q_{\rm f}, q_{\rm in};\, \Delta t) = \frac{1}{\sqrt{2\pi i \Delta t}} \exp\left\{\frac{i(q_{\rm f} - q_{\rm in})^2}{2\Delta t}\right\}.} \tag{6.9}$$

An exact analytical expression for the kernel can also be derived for the Hamiltonian of the harmonic oscillator. It reads [68]

$$\begin{aligned}\mathcal{K}^{\rm osc}(q_{\rm f}, q_{\rm in};\, \Delta t) &= \sqrt{\frac{\omega}{2\pi i \sin(\omega\Delta t)}} \\ &\times \exp\left\{\frac{i\omega}{2\sin(\omega\Delta t)}\left[(q_{\rm f}^2 + q_{\rm in}^2)\cos(\omega\Delta t) - 2q_{\rm f}q_{\rm in}\right]\right\}.\end{aligned} \tag{6.10}$$

But how to solve the problem in the general case?

We note first that the matrix element (6.3) satisfies the completeness relation

$$\langle q_{\rm f}, t_{\rm f} | q_{\rm in}, t_{\rm in}\rangle = \int_{-\infty}^{\infty} dq_* \, \langle q_{\rm f}, t_{\rm f} \,|\, q_*, t_*\rangle \, \langle q_*, t_* \,|\, q_{\rm in}, t_{\rm in}\rangle\,, \tag{6.11}$$

where $t_*$ is any time moment between $t_{\rm in}$ and $t_{\rm f}$. The relation (6.11) has a transparent physical meaning: the probability amplitude for the particle to go from the point $q_{\rm in}$ at the initial moment $t_{\rm in}$ to the point $q_{\rm f}$ at the final moment $t_{\rm f}$ is given by the convolution of the probability amplitudes to go first to an intermediate point $q_*$ at $t_*$ and then to the final point.[2]

[2]The reader-physicist may recall the Huygens–Fresnel principle in optics, which tells basically the same thing but for the amplitude of an electromagnetic wave. The analogy between quantum mechanics and optics is very deep, though it is, of course, outside the scope of our book and we do not pursue it here.

We divide now the time interval $\Delta t$ in a large number $n$ of equal tiny time slices and write

$$\langle q_\mathrm{f}, t_\mathrm{f} | q_\mathrm{in}, t_\mathrm{in}\rangle = \int_{-\infty}^{\infty} \prod_{j=1}^{n-1} dq_j \ \langle q_\mathrm{f}, t_\mathrm{f} | q_{n-1}, t_\mathrm{f} - \epsilon\rangle \cdots \langle q_1, t_\mathrm{in} + \epsilon | q_\mathrm{in}, t_\mathrm{in}\rangle \tag{6.12}$$

with $\epsilon = \Delta t/n = (t_\mathrm{f} - t_\mathrm{in})/n$.

The integrand involves the product of a large number of factors. To evaluate each such factor, we use (6.2) and (6.3). For example,

$$\begin{aligned}\langle q_1, t_\mathrm{in} + \epsilon | q_\mathrm{in}, t_\mathrm{in}\rangle &= \langle q_1 | \exp\{-i\epsilon \hat{H}\} | q_\mathrm{in}\rangle \\ &= \int \frac{dp}{2\pi} \langle q_1 | p\rangle \langle p | e^{-i\epsilon\hat{H}} | q_\mathrm{in}\rangle. \end{aligned} \tag{6.13}$$

We used here the completeness relation again, inserting the intermediate states $|p\rangle$ that represent the eigenstates of the momentum operator $\hat{p} = -i\partial/\partial q$. They are plane waves:

$$\langle q_1 | p\rangle = \Psi_p(q_1) = e^{ipq_1}, \quad \langle p | q_\mathrm{in}\rangle = \overline{\Psi_p(q_\mathrm{in})} = e^{-ipq_\mathrm{in}}. \tag{6.14}$$

The factor $1/2\pi$ in the measure provides the correct normalization:

$$\langle q_1 | q_\mathrm{in}\rangle = \int \frac{dp}{2\pi} e^{ip(q_1 - q_\mathrm{in})} = \delta(q_1 - q_\mathrm{in}).$$

Suppose that the Hamiltonian has a simple form

$$\hat{H} = \frac{\hat{p}^2}{2} + V(q). \tag{6.15}$$

We may then write for small $\epsilon$

$$\langle p | e^{-i\epsilon\hat{H}} | q_\mathrm{in}\rangle \approx e^{-i\epsilon H(p, q_\mathrm{in})} \langle p | q_\mathrm{in}\rangle = e^{-ipq_\mathrm{in} - i\epsilon H(p, q_\mathrm{in})}, \tag{6.16}$$

where $H(p, q)$ is the classical Hamiltonian. Plugging it in (6.13), we derive

$$\langle q_1, t_\mathrm{in} + \epsilon | q_\mathrm{in}, t_\mathrm{in}\rangle \approx \int \frac{dp}{2\pi} \exp\{ip(q_1 - q_\mathrm{in}) - iH(p, q_\mathrm{in})\epsilon\}. \tag{6.17}$$

For a generic Hamiltonian mixing the coordinates and momenta, the situation is more complicated. As we discussed earlier, the order in which $\hat{p}$ and $q$ enter $\hat{H}$ matters. This ordering ambiguity matches the ambiguity in the path integral definition. Generically, it matters whether we write $H(p, q_\mathrm{in})$, $H(p, q_1)$ or maybe $H\left(p, \frac{q_1 + q_\mathrm{in}}{2}\right)$ in the integrand in Eq. (6.17). For example, the relation (6.16) holds for $\hat{H} = \hat{p}^2 q^2$ but not for $\hat{H} = q^2 \hat{p}^2$.

We address the reader to Ref. [69] where this question is thouroughly studied and pedagogically explained. We see later, however, that this ordering ambiguity is irrevant for the final result we are interested in — the integral representation of the Witten index.

Doing the same transformation for each factor in (6.12) and sending $n$ to infinity, we obtain the evolution kernel at finite time interval in the following form:

$$\begin{aligned}\mathcal{K}(q_{\rm f}, q_{\rm in}; t_{\rm f} - t_{\rm in}) = \lim_{n\to\infty} \int & \exp\{ip_n(q_{\rm f} - q_{n-1}) + \cdots + ip_1(q_1 - q_{\rm in}) \\ & - i\epsilon[H(p_n, q_{n-1}) + \cdots + H(p_1, q_{\rm in})]\} \\ & \times \frac{dp_n}{2\pi} \frac{dp_{n-1}dq_{n-1}}{2\pi} \cdots \frac{dp_1 dq_1}{2\pi}. \end{aligned} \tag{6.18}$$

Trading $\epsilon$ for $dt$, this can be formally written as

$$\boxed{\mathcal{K}(q_{\rm f}, q_{\rm in}; t_{\rm f} - t_{\rm in}) = \int \exp\left\{i \int_{t_{\rm in}}^{t_{\rm f}} [p\dot{q} - H(p,q)]dt\right\} \prod_t \frac{dp(t)dq(t)}{2\pi}} \tag{6.19}$$

with the boundary conditions

$$q(t_{\rm in}) = q_{\rm in}, \quad q(t_{\rm f}) = q_{\rm f}, \tag{6.20}$$

while the function $p(t)$ is quite arbitrary and is integrated over. For the Hamiltonian (6.15), the integral over all $dp_n$ in Eq. (6.18) is Gaussian and can be easily done:

$$\begin{aligned}\mathcal{K}(q_{\rm f}, q_{\rm in}; t_{\rm f} - t_{\rm in}) &= \mathcal{N} \lim_{n\to\infty} \int \prod_{j=1}^{n-1} dq_j \\ &\times \exp\left\{i\left[\frac{(q_{\rm f} - q_{n-1})^2}{2\epsilon} + \cdots + \frac{(q_1 - q_{\rm in})^2}{2\epsilon} - \epsilon V(q_{n-1}) - \cdots - \epsilon V(q_{\rm in})\right]\right\} \\ &\propto \int \prod_t dq(t) \exp\left\{i \int_{t_{\rm in}}^{t_{\rm f}} \left[\frac{\dot{q}^2}{2} - V(q)\right] dt\right\}. \end{aligned} \tag{6.21}$$

An infinite normalization factor $\mathcal{N}$ plays no role for our purposes. The third line in Eq. (6.21) was derived from the second one by replacing $q_1 - q_{in} \to \epsilon\dot{q}(t_{\rm in})$, etc. The representation (6.21) is the original Feynman form of the path integral where the integral is done over the trajectories in configuration space rather than in phase space as it is the case in (6.19). The boundary conditions (6.20) are assumed. The integral in the exponent is just the action $S = \int L dt$ on a trajectory $q(t)$ connecting the points $q_{\rm in}, t_{\rm in}$ and $q_{\rm f}, t_{\rm f}$.

The (semi-)classical limit of quantum theory corresponds to large action (in units of $\hbar$). In this case, the main contribution to the integral (6.21) stems from the stationary points (minima) of the action where many trajectories are added up with nearly identical phase factors. That is how the classical principle of least action emerges.

The path integral has been now used in physics for three quarters of a century. But not in mathematics — mathematicians still do not like it and consider it not to be a rigorously defined object. To be more precise, they have nothing against the *Euclidean* path integral where the time is considered to be imaginary, $t = -i\tau$. In that case, the function to be integrated in (6.21) goes over to

$$\exp\left\{-\int d\tau \left[\frac{1}{2}\left(\frac{dq}{d\tau}\right)^2 + V(q)\right]\right\}. \tag{6.22}$$

If the potential grows at large $q$, the contribution of the trajectories with wildly deviating $q(\tau)$ is exponentially suppressed. Mathematicians call such integrals the integrals with *Wiener measure* and can work with them. A good news for a reader-mathematician is that we will also only work with Euclidean path integrals in this book!

It is not so relevant for us, but we can still mention that the main problem with the real-time path integral is that it is difficult to prove the existence of the limit of a finite-dimensional integral (6.12), when the number of points goes to infinity and the time interval $\epsilon$ between the adjacent points goes to zero.

I am not myself a mathematician and cannot even try to prove it. The only thing that I can say is that, for simple-minded but nontrivial quantum mechanical problems like the anharmonic oscillator, the real-time path integrals were calculated *numerically* and the existence of the limit $\epsilon \to 0$ was established *experimentally*, if you will [70].[3]

---

[3]In this paper, the Fourier image

$$G(q, E) = \int_0^\infty dT\, e^{iET}\, \mathcal{K}(q, q; T) \tag{6.23}$$

of the evolution operator (6.21) with just 2–3 intermediate points was evaluated and the peaks at the right places — at the eigenvalues $E_k$ of the Hamiltonian — were observed, in agreement with the spectral decomposition (6.5). The actual calculations were made

Let us concentrate on the Euclidean path integral (6.22). In the Minkowski case, we were interested with the kernel (6.3) of the evolution operator that depended on the initial and final coordinates and the time interval $\Delta t$. But the object of principal interest for us in this book is the *partition function* and especially its supersymmetric generalization to be discussed soon.

**Definition 6.1.** The partition function of a quantum system is defined as the following sum over the spectrum of its Hamiltonian:

$$Z(\beta) = \sum_k e^{-\beta E_k}, \tag{6.24}$$

where $\beta$ is the parameter whose physical interpretation is the inverse temperature, $\beta = 1/T$.

Note that for large enough $\beta$ (small enough temperatures), the main contribution in the sum (6.24) comes from the ground state in the spectrum, while the contribution of the excited states is suppressed.

**Theorem 6.2.** *The partition function is given by the integral*

$$\boxed{Z = \int dq\, \mathcal{K}(q, q; -i\beta),} \tag{6.25}$$

*where* $\mathcal{K}(q, q; -i\beta)$ *is the Euclidean evolution operator over the imaginary time* $-i\beta$ *with coinciding initial and final points.*

**Proof.** Simply substitute $-i\beta$ for $\Delta t$ in Eq. (6.5), set $q_{\text{in}} = q_{\text{f}} \to q$ and integrate over $q$. □

Another name for the Euclidean evolution operator $\mathcal{K}(q_{\text{f}}, q_{\text{in}}; -i\beta)$ is *heat kernel.* It is called so because it represents the fundamental solution of the *diffusion* equation or *heat* equation in the same way as $\mathcal{K}(q_{\text{f}}, q_{\text{in}}; \Delta t)$ represents the fundamental solution of the Schrödinger equation.

## 6.2 Grassmann evolution kernel

We now go over to the systems including Grassmann dynamical variables. As a toy model, consider first the Grassmann oscillator of Section 4.4. It is

---

assuming that the parameter $E$ in (6.23) has a small positive imaginary part (a little step in the Euclidean direction). This makes the integrals well defined and convergent.

described by a single holomorphic Grassmann variable $\psi$, and the Hilbert space includes only two states: the state $|1\rangle$ with $\Psi(\psi) = 1$ and the state $|\psi\rangle$ with $\Psi(\psi) = \psi$. To proceed, we need to define the inner product in this space. We should have $\langle 1|1\rangle = \langle\psi|\psi\rangle = 1$ and $\langle 1|\psi\rangle = \langle\psi|1\rangle = 0$. This follows from the nice formula

$$\langle f|g\rangle = \int d\psi d\bar{\psi}\, e^{-\psi\bar{\psi}}\, \overline{f(\psi)} g(\psi). \tag{6.26}$$

As we see, the measure includes now the factor $e^{-\psi\bar{\psi}}$. Its appearance is not so surprising — an exponential factor in the measure arises also for the ordinary oscillator in the holomorphic representation. The states in the first line in Eq. (5.12) form an orthonormal basis with the measure[4]

$$d\mu = \frac{dad\bar{a}}{2\pi}\, e^{-a\bar{a}}. \tag{6.27}$$

We later discuss the systems including many Grassmann dynamical variables $\psi_\alpha$, in which case the fermion measure is

$$d\mu_G = \prod_\alpha d\psi_\alpha d\bar{\psi}_\alpha\, e^{-\psi_\alpha\bar{\psi}_\alpha}. \tag{6.28}$$

Going back to the Grassmann oscillator, the formal solution to the Schrödinger equation that describes its dynamics is

$$\Psi(\psi_{\rm f}, t_{\rm f}) = \int d\psi_{\rm in} d\bar{\psi}_{\rm in}\, e^{-\psi_{\rm in}\bar{\psi}_{\rm in}}\, \mathcal{K}(\psi_{\rm f}, \bar{\psi}_{\rm in}; \Delta t)\, \Psi(\psi_{\rm in}, t_{\rm in}) \tag{6.29}$$

with the kernel

$$\mathcal{K}(\psi_{\rm f}, \bar{\psi}_{\rm in}; \Delta t) = \langle f|e^{-i\hat{H}\Delta t}|i\rangle$$

given by the spectral decomposition (a generalization of Theorem 6.1 for Grassmann systems):

$$\boxed{\mathcal{K}(\psi_{\rm f}, \bar{\psi}_{\rm in}; \Delta t) = \sum_k \Psi_k(\psi_{\rm f}) \overline{\Psi_k(\psi_{\rm in})} e^{-iE_k\Delta t}.} \tag{6.30}$$

The integration over the initial phase space in Eq. (6.29) is done with the measure prescribed in (6.26).

[4]We use the convention $dad\bar{a} = dxdy$, where $a = (x + iy)/\sqrt{2}$ — cf. Eq. (2.38).

In our case, the sum in (6.30) includes only two terms with $E_0 = 0$ and $E_1 = \omega$. We derive

$$\mathcal{K}(\psi_\mathrm{f}, \bar{\psi}_\mathrm{in}; \Delta t) = 1 + \psi_\mathrm{f}\bar{\psi}_\mathrm{in} e^{-i\omega\Delta t}. \tag{6.31}$$

The Euclidean evolution operator is given by (6.31) where $\Delta t$ is substituted for $-i\beta$:

$$\mathcal{K}(\psi_\mathrm{f}, \bar{\psi}_\mathrm{in}; -i\beta) = 1 + \psi_\mathrm{f}\bar{\psi}_\mathrm{in} e^{-\omega\beta}. \tag{6.32}$$

Consider a natural generalization of (6.25):

$$\tilde{Z}(\beta) = \int d\psi d\bar{\psi}\, e^{-\psi\bar{\psi}}\, \mathcal{K}(\psi, \bar{\psi}; -i\beta), \tag{6.33}$$

where the measure (6.26) was used again. Doing the integral, we obtain

$$\tilde{Z}(\beta) = 1 - e^{-\beta\omega}. \tag{6.34}$$

This resembles the partition function (6.24), but the second term, the contribution of the excited oscillator state, enters with the negative sign!

Remember this remarkable fact and look again at the kernel (6.32). For $\beta = 0$ (when $\psi_\mathrm{in}$ and $\psi_\mathrm{f}$ coincide), it is reduced to $\mathcal{K}(\psi, \bar{\psi}; 0) = 1 + \psi\bar{\psi} = e^{\psi\bar{\psi}}$. For small $\beta \ll \omega^{-1}$, the kernel can be written as

$$\mathcal{K}(\psi_\mathrm{f}, \bar{\psi}_\mathrm{in}; -i\beta) \approx e^{\psi_\mathrm{f}\bar{\psi}_\mathrm{in} - \beta H(\psi_\mathrm{f}, \bar{\psi}_\mathrm{in})}, \tag{6.35}$$

where $H(\psi, \bar{\psi})$ is the classical Hamiltonian of the Grassmann oscillator (4.55).

The sign $\approx$ in Eq. (6.35) means the following. For small $\beta$, we keep in $\mathcal{K}(\psi_\mathrm{f}, \bar{\psi}_\mathrm{in}; -i\beta)$ the terms of order $\sim \beta\psi\bar{\psi}$ but neglect the terms of order $\sim \beta^{n\geq 2}\psi\bar{\psi}$ (and would neglect the terms of order $\sim \beta^{n\geq 1}$ not involving the factor $\psi\bar{\psi}$, were such terms present). The meaning of this selective neglection will be clarified by the end of this chapter: we are going to see that the neglected terms are irrelevant for the Witten index calculation. See also the discussion on p. 384.

The result (6.35) can be generalized for systems with an arbitrary number of Grassmann variables $\psi_{\alpha=1,\ldots,n}$. The inner product is then defined with the measure (6.28).

**Definition 6.2.** A *normal ordered* operator $\hat{A}_N(\psi_\alpha, \hat{\bar{\psi}}_\alpha)$ is the operator where all the factors $\psi_\alpha$ stay on the left and the factors $\hat{\bar{\psi}}_\alpha = \partial/\partial\psi_\alpha$ stay on the right.

It is clear that any operator can be brought to the normal ordered form by pulling the factors $\hat{\bar{\psi}}_\alpha$ to the right and using $\{\hat{\bar{\psi}}_\alpha, \psi_\beta\} = \delta_{\alpha\beta}$.

**Definition 6.3.** The function $A_N$ obtained from $\hat{A}_N$ by lifting the hats is called the *normal symbol* of the operator $\hat{A}$.

Thus, the normal symbol of $\psi_\alpha \hat{\bar{\psi}}_\beta$ is $\psi_\alpha \bar{\psi}_\beta$ and the normal symbol of $\hat{\bar{\psi}}_\beta \psi_\alpha$ is $\delta_{\alpha\beta} - \psi_\alpha \bar{\psi}_\beta$.

**Theorem 6.3.** *The kernel $\mathcal{K}_A(\psi_\alpha, \bar{\psi}_\alpha)$ of any operator $\hat{A}$, defined according to*[5]

$$(\hat{A}\Psi)(\psi) = \int \left( \prod_\beta d\chi_\beta d\bar{\chi}_\beta \, e^{-\chi_\beta \bar{\chi}_\beta} \right) \mathcal{K}_A(\psi_\alpha, \bar{\chi}_\alpha) \, \Psi(\chi), \tag{6.36}$$

*is related to its normal symbol $A_N(\psi_\alpha, \bar{\psi}_\alpha)$ by the formula*

$$\mathcal{K}_A(\psi_\alpha, \bar{\psi}_\alpha) = \prod_\beta e^{\psi_\beta \bar{\psi}_\beta} A_N(\psi_\alpha, \bar{\psi}_\alpha). \tag{6.37}$$

**Proof.** A normal ordered operator $\hat{A}$ can be decomposed as

$$\hat{A} = C(\psi) + A_\alpha(\psi) \hat{\bar{\psi}}_\alpha + A_{\alpha\beta}(\psi) \hat{\bar{\psi}}_\alpha \hat{\bar{\psi}}_\beta + \cdots \tag{6.38}$$

In the simplest case, the operator does not depend on $\hat{\bar{\psi}}$ and its action boils down to multiplication by $C(\psi)$. Then, to prove the theorem, we have to prove that

$$\int \left( \prod_\beta d\chi_\beta d\bar{\chi}_\beta e^{-\chi_\beta \bar{\chi}_\beta} e^{\psi_\beta \bar{\chi}_\beta} \right) C(\psi) \Psi(\chi) = C(\psi) \Psi(\psi) \tag{6.39}$$

or

$$\int \left( \prod_\beta d\chi_\beta d\bar{\chi}_\beta \, e^{-\chi_\beta \bar{\chi}_\beta} e^{\psi_\beta \bar{\chi}_\beta} \right) \Psi(\chi) = \Psi(\psi) \tag{6.40}$$

for any wave function $\Psi$. And this is true for any monomial, like $\Psi(\chi) = \chi_1 \chi_3 \chi_7$, due to the properties

$$\int d\chi d\bar{\chi} \, e^{(\psi - \chi)\bar{\chi}} = 1, \quad \int d\chi d\bar{\chi} \, e^{(\psi - \chi)\bar{\chi}} \chi = \psi. \tag{6.41}$$

[5] Cf. Eq. (6.29) defining the kernel of the evolution operator.

Consider now the second term $A_\alpha(\psi)\hat{\bar{\psi}}_\alpha$ in (6.38). We have to prove that

$$\int \left( \prod_\beta d\chi_\beta d\bar{\chi}_\beta \, e^{(\psi_\beta - \chi_\beta)\bar{\chi}_\beta} \right) A_\alpha(\psi)\bar{\chi}_\alpha \Psi(\chi) = A_\alpha(\psi) \frac{\partial}{\partial \psi_\alpha} \Psi(\psi) \quad (6.42)$$

or

$$\int \left( \prod_\beta d\chi_\beta d\bar{\chi}_\beta \, e^{(\psi_\beta - \chi_\beta)\bar{\chi}_\beta} \right) \bar{\chi}_\alpha \Psi(\chi) = \frac{\partial}{\partial \psi_\alpha} \Psi(\psi). \quad (6.43)$$

To prove the validity of the latter relation, it is sufficient to observe that, for the integral not to vanish, the factor $\bar{\chi}_\alpha$ in the integrand should be paired with some factor $\chi_\gamma$ in the expansion of $\Psi(\chi)$. And such a pairing amounts to differentiating $\partial\Psi(\chi)/\partial\chi_\alpha$. Using (6.40) for the function $\partial\Psi(\psi)/\partial\psi_\alpha$, we derive (6.43).

The same reasoning works for the higher terms of the expansion (6.38). For example, the validity of the theorem for the term including two operators $\hat{\bar{\psi}}$ follows from the relation

$$\int \left( \prod_\beta d\chi_\beta d\bar{\chi}_\beta \, e^{(\psi_\beta - \chi_\beta)\bar{\chi}_\beta} \right) \bar{\chi}_\alpha \bar{\chi}_\gamma \, \Psi(\chi) = \frac{\partial^2}{\partial \psi_\alpha \partial \psi_\gamma} \Psi(\psi). \quad (6.44)$$

□

Applying this theorem to the Euclidean evolution operator $e^{-\beta \hat{H}}$, we derive

$$\mathcal{K}(\psi^{\rm f}_\alpha, \bar{\psi}^{\rm in}_\alpha; -i\beta) = \left( \prod_\alpha \exp\{\psi^{\rm f}_\alpha \bar{\psi}^{\rm in}_\alpha\} \right) \left( e^{-\beta \hat{H}} \right)_N . \quad (6.45)$$

Now, *for small* $\beta$, the normal symbol of the exponential can be approximated by the exponential $e^{-\beta H}$ of the classical Hamiltonian. The latter can be chosen as the normal symbol of the quantum one, but it is not necessary, and actually we cannot make a particular choice of $H$ — it is beyond the validity of our approximation. We obtain

$$\mathcal{K}(\psi^{\rm f}_\alpha, \bar{\psi}^{\rm in}_\alpha; -i\beta) \approx \left( \prod_\alpha \exp\{\psi^{\rm f}_\alpha \bar{\psi}^{\rm in}_\alpha\} \right) e^{-\beta H(\psi^{\rm f}_\alpha, \bar{\psi}^{\rm in}_\alpha)}. \quad (6.46)$$

The reader may look again at Eq. (6.16) for the bosonic infinitesimal evolution kernel. It could also be expressed in the language of symbols. In close analogy with the theorem just proved, one can show that the matrix element $\langle p|\hat{A}|q\rangle$ of an arbitrary operator $\hat{A}$ coincides with the product of the factor $e^{-ipq}$ and the so-called *p-q symbol* [69] of $\hat{A}$. And for small $\epsilon$, the p-q symbol of the evolution operator can be approximated by the exponential $\exp\{-i\epsilon H(p,q)\}$, where $H(p,q)$ is the p-q symbol of $\hat{H}$. We arrive at (6.16).

## 6.3 Witten index

Consider now an SQM system including both ordinary variables $q_j$ and Grassmann variables $\psi_\alpha$. Consider the infinitesimal Euclidean evolution kernel

$$\mathcal{K}_{\text{inf}}(q_j, \psi_\alpha; q_j, \bar{\psi}_\alpha; -i\beta) = \langle q_j, \psi_\alpha; -i\beta | q_j, \psi_\alpha; 0\rangle$$

with the same values of the variables at $t = 0$ and $t = -i\beta$. Combining the results (6.17) (with $\epsilon \to -i\beta$ ) and (6.46), we derive

$$\mathcal{K}_{\text{inf}}(q_j, \psi_\alpha; q_j, \bar{\psi}_\alpha; -i\beta) \approx e^{\psi_\alpha \bar{\psi}_\alpha} \int \prod_j \frac{dp_j}{2\pi} e^{-\beta H(p_j, q_j; \psi_\alpha, \bar{\psi}_\alpha)}, \quad (6.47)$$

where $H(p_j, q_j; \psi_\alpha, \bar{\psi}_\alpha)$ is the classical Hamiltonian in *some* ordering prescription.

The expression above is written in the small $\beta$ approximation. The evolution operator at finite $\beta$ may be calculated as a path integral involving the convolution of an infinite number of the infinitesimal evolution operators:

$$\begin{aligned}
&\mathcal{K}(q_j, \psi_\alpha; q_j, \bar{\psi}_\alpha; -i\beta) \\
&= \lim_{n\to\infty} \int \prod_{m=1}^{n-1} \left( \prod_j dq_j^{(m)} \prod_\alpha d\psi_\alpha^{(m)} d\bar{\psi}_\alpha^{(m)} e^{-\psi_\alpha^{(m)} \bar{\psi}_\alpha^{(m)}} \right) \\
&\times \mathcal{K}_{\text{inf}} \left( q_j, \psi_\alpha; q_j^{(1)}, \bar{\psi}_\alpha^{(1)}; \frac{-i\beta}{n} \right) \cdots \\
&\times \mathcal{K}_{\text{inf}} \left( q_j^{(n-1)}, \psi_\alpha^{(n-1)}; q_j, \bar{\psi}_\alpha^{(n-1)}; \frac{-i\beta}{n} \right).
\end{aligned} \quad (6.48)$$

But a very good news is that in most cases (though not in *all* cases — see the discussion in Chapter 15), we do not need to evaluate this complicated path integral. For our present purposes, it is sufficient to know the expression (6.47) for the evolution kernel at small $\beta$ and its spectral decomposition at an arbitrary $\beta$.

The spectral decomposition of the bosonic evolution kernel was written in Eq. (6.5) and, for the kernel in the toy Grassmann oscillator model, in

Eq. (6.30). Performing an obvious generalization and setting also "in" = "f", we may write

$$\mathcal{K}(q_j, \psi_\alpha; q_j, \bar{\psi}_\alpha; -i\beta) = \sum_k \Psi_k(q_j, \psi_\alpha)\overline{\Psi_k(q_j, \psi_\alpha)}e^{-\beta E_k}. \tag{6.49}$$

As a generalization of (6.33), we consider the integral

$$\tilde{Z}(\beta) = \int \prod_j dq_j \prod_\alpha d\psi_\alpha d\bar{\psi}_\alpha \, e^{-\psi_\alpha \bar{\psi}_\alpha} \, \mathcal{K}(q_j, \psi_\alpha; q_j, \bar{\psi}_\alpha; -i\beta). \tag{6.50}$$

We substitute there the spectral decomposition (6.49) and, bearing in mind the normalization condition $\langle \Psi_k | \Psi_k \rangle = 1$ with the measure (6.28), we derive

$$\boxed{\tilde{Z}(\beta) = \sum_k \eta_k e^{-\beta E_k},} \tag{6.51}$$

where $\eta_k = 1$ for the bosonic states with Grassmann-even wave functions $\Psi_k$ and $\eta_k = -1$ for the fermionic states with Grassmann-odd wave functions. The quantity $\tilde{Z}$ is called the *supersymmetric partition function* or simply the *Witten index*.[6]

The path integral (6.50) for $\tilde{Z}$ can be presented in the following symbolic form [cf. (6.21)]:

$$\tilde{Z}(\beta) = \int \prod_{j\tau} dq_j(\tau) \prod_{\alpha\tau} d\psi_\alpha(\tau) d\bar{\psi}_\alpha(\tau)$$
$$\times \exp\left\{ -\int_0^\beta L_E[q_j(\tau), \psi_\alpha(\tau), \bar{\psi}_\alpha(\tau)]d\tau \right\}, \tag{6.52}$$

where $L_E$ is the Euclidean Lagrangian and both the ordinary and Grassmann variables satisfy the *periodic* boundary conditions:

$$q_j(\beta) = q_j(0); \quad \psi_\alpha(\beta) = \psi_\alpha(0); \quad \bar{\psi}_\alpha(\beta) = \bar{\psi}_\alpha(0). \tag{6.53}$$

The reader may wonder whether it is possible to define in a similar way the conventional partition function (6.24) where all the states, both bosonic

[6]We are going to see in Chapter 15 that the notion of Witten index is closely related to Atiyah–Singer index introduced by mathematicians in the 1960s [66]. And the first (to the best of my knowledge) mathematician who considered a similar object (the so-called *equivariant index*) was Henri Cartan [71].

and fermionic, contribute in the sum with the positive sign. The answer is affirmative. One just has to write

$$Z(\beta) = \int \prod_j dq_j \prod_\alpha d\psi_\alpha d\bar\psi_\alpha \, e^{-\psi_\alpha \bar\psi_\alpha} \, \mathcal{K}(q_j, \psi_\alpha; q_j, -\bar\psi_\alpha; -i\beta). \quad (6.54)$$

The minus before the second Grassmann argument in $\mathcal{K}$ brings about an extra minus in the spectral sum for Grassmann-odd wave functions, which compensates the factor $\eta_k = -1$ in (6.51).[7]

Going back to the alternating sum (6.51), it has interesting properties. You remember that a supersymmetric quantum system is characterized by the double degeneracy of all excited states — for any bosonic state, there is a fermionic superpartner with the same energy. This means that the excited states do not give any contribution to the sum (6.51), only the ground alias vacuum states having zero energy, which are not necessarily paired, contribute. And if $E_k = 0$, $e^{-\beta E_k} = 1$. We see that $\tilde Z(\beta)$ does *not* depend on $\beta$ and is given by a remarkable formula [61, 72]

$$\boxed{\tilde Z(\beta) \equiv I_W = n_B^{E=0} - n_F^{E=0},} \quad (6.55)$$

where $n_B^{E=0}$ is the number of bosonic vacuum states and $n_F^{E=0}$ is the number of fermionic vacuum states. Obviously, the Witten index is an integer number. But if $\tilde Z(\beta)$ does not depend on $\beta$, one may calculate it for any $\beta$, and it is convenient to do so when $\beta$ is small. In this case, we can substitute the approximate expression (6.47) for the integrand in (6.50) and calculate the integral. We obtain [73]

$$\boxed{I_W = \lim_{\beta \to 0} \int \prod_j \frac{dp_j dq_j}{2\pi} \prod_\alpha d\psi_\alpha d\bar\psi_\alpha \, e^{-\beta H(p_j, q_j; \bar\psi_\alpha, \psi_\alpha)}.} \quad (6.56)$$

This is an ordinary finite-dimensional integral, and it can be evaluated analytically for many SQM systems.

One can also represent the result (6.56) in an alternative form when the Grassmann dynamical variables are not introduced, but the supersymmetric

---

[7] It is not so relevant for us, but one can mention that the conventional partition function is interesting for a physicist when they study thermal properties of the system. For a finite $\beta$, the evolution kernel and the partition function are described by a path integral. The appearance of the extra minus sign in the second argument of $\mathcal{K}$ for the fermionic but not bosonic variables means that one should impose in the path integral (6.52) periodic boundary conditions in imaginary time for the bosonic variables and antiperiodic ones for the fermionic variables. Physicists say that the symmetry between bosons and fermions, i.e. supersymmetry, is broken by finite temperature effects.

Hamiltonian has matrix nature. For example, for the Witten model, we may represent the classical Hamiltonian as

$$\hat{H} = \frac{P^2 + [W'(x)]^2}{2} + \frac{1}{2}W''(x)\sigma^3. \tag{6.57}$$

Here, the hat over $H$ only signifies that $\hat{H}$ is a matrix, but it is not the full-scale quantum Hamiltonian representing a differential operator. The momentum $P$ is a classical phase space variable.

In this case, the Euclidean evolution kernel also has matrix nature. The functional trace of this kernel including the ordinary matrix trace and the integration over coordinates, as in (6.54), gives us the partition function of the system:

$$Z(\beta) = \int \prod_j dq_j \, \mathrm{Tr}\{\hat{\mathcal{K}}(q_j; q_j; -i\beta)\}. \tag{6.58}$$

For small $\beta$, the infinitesimal matrix evolution operator is given by the expression similar to (6.17):

$$\hat{\mathcal{K}}(q_j; q_j; -i\beta) \approx \int \prod_j \frac{dp_j}{2\pi} \exp\{-\beta \hat{H}(p_j, q_j)\}. \tag{6.59}$$

To find the supersymmetric partition function, we have to recall that the spectrum of a supersymmetric Hamiltonian includes two towers of degenerate states, as in Fig. 5.2. We introduce then the operator $(-1)^{\hat{F}}$ which commutes with the Hamiltonian and whose eigenvalues are $+1$ for one tower and $-1$ for another tower.

To understand the commonly used notation $(-1)^{\hat{F}}$ for this operator, one should go back to the Grassmann representation. Then, $\hat{F}$ is the operator of the *fermion charge*:

$$\hat{F} = \psi_\alpha \hat{\bar{\psi}}_\alpha. \tag{6.60}$$

It simply counts the number of the Grassmann factors in the wave function. The operator $\hat{F}$ may commute with the Hamiltonian, and in practice it almost always does, but it is not necessary. The property $[(-1)^{\hat{F}}, \hat{H}]$ always holds, however. Indeed, the supercharge operators always include an odd number of Grassmann factors $\psi$ or $\hat{\bar{\psi}}$, and they transform the states with even $F$ to the states with odd $F$ and vice versa. Thus, in the Grassmann language, the two towers of states, the eigenstates of the Hamiltonian mentioned in the previous paragraph, are the states of even and odd fermionic charge. The action of $\hat{H}$ does not change $(-1)^F$.

It is natural to call bosonic the states with even $F$ and fermionic the states with odd $F$.

We go back to the matrix formulation. The supersymmetric partition function can be represented as

$$\tilde{Z}(\beta) = \int \prod_j dq_j \,\mathrm{Tr}\{(-1)^{\hat{F}}_{\mathrm{matr}} \hat{\mathcal{K}}(q_j; q_j; -i\beta)\}, \tag{6.61}$$

where $[(-1)^{\hat{F}}]_{\mathrm{matr}}$ is the matrix representation of the fermion parity operator defined above. As we see a bit later, in the conventions spelled out in Eq. (4.68), $(-1)^{\hat{F}} \equiv -\sigma^3$ both for the Witten model and the Landau model.

Bearing in mind (6.59), the formula (6.56) acquires the form

$$I_W = \lim_{\beta \to 0} \int \prod_j \frac{dp_j dq_j}{2\pi} \,\mathrm{Tr}\left\{(-1)^{\hat{F}}_{\mathrm{matr}}\, e^{-\beta \hat{H}(p_j, q_j)}\right\}. \tag{6.62}$$

The importance of the Witten index notion is the following. Suppose that we have a complicated supersymmetric system (not necessarily an SQM system, it can be a field theory system) and we want to know whether the spectrum of this system involves supersymmetric zero-energy states annihilated by the action of supercharges or not. In other words, we want to know whether supersymmetry is spontaneously broken or not. To answer this question by explicitly solving the Schrödinger equation or the equation $\hat{Q}\Psi = \hat{\bar{Q}}\Psi = 0$ may be a very difficult task.

But there is an alternative to this *brute force* approach: one can determine $I_W$ by calculating the integral (6.56). If this integral is still complicated in the theory in question, one can modify the theory by a *smooth* change of parameters. The Witten index is an integer, and it cannot change abrubtly under a smooth parameter change. For example, one can modify a field theory by putting it into a finite spatial box and then sending the size of the box to zero so that only zero Fourrier modes of the fields are left, and the problem is reduced to a SQM problem [72].

We hasten to comment here that the method just described has its limits of applicability:

- To begin with, the very notion of the Witten index applies only to the systems with discrete spectrum. As we already mentioned at the end of Section 4.2, only such systems are mathematically well defined. If the spectrum is continuous and the states are not normalizable, most methods of the functional analysis do not work. And even a physicist who is brave enough to consider systems with continuous spectrum (like the system describing free motion along a line) always has in mind a finite-volume regularization making the spectrum discrete. For systems with continuous spectrum, the integrals (6.56) have often fractional values, which does not have much sense.[8]

- Even for the systems with discrete spectrum, the integral (6.56) gives in some cases wrong fractional numbers. The reason for this is that the approximation (6.47) does not work there and *one-loop effects* come into play. We meet and treat such systems in Chapter 15.

Once $I_W$ is calculated, we can answer the question on whether supersymmetry is spontaneously broken or not. Strictly speaking, we can only give a definite answer if $I_W \neq 0$. Then, a supersymmetric vacuum exists and supersymmetry is not broken. And if $I_W = 0$, one cannot issue a judgement: $n_B^{E=0}$ and $n_F^{E=0}$ may still be nonzero but equal. In most cases, however, the vanishing of $I_W$ signalizes also the vanishing of $n_B^{E=0} = n_F^{E=0}$. If these numbers do not vanish, there should be a special reason for that. And in such a case, it is usually possible to define a modified elaborated version of $I_W$ that does not vanish.

### 6.3.1 *Examples*

We now show how this general machinery works for the simple SQM models discussed in the previous chapter. Consider first the Witten model with the classical Hamiltonian

$$H = \frac{P^2 + [W'(x)]^2}{2} + W''(x)\psi\bar{\psi}. \tag{6.63}$$

Substitute this into (6.56). Integrating over $d\psi d\bar{\psi}$ and over momenta, we derive

$$I_W = \lim_{\beta \to 0} \sqrt{\frac{\beta}{2\pi}} \int_{-\infty}^{\infty} dx\, W''(x) \exp\left\{-\frac{\beta[W'(x)]^2}{2}\right\}. \tag{6.64}$$

[8]There are certain rather complicated supersymmetric *gauge* models where the spectrum is continuous, but on top of that, ground zero-energy normalizable states exist and Witten index can still be defined [74] (see Chapter 9 of Ref. [75] for the review). These models are rather interesting (incidentally, they provide an example of systems where the fermion charge does not commute with the Hamiltonian), but they have no application to geometry and are beyond the scope of this book.

Changing the variable $y = W'(x)$, we obtain

$$I_W = \lim_{\beta \to 0} \sqrt{\frac{\beta}{2\pi}} \int_{W'(-\infty)}^{W'(\infty)} dy \exp\left\{-\frac{\beta y^2}{2}\right\}. \tag{6.65}$$

We are interested only in the systems where the potential $U(x) = [W'(x)]^2/2$ grows at infinity. Otherwise, the spectrum of the Hamiltonian would not be discrete and the notion of the index would not be well defined. Let $W(x)$ be a polynomial. There are three possibilities:

(1) The highest power in $W(x)$ is odd. Then, $W'(\infty) = W'(-\infty)$, the limits in the integral (6.65) coincide and $I_W = 0$. This signalizes the spontaneous supersymmetry breaking, which agrees with the explicit analysis of the preceding chapter.
(2) The highest power in $W(x)$ is even and the coefficient in front is positive. Then, $W'(\pm\infty) = \pm\infty$ and the integral in (6.65) does not depend on $\beta$, giving $I_W = 1$ — there is one bosonic zero-energy eigenstate in the spectrum[9] in agreement with the explicit analysis in Section 5.4.
(3) The higher power in $W(x)$ is even and the coefficient in front is negative. Then, $W'(\pm\infty) = \mp\infty$ and the integral (6.65) gives $I_W = -1$: there is one fermionic zero-energy eigenstate in the spectrum. Again, this agrees with Section 5.4.

Consider now the generalized Landau problem when the magnetic field $B$ has the form (5.18). We set $m = c = 1, e = -1$. The classical Hamiltonian then reads

$$H = \frac{1}{2}(P_x + A_x)^2 + \frac{1}{2}(P_y + A_y)^2 + B(x,y)\psi\bar{\psi}, \tag{6.66}$$

where $A_{x,y}$ are the components of the vector potential. We have

$$\begin{aligned} I_W = \lim_{\beta \to 0} \int \frac{dP_x dx}{2\pi} \frac{dP_y dy}{2\pi} d\psi d\bar{\psi} \\ \times \exp\left\{-\frac{\beta}{2}[(P_x + A_x)^2 + (P_y + A_y)^2] - \beta B(x,y)\psi\bar{\psi}\right\}. \end{aligned} \tag{6.67}$$

[9] Strictly speaking, we can only say that $n_B^{E=0} - n_F^{E=0} = 1$, which does not exclude the possibility for the system to have e.g. two bosonic and one fermionic vacua. But one can make here the same comment that we did when discussing the case $I_W = 0$: there should be a special reason for the existence of extra vacuum states not dictated by the index considerations. For the Witten model, there is no such special reason.

Doing the integral over $dP_x dP_y d\psi d\bar{\psi}$, we obtain

$$\boxed{I_W = \frac{1}{2\pi}\int dxdy\, B(x,y) = \frac{\Phi}{2\pi},} \tag{6.68}$$

where $\Phi$ is the magnetic flux. We have reproduced thereby the relation (5.25) with an additional specification: the sign of the index coincides with the sign of the flux. And this conforms with the analysis of Section 5.3: if $B > 0$, the zero-energy states are spin-down in the spinorial language and bosonic in the Grassmannian language.

It is instructive to reproduce the result (6.68) using the matrix formula (6.62). In this case,

$$[(-1)^{\hat{F}}]_{\text{matr}} = -\sigma_3. \tag{6.69}$$

Indeed, as we know from Eq. (4.68), the operator $\psi$ goes over to $\sigma^+$ and the operator $\hat{\bar{\psi}}$ goes over to $\sigma^-$ in the matrix formulation. Then, $\hat{F} = \psi\hat{\bar{\psi}}$ goes to

$$\sigma^+\sigma^- = \begin{pmatrix} 1 & 0 \\ 0 & 0 \end{pmatrix}.$$

Trading $(-1)^{\hat{F}}$ for $e^{i\pi\hat{F}}$, expanding and using the property $\hat{F}^2 = \hat{F}^3 = \cdots = \hat{F}$, we derive $e^{i\pi\hat{F}} = \mathbb{1} + (e^{i\pi} - 1)\hat{F}$ or

$$(-1)^{\hat{F}} = \begin{pmatrix} 1 & 0 \\ 0 & 1 \end{pmatrix} - 2\begin{pmatrix} 1 & 0 \\ 0 & 0 \end{pmatrix} = \begin{pmatrix} -1 & 0 \\ 0 & 1 \end{pmatrix} = -\sigma_3. \tag{6.70}$$

We derive

$$I_W = \lim_{\beta\to 0}\frac{1}{2\pi\beta}\int dxdy\,\mathrm{Tr}\left\{-\sigma^3 e^{-\beta\sigma^3 B(x,y)/2}\right\}, \tag{6.71}$$

which coincides with (6.68).

By definition, the index $I_W$ is an integer number. It follows that the flux $\Phi$ should be quantized, $\Phi = 2\pi n$. In Section 14.2.1, we are going to see what is wrong with the systems where $\Phi/2\pi$ is not an integer. In contrast to what is usually assumed, one *can* in this case define the spectral problem for the Hamiltonian, but this Hamiltonian is not supersymmetric anymore.

For the most part of this book, we assume, when talking about the gauge fields, that the flux (6.68) and its generalizations for more complicated systems *are* quantized.

The reader may have noted that in the examples considered above, the expressions (6.64), (6.67) do not depend on $\beta$ and taking the limit $\beta \to 0$ was superfluous.[10] Actually, this was due to the clever choice of the classical Hamiltonians (6.63), (6.66) which are given (with the factor $-i$) by the Poisson bracket of the classical supercharges obtained from the quantum expressions like in Eq. (5.27) by replacing $\hat{P} \to P$. In other words, the classical supersymmetry algebra (5.43) holds.

In the expressions (5.27), the coordinates and momenta are not entangled and there is no ordering problem. We mentioned at the end of Chapter 5 that, in the complicated systems where such a problem exists (we meet these systems in the third part of this book), classical supercharges should be taken as the Weyl symbols of the quantum ones. We can formulate a **conjecture:**

*Let $\hat{Q}, \hat{\bar{Q}}$ and $\hat{H}$ be quantum operators satisfying the supersymmetry algebra (5.1) and expressed via $\psi_\alpha$ and $\hat{\bar{\psi}}^\alpha$. Let the spectrum of $\hat{H}$ be discrete. Let $Q$ and $\bar{Q}$ be the Weyl symbols of $\hat{Q}$ and $\hat{\bar{Q}}$ and let the Poisson bracket $\{Q, Q\}_P$ and hence $\{\bar{Q}, \bar{Q}\}_P$ vanish.[11] Define the classical Hamiltonian as $H = -i\{Q, \bar{Q}\}_P$. Then, the integral in (6.56) does not depend on $\beta$.*

We do not know a general proof of this statement, but it holds in all the SQM systems that we know.

If $H$ is chosen sloppily, but the quantum Hamiltonian is still derived from $H$ by *some* ordering prescription, the integral in (6.56) may depend on $\beta$, but its value in the limit $\beta \to 0$ should be the same in all definitions. Again, we cannot give a mathematical proof of this assertion, but a heuristic physical reasoning is the following. The ordering ambiguities is a quantum effect. On the other hand, small $\beta$ correspond to high temperatures when quantum effects are not important and the dynamics is classical. Then, the integral in (6.56) should not be sensitive in this limit to a particular way to resolve these ambiguities.

What we can do is to illustrate this general property in a particular example. Let us concentrate on Witten's model. We note the following:

- As the coordinates and momenta are not entangled, the Weyl symbols of the quantum supercharges (5.27) have the same functional form as in (5.27).

[10] This was true in the Grassmann formulation. On the other hand, the RHS in (6.71) *does* depend on $\beta$ and taking the limit $\beta \to 0$ is essential.

[11] This reservation is necessary to exclude the systems suffering from the classical supersymmetry anomaly, like the system (5) of Ref. [63] (see the discussion on pp. 119, 330).

- The classical Hamiltonian (6.63) coincides with $-i\{Q, \bar{Q}\}_P$. [One can also note that it coincides in this case with the Weyl symbol of the quantum Hamiltonian (5.28), but it is not so for more complicated systems.]
- With this choice, the integral for the index is $\beta$-independent, indeed.

If another form of $H$ were chosen — for example, the normal symbol of (5.28) instead of the Weyl one — the integrand in (6.64) would acquire the extra factor $\exp\{\beta W''(x)/2\}$, and the integral would depend on $\beta$.

Note, however, that, for the motion to be finite and the spectrum discrete (otherwise, the index is ill-defined), the potential $[W'(x)]^2/2$ and hence the function $W'(x)$ should grow at infinity. Then, for small $\beta$, the integral in (6.64) is saturated by large $x$. And under these conditions, $[W'(x)]^2/2$ dominates over $W''(x)/2$ and the factor $\exp\{\beta W''(x)/2\}$ is irrelevant — it does not affect the integral for the index in the limit $\beta \to 0$.

# Chapter 7

# Superspace and Superfields

Supersymmetric Lagrangians (5.46), (5.48), and (5.52) constructed in Chapter 5 are invariant up to a total time derivative under the supersymmetry transformations (5.47), (5.49), and (5.53). These are the transformations mixing the ordinary and Grassmann variables, which can be compared to rotations mixing different vector components. In the latter case, we know that the description of a vector as a set of its components is possible, but it is much more convenient to use the vector notation and to write instead of many different scalar equations a single vector equation, like $\boldsymbol{B} = \boldsymbol{\nabla} \times \boldsymbol{A}$ instead of $B_x = \partial_y A_z - \partial_z A_y$, etc.

One can ask, "Is there an analog of such vector notation in supersymmetry? Can one introduce a single object including the ordinary and Grassmann variables as different components, so that the supersymmetric actions and equations of motion can be written in a more compact and transparent way?" The answer is positive. Such an object exists, it is called *superfield* — a function depending on *superspace* coordinates.

At this point, we should complain again on the universally accepted terminology. Historically, the concept of superspace was first introduced for supersymmetric field theories [76]. In this book, we are interested only in supersymmetric quantum *mechanics*, and the objects that we are about to introduce should better be called dynamical *supervariables* depending on *supertime*. But people do not do so. They still use the terms "superfield" and "superspace" (adding sometimes a specification "one-dimensional"). And that is what we also do.

## 7.1 $\mathcal{N} = 1$

**Definition 7.1.** The simplest $\mathcal{N} = 1$ one-dimensional superspace includes on top of time $t$ a single real Grassmann-odd variable $\Theta = \bar{\Theta}$. The transformations

$$\begin{aligned} \Theta &\to \Theta + E, \\ t &\to t + iE\Theta, \end{aligned} \tag{7.1}$$

with a real Grassmann-odd $E$ play there the role of "rotations".

Consider the composition of two different transformations (7.1). We obtain

$$\begin{gathered} \Theta_1 = \Theta + E_1, \quad \Theta_2 = \Theta_1 + E_2 = \Theta + E_1 + E_2, \\ t_1 = t + iE_1\Theta, \quad t_2 = t_1 + iE_2\Theta_1 = t + i(E_1 + E_2)\Theta + iE_2E_1. \end{gathered} \tag{7.2}$$

In other words, the composition of two supertransformations characterized by the parameters $E_1$ and $E_2$ is a supertransformation with the parameter $E_1 + E_2$ supplemented by a shift of time by a real Grassmann-even parameter $\alpha = iE_2E_1$. ($\bar{\alpha} = -iE_1E_2 = iE_2E_1 = \alpha$.)

**Theorem 7.1.** *The set of all transformations* (7.1) *supplemented by time shifts $t \to t + \alpha$ and characterized by a Grassmann-even parameter $\alpha$ and a Grassmann-odd parameter $E$ form a group.*

**Proof.** We leave it to the reader. □

Such groups characterized by both Grassmann-even and Grassmann-odd parameters and acting on superspaces are called *supergroups*.

Calling an element of this supergroup $g(\alpha, E)$, the law (7.1) can be rewritten as $g(0, E_2)g(0, E_1) = g(iE_2E_1, E_2 + E_1)$. We see that this group is not commutative. The group commutator

$$g(0, E_2)g(0, E_1)g(0, -E_2)g(0, -E_1) = g(2iE_2E_1, 0): \quad \begin{cases} \Theta \to \Theta \\ t \to t + 2iE_2E_1 \end{cases}$$

is not trivial and amounts to a pure time shift.

A group element is the exponential of a linear combination of its generators. In our case, we may write

$$g(\alpha, E) = \exp\{-i\alpha\hat{H} + E\hat{\mathcal{Q}})\}, \tag{7.3}$$

where

$$\hat{H} = i\partial/\partial t \tag{7.4}$$

is the generator of time shifts, i.e. the Hamiltonian, and

$$\boxed{\hat{\mathcal{Q}} = \frac{\partial}{\partial \Theta} + i\Theta \frac{\partial}{\partial t}} \tag{7.5}$$

is none other than the *supercharge*! Indeed, the relations

$$\hat{\mathcal{Q}}^2 = \hat{H} \quad \text{and hence} \quad [\hat{\mathcal{Q}}, \hat{H}] = 0 \tag{7.6}$$

(the $\mathcal{N} = 1$ superalgebra) hold. This algebra, a simplified version of (5.3), is called the $\mathcal{N} = 1$ supersymmetry algebra. Equations (7.4) and (7.5) provide a realization of this algebra via differential operators in one-dimensional superspace.

By the same token as the symmetry under time shifts brings about, for a particular dynamical system, a particular classical Hamiltonian and its quantum counterpart, the symmetry under the $\mathcal{N} = 1$ supersymmetry transformations (7.1) brings about, by Noether's theorem, a particular classical conserving supercharge and its quantum counterpart. The quantum Hamiltonian and the supercharge thus derived (we see in many examples how this general procedure works in the third part of our book) must obey the same algebra as the generators (7.4) and (7.5).

If the Hamiltonian $\hat{H}$ acting in the Hilbert space of our quantum system is positive definite (and, if the spectrum of the Hamiltonian is bounded from below, which we assume, one can always make it positive definite by adding an appropriate constant), the algebra (7.6) dictates that the spectrum of $\hat{\mathcal{Q}}$ is real, i.e. $\hat{\mathcal{Q}}$ is Hermitian. This applies e.g. to the operator $\boldsymbol{\sigma} \cdot (\hat{\boldsymbol{P}} - e\boldsymbol{A}/c)$ whose square gives the Pauli Hamiltonian (5.14).

Now, we saw in Chapter 5 that the algebra (7.6) treated as a symmetry algebra of a dynamical system is too small and too poor to entail nontrivial dynamical consequencies. We are still discussing in detail the $\mathcal{N} = 1$ case by two reasons:

(1) It is the simplest possible superspace construction. Having understood this, it will be easy to understand everything that follows.
(2) Though $\mathcal{N} = 1$ dynamical systems not involving some extra symmetries are devoid of nontrivial physical content, $\mathcal{N} = 1$ superfields represent a convenient technical tool that is useful to keep in theorist's kit. For example, $\mathcal{N} = 1, D = 1$ superfields were used in Ref. [77] to describe one-dimensional supergravity. We use this tool in this chapter to give an alternative superfield description of Landau's problem, then in Chapter 9 to describe in the supersymmetric language generic complex geometry and in Chapter 13 to write a generic bi-HKT SQM

action[1] and explore the conditions when it is reduced to the HKT one (see Theorems 13.2 and 13.4).

To derive in the superspace approach particular Lagrangians (and then the corresponding Hamiltonians and supercharges) as functions of dynamical variables, we have to introduce supervariables, alias superfields.

**Definition 7.2.** A $D = 1$ $\mathcal{N} = 1$ superfield is a function

$$\boxed{\mathcal{X}(t, \Theta) = x(t) + i\Theta\Psi(t).} \tag{7.7}$$

The ordinary dynamical variable $x(t)$ and the Grassmann dynamical variable $\Psi(t)$ are its *bosonic* and *fermionic* components.[2]

The transformations (7.1) of the superspace coordinates induce the transformation

$$\delta\mathcal{X} = E\hat{Q}\mathcal{X} \tag{7.8}$$

or in components:

$$\begin{aligned} \delta x(t) &= iE\Psi(t), \\ \delta\Psi(t) &= -E\dot{x}(t). \end{aligned} \tag{7.9}$$

Generically, $x(t)$ and $\Psi(t)$ may be complex, but one can also impose a constraint $\mathcal{X} = \bar{\mathcal{X}}$ which makes them real[3]. This constraint is compatible with the transformation law (7.9). One can say that the real superfield $\mathcal{X}(t, \Theta)$ represents an *irreducible representation* of the superalgebra (7.6).[4]

**Theorem 7.2.** *Let $\mathcal{X}_M$ be a set of real superfields that all vanish at $t = \pm\infty$. The functional*[5]

$$S = \int d\Theta \int_{-\infty}^{\infty} dt\, F(\mathcal{X}_M) \tag{7.10}$$

*is invariant under the transformations* (7.1).

---

[1]A reader should wait till Chapter 10 to learn what "bi-HKT" means.

[2]We keep using this physical terminology, though one cannot really talk about bosons and fermions in mechanical systems; one should go over to field theory for that.

[3]That is why we have chosen the capital $\Psi$ for the Grassmann component of $\mathcal{X}$ — to distinguish it from the complex fermion variables entering the expansion (7.25), etc. Likewise, we use Greek capitals for the $\mathcal{N} = 1$ superspace coordinate $\Theta$ and the $\mathcal{N} = 1$ transformation parameter $E$.

[4]Note that we are dealing here with an infinite-dimensional representation. Finite-dimensional representations of the superalgebras were discussed in Chapter 5: a doublet of denenerate states for $\mathcal{N} = 2$ algebra, a quartet of denenerate states for $\mathcal{N} = 4$ algebra, etc.

[5]The Berezin integral $\int d\Theta$ entering (7.10) was defined in Eq. (4.46).

**Proof.** If $\mathcal{X}_M$ are superfields, so is $F(\mathcal{X}_M)$ — it also transforms as in (7.8). We derive

$$\delta S = E \int d\Theta dt \left( \frac{\partial F}{\partial \Theta} + i\Theta \frac{\partial F}{\partial t} \right) = 0. \tag{7.11}$$

Indeed, the second term includes a time integral of a total time derivative and vanishes. The first term vanishes too due to the property (4.48). □

Suppose that we have only one real superfield $\mathcal{X}$ at our disposal. Substituting (7.7) into (7.10) and integrating over $d\Theta$, using (4.47), we express the functional (7.10) in components:

$$S = i \int F'[x(t)]\, \Psi(t)\, dt. \tag{7.12}$$

This expression involves no time derivatives and is too simple to play the role of an action functional of a dynamical system. More interesting expressions can be derived if we introduce the operator

$$\mathcal{D} = \frac{\partial}{\partial \Theta} - i\Theta \frac{\partial}{\partial t} \tag{7.13}$$

called the *supersymmetric covariant derivative*. It has a nice property:

**Theorem 7.3.** *If $\mathcal{X}$ is a real superfield, $i\mathcal{D}\mathcal{X}$ is also a real superfield — it is real and transforms in the same way as $\mathcal{X}$.*

**Proof.** The operators (7.5) and (7.13) anticommute. The infinitesimal variation of $\mathcal{X}$ under (7.1) is by definition $\delta\mathcal{X} = E\hat{Q}\mathcal{X}$. The variation of $\mathcal{D}\mathcal{X}$ is then

$$\delta(\mathcal{D}\mathcal{X}) = \mathcal{D}(\delta\mathcal{X}) = \mathcal{D}(E\hat{Q}\mathcal{X}) = E\hat{Q}(\mathcal{D}\mathcal{X}) \tag{7.14}$$

— the same law as for $\mathcal{X}$. The reality of $i\mathcal{D}\mathcal{X} = -\Psi + \Theta\dot{x}$ is manifest.[6] □

[6] A distinguishing feature of $i\mathcal{D}\mathcal{X}$ is, however, the Grassmann nature of its lowest $\Theta$-independent component. In fact, besides the bosonic superfields (7.7), one can also consider the Grassmann superfields of the type

$$\Phi(t, \Theta) = \phi(t) + \Theta F(t), \tag{7.15}$$

where $\phi(t)$ is a real fermionic and $F(t)$ a real bosonic variable. We meet such superfields in Chapters 8 and 10.

The time derivative of $\mathcal{X}$ is also a superfield, of course. Algebraically, that follows from $[\hat{\mathcal{Q}}, \hat{H}] = 0$ or else from the relation $\mathcal{D}^2 = -i\partial/\partial t$ and from the theorem just proven. Thus, the functional[7]

$$S = \int d\Theta dt\, F(\mathcal{X}_M, \dot{\mathcal{X}}_M, \mathcal{D}\mathcal{X}_M) \tag{7.16}$$

is a supersymmetric invariant. The simplest example giving a truly *dynamical* system is

$$\boxed{S = \frac{i}{2}\int d\Theta dt\, \dot{\mathcal{X}}\mathcal{D}\mathcal{X}.} \tag{7.17}$$

Substituting there (7.7) and integrating by $d\Theta$, we obtain the system with the Lagrangian

$$L = \frac{\dot{x}^2}{2} - \frac{i}{2}\dot{\Psi}\Psi. \tag{7.18}$$

Its bosonic part describes free motion along the $x$-axis.

## 7.2 $\mathcal{N} = 2$

Consider now a superspace including $t$ and a *complex* Grassmann variable $\theta$. The supersymmetry transformations are now defined as

$$\begin{aligned} &\theta \to \theta + \epsilon, \\ &\bar{\theta} \to \bar{\theta} + \bar{\epsilon}, \\ &t \to t + i(\epsilon\bar{\theta} - \theta\bar{\epsilon}) \end{aligned} \tag{7.19}$$

with complex Grassmann $\epsilon$. An element of the supergroup is convenient to write as

$$g(\alpha, \epsilon, \bar{\epsilon}) = \exp\{-i\alpha\hat{H} + \sqrt{2}(\epsilon\hat{Q} + \bar{\epsilon}\hat{\bar{Q}})\}, \tag{7.20}$$

where

$$\boxed{\hat{Q} = \frac{1}{\sqrt{2}}\left(\frac{\partial}{\partial\theta} + i\bar{\theta}\frac{\partial}{\partial t}\right), \quad \hat{\bar{Q}} = \frac{1}{\sqrt{2}}\left(\frac{\partial}{\partial\bar{\theta}} + i\theta\frac{\partial}{\partial t}\right)} \tag{7.21}$$

and $\hat{H}$ is still $i\partial/\partial t$.

---

[7]We will not display the limits for the time integral anymore. Heretofore, speaking about the invariance of the Lagrangian, we will always mean *invariant up to a total derivative* and, speaking about the invariance of the action, we will always mean *invariant up to a boundary term*.

The supercharges (7.21) satisfy the algebra (5.1), which means that we are on the right track to describe in the superspace approach the supersymmetric models of the previous chapter.

We also introduce the covariant derivatives

$$D = \frac{\partial}{\partial \theta} - i\bar{\theta}\frac{\partial}{\partial t}, \quad \bar{D} = -\frac{\partial}{\partial \bar{\theta}} + i\theta\frac{\partial}{\partial t}. \tag{7.22}$$

$D$ and $\bar{D}$ are nilpotent and anticommute with both $\hat{Q}$ and $\hat{\bar{Q}}$.

A generic $\mathcal{N} = 2$ superfield reads

$$Z = z + \theta\psi + \bar{\theta}\chi + \theta\bar{\theta}F, \tag{7.23}$$

where $z$ and $F$ are complex bosonic and $\psi$ and $\chi$ are complex fermionic variables. Note that the number of bosonic and fermionic variables in the expansion (7.23) is equal.

**Theorem 7.4.** *The latter property holds for all superfields.*

**Proof.** Consider an arbitrary superfield $\Phi(t, \theta_\alpha)$ depending on $t$ and several real Grassmann coordinates $\theta_{\alpha=1,\ldots,n}$. It is clear that its $\theta$ expansion gives altogether $2^n$ components ($2^{n+1}$ real components if $\Phi$ is complex). A half of them multiply Grassmann-even monomials $\theta_{\alpha_1} \cdots \theta_{\alpha_{2k}}$ and another half multiply Grassmann-odd monomials. □

As was the case for $\mathcal{N} = 1$, the supertransformations (7.19) induce the supertransformations of the superfields that can be expressed in a form:

$$\delta Z = \sqrt{2}(\epsilon\hat{Q} + \bar{\epsilon}\hat{\bar{Q}})Z. \tag{7.24}$$

However, (7.23) is not an irreducible representation of the superalgebra (5.1). Irreducible ones can be constructed by imposing certain constraints on $Z$. To begin with, we can impose the constraint $Z = \bar{Z}$. This gives a real $\mathcal{N} = 2$ superfield:

$$\boxed{X = x + \theta\psi + \bar{\psi}\bar{\theta} + \theta\bar{\theta}F} \tag{7.25}$$

with real $x, F$. A real superfield stays real after a supertransformation (7.24).

The superfield $X$ has the components at the three levels: (i) the $\theta$-independent bosonic component $x$, (ii) the fermion components $\psi$ and $\bar{\psi}$ that appear in the linear terms of the $\theta$ expansion and (iii) the bosonic component $F$ in the quadratic term. Using the convenient notation suggested in Ref. [78], we can attribute the notation (**1**, **2**, **1**) to this superfield, where the numbers count the real degrees of freedom in the $\theta$-independent, linear and quadratic sectors.[8]

**Theorem 7.5.** *The superfield* (7.25) *is irreducible. One cannot impose on* $X$ *additional constraints compatible with the supersymmetry.*

**Proof.** It follows from the fact that any superfield representation of the $\mathcal{N} = 2$ supersymmetry must have at least 2 bosonic and 2 fermionic components. Indeed, take any bosonic component and act on it by two different real supercharges available in the algebra; you obtain two different fermionic components. And starting from a fermionic component, we obtain two different bosonic ones. The superfield (7.25) saturates this minimum, and the number of components cannot be reduced further. □

**Remark:** The same concerns the chiral superfields (7.28) and (7.29) discussed in the following. They are also irreducible.

The supertransformations of the components of (7.25) induced by (7.24) read

$$\begin{aligned}
\delta x &= \epsilon\psi - \bar{\epsilon}\bar{\psi}, \\
\delta\psi &= \bar{\epsilon}(F - i\dot{x}), \\
\delta\bar{\psi} &= \epsilon(F + i\dot{x}), \\
\delta F &= -i(\epsilon\dot{\psi} + \bar{\epsilon}\dot{\bar{\psi}}).
\end{aligned} \tag{7.26}$$

One can observe the pattern: the variation of a certain component of $X$ involves a higher component ($\psi$ and $\bar{\psi}$ for $\delta x$, $F$ for $\delta\psi$ and $\delta\bar{\psi}$) and the time derivative of a lower component. This pattern also holds for more complicated superfields to be considered later. In particular, the supersymmetric variation of the highest superfield component is always a total time derivative.

For $\mathcal{N} = 1$, the only irreducible bosonic superfield was the real superfield (7.7). But the case $\mathcal{N} = 2$ is more rich, and besides the real superfield (7.25), there exist also two other irreducible infinite-dimensional representations of the supersymmetry algebra.

**Definition 7.3.** A *holomorphic chiral* $\mathcal{N} = 2$ *superfield* $Z$ *is a generic superfield* (7.23) *satisfying the constraint* $\bar{D}Z = 0$. *An antiholomorphic*

[8] And the $\mathcal{N} = 1$ superfield (7.7) is, obviously, $(\mathbf{1},\ \mathbf{1})$ in these terms!

*chiral* $\mathcal{N} = 2$ superfield $\bar{Z}$ is a generic superfield (7.23) satisfying the constraint[9] $D\bar{Z} = 0$.

Bearing in mind that $\{\bar{D}, \hat{Q}\} = \{\bar{D}, \hat{\bar{Q}}\} = 0$, one immediately sees that the variation (7.24) is holomorphic chiral if $Z$ is holomorphic chiral. The same is true for antiholomorphic fields. The condition $\bar{D}Z = 0$ kills a half of the degrees of freedom. A generic solution of this constraint is[10]

$$\boxed{Z = z + \sqrt{2}\,\theta\psi - i\theta\bar{\theta}\dot{z}.} \tag{7.27}$$

We see that the term $\propto \bar{\theta}$ is now absent and the term $\propto \theta\bar{\theta}$ is not independent any more but is given by the time derivative of the lowest term.

A convenient technical trick is to introduce a "holomorphic time"

$$t_L = t - i\theta\bar{\theta}.$$

It transforms under supersymmetry as $t_L \to t_L - 2i\theta\bar{\epsilon}$ and does not "feel" the presence of $\bar{\theta}$. The covariant derivative $\bar{D}$ reduces in this frame simply to $-\partial/\partial\bar{\theta}$ and the solutions to the equation $\bar{D}Z = 0$ have no dependence on $\bar{\theta}$:

$$Z = Z(t_L, \theta) = z(t_L) + \sqrt{2}\,\theta\psi(t_L). \tag{7.28}$$

Similarly,

$$\bar{Z} = \bar{Z}(t_R, \bar{\theta}) = \bar{z}(t_R) - \sqrt{2}\,\bar{\theta}\bar{\psi}(t_R) \tag{7.29}$$

with $t_R = \overline{t_L} = t + i\theta\bar{\theta}$.

**Definition 7.4.** The set of coordinates $(t_L, \theta)$ describes the *holomorphic chiral* $\mathcal{N} = 2$ *superspace* and the set $(t_R, \bar{\theta})$ describes the *antiholomorphic chiral* $\mathcal{N} = 2$ *superspace.*

The lowest bosonic term in the expansion (7.28) or (7.29) includes two real degrees of freedom, the linear in $\theta$ or $\bar{\theta}$ term gives two real fermionic degrees of freedom, and there is no quadratic term. These are **(2, 2, 0)** superfields.

---

[9]In the slang used by the physicists, these fields are called "left chiral" and "right chiral". This terminology came from field theories, where the words "left" and "right" have a physical meaning: the particle representing an elementary excitation of a chiral field is not symmetric under mirror reflection, like a left or right hand (the word *chiral* comes from the Greek root χέρι meaning "hand"). But the notions of left and right make, of course, little sense for mechanical systems, and we try not to use these words, allowing, however, for their traces in the notation $t_L$ and $t_R$ below.

[10]The factor $\sqrt{2}$ is a convenient convention.

The supertransformations of the components induced by (7.24) read

$$\delta z = \sqrt{2}\,\epsilon\psi, \quad \delta\psi = -i\sqrt{2}\,\bar{\epsilon}\dot{z}$$
$$\delta\bar{z} = -\sqrt{2}\,\bar{\epsilon}\bar{\psi}, \quad \delta\bar{\psi} = i\sqrt{2}\,\epsilon\dot{\bar{z}}. \tag{7.30}$$

Besides the real superfield (7.25) and the chiral superfields (7.28), (7.29), there are no other irreducible infinite-dimensional representations of the $\mathcal{N} = 2$ supersymmetry algebra. This is a nontrivial statement. As we learn in Chapter 9 (see the discussion after Theorem 9.4), it is in fact a corollary of Theorem 2.6 — the Newlander–Nirenberg theorem. Note that the transformations (7.30) coincide up an irrelevant factor $\sqrt{2}$ with the supertransformations (5.53) of the variables describing the Landau problem!

To derive its Lagrangian in the superspace approach, we need to write a superfield action. According to a generalization of Theorem 7.2 for $\mathcal{N} = 2$ (there are similar generalizations for higher $\mathcal{N}$), any full superspace integral of a generic superfield is a supersymmetric invariant. For the action to be real, this superfield should be real too.[11]

There are many ways to cook up a real superfield out of the chiral superfield (7.27) that we have at our disposal. One can simply multiply $Z$ by $\bar{Z}$ and the product $Z\bar{Z}$ is obviously real. But to obtain a nontrivial dynamics, we should also get the covariant derivatives $D$ and $\bar{D}$ out of our tool kit. As was also the case for $\mathcal{N} = 1$, they make a superfield out of a superfield. But the nature of the superfield changes: due to the nilpotency of $D$ and $\bar{D}$, $DZ$ is antiholomorphic chiral and $\bar{D}\bar{Z}$ is holomorphic chiral for any $Z$. One can be convinced that the action[12]

$$\boxed{S = \int d\bar{\theta}d\theta dt \left(\frac{1}{4}\bar{D}\bar{Z}\,DZ - \frac{B}{2}Z\bar{Z}\right)} \tag{7.31}$$

does the job and the component Lagrangian derived from it coincides up to a total derivative with (5.52).

Equations (5.52) and (7.31) are two different ways of representing the same supersymmetric system. In the first case, the Lagrangian is expressed

[11] We hasten to comment that there is an alternative: one can write the integral of a chiral superfield over the chiral superspace and add the complex conjugate, as we as we will do in Eq. (7.50) later. But one would not obtain the Landau Hamiltonian in this way.

[12] Note that the integral $\int d\bar{\theta}d\theta\, X$ coincides with the highest component of the superfield (7.25). This is how the construction of supersymmetric actions is described in the textbook [79]: construct a real superfield out of the basic superfields in your kit and then pick up the highest component in its $\theta$ expansion.

in components, and to show that it is supersymmetric, one has to check the invariance of the action under the transformations (5.53) explicitly. On the other hand, supersymmetry of the action (7.31) is a corollary of Theorem 7.2 (its generalization for $\mathcal{N} = 2$). The law of transformations of components follows from the transformation law (7.19) of the superspace coordinates. One may but does not need to check anything. The superspace approach gives one, if you will, an "industrial" method to write supersymmetric Lagrangians without much effort, like the calculus deleloped by Newton and Leibniz allowed one to calculate the volume of a ball — the problem first solved by Archimedes by an ingenious geometric construction — in two lines.

Besides the component description and $\mathcal{N} = 2$ description, Landau's problem also admits a third intermediate way of its description. We pose

$$z = \frac{x_1 + ix_2}{\sqrt{2}}, \quad \psi = \frac{\Psi_1 + i\Psi_2}{\sqrt{2}}, \quad \text{and} \quad \theta = i\frac{\Theta_1 + i\Theta_2}{\sqrt{2}} \tag{7.32}$$

with real $\Psi_{1,2}, \Theta_{1,2}$ to express (7.27) and its conjugate in the form

$$\begin{aligned} Z &= \frac{1}{\sqrt{2}}(1 + i\Theta_2\mathcal{D}_1)(\mathcal{X}_1 + i\mathcal{X}_2), \\ \bar{Z} &= \frac{1}{\sqrt{2}}(1 - i\Theta_2\mathcal{D}_1)(\mathcal{X}_1 - i\mathcal{X}_2), \end{aligned} \tag{7.33}$$

where $\mathcal{X}_{1,2} = x_{1,2} + i\Theta_1\Psi_{1,2}$ are $\mathcal{N} = 1$ superfields and $\mathcal{D}_1$ is the $\mathcal{N} = 1$ covariant derivative (7.13).

Replacing in (7.31) $d\bar{\theta}d\theta$ by[13] $-id\Theta_1 d\Theta_2$, integrating it over $d\Theta_2$ and trading $\Theta_1$ for $\Theta$, we may express the action as

$$S = \frac{i}{2}\int d\Theta dt\, \dot{\mathcal{X}}_M \mathcal{D}\mathcal{X}_M - \frac{iB}{2}\varepsilon_{MN}\int d\Theta dt\, \mathcal{X}_M \mathcal{D}\mathcal{X}_N. \tag{7.34}$$

Going down to components, we obtain

$$L = \frac{1}{2}\dot{x}_M^2 - \frac{i}{2}\dot{\Psi}_M\Psi_M + \frac{B}{2}\varepsilon_{MN}(\dot{x}_M x_N + i\Psi_M\Psi_N), \tag{7.35}$$

which is equivalent to (5.52).

In the expression (7.34), $\Theta$ is real, and the transformation (7.9) of the $\mathcal{N} = 1$ superfield components gives *one* of two real supersymmetry transformations that appear in (7.30): $E = \sqrt{2}\,\mathrm{Im}(\epsilon)$. And *another* transformation

[13]Note that, if $\theta = i(\Theta_1 + i\Theta_2)/\sqrt{2}$, then we must define $d\theta = -i(d\Theta_1 - id\Theta_2)/\sqrt{2}$ to harmonize the properties $\int d\theta\, \theta = 1$ and $\int d\Theta_{1,2}\, \Theta_{1,2} = 1$. Also, $D = -i(\mathcal{D}_1 - i\mathcal{D}_2)/\sqrt{2}$.

with the real Grassmann parameter $\eta$ (the real part of $\sqrt{2}\,\epsilon$) is realized as

$$\delta \mathcal{X}_M = \eta\, \varepsilon_{MN} \mathcal{D}\mathcal{X}_N. \tag{7.36}$$

One can be explicitly convinced that (7.36) leaves the action (7.34) invariant, using $\mathcal{D}^2 = -i\partial/\partial t$.

Rather often, such a mixed description, when a part of supersymmetries is realized by superspace transformations and another part as a transformation of superfields of low rank, turns out convenient. We meet many such constructions later.

We concentrated in the last couple of pages on the Landau problem with a homogeneous magnetic field. But one could also replace $-BZ\bar{Z}/2$ in (7.31) by any function $\mathcal{H}(\bar{Z}, Z)$ to obtain a system with a nonhomogeneous magnetic field

$$B(z, \bar{z}) = -2\frac{\partial^2 \mathcal{H}}{\partial z \partial \bar{z}}. \tag{7.37}$$

The Hamiltonian of this system was written before in Eq. (6.66).

To derive the Lagrangian (5.48) for the Witten model, one has to take the real $\mathcal{N} = 2$ superfield (7.25) and write

$$\boxed{\begin{aligned} L &= \int d\bar{\theta} d\theta \left[\frac{1}{2}\bar{D}X DX - W(X)\right] \\ &= \frac{1}{2}(\dot{x}^2 + F^2) - i\dot{\psi}\bar{\psi} - W''(x)\psi\bar{\psi} - FW'(x). \end{aligned}} \tag{7.38}$$

This resembles (5.48) but does not coincide with it completely: the Lagrangian $L$ depends on an extra variable $F$, which was absent in (5.48). One can note, however, that $F$ is an *auxiliary* nondynamical variable: it enters the Lagrangian without derivatives and the corresponding equation of motion is simply

$$\frac{\partial L}{\partial F} = F - W'(x) = 0. \tag{7.39}$$

It has a trivial solution $F = W'(x)$. If substituting this solution in (7.38), we reproduce (5.48). When doing the same for the supertransformation (7.26), we reproduce (5.49).

One can mention in parentheses at this point that $F$ is in this case auxiliary and can be algebraically excluded for the Lagrangian (7.38), but if one writes a more complicated action including extra time derivative, like

$$S = \int d\bar{\theta} d\theta dt \, DX\bar{D}\dot{X}, \tag{7.40}$$

the component Lagrangian *would* depend on $\dot{F}$ [58].

A final comment in this section concerns the supersymmetric oscillator. In the previous chapter, we described it first using the holomorphic bosonic variables $a, \bar{a}$ and then noted that the same system, formulated in the ordinary coordinate representation, has the Hamiltonian (5.29) and represents a particular case of Witten's model. It is this second formulation which can also be derived in the superspace approach. We do not know how to directly derive in this way the Lagrangian (5.46) involving bosonic holomorphic variables. The problem is that the supersymmetric transformations (5.47) do not involve time derivatives, as is always the case for the transformation law of superfield components.

## 7.3 $\mathcal{N} = 4$

Consider the superspace including $t$ and *two* complex Grassmann variables $\theta_j$ and their conjugates $\bar{\theta}^j$. They transform under supersymmetry as

$$\begin{aligned} \theta_j &\to \theta_j + \epsilon_j, \\ \bar{\theta}^j &\to \bar{\theta}^j + \bar{\epsilon}^j, \\ t &\to t + i(\epsilon_j \bar{\theta}^j + \bar{\epsilon}^j \theta_j). \end{aligned} \tag{7.41}$$

The generators of these transformations

$$\begin{aligned} \hat{Q}^j &= \frac{1}{\sqrt{2}} \left( \frac{\partial}{\partial \theta_j} + i\bar{\theta}^j \frac{\partial}{\partial t} \right), \\ \hat{\bar{Q}}_j &= \frac{1}{\sqrt{2}} \left( \frac{\partial}{\partial \bar{\theta}^j} + i\theta_j \frac{\partial}{\partial t} \right) \end{aligned} \tag{7.42}$$

satisfy the extended supersymmetry algebra

$$\{Q^j, Q^k\} = \{\bar{Q}_j, \bar{Q}_k\} = 0, \quad \{Q^j, \bar{Q}_k\} = i\delta^j_k \frac{\partial}{\partial t}, \tag{7.43}$$

as in (5.36). The covariant derivatives

$$\begin{aligned} D^j &= \frac{\partial}{\partial \theta_j} - i\bar{\theta}^j \frac{\partial}{\partial t}, \\ \bar{D}_j &= -\frac{\partial}{\partial \bar{\theta}^j} + i\theta_j \frac{\partial}{\partial t} \end{aligned} \tag{7.44}$$

satisfy a similar algebra,

$$\{D^j, D^k\} = \{\bar{D}_j, \bar{D}_k\} = 0, \quad \{D^j, \bar{D}_k\} = 2i\delta^j_k \frac{\partial}{\partial t}, \tag{7.45}$$

and anticommute with all the supercharges.

A generic complex $\mathcal{N} = 4$ superfield includes many terms in its $\theta$ expansion (up to $\sim \theta^2\bar{\theta}^2$) and 16 + 16 real bosonic and fermionic components. This representation of the algebra (5.36) is far from being irreducible. In $\mathcal{N} = 1$ superspace, we had only one superfield (7.7). In $\mathcal{N} = 2$ superspace, there are irreducible superfields of two types: real and chiral. In $\mathcal{N} = 4$ superspace, there are several types of irreducible superfields.

### 7.3.1 *Superfield* (2, 4, 2)

The direct generalizations of the $\mathcal{N} = 2$ chiral superfields discussed in the previous section are $\mathcal{N} = 4$ chiral superfields.

**Definition 7.5.** A holomorphic chiral $\mathcal{N} = 4$ superfield $\mathcal{Z}$ is a generic $\mathcal{N} = 4$ superfield satisfying the constraint $\bar{D}_j\mathcal{Z} = 0$. An antiholomorphic chiral field $\bar{\mathcal{Z}}$ satisfies the constraint $D^j\bar{\mathcal{Z}} = 0$.

If we introduce the holomorphic time

$$t_L = t - i\theta_j\bar{\theta}^j, \tag{7.46}$$

the constraint $\bar{D}_j\mathcal{Z} = 0$ becomes

$$\partial\mathcal{Z}/\partial\bar{\theta}^j\big|_{\text{fixed } t_L} = 0.$$

In this frame, the holomorphic chiral superfield does not depend on $\bar{\theta}^j$ and has only three terms in its expansion:

$$\boxed{\mathcal{Z}(t_L, \theta_j) = z(t_L) + \sqrt{2}\,\theta_j\psi^j(t_L) + 2\,\theta_1\theta_2 F(t_L).} \tag{7.47}$$

By the same token, an antiholomorphic chiral superfield can be represented as[14]

$$\bar{\mathcal{Z}}(t_R, \bar{\theta}^j) = \bar{z}(t_R) + \sqrt{2}\,\bar{\theta}^j\bar{\psi}_j(t_R) + 2\,\bar{\theta}^2\bar{\theta}^1\bar{F}(t_R), \tag{7.48}$$

where

$$t_R = \overline{t_L} = t + i\theta_j\bar{\theta}^j, \tag{7.49}$$

[14]Recall that the indices are raised and lowered according to (5.37). The complex conjugate of $\psi_j$ is denoted as $\bar{\psi}^j$ — in the same way as for $\theta_j$. It follows that $\bar{\theta}_j = -\overline{\theta^j}$ and $\bar{\psi}_j(t_R) = -\overline{\psi^j(t_L)}$. Note that $A_jB^j = -A^jB_j$.

Now, $z(t_L)$ has two real bosonic components, there are also two real bosonic components in the quadratic term $\sim \theta^2$, and the linear in $\theta$ term provides us with four real fermionic components. The superfields (7.47) and (7.48) have the type $(\mathbf{2}, \mathbf{4}, \mathbf{2})$.

The field (7.47) lives in the holomorphic chiral $\mathcal{N} = 4$ superspace and the field (7.48) in the antiholomorphic chiral $\mathcal{N} = 4$ superspace.

Consider the system

$$\boxed{\begin{aligned} S = \frac{1}{4}\int d^2\bar{\theta}d^2\theta dt\, \bar{\mathcal{Z}}(t_R, \bar{\theta})\, \mathcal{Z}(t_L, \theta) \\ + \frac{1}{2}\int d^2\theta dt_L\, W(\mathcal{Z}) + \frac{1}{2}\int d^2\bar{\theta} dt_R\, \overline{W(\mathcal{Z})}, \end{aligned}} \tag{7.50}$$

with

$$d^2\theta = d\theta_1 d\theta_2, \quad d^2\bar{\theta} = d\bar{\theta}^2 d\bar{\theta}^1. \tag{7.51}$$

$W(\mathcal{Z})$ is an arbitrary holomorphic function. The first term in (7.50) involves the integral over the whole superspace and the second and the third terms involve the integrals over the chiral superspaces $(t_L, \theta)$ and $(t_R, \bar{\theta})$. All these integrals are supersymmetric invariants.

Substituting Eqs. (7.46)–(7.49) in this expression, integrating over $d^2\theta$ and $d^2\bar{\theta}$ (doing so for the first term, one has to keep in mind that $\mathcal{Z}$ and $\bar{\mathcal{Z}}$ depend on the different arguments and one should expand $z(t_L) = z - i\theta_j\bar{\theta}^j\dot{z} - (\theta_j\bar{\theta}^j)^2\ddot{z}/2$ and similarly for other variables) and omitting some total time derivatives, we obtain

$$\begin{aligned} L = \dot{z}\dot{\bar{z}} + F\bar{F} + \frac{i}{2}(\bar{\psi}^j\dot{\psi}_j - \dot{\bar{\psi}}^j\psi_j) \\ - FW'(z) - \bar{F}\overline{W'(z)} + W''(z)\psi_1\psi_2 + \overline{W''(z)}\,\bar{\psi}^2\bar{\psi}^1. \end{aligned} \tag{7.52}$$

We see that the variables $F$ and $\bar{F}$ are not dynamical and one can exclude them by substituting the solutions $F = \overline{W'(z)}$, $\bar{F} = W'(z)$ of the corresponding (purely algebraic) equations of motion in the Lagrangian. Doing so and going over to the classical and then quantum Hamiltonian, we derive

$$\hat{H} = \hat{\pi}\hat{\pi}^\dagger + W'(z)\overline{W'(z)} - W''(z)\psi_1\psi_2 - \overline{W''(z)}\,\hat{\bar{\psi}}^2\hat{\bar{\psi}}^1. \tag{7.53}$$

When the superpotential $W(\mathcal{Z}) = \mathcal{Z}^3/3$ is chosen, this Hamiltonian coincides with the Hamiltonian (5.39) (our first example of an extended supersymmetric system). In fact, the system just considered is none other than the SQM version of the *Wess–Zumino* supersymmetric field theory system [80].

To define the chiral $\mathcal{N} = 4$ superfields, we imposed the constraints $\bar{D}_j \mathcal{Z} = 0$ or $D^j \bar{\mathcal{Z}} = 0$. But one could also impose a "mixed" constraint like

$$D^2 \mathcal{Z} = \bar{D}_1 \mathcal{Z} = 0. \tag{7.54}$$

A superfield satisfying the constraint (7.54) can be called a "twisted" or "mirror" chiral superfield. If the action of the model includes only such mirror fields and their complex congugates, it does not really matter — just redefine $\theta_2 \leftrightarrow \bar{\theta}^2$ and $\mathcal{Z}$ becomes a conventional holomorphic chiral superfield. Nontrivial models appear when the action includes *both* ordinary and mirror field. Such models were first constructed in the field theory context in Ref. [81]. Different SQM models including both ordinary and mirror (**2**, **4**, **2**) superfields were studied in Ref. [82]. We discuss them in Chapter 10.

### 7.3.2 *Superfields* (**3**, **4**, **1**) *and* (**4**, **4**, **0**)

The number of real fermionic components in an irreducible $\mathcal{N} = 4$ superfield is equal to 4. The same concerns the bosonic components, but not all of them are always dynamical. In the example that we have just considered, there were 2 dynamical and 2 auxiliary bosonic degrees of freedom. Superfields with a different dynamic-auxiliary decomposition can also be constructed. We discuss two such superfields: the superfield (**3**, **4**, **1**) and the superfield (**4**, **4**, **0**).

Take a generic $\mathcal{N} = 4$ superfield $V$ and impose the condition of reality, $V = \bar{V}$. We are left at this stage with 8 bosonic and 8 fermionic real components. Consider now the superfield:

$$\boxed{\Phi_{jk} = i(D_j \bar{D}_k + D_k \bar{D}_j)V.} \tag{7.55}$$

It satisfies the pseudoreality condition,

$$\overline{\Phi_{jk}} = \varepsilon^{jl} \varepsilon^{kp} \Phi_{lp}, \tag{7.56}$$

and also the constraint

$$D_p \Phi_{jk} + D_j \Phi_{kp} + D_k \Phi_{pj} = 0. \tag{7.57}$$

Inversely, one can show that any symmetric in $j \leftrightarrow k$ complex $\mathcal{N} = 4$ superfield $\Phi_{jk}$ satisfying the constraints (7.56) and (7.57) can be represented in the form (7.55).

A remarkable fact is that $\Phi_{jk}$ does not change under the transformation

$$V \to V + \mathcal{Z} + \bar{\mathcal{Z}}, \tag{7.58}$$

where $\mathcal{Z}$ is a chiral (**2**, **4**, **2**) superfield of the preceding subsection. This follows from $\bar{D}_j\mathcal{Z} = D_j\bar{\mathcal{Z}} = 0$ and $\{D_j, \bar{D}_k\} = 2i\varepsilon_{jk}\,\partial/\partial t$. The chiral superfield $\mathcal{Z}$ involves, as we know, 4 bosonic and 4 fermionic components. To count the independent components in $\Phi_{jk}$, one has to *subtract* these numbers from 8+8 components in $V$. One is thus left with 4+4 components meaning that $\Phi_{jk}$ is an irreducible superfield.

In fact, the transformation (7.58) is nothing but a *supersymmetric gauge transformation.* A reader-physicist who already has some notions about supersymmetry met this transformation (its field theory version) e.g. in Chapter 6 of the standard textbook in supersymmetry [79]. Actually, the one-dimensional superfield V can be obtained [83] by dimensional reduction from the real four-dimensional superfield $V$ discussed there and used to construct supersymmetric gauge theories.

It is thus not accidental that the first example of an SQM system based on the superfield $\Phi_{jk}$ was found (in the component form) as an effective Hamiltonian of a four-dimensional supersymmetric gauge theory put in a small spatial box [84]. Its superfield description was worked out in Refs. [85,86].

To mark out the independent components explicitly, one can use the freedom (7.58) and bring $V$ in the form (the so-called *Wess–Zumino gauge*):

$$V = -\frac{1}{2}A_M\,\bar{\theta}\sigma_M\theta + (\psi\theta + \bar{\psi}\bar{\theta})\bar{\theta}\theta + F\,\theta_1\theta_2\bar{\theta}^2\bar{\theta}^1 \tag{7.59}$$

with real $A_{M=1,2,3}$ and $F$ and complex $\psi_j$. The shorthands in the equation above are spelled out as

$$\bar{\theta}\sigma_M\theta \equiv \bar{\theta}^j(\sigma_M)_j{}^k\theta_k, \quad \bar{\theta}\theta \equiv \bar{\theta}^j\theta_j, \quad \psi\theta \equiv \psi^j\theta_j, \quad \bar{\psi}\bar{\theta} \equiv \bar{\psi}_j\bar{\theta}^j. \tag{7.60}$$

It is now convenient to trade the pseudoreal superfields $\Phi_{jk}$ for the triplet of real superfields[15]

$$\Phi_M = \frac{i}{2}(\sigma_M)^{jk}\Phi_{jk}, \quad \Phi_{jk} = i(\sigma_M)_{jk}\Phi_M \tag{7.61}$$

with $(\sigma_M)^{jk}$ and $(\sigma_M)_{jk}$ given in (3.18). In these terms, the constraint (7.57) acquires the form

$$D_s\Phi_M = \frac{i}{2}\varepsilon_{MNP}(\sigma_P)_s{}^jD_j\Phi_N. \tag{7.62}$$

[15]They correspond to *superconnections* $\Gamma_\mu$ in the four-dimensional theory (see e.g. Ref. [87]). In four dimensions, $\Gamma_\mu$ depends on the gauge, but its one-dimensional counterpart $\Phi_M$ is gauge invariant.

The solution to (7.62) is

$$D_s \Phi_M = (\sigma_M)_{sq} \bar{\Xi}^q \tag{7.63}$$

with arbitrary $\bar{\Xi}^q$. In other words, the constraint (7.57) eliminates the spin 3/2 part in $D_s \Phi_M$ (whatever spin in one dimensions means).

The bar in $\bar{\Xi}^q$ indicates that we are dealing with a pair of antiholomorphic chiral superfields. Antiholomorphicity follows from the definitions (7.55) and (7.61). It is not so much an issue, however. The identity

$$\bar{D}^s \Phi_M = (\sigma_M)^{sq} \Xi_q \tag{7.64}$$

with holomorphic $\Xi_q$ also holds. $\Phi_M$ kind of straddles the middle stretch of the seesaw [88] between $\Xi_q$ and $\bar{\Xi}^q$ — one-dimensional counterparts of the well-known four-dimensional superfields $\mathcal{W}_\alpha$ and $\bar{\mathcal{W}}_{\dot{\alpha}}$.

The component expansion of $\Phi_M$ satisfying the constraint (7.62) reads

$$\begin{aligned} \Phi_M = {} & A_M + \psi\sigma_M\theta - \bar{\theta}\sigma_M\bar{\psi} + \varepsilon_{MNP}\dot{A}_N\,\bar{\theta}\sigma_P\theta + F\,\bar{\theta}\sigma_M\theta \\ & + i(\bar{\theta}\sigma_M\theta)(\dot{\psi}\theta - \dot{\bar{\psi}}\bar{\theta}) + \ddot{A}_M\,\theta_1\theta_2\bar{\theta}^2\bar{\theta}^1, \end{aligned} \tag{7.65}$$

where $\psi\sigma_M\theta = \psi^j(\sigma_M)_j{}^k\theta_k$ and $\bar{\theta}\sigma_M\bar{\psi} = \bar{\theta}^j(\sigma_M)_j{}^k\bar{\psi}_k$. The simplest action

$$\boxed{S = -\frac{1}{12}\int d^2\bar{\theta}d^2\theta dt\,(\Phi_M)^2} \tag{7.66}$$

gives the component Lagrangian

$$L = \frac{1}{2}(\dot{A}_M)^2 + \frac{i}{2}(\bar{\psi}\dot{\psi} - \dot{\bar{\psi}}\psi) + \frac{F^2}{2}. \tag{7.67}$$

We observe the presence of the complex fermion doublet, of three dynamical variables $A_M$ (and the bosonic part of the Lagrangian describes free three-dimensional motion) and of an auxiliary variable $F$, which can be excluded thanks to the equation of motion $F = 0$.

We postpone the discussion of more complicated models based on the superfield $\Phi_M$ to Section 10.2.

To define the irreducible superfield[16] (**4**, **4**, **0**), we take a couple of holomorphic chiral $\mathcal{N} = 4$ superfields $\mathcal{Z}^A$ and impose on top of the constraints $\bar{D}_j\mathcal{Z}^A = 0$ the constraint[17]

$$\boxed{D_j\mathcal{Z}^A - \varepsilon^{AB}\,\bar{D}_j\bar{\mathcal{Z}}^B = 0.} \tag{7.68}$$

---

[16]It was first introduced in Ref. [89]. In our description, we follow the approach of Ref. [82].

[17]Here, the convention $\varepsilon^{12} = 1$ is chosen. In contrast to the Latin indices in Eq. (7.41), etc., the position of the indices $A, B$ plays no role.

This constraint kills a half of the degrees of freedom, and we are left once again with 4 bosonic and 4 fermionic variables. The solution of (7.68) may be presented as

$$\begin{aligned} \mathcal{Z}^A &= Z^A + \theta_2 \varepsilon^{AB} \bar{D}_1 \bar{Z}^B - i\theta_2 \bar{\theta}^2 \dot{Z}^A, \\ \bar{\mathcal{Z}}^A &= \bar{Z}^A - \bar{\theta}^2 \varepsilon^{AB} D^1 Z^B + i\theta_2 \bar{\theta}^2 \dot{\bar{Z}}^A, \end{aligned} \tag{7.69}$$

where $Z^A$ are the holomorphic chiral $\mathcal{N} = 2$ superfields (7.27) living in the subspace $(t; \theta_1, \bar{\theta}^1)$. At the level of components, we have two complex bosonic variables $z^A$ and two complex fermionic variables $\psi^A$.

The simplest invariant action reads

$$S = \frac{1}{8} \int d^2\bar{\theta} d^2\theta dt \, \mathcal{Z}^A \bar{\mathcal{Z}}^A. \tag{7.70}$$

The corresponding component Lagrangian is very simple:

$$L = \dot{z}^A \overline{\dot{z}^A} + \frac{i}{2}(\bar{\psi}^A \dot{\psi}^A - \dot{\bar{\psi}}^A \psi^A). \tag{7.71}$$

It does not involve any auxiliary variables.[18]

Similar to what we had for (**2**, **4**, **2**), one can also define a *mirror* (**4**, **4**, **0**) superfield by interchanging $\theta_2 \leftrightarrow \bar{\theta}^2$ in the constraints.

## 7.4 Harmonic superspace

The goal of this book is to describe geometric structures of the smooth manifolds using supersymmetric methods. And the methods discussed so far have been sufficient to describe (i) generic manifolds with their de Rham complex, (ii) complex manifolds with their Dolbeault complex, (iii) Kähler manifolds and (iv) *Obata-flat* HKT manifolds. In addition, they will allow us to describe some *new* geometrical structures not noted by mathematicians heretofore (see Chapters 10–12). But these methods turn out to be not sufficient to describe in a complete way generic HKT manifolds and hyperkähler manifolds. To do so, we need to learn an elaboration of the superspace technique developed in Dubna — the *harmonic superspace* technique [90].

We saw in the preceding section that the higher the $\mathcal{N}$ in the supersymmetry algebra (5.36), the more intricate is its superfield description. The problem is to find a set of constraints that the superfield should obey to be

[18]The auxiliary variables that were present in the expansion (7.47) of the (**2**, **4**, **2**) superfields $\mathcal{Z}^A$ have been eliminated by the constraints (7.68).

an irreducible representation of (5.36). We managed to do so for $\mathcal{N} = 1$, $\mathcal{N} = 2$ and $\mathcal{N} = 4$, but this direct approach does not work for $\mathcal{N} = 8$. An irreducible $\mathcal{N} = 8$ superfield must include 8 ordinary and 8 Grassmann real components, but a generic $\mathcal{N} = 8$ superfield includes 256+256 components, and a set of constraints in the ordinary $\mathcal{N} = 8$ superspace reducing it to 8+8 is not known. In addition, already at the $\mathcal{N} = 4$ level, we discover in Chapters 10 and 13 that the constraints $\bar{D}_j \mathcal{Z}^A = 0$ supplemented by (7.68) that we used to define a (**4**, **4**, **0**) superfield are too rigid. They are not general enough to describe interesting geometry.

Harmonic superspace was originally invented to describe extended supersymmetric field theories. The adaptation of this technique to SQM systems was worked out in Ref. [91], and we mostly (though not always) follow their conventions. Let us harmonize first the $\mathcal{N} = 4$ superspace and leave the $\mathcal{N} = 8$ superspace for the dessert — the complications that will show up in that case are not so substantial.

The harmonic superspace technique capitalizes on the presence of the $R$-symmetry that rotates complex supercharges and the associated fermion variables. [We discussed it above simultaneously with introducing the extended supersymmetry algebra (5.36).] In our case, the group is $U(2)$. Its $U(1)$ subgroup is present also in the case of the mininal $\mathcal{N} = 2$ supersymmetry. The new relevant part is $SU(2)$. The main idea is to *extend* the $\mathcal{N} = 4$ superspace $(t; \theta_j, \bar{\theta}^j)$ by adding extra commuting coordinates parameterizing this $SU(2)$ group.

For our purposes, the most convenient way of parameterization is introducing a complex doublet $u^{+(j=1,2)}$ of unit length:

$$u^{+j}\overline{u^{+j}} \stackrel{\text{def}}{=} u^{+j} u_j^- = 1. \tag{7.72}$$

One may write an explicit expression of $u_j^+ = \varepsilon_{jk} u^{+k}$ via the three Euler angles:

$$u_j^+ = e^{i\psi/2} \begin{pmatrix} \cos(\theta/2)\, e^{i\phi/2} \\ i \sin(\theta/2)\, e^{-i\phi/2} \end{pmatrix}, \tag{7.73}$$

$0 \le \theta \le \pi, 0 \le \phi \le 2\pi, 0 \le \psi \le 4\pi$. Then,

$$u_j^- = e^{-i\psi/2} \begin{pmatrix} i \sin(\theta/2)\, e^{i\phi/2} \\ \cos(\theta/2)\, e^{-i\phi/2} \end{pmatrix}. \tag{7.74}$$

The matrix

$$U = \begin{pmatrix} u_1^+ & u_1^- \\ u_2^+ & u_2^- \end{pmatrix} \tag{7.75}$$

belongs to $SU(2)$ so that the property

$$u_j^+ u_k^- - u_j^- u_k^+ = \varepsilon_{jk} \tag{7.76}$$

holds.

Introduce now the odd coordinates $\theta^\pm = \theta^j u_j^\pm$, $\bar\theta^\pm = \bar\theta^j u_j^\pm$. They are $SU(2)$ invariants. Note that $\bar\theta^\pm$ are not conjugates of $\theta^\pm$. We have instead

$$\overline{\theta^+} = -\bar\theta^-, \quad \overline{\theta^-} = \bar\theta^+. \tag{7.77}$$

We also use in the following a special *pseudoconjugation* operation — a combination of the ordinary complex conjugation and the following transformation of the harmonics[19]: $u_j^+ \to u_j^-, u_j^- \to -u_j^+$. This transformation acts as

$$\widetilde{\theta^\pm} = \bar\theta^\pm, \quad \widetilde{\bar\theta^\pm} = -\theta^\pm, \quad \widetilde{u_j^\pm} = u^{\pm j}. \tag{7.78}$$

**Definition 7.6.** The $\mathcal{N} = 4$ one-dimensional harmonic superspace is a superspace with the coordinates

$$t;\ \theta^+, \bar\theta^+, \theta^-, \bar\theta^-;\ u_j^\pm. \tag{7.79}$$

The supertransformations of coordinates (7.41) induce the transformations

$$\begin{aligned} t &\to t + i(\epsilon^- \bar\theta^+ - \epsilon^+ \bar\theta^- + \bar\epsilon^+ \theta^- - \bar\epsilon^- \theta^+), \\ \theta^\pm &\to \theta^\pm + \epsilon^\pm, \\ \bar\theta^\pm &\to \bar\theta^\pm + \bar\epsilon^\pm \end{aligned} \tag{7.80}$$

with $\epsilon^\pm = \epsilon^j u_j^\pm$ and $\bar\epsilon^\pm = \bar\epsilon^j u_j^\pm$. A generic harmonic superfield is a function in this superspace:

$$\boxed{\Phi = \phi(t,u) + \ldots + D(t,u)\, \bar\theta^+ \theta^+ \bar\theta^- \theta^-.} \tag{7.81}$$

The supercoordinate transformations (7.80) generate the transformations of the components of (7.81).

---

[19]It may be interpreted as the antipodal reflection on the 2-sphere $SU(2)/U(1)$: $\theta \to \pi - \theta, \phi \to \phi + \pi$, supplemented by the $U(1)$ reflection $\psi \to -\psi$, in the parameterization (7.73).

As in the ordinary superspace, one defines the covariant derivatives. In this case, there are covariant derivatives of two types. First, we have *Grassmann* covariant derivatives:

$$\begin{aligned} D^+ &= \frac{\partial}{\partial\theta^-} - i\bar{\theta}^+\frac{\partial}{\partial t}, \quad D^- = -\frac{\partial}{\partial\theta^+} - i\bar{\theta}^-\frac{\partial}{\partial t}, \\ \bar{D}^+ &= -\frac{\partial}{\partial\bar{\theta}^-} - i\theta^+\frac{\partial}{\partial t}, \quad \bar{D}^- = \frac{\partial}{\partial\bar{\theta}^+} - i\theta^-\frac{\partial}{\partial t}. \end{aligned} \tag{7.82}$$

Second, there are *harmonic* covariant derivatives:

$$\begin{aligned} D^0 &= \partial^0 + \theta^+\frac{\partial}{\partial\theta^+} + \bar{\theta}^+\frac{\partial}{\partial\bar{\theta}^+} - \theta^-\frac{\partial}{\partial\theta^-} - \bar{\theta}^-\frac{\partial}{\partial\bar{\theta}^-}, \\ D^{++} &= \partial^{++} + \theta^+\frac{\partial}{\partial\theta^-} + \bar{\theta}^+\frac{\partial}{\partial\bar{\theta}^-}, \\ D^{--} &= \partial^{--} + \theta^-\frac{\partial}{\partial\theta^+} + \bar{\theta}^-\frac{\partial}{\partial\bar{\theta}^+}, \end{aligned} \tag{7.83}$$

where

$$\partial^0 = u_j^+\frac{\partial}{\partial u_j^+} - u_j^-\frac{\partial}{\partial u_j^-}, \quad \partial^{++} = u_j^+\frac{\partial}{\partial u_j^-}, \quad \partial^{--} = u_j^-\frac{\partial}{\partial u_j^+}. \tag{7.84}$$

They satisfy the $su(2)$ algebra

$$[D^{++}, D^{--}] = D^0, \quad [D^0, D^{\pm\pm}] = \pm 2D^{\pm\pm}, \tag{7.85}$$

the same as the "short" derivatives (7.84). The other nonzero (anti)commutators involving the covariant derivatives are

$$\begin{aligned} &[D^{\pm\pm},\, D^{\mp}] = D^{\pm}, \quad [D^{\pm\pm},\, \bar{D}^{\mp}] = \bar{D}^{\pm} \\ &\{D^-,\, \bar{D}^+\} = -\{D^+,\, \bar{D}^-\} = 2i\frac{\partial}{\partial t}. \end{aligned} \tag{7.86}$$

Note that all the Grassmann covariant derivatives (7.82) anticommute and the harmonic covariant derivatives (7.83) commute with the supercharges — generators of the transformations (7.80):

$$Q^+ = \frac{\partial}{\partial\theta^-} + i\bar{\theta}^+\frac{\partial}{\partial t}, \quad \text{etc.}$$

**Corollary.** If $\Phi$ is a harmonic superfield (7.81), then $D\Phi$ is also a harmonic superfield with the same transformation law for the components, where $D$ is any covariant derivative in (7.82) and (7.83) (cf. Theorem 7.3).

The operator $D^0$ is called the *harmonic charge.* If the parameterization (7.73) for $u_j^+$ is chosen, its short version is expressed as $\partial^0 = -2i\partial/(\partial\psi)$. The variables $u_j^+$ as well as $\theta^+$ and $\bar{\theta}^+$ have the harmonic charge +1. And the variables $u_j^-, \theta^-, \bar{\theta}^-$ carry charge $-1$. All the harmonic superfields that we are going to deal with have a definite integer value of this charge. The operator $D^{++}$ acting on a superfield of harmonic charge $q$ produces a superfield of charge $q+2$. The operator $D^{--}$ acting on a superfield of charge $q$ produces a superfield of charge $q-2$. For any expression $X^q$ carrying a definite harmonic charge $q$, the property

$$\widetilde{\widetilde{X^q}} = (-1)^q X^q \tag{7.87}$$

holds.

The integral $\int dt\, d^2\theta^+ d^2\theta^- \, \Phi$ is invariant under the transformations (7.80). It still depends on $u^\pm$, however, and to get rid of this dependence and to write down an invariant action, one has to integrate over the harmonics, in other words — over the group $SU(2)$ that they parameterize. We normalize the group volume $\int du$ to 1. For any function of harmonics $\Phi^{(q)}$ having a nonzero harmonic charge $q$, the integral $\int \Phi^q(u)\, du$ vanishes. And the integrals like $\int u_j^+ u_k^- \, du$ or $\int u_j^+ u_k^+ u_l^- u_p^- \, du$ must be expressed via the $SU(2)$ invariant tensor $\varepsilon_{jk}$. Using the properties (7.72) and (7.76) and the notrmalization $\int du = 1$, this gives

$$\int u_j^+ u_k^- \, du = \frac{1}{2}\varepsilon_{jk},$$

$$\int u_j^+ u_k^+ u_l^- u_p^- \, du = \frac{1}{6}(\varepsilon_{jl}\varepsilon_{kp} + \varepsilon_{jp}\varepsilon_{kl}), \quad \text{etc.} \tag{7.88}$$

But why is it convenient to extend the $\mathcal{N} = 4$ superspace in such a way? The harmonic superspace described above seems to be a rather complicated construction. Why not stay with the standard $\mathcal{N} = 4$ superspace of the previous section?

The point is that the harmonic superspace approach allows one to find simple constraints to be imposed on (7.81) that single out irreducible representations of the superalgebra. These constrained irreducible superfields are much more handleable than generic superfields (7.81) and also much more useful: we use them in the third part of this book to describe nontrivial geometric structures on the manifolds.

To begin with, we impose the constraints

$$D^+\Phi = \bar{D}^+\Phi = 0. \tag{7.89}$$

The superfields $\Phi$ satisfying this constraint are called the *Grassmann-analytic* or *G-analytic* superfields. They have many common features with the chiral $\mathcal{N} = 2$ superfields (7.28) and (7.29). As we saw, a holomorphic chiral $\mathcal{N} = 2$ superfield lives in fact in the reduced holomorphic chiral superspace that does not include $\bar{\theta}$ but only $\theta$ and the holomorphic time $t_L = t - i\theta\bar{\theta}$. Similarly, a G-analytic superfield lives in fact in the reduced *analytic superspace*

$$(t_A;\ \theta^+, \bar{\theta}^+;\ u) \equiv (\zeta, u) \tag{7.90}$$

including the harmonics, the *analytic time*

$$\boxed{t_A = t + i(\theta^+\bar{\theta}^- + \theta^-\bar{\theta}^+),} \tag{7.91}$$

which is annihilated under the action of $D^+$ or $\bar{D}^+$, and the odd coordinates $\theta^+$ and $\bar{\theta}^+$ but not $\theta^-$ and $\bar{\theta}^-$. As a result, the $\theta$-expansion for the G-analytic superfields is much shorter than for the generic ones. Note that $t_A$ is invariant under pseudoconjugation.

Indeed, in the analytic basis, the Grassmann covariant derivatives acquire the form

$$\begin{aligned} D^+ &= \frac{\partial}{\partial\theta^-}, \quad D^- = -\frac{\partial}{\partial\theta^+} - 2i\bar{\theta}^- \frac{\partial}{\partial t_A}, \\ \bar{D}^+ &= -\frac{\partial}{\partial\bar{\theta}^-}, \quad \bar{D}^- = \frac{\partial}{\partial\bar{\theta}^+} - 2i\theta^- \frac{\partial}{\partial t_A}. \end{aligned} \tag{7.92}$$

And we see that the constraints (7.89) boil down in the analytic basis to the condition that $\Phi$ does not depend on $\theta^-$ and $\bar{\theta}^-$.

The analytic space coordinates transform as

$$\begin{aligned} t_A &\to t_A + 2i(\epsilon^-\bar{\theta}^+ - \bar{\epsilon}^-\theta^+), \\ \theta^+ &\to \theta^+ + \epsilon^+, \\ \bar{\theta}^+ &\to \bar{\theta}^+ + \bar{\epsilon}^+, \end{aligned} \tag{7.93}$$

where $\epsilon^\pm = \epsilon^j u_j^\pm$ and $\bar{\epsilon}^\pm = \bar{\epsilon}^j u_j^\pm$.

These transformations do not include $\theta^-$ or $\bar{\theta}^-$, and hence the analytic superspace (7.90) represents an invariant subspace of the full harmonic superspace (7.79).

The fact that the analytic time is invariant under the pseudoconjugation defined in (7.78) (but not under the ordinary complex conjugation) together with the fact that $\widetilde{\theta^+} = \bar{\theta}^+$ and $\widetilde{\bar{\theta}^+} = -\theta^+$ mean that the pseudoconjugation leaves the whole analytic superspace intact.

The most important for geometric applications object is the G-analytic superfield $q^+(\zeta, u)$ of harmonic charge $+1$. Its $\theta$-expansion is short, as promised (cf. the short expansions (7.28) and (7.29) for the chiral superfields):

$$q^+(\zeta, u) = f^+(t_A, u) + \theta^+ \chi(t_A, u) + \bar{\theta}^+ \kappa(t_A, u) + \theta^+ \bar{\theta}^+ A^-(t_A, u), \tag{7.94}$$

where $f^+(t_A, u)$ and $A^-(t_A, u)$ are Grassmann-even variables of harmonic charge $+1$ and $-1$, respectively, and $\chi(t_A, u)$ and $\kappa(t_A, u)$ are Grassmann-odd variables of zero harmonic charge.

We have decreased the number of components in the $\theta$-expansion, compared to the ordinary (non harmonic) superspace description, but had to pay, as it seems, a heavy price for that: $f^+(t_A, u)$, etc. not only depend on (analytic) time but also represent infinite series in $u^+$ and $u^-$. It is convenient to expand over the symmetrized products of the harmonics. For example,

$$f^+(t_A, u) = f^j(t_A)\, u_j^+ + A^{(ijk)}(t_A)\, u_i^+ u_j^+ u_k^- + \ldots . \tag{7.95}$$

Then, the different terms of this expansion are orthogonal to each other with respect to the measure (7.88). It seems that we have to deal with an infinite number of ordinary and Grassmann time-dependent variables that multiply an infinite number of monomials in these series.

A remarkable fact, however, is that one can eliminate almost all of them by imposing a constraint

$$D^{++} q^+ = 0. \tag{7.96}$$

Indeed, the harmonic derivatives in the analytical basis read

$$\begin{aligned} D^{++} &= \partial^{++} + 2i\theta^+\bar{\theta}^+ \frac{\partial}{\partial t_A} + \theta^+ \frac{\partial}{\partial \theta^-} + \bar{\theta}^+ \frac{\partial}{\partial \bar{\theta}^-}, \\ D^{--} &= \partial^{--} + 2i\theta^-\bar{\theta}^- \frac{\partial}{\partial t_A} + \theta^- \frac{\partial}{\partial \theta^+} + \bar{\theta}^- \frac{\partial}{\partial \bar{\theta}^+}. \end{aligned} \tag{7.97}$$

Expanding the LHS of (7.96) in $\theta$ and $u$, one obtains an infinite number of conditions. And it turns out that this infinity *almost* matches the infinity of the terms in the expansion of $q^+(\zeta, u)$. A general solution to the constraint (7.96) is

$$q^+(\zeta, u) = f^j(t_A) u_j^+ + \theta^+ \chi(t_A) + \bar{\theta}^+ \kappa(t_A) - 2i\theta^+\bar{\theta}^+ \dot{f}^j(t_A) u_j^- . \tag{7.98}$$

We are left with only two complex bosonic variables $f^j(t_A)$ and two complex fermionic variables $\chi(t_A)$ and $\kappa(t_A)$. This set of components matches the set of components of the irreducible $\mathcal{N} = 4$ superfield $(\mathbf{4},\ \mathbf{4},\ \mathbf{0})$ discussed above!

A legitimate question is why have we put so much attention to a rather nontrivial and complicated way to rederive the $(\mathbf{4},\ \mathbf{4},\ \mathbf{0})$ superfield, which we have already derived earlier without breaking so much sweat. The answer is the following. Yes, for the *linear* multiplet $q^+$ satisfying the constraint (7.96), the harmonic description is not at all necessary. All the results coincide with the results that can be derived in a traditional way, which is simpler. But only the harmonic description allows one to treat the so-called *nonlinear* multiplets[20] satisfying the nonlinear constraints [92]

$$\boxed{D^{++} q^{+a} = \mathcal{L}^{+3a}(q^{+b}, u),} \tag{7.99}$$

where $\mathcal{L}^{+3a}$ is a set of $2n$ arbitrary functions carrying harmonic charge $+3$ of $2n$ G-analytic superfields $q^{+b}$ and harmonics.[21] This *is* the way to construct a SQM model that describes a generic HKT manifold of complex dimension $2n$. In addition, only the harmonic superspace description allows one to incorporate in the construction external gauge fields. We continue this discussion in Chapter 13.

Another interesting $\mathcal{N} = 4$ harmonic superfield is the G-analytic superfield $V^{++}(\zeta, u)$ of harmonic charge $+2$. To eliminate higher-order terms in the harmonic expansion, we impose the constraint

$$D^{++} V^{++} = 0. \tag{7.102}$$

To make $V^{++}$ an irreducible representation of the superalgebra, we also require

$$V^{++} = \widetilde{V^{++}}. \tag{7.103}$$

[20]Sometimes, it is the multiplet $(\mathbf{3},\ \mathbf{4},\ \mathbf{1})$ that is called linear. But we reserve the names "linear" and "nonlinear" for the multiplets $(\mathbf{4},\ \mathbf{4},\ \mathbf{0})$ of different kind.

[21]The variables $q^{+a}$ satisfy the additional pseudoreality constraint

$$\widetilde{q^{+a}} = -\Omega_{ab} q^{+b} \equiv -q_a^+, \tag{7.100}$$

where $\Omega_{ab}$ is the symplectic matrix (satisfying $\Omega_{ab} = -\Omega_{ba}, \Omega^2 = -\mathbb{1}$) and $\tilde{}$ is the pseudoconjugation. This constraint is equivalent to the constraint

$$\overline{q^{ja}} = \varepsilon_{jk} \Omega_{ab} q^{kb} \equiv q_{aj}, \tag{7.101}$$

which we already met in Eq. (3.16). Note that the orders of the indices upstairs and downstairs are opposite — the convention that comes from (3.13).

A general solution to the analyticity constraints $D^+V^{++} = \bar{D}^+V^{++} = 0$ and the constraints (7.102) and (7.103) reads [91]

$$V^{++} = v^{(jk)}(t_A)u_j^+u_k^+ + i\theta^+\psi^j(t_A)u_j^+ + i\bar{\theta}^+\bar{\psi}^j(t_A)u_j^+ \\ + i\theta^+\bar{\theta}^+[F(t_A) - 2\dot{v}^{(jk)}(t_A)u_j^+u_k^-], \tag{7.104}$$

where $F$ is a real function of $t_A$ and $v_{jk} = v_{kj}$ is a set of functions satisfying the pseudoreality constraint like in (7.56). As we have seen, such a set can be described by three real independent functions.

We see that the physical component content of the superfield (7.104) is the same as in (7.65). In other words, $V^{++}$ gives a harmonic description of the **(3, 4, 1)** multiplet.

In the harmonic approach, there are two ways to write down invariant actions. Either we integrate a generic superfield $\Phi$ over the full superspace,

$$S = \int dt\, d^2\theta^+ d^2\theta^- du\, \Phi = \int dt\, d\bar{\theta}^+ d\theta^+ d\bar{\theta}^- d\theta^- du\, \Phi, \tag{7.105}$$

or else we can integrate a G-analytic superfield $\Phi_G^{++}$ of double harmonic charge over the analytic superspace[22]:

$$S = \int dt_A\, d\bar{\theta}^+ d\theta^+ du\, \Phi_G^{++}. \tag{7.106}$$

**Remark:** If we want the action to be real, we have to require $\Phi$ to be real and $\Phi_G^{++}$ to be pseudoreal — invariant under the pseudoconjugation [this immediately follows from (7.77) and (7.78)].

The simplest nontrivial invariant action involving $V^{++}$ reads

$$S \propto \int dt\, d^2\theta^+ d^2\theta^- du\, (D^{--}V^{++})^2. \tag{7.107}$$

Going down to components, we reproduce up to irrelevant coefficients the free Lagrangian (7.67).

We go over now to $\mathcal{N} = 8$ harmonic superspace. An ordinary $\mathcal{N} = 8$ superspace includes four complex odd coordinates, which we denote $\theta_{j\alpha}$, with both $j$ and $\alpha$ taking two values. Both indices are lifted under complex conjugation: $\theta_{j\alpha} \to \bar{\theta}^{j\alpha}$. To harmonize it, we consider the projections

[22]The symbols $d\bar{\theta}^+, d\theta^+$ carry harmonic charge $-1$: integration over $d\theta^+$ is equivalent to taking the derivative $\partial/\partial\theta^+$.

$\theta^\pm_\alpha = \theta^j_\alpha u^\pm_j$ and $\bar\theta^{\pm\alpha} = \bar\theta^{j\alpha} u^\pm_j$. We then introduce the analytic time $t_A$ given by the same formula (7.91) as for $\mathcal{N} = 4$ but with an extra summation over $\alpha$. The $\mathcal{N} = 8$ analytic superspace includes the variables $t_A, \theta^+_\alpha, \bar\theta^{+\alpha}$ and $u$.

Consider now a G-analytic superfield

$$\boxed{Q^+(\zeta, u) = F^+(t_A, u) + \theta^+_\alpha \chi^\alpha(t_A, u) + \ldots + (\theta^+_\alpha \bar\theta^{+\alpha})^2 P^{-3}(t_A, u).} \tag{7.108}$$

Take $2n$ such superfields $Q^{+a}$ and impose on them the pseudoreality constraint (7.100) — the *only* constraint that one has to impose in this case. Consider then the superfield action (see Chapter 5 of Ref. [90])

$$\boxed{S \propto \int dt_A \, du \, d^2\bar\theta^+ d^2\theta^+ \left[\frac{1}{2}\, Q^+_a D^{++} Q^{+a} + \mathcal{L}^{+4}(Q^{+b}, u)\right]} \tag{7.109}$$

with

$$D^{++} = \partial^{++} + 2i\theta^+_\alpha \bar\theta^{+\alpha} \frac{\partial}{\partial t_A}. \tag{7.110}$$

Here, $\mathcal{L}^{+4}$ is an arbitrary real function of $Q^{+a}$ and $u^\pm$ that carries harmonic charge $+4$.

The superfields $Q^{+a}(\zeta, u)$ involve in their harmonic expansion an infinity of Grassmann-even and Grassmann-odd variables. How to deal with them and what dynamics does the action (7.109) describe?

It turns out that almost all these variables have a non dynamical auxiliary nature. They are algebraically determined by the equations of motion following from (7.109). The latter can be compactly written in the superfield form:

$$D^{++} Q^{+a} = \Omega^{ab} \frac{\partial \mathcal{L}^{+4}}{\partial Q^{+b}}. \tag{7.111}$$

By going down to components and resolving a set of equations following from (7.111),[23] we are left with $4n$ real bosonic and $8n$ real fermionic dynamical variables. Plugging in (7.109) the solutions for the auxiliary variables, we derive a dynamical SQM model describing the motion along a $4n$-dimensional hyperkähler manifold.

We discuss in detail the hyperkähler and HKT models in Chapter 13.

[23] These equations are rather intricate and can be *analytically* resolved only in some special cases.

# PART 3
# SYNTHESIS

# Chapter 8

# Supersymmetric Description of the de Rham Complex

## 8.1 Basic structures

In Chapters 5–7, we presented the simplest SQM models. Now we begin the discussion of more complicated models describing the motion along curved manifolds. By some historical reasons, physicists call them the *supersymmetric sigma models.* The study of these models will allow us to reproduce and elaborate the geometric constructions of the first part.

**Definition 8.1.** An ordinary mechanical[1] sigma model is a dynamical system including $D$ dynamical variables $x^M(t)$ with the Lagrangian

$$L = \frac{1}{2} g_{MN}(x)\, \dot{x}^M \dot{x}^N. \tag{8.2}$$

Here $x^M$ have the meaning of the coordinates of a $D$-dimensional manifold and $g_{MN}(x)$ is its metric tensor. The equations of motion derived from (8.2) coincide with (4.1) and describe the motion along the geodesic lines.

We will consider in this and the subsequent chapters many different supersymmetric sigma models including, besides $x^M(t)$, Grassmann dynamical variables.

---

[1] Physicists also discuss sigma models involving not a finite number of dynamical variables $x^M(t)$, but the fields $x^M(t, \boldsymbol{\xi})$, where the components of $\boldsymbol{\xi}$ are the flat spatial coordinates that have nothing to do with $x^M$. Such models are characterized by the actions like

$$S = \frac{1}{2} \int dt d\boldsymbol{\xi}\, g_{MN}(x)\, \partial_\mu x^M \partial^\mu x^N \tag{8.1}$$

with $\partial_\mu = (\partial_t, \partial_{\boldsymbol{\xi}})$.

We had to mention that to give the reader a proper perspective, but, with rare exceptions, we will not consider field-theory models further.

For example, one can take a set of $D$ $\mathcal{N} = 1$ superfields (7.7) and generalize (7.17) by writing

$$L = \frac{i}{2} \int d\Theta \, g_{MN}(\mathcal{X}) \, \dot{\mathcal{X}}^M \mathcal{D} \mathcal{X}^N. \tag{8.3}$$

In components, this gives

$$L = \frac{1}{2} g_{MN}(x)(\dot{x}^M \dot{x}^N + i\Psi^M \boldsymbol{\nabla} \Psi^N), \tag{8.4}$$

where

$$\boldsymbol{\nabla} \Psi^M = \dot{\Psi}^M + \Gamma^M_{PS} \dot{x}^P \Psi^S \tag{8.5}$$

and $\Gamma^M_{PS}$ are the Christoffel symbols (1.17). (The operator $\boldsymbol{\nabla}$ is a covariant time derivative: if $x^M$ and $\Psi^M$ are vectors, $\boldsymbol{\nabla}\Psi^M$ is also a vector.) Theorem 7.2 assures that the Lagrangian (8.4) may only shift under the supertransformations (7.9) for all $(x^M, \Psi^M)$ by a total time derivative.

However, as we mentioned before, $\mathcal{N} = 1$ supersymmetry is in fact not a real symmetry — it does not entail dynamical consequences like the double degeneracy of the spectrum. We will see later that for *complex* manifolds the Lagrangian (8.4) enjoys an additional "hidden" supersymmetry [cf. Eq. (7.36)]. But in a general case, the study of the system (8.4) will not help us to elucidate geometric properties of the manifold, and we have to consider the models with stronger supersymmetries starting from $\mathcal{N} = 2$.

Well, if we want to have $\mathcal{N} = 2$ supersymmetry, the best way to proceed is to express the Lagrangian in terms of $\mathcal{N} = 2$ superfields. For a generic real manifold, the only option is to take $D$ real superfields (7.25), with the dynamical bosonic components coinciding with the coordinates $x^M$. To supersymmetrize (8.2), we write the Lagrangian [93]

$$\boxed{L = \frac{1}{2} \int d\bar{\theta} d\theta \, g_{MN}(X) \bar{D} X^M D X^N,} \tag{8.6}$$

with $\mathcal{N} = 2$ covariant derivatives (7.22).

In components, this gives

$$L = \frac{1}{2} g_{MN}(x) \left[ \dot{x}^M \dot{x}^N + F^M F^N + i(\bar{\psi}^M \boldsymbol{\nabla} \psi^N - \boldsymbol{\nabla} \bar{\psi}^M \psi^N) \right]$$
$$+ \Gamma_{M,PQ} F^M \psi^P \bar{\psi}^Q + \frac{1}{2} (\partial_P \partial_Q g_{MN}) \, \bar{\psi}^P \bar{\psi}^M \psi^Q \psi^N. \tag{8.7}$$

The auxiliary variables $F^M$ enter the Lagrangian without derivatives and the corresponding equations of motion $\partial L/\partial F^M = 0$ are purely algebraic and can be resolved for $F^M$. The solution reads

$$F^M = -\Gamma^M_{PQ}\psi^P\bar{\psi}^Q. \tag{8.8}$$

Plugging this into (8.7), we derive

$$\boxed{\begin{aligned} L = {}& \frac{1}{2}g_{MN}\left[\dot{x}^M\dot{x}^N + i(\bar{\psi}^M\boldsymbol{\nabla}\psi^N - \boldsymbol{\nabla}\bar{\psi}^M\psi^N)\right] \\ & - \frac{1}{4}R_{PMQN}\,\bar{\psi}^P\bar{\psi}^M\psi^Q\psi^N, \end{aligned}} \tag{8.9}$$

where $R_{PMQN}$ is the Riemann tensor.

Using the $\mathcal{N} = 2$ superfields was the best option, but one also could express the Lagrangian of the model in terms of $\mathcal{N} = 1$ superfields [94]. Consider along with the set of bosonic real $\mathcal{N} = 1$ superfields $\mathcal{X}^M = x^M + i\Theta\Psi^M$ a set of $D$ *Grassmann* real superfields (7.15): $\Phi^M = \phi^M + \Theta F^M$. Then the Lagrangian

$$L = \frac{1}{2}\int d\Theta\, g_{MN}(\mathcal{X})\left\{i\dot{\mathcal{X}}^M\mathcal{D}\mathcal{X}^N + \Phi^M[\mathcal{D}\Phi^N + \Gamma^N_{PS}(\mathcal{X})\mathcal{D}\mathcal{X}^P\Phi^S]\right\}, \tag{8.10}$$

being expressed in components, gives the same expression (8.7) after the identification $\psi^M = (\Psi^M - i\phi^M)/\sqrt{2}$. Besides the explicit $\mathcal{N} = 1$ supersymmetry mixing the components of $\mathcal{X}^M$ and $\Phi^M$ independently, the Lagrangian (8.10) enjoys the extra $\mathcal{N} = 1$ symmetry under the transformations

$$\begin{aligned} \delta\mathcal{X}^M &= -i\eta\Phi^M, \\ \delta\Phi^M &= \eta\dot{\mathcal{X}}^M, \end{aligned} \tag{8.11}$$

with real $\eta$. These transformations correspond to the transformations (7.26) with $\epsilon = \bar{\epsilon} = \eta/\sqrt{2}$.

**Note the rule of thumb:** the farther in the $\theta$ expansion is placed a superfield component, the less is the degree of the time derivative with which it enters the component Lagrangian. The lowest component of $X^M$ gives the term $\propto \dot{x}^2$, the middle component gives the structure $\propto \dot{\psi}$, and the auxiliary field $F$ enters without derivatives. This is quite natural. For the highest component of $X^M$, it is only the terms $\propto \partial/\partial\theta$ in $D$ and $\partial/\partial\bar{\theta}$ in $\bar{D}$ that work, while for the lower components, the terms $\propto \bar{\theta}\partial/\partial t$ and

$\propto \theta\partial/\partial t$ in the covariant derivatives contribute: for the lower components we need to increase the number of the factors $\theta$ in the integrand to provide a nonzero contribution in the superspace integral rather than diminish them. This brings about time derivatives in the component Lagrangian.

As was briefly mentioned in Chapter 7, it is also possible to write the structures with extra derivatives like

$$\int d\bar{\theta}d\theta\, \kappa_{MN}(X)\bar{D}X^M \frac{d}{dt}DX^N \tag{8.12}$$

in the superfield Lagrangian; then the component $F$ would enter with a time derivative. But in that case, the lowest component $x$ would enter with still higher derivatives $\propto \dot{x}\ddot{x}$. This gives interesting models including "benign ghosts" [58], but they do not represent an interest for the problems of differential geometry. We will not consider the systems with higher derivatives in our book.

Our next task is to construct the supercharges in this model. A general method to derive conserved supercharges in any Lagrangian system is provided by the Noether theorem. It follows from (7.26) and (8.8) that the Lagrangian (8.9) is invariant under the following nonlinear supersymmetry transformations:

$$\begin{aligned} \delta x^M &= \epsilon\psi^M + \bar{\psi}^M\bar{\epsilon}, \\ \delta\psi^M &= -\bar{\epsilon}\left(i\dot{x}^M + \Gamma^M_{PQ}\psi^P\bar{\psi}^Q\right), \quad \delta\bar{\psi}^M = \epsilon\left(i\dot{x}^M - \Gamma^M_{PQ}\psi^P\bar{\psi}^Q\right). \end{aligned} \tag{8.13}$$

From this and using the general Theorem 4.6, we arrive after some calculations at a very simple result [95, 96]:

$$Q = \frac{1}{\sqrt{2}} g_{MN}\psi^M\dot{x}^N, \quad \bar{Q} = \frac{1}{\sqrt{2}} g_{MN}\bar{\psi}^M\dot{x}^N, \tag{8.14}$$

where the factor $1/\sqrt{2}$ is introduced for further convenience.

A very good exercise for a good student is to derive the classical equations of motion from the Lagrangian (8.9) and verify that the supercharges (8.14) conserve indeed.[2]

[2]Frankly, I do not know whether anybody gave himself a trouble to make this test. I did not.

The expressions for the supercharges in terms of coordinates and canonical momenta,

$$\Pi_M = g_{MN}\dot{x}^N + \frac{i}{2}[\partial_Q g_{PM} - \partial_P g_{QM}]\psi^P\bar{\psi}^Q, \tag{8.15}$$

are

$$\begin{aligned} Q &= \frac{1}{\sqrt{2}}\psi^M\left[\Pi_M + \frac{i}{2}\Gamma_{M,PQ}\,\psi^P\bar{\psi}^Q\right] = \frac{1}{\sqrt{2}}\psi^M\left[\Pi_M - \frac{i}{2}\partial_M g_{PQ}\,\psi^P\bar{\psi}^Q\right], \\ \bar{Q} &= \frac{1}{\sqrt{2}}\bar{\psi}^M\left[\Pi_M - \frac{i}{2}\Gamma_{M,PQ}\,\psi^P\bar{\psi}^Q\right] = \frac{1}{\sqrt{2}}\bar{\psi}^M\left[\Pi_M + \frac{i}{2}\partial_M g_{PQ}\,\psi^P\bar{\psi}^Q\right]. \end{aligned} \tag{8.16}$$

The classical Hamiltonian is obtained by the Legendre transformation of the Lagrangian (8.9). It reads

$$H = \frac{g^{MN}}{2}\mathcal{P}_M\mathcal{P}_N + \frac{1}{4}R_{PMQN}\,\bar{\psi}^P\bar{\psi}^M\psi^Q\psi^N, \tag{8.17}$$

where

$$\mathcal{P}_M = \Pi_M + \frac{i}{2}(\partial_P g_{QM} - \partial_Q g_{PM})\psi^P\bar{\psi}^Q.$$

For some purposes (in particular, for quantization, which is our next task), it is more convenient to express the classical supercharges and the Hamiltonian in terms of the tangent space fermion variables $\psi_A = e_{AM}\psi^M$. Then the supercharges acquire the following form [95, 96]:

$$\begin{aligned} Q &= \frac{1}{\sqrt{2}}\,\psi_C\,e_C^M\,(P_M - i\omega_{AB,M}\,\psi_A\bar{\psi}_B), \\ \bar{Q} &= \frac{1}{\sqrt{2}}\,\bar{\psi}_C\,e_C^M\,(P_M - i\omega_{AB,M}\,\psi_A\bar{\psi}_B), \end{aligned} \tag{8.18}$$

where $\omega_{AB,M}$ is the spin connection. When comparing (8.16) with (8.18), it is necessary to have in mind that the canonical momenta $\Pi_M$ and $P_M$ do *not* coincide. $\Pi_M$ is given by the variation of the Lagrangian with respect to $\dot{x}^M$ when $\psi^M$ are kept fixed, while $P_M$ is this variation taken with fixed $\psi_A$. The difference stems from the terms including the time derivatives of the fermion variables [97, 98]:

$$P_M = \Pi_M + \frac{\partial \dot\psi^Q}{\partial \dot x_M}\frac{\partial L}{\partial \dot\psi^Q} + \frac{\partial \dot{\bar\psi}^Q}{\partial \dot x_M}\frac{\partial L}{\partial \dot{\bar\psi}^Q}$$
$$= \Pi_M + \frac{i}{2}[(\partial_M e^Q_B)\, e_{CQ} - (\partial_M e^Q_C)\, e_{BQ}]\psi_C\bar\psi_B. \qquad (8.19)$$

The only nonzero Poisson brackets in the basis $(x^M, P_M; \psi_A, \bar\psi_A)$ are

$$\{P_M, x^N\}_P = \delta^N_M, \quad \{\psi_A, \bar\psi_B\}_P = i\delta_{AB}, \qquad (8.20)$$

whereas in the basis $(x^M, \Pi_M; \psi^M, \bar\psi^M)$, the Poisson brackets are more complicated:[3]

$$\{\Pi_N, x^M\}_P = \delta^M_N, \quad \{\psi^M, \bar\psi^N\}_P = ig^{MN},$$
$$\{\Pi_M, \psi^N\}_P = \frac{1}{2}\psi_Q\,\partial_M g^{NQ}, \quad \{\Pi_M, \bar\psi^N\}_P = \frac{1}{2}\bar\psi_Q\,\partial_M g^{NQ}. \qquad (8.22)$$

Thus, the basis $(x^M, P_M; \psi_A, \bar\psi_A)$ is orthogonal with respect to the symplectic Poisson bracket structure in contrast to the basis $(x^M, \Pi_M; \psi^M, \bar\psi^M)$. In the former case, the Poisson bracket of arbitrary phase space functions has a canonical form (4.59).

The Poisson bracket $-i\{Q, \bar Q\}_P$ gives the classical Hamiltonian

$$H = \frac{1}{2}g^{MN}(P_M - i\omega_{AB,M}\psi_A\bar\psi_B)(P_N - i\omega_{CD,N}\psi_C\bar\psi_D)$$
$$+ \frac{1}{4}R_{ABCD}\,\bar\psi_A\bar\psi_B\psi_C\psi_D, \qquad (8.23)$$

which coincides with (8.17).

To quantize this system, we take the classical supercharges (8.18) and apply the recipe suggested in Ref. [52] and discussed earlier on p. 119. Namely, we order the classical supercharges using the symmetric (or antisymmetric for the products of Grassmann-odd variables) *Weyl prescription*

---

[3] For example,

$$\{\Pi_M, \psi^N\}_P = (\partial_M e^N_A)\,\psi_A + \frac{i}{2}e^N_A\left[(\partial_M e^Q_C)\,e_{BQ} - (\partial_M e^Q_B)\,e_{CQ}\right]\{\psi_C\bar\psi_B, \psi_A\}_P$$
$$= (\partial_M e^N_A)\,\psi_A - \frac{1}{2}e^N_A\left[(\partial_M e^Q_C)\,e_{AQ} - (\partial_M e^Q_A)\,e_{CQ}\right]\psi_C$$
$$= \frac{1}{2}(\partial_M e^N_A)\psi_A + \frac{1}{2}e^N_A(\partial_M e^Q_A)\,\psi_Q = \frac{1}{2}\psi_Q\partial_M g^{NQ}. \qquad (8.21)$$

and then replace $P_M \to \hat{P}_M = -i\partial/\partial x^M$ and $\bar\psi_A \to \hat{\bar\psi}_A = \partial/\partial\psi_A$. In particular, we have to replace

$$e^M_C P_M \to \frac{1}{2}(e^M_C \hat{P}_M + \hat{P}_M e^M_C) = e^M_C \hat{P}_M - \frac{i}{2}(\partial_M e^M_C),$$

$$\psi_C\psi_A\bar\psi_B \to \frac{1}{6}\left(2\psi_C\psi_A\frac{\partial}{\partial\psi_B} + 2\frac{\partial}{\partial\psi_B}\psi_C\psi_A + \psi_A\frac{\partial}{\partial\psi_B}\psi_C - \psi_C\frac{\partial}{\partial\psi_B}\psi_A\right)$$

$$= \psi_C\psi_A\frac{\partial}{\partial\psi_B} + \frac{1}{2}(\delta_{BC}\psi_A - \delta_{BA}\psi_C). \tag{8.24}$$

After some calculations using the expression (1.58) for the spin connection (with $G^N_{MP} \to \Gamma^N_{MP}$) and the identity $\Gamma^N_{NM} = \frac{1}{2}\partial_M \ln g$, where $g = \det(g_{MN})$, we obtain

$$\hat{Q} = -\frac{i}{\sqrt{2}}\psi^M\left[\frac{\partial}{\partial x^M} + \omega_{AB,M}\,\psi_A\frac{\partial}{\partial\psi_B} - \frac{1}{4}\,\partial_M \ln g\right],$$

$$\hat{\bar{Q}} = -\frac{i}{\sqrt{2}}e^M_C\frac{\partial}{\partial\psi_C}\left[\frac{\partial}{\partial x^M} + \omega_{AB,M}\,\psi_A\frac{\partial}{\partial\psi_B} - \frac{1}{4}\,\partial_M \ln g\right]. \tag{8.25}$$

These supercharges are Hermitially conjugate to each other in the Hilbert space involving the wave functions

$$\Psi(x^M, \psi_A) = \Psi_0(x^M) + \Psi_A(x^M)\psi_A + \cdots \tag{8.26}$$

with the complex-valued coefficients and endowed with the inner product with the flat measure [see Eq. (6.26)]:

$$d\mu_{\text{flat}} \sim \prod_M dx^M \prod_A d\psi_A d\bar\psi_A e^{-\psi_A\bar\psi_A}. \tag{8.27}$$

If we want to make a contact with geometry, we have to derive the expressions for the *covariant* operators acting in the Hilbert space with the measure involving the extra factor $\sqrt{g}$. The wave functions in the covariant Hilbert space are related to the wave functions in the Hilbert space with the flat measure as $\Psi^{\text{cov}} = g^{-1/4}\Psi^{\text{flat}}$. Accordingly, the covariant supercharges are

$$(\hat{Q}^{\text{cov}}, \hat{\bar{Q}}^{\text{cov}}) = g^{-1/4}\,(\hat{Q}^{\text{flat}}, \hat{\bar{Q}}^{\text{flat}})g^{1/4}. \tag{8.28}$$

This similarity transformation kills the last terms in (8.25) and we finally derive

$$\boxed{\begin{aligned}\hat{Q}^{\text{cov}} &= -\frac{i}{\sqrt{2}}\psi^M\left[\frac{\partial}{\partial x^M} + \omega_{AB,M}\,\psi_A\frac{\partial}{\partial\psi_B}\right],\\ \hat{\bar{Q}}^{\text{cov}} &= -\frac{i}{\sqrt{2}}e^M_C\frac{\partial}{\partial\psi_C}\left[\frac{\partial}{\partial x^M} + \omega_{AB,M}\,\psi_A\frac{\partial}{\partial\psi_B}\right],\end{aligned}} \tag{8.29}$$

where the operator

$$D_M = \frac{\partial}{\partial x^M} + \omega_{AB,M}\psi_A \frac{\partial}{\partial \psi_B} \tag{8.30}$$

acts on the coefficients in the expansion (8.26) as the spin part in the covariant derivative (1.55).

The operators (8.29) are nilpotent. The anticommutator $\{\hat{Q}^{\rm cov}, \hat{\bar{Q}}^{\rm cov}\}$ gives the covariant quantum Hamiltonian, which has the form[4] [96]

$$\boxed{\hat{H}^{\rm cov} = -\frac{1}{2\sqrt{g}} D_M \sqrt{g}\, g^{MN} D_N - \frac{1}{2} R_{ABCD}\, \hat{\bar{\psi}}_A \psi_B \hat{\bar{\psi}}_C \psi_D.} \tag{8.31}$$

In the first term in (8.31), the reader will recognize the Laplace–Beltrami operator (1.38), with $D_M$ including also the Grassmann part.

The operator $D_M$ and the whole Hamiltonian (8.31) commute with the fermion charge operator, $\hat{F} = \psi_A \hat{\bar{\psi}}_A$, that counts the number of Grassmann factors in the wave function. Hence, the fermion charge is conserved in this model.

Note that the quantum Hamiltonian is *not* obtained by the procedure outlined above (Weyl ordering supplemented by the similarity transformation from the flat to covariant measure). If you want to preserve a symmetry (in our case, supersymmetry) at the quantum level, it is the *symmetry generators* (the quantum supercharges in our case) that should be derived this way.

We are ready now to prove the key theorem.

**Theorem 8.1.** *The action of the supercharges $\hat{Q}^{\rm cov}$ and $\hat{\bar{Q}}^{\rm cov}$ on the wave functions $\Psi(x^M, \psi_A)$ is isomorphic to the action of the exterior derivative operator $d$ and its conjugate $d^\dagger$ in the de Rham complex.*

**Proof.** Consider a wave function of fermion charge $p$. Represent it as

$$\Psi_p = A_{M_1 \ldots M_p}(x)\, \psi^{M_1} \cdots \psi^{M_p}, \tag{8.32}$$

[4]The different coefficients and signs before the last term in (8.23) and (8.31) are not a mistake. In the classical limit, (8.31) is reduced to (8.23), bearing in mind the cyclic property of the Riemann tensor [the second line in (1.23)].

where $\psi^M = e_A^M \psi_A$ and $A_{M_1 \ldots M_p}(x)$ is totally antisymmetric. There are two terms in $\hat{Q}^{\mathrm{cov}}$. The action of the derivative term is

$$-\frac{i}{\sqrt{2}}\psi^M \frac{\partial}{\partial x^M}\Psi_p = -\frac{i}{\sqrt{2}}\left(\partial_M A_{M_1 \ldots M_p}\right)\psi^M \psi^{M_1} \cdots \psi^{M_p} - \frac{ip}{\sqrt{2}} A_{M_1 \ldots M_p}\psi^M \left(\partial_M e_A^{M_1}\right)\left(e_{AN}\psi^N\right)\psi^{M_2} \cdots \psi^{M_p}. \tag{8.33}$$

The action of the term $\propto \omega_{AB,M}$ gives

$$-\frac{ip}{\sqrt{2}}\psi^M \omega_{AB,M}\left(e_{AN}\psi^N\right) e_B^{M_1} A_{M_1 \ldots M_p}\psi^{M_2} \cdots \psi^{M_p}.$$

Adding this to the second term in (8.33), we obtain an expression involving the factor

$$e_{AN}(\partial_M e_A^{M_1} + \omega_{AB,M} e_B^{M_1})\psi^M \psi^N.$$

Using (1.55) and (1.57), we represent it as

$$-e_{AN}\Gamma^{M_1}_{MP} e_A^P \psi^M \psi^N = -\Gamma^{M_1}_{MN}\psi^M \psi^N = 0.$$

We are left with the result

$$\hat{Q}^{\mathrm{cov}}\Psi_p = -\frac{i}{\sqrt{2}}\left(\partial_M A_{M_1 \ldots M_p}\right)\psi^M \psi^{M_1} \cdots \psi^{M_p}, \tag{8.34}$$

which has the same form, up to the factor $-i/\sqrt{2}$, as Eq. (1.32) defining the action of $d$. One only has to map $\psi^M \leftrightarrow dx^M$.

The operator $d^\dagger$ is adjoint to $d$ with the measure (1.29) including the factor $\sqrt{g}$, and the operator $\hat{\bar{Q}}^{\mathrm{cov}}$ is adjoint to $\hat{Q}^{\mathrm{cov}}$ with a similar covariant measure. Thus, if $\hat{Q}^{\mathrm{cov}}$ is isomorphic to $d$, $\hat{\bar{Q}}^{\mathrm{cov}}$ must be isomorphic to $d^\dagger$. □

## 8.2 Euler characteristic

We have succeeded in translating the content of Section 1.2 into supersymmetric language by mapping the de Rham complex of differential forms to the Hilbert space of a certain SQM system. A legitimate question is: what have we gained by that? Was the traditional language of differential forms not good enough for working with this system?

There are two answers to this question. Firstly, one can also derive in this way certain *new* mathematical results; we will display them later. Secondly, many known mathematical results can be derived *simpler* by using supersymmetry. As an example, we will prove an important theorem here.

**Theorem 8.2 (Hodge).** *Any differential form $\alpha$ can be uniquely presented as a sum of an exact, coexact and a harmonic form,*

$$\alpha = d\beta + d^\dagger \gamma + h. \tag{8.35}$$

**Proof.** This immediately follows from the general properties of the spectrum of a SQM system established in Chapter 5: it consists of degenerate doublets interrelated by the action of $\hat{Q}$ and $\hat{\bar{Q}} \equiv \hat{Q}^\dagger$ and of some unpaired zero-energy states, as shown in Fig. 5.2. Exact forms map to the right tower of states in this figure, coexact forms map to the left tower, and harmonic forms to the zero-energy states at the bottom. □

A related classical mathematical notion that shines differently, when looking through the supersymmetric glasses, is the *de Rham cohomology*. We already mentioned it on p. 15. It is the space of all closed forms, i.e. the forms satisfying the condition $d\alpha_p = 0$, factorized over the space of exact forms that can be represented as $\alpha_p = d\beta_{p-1}$. For a given degree $p$, the coset *closed/exact* represents a vector space $H^p_{\text{dR}}(\mathcal{M})$.

**Definition 8.2.** The *Betti number* $b_p$ is the dimension of $H^p_{\text{dR}}$.

**Definition 8.3.** The *Euler characteristic* of the manifold is the alternating sum

$$\boxed{\chi(\mathcal{M}) = \sum_p (-1)^p \, b_p(\mathcal{M}).} \tag{8.36}$$

I give two elementary examples here:

(1) For $S^2$, $b_0 = 1$ (a closed but not exact 0-form is a real constant $C$) and $H^0_{\text{dR}}(S^2) = \mathbb{R}$. Similarly, $b_2 = 1$: a closed but not exact 2-form is a real constant multiplied by $dx^1 \wedge dx^2$. There are no closed but not exact 1-forms and $b_1 = 0$. The Euler characteristic is

$$\chi(S^2) = 1 - 0 + 1 = 2. \tag{8.37}$$

(2) For $T^2$, $b_0 = b_2 = 1$, as for the sphere,[5] but there exist now closed but not exact 1-forms,[6] $\alpha_1 = Adx^1 + Bdx^2$. Then $H^2_{\text{dR}}(T^2) = \mathbb{R}^2$ and

[5] Actually, for any manifold $\mathcal{M}$ of dimension $D$, $b_0 = b_D = 1$.

[6] The form $dx^1$ is not exact because $x^1$ is not an admissible 0-form — it is not uniquely defined on the torus.

$b_1 = 2$. The Euler characteristic is

$$\chi(T^2) = 1 - 2 + 1 = 0. \tag{8.38}$$

**Remark:** It is a proper place to say that, when we are discussing forms, the de Rham and other complexes, we always have in mind an important restriction. We always assume that our space of forms is endowed by a well-defined and finite inner product (1.29). Mathematicians also do that, but for us, it is especially necessary. If the inner product is not defined, we cannot map the complex to the Hilbert space of an SQM system, and the whole ideology of this book fails. The evaluation of the Euler characteristic above also implies the existence of the inner product. If this restriction is lifted, the evaluation is not correct. Indeed, consider the form $d\phi$ on $S^2$. It seems to have the same properties as the form $dx^1$ on $T^2$, being closed but not exact. But the norm (1.29) for this form diverges $\sim \int d\theta/|\sin\theta|$, and it is not admissible for this reason. (On the other hand, the form $d\theta$ is closed, has a finite norm, but it is also *exact* ($\theta$ is a legitimate 0-form on $S^2$) and hence does not contribute in $b_1$.)

The Betti numbers and hence the Euler characteristic are topological invariants. We will show now how the latter can be calculated in the SQM framework. The key observation is as follows.

**Theorem 8.3.** *The Euler characteristic of a smooth manifold coincides with the Witten index of the supersymmetric sigma model* (8.3).

**Proof.** A look at Fig. 5.2 makes it evident that any eigenstate $\Psi$ of an SQM Hamiltonian that is annihilated by $\hat{Q}$, but is not representable as $\Psi = \hat{Q}\Psi'$, does not belong to a supersymmetric doublet of states carrying nonzero energy. Thus, $b_p$ counts the number of zero-energy states of the Hamiltonian (8.31) having fermion charge $p$. If $p$ is even, these are bosonic states, and if $p$ is odd, these are fermionic states. Then the sum (8.36) coincides with the definition (6.55) of the Witten index for this system. □

In Chapter 6, we explained how one can calculate the Witten index of an SQM system as a phase-space integral (6.56) to which the functional integral (6.50) is reduced in the small $\beta$ limit.[7] Substituting there the

[7]We already noted in Chapter 6 and will demonstrate it Chapter 15 that this recipe works in many, but not in *all* cases. But in this case, it *works*.

Hamiltonian (8.23) and integrating over the momenta, we obtain the integral representation of the Euler characteristic:

$$\chi = \frac{1}{(2\pi\beta)^{D/2}} \int d^D x \sqrt{g} \int \prod_{A=1}^{D} d\psi_A d\bar{\psi}_A \exp\left\{-\frac{\beta}{4} R_{ABCD}\, \bar{\psi}_A \bar{\psi}_B \psi_C \psi_D\right\}. \tag{8.39}$$

The first immediate observation is that the fermion integral vanishes when $D$ is odd. Indeed, the Euler characteristic of any odd-dimensional manifold vanishes, as follows from the definition (8.36) and from the identity $b_p = b_{D-p}$, which in turn stems from the fact that $p$-forms and $(D-p)$-forms are interrelated by the duality transformation $\star$.

If $D$ is even and $R_{ABCD} \neq 0$, the fermion integral does not generically vanish. A nonzero contribution stems from the $(D/2)$th term in the expansion of the exponential. The factor $\beta^{D/2}$ cancels the factor $\beta^{-D/2}$ in front of the integral in (8.39) that came from the momentum integrations. In the simplest two-dimensional case, we obtain

$$\chi_2 = \frac{1}{2\pi} \int d^2 x \sqrt{g}\, \mathbb{R}, \tag{8.40}$$

where $\mathbb{R} = R_{1212}$ is the Gaussian curvature. This is none other than the very well-known *Gauss–Bonnet formula.* For an arbitrary $D = 2n$, we derive the generalized Gauss–Bonnet formula:

$$\boxed{\chi_{2n} = \frac{1}{(8\pi)^n n!} \int d^D x \sqrt{g}\, \varepsilon_{A_1 B_1 \ldots A_n B_n} \varepsilon_{C_1 D_1 \ldots C_n D_n} \prod_{k=1}^{n} R_{A_k B_k C_k D_k}.} \tag{8.41}$$

In four dimensions, this can be presented as

$$\chi_4 = \frac{1}{32\pi^2} \int d^4 x \sqrt{g}\, R_{ABCD} \tilde{R}_{ABCD}, \tag{8.42}$$

where

$$\tilde{R}_{ABCD} = \frac{1}{4} \varepsilon_{ABEF}\, \varepsilon_{CDGH} R_{EFGH} \tag{8.43}$$

is the dual Riemann tensor.

The formula (8.41) is the second example of an *index theorem* — an integral representation for a topological invariant — that we meet in our book (the first such example was the magnetic flux (6.68)). More examples will be given in Chapters 14 and 15. The generalized Gauss–Bonnet formula was first derived by Chern back in 1942 [99]. To prove it, Chern used traditional mathematical methods. Its supersymmetric proof was given in 1983–1984 in Refs. [67]. As you have seen, this proof is very simple.[8]

## 8.3 Deformations □ Morse theory □ Quasitorsions

The Lagrangian (8.6) is not the most general supersymmetric Lagrangian that one can construct with the $\mathcal{N} = 2$ $(\mathbf{1}, \mathbf{2}, \mathbf{1})$ superfields $X^M(t, \theta, \bar\theta)$. By the same token as in (7.38), one can add the superpotential term

$$L_{\text{pot}} = -\int d\bar\theta d\theta \, W(X^M). \tag{8.44}$$

Adding this to (8.6) and performing the same standard manipulations as for the undeformed Lagrangian (8.6), which we will not describe in detail anymore, we arrive at the following form of the quantum supercharges:

$$\boxed{\begin{aligned} \hat Q_W &= \hat Q_0 - \frac{i}{\sqrt2}\,\partial_M W \psi^M = e^{-W} \hat Q_0 \, e^W, \\ \hat{\bar Q}_W &= \hat{\bar Q}_0 + \frac{i}{\sqrt2}\, e^M_A (\partial_M W) \frac{\partial}{\partial \psi_A} = e^W \hat{\bar Q}_0 \, e^{-W}, \end{aligned}} \tag{8.45}$$

where $\hat Q_0$ and $\hat{\bar Q}_0$ are the undeformed covariant quantum supercharges (8.29).

In the language of the differential forms, this is translated as replacing the operators $d$ and $d^\dagger$ by $d_W$ and $d_W^\dagger$ defined as

$$\begin{aligned} d_W \alpha &= d\alpha + dW \wedge \alpha, \\ d_W^\dagger \alpha &= d^\dagger \alpha - \langle dW, \alpha \rangle, \end{aligned} \tag{8.46}$$

where $\langle dW, \alpha\rangle$ stands for the interior product,

$$\langle dW, \alpha \rangle = p (\partial_{M_1} W) \alpha^{M_1}{}_{M_2 \ldots M_p} dx^{M_2} \wedge \cdots \wedge dx^{M_p}, \tag{8.47}$$

where $p$ is the degree of $\alpha$ (see Definition 1.9).The supercharges (8.45) are nilpotent as the supercharges (8.29) are. The anticommutator $\{\hat Q_W, \hat{\bar Q}_W\}$

---

[8] Of course, to appreciate its simplicity, one had first to learn the basics of supersymmetric dynamics outlined in Chapters. 5–7. Any new method requires some time and effort to learn. But having done this, a scholar is rewarded by a rich harvest of interesting and nontrivial results.

gives the quantum Hamiltonian:

$$\hat{H}_W = \hat{H}_0 + \frac{1}{2}\partial_M W\,\partial^M W + \frac{1}{2}\partial_M\partial_N W(\psi^M\hat{\bar{\psi}}^N - \hat{\bar{\psi}}^N\psi^M), \quad (8.48)$$

where $\hat{H}_0$ was written in (8.31).

Similarly, the Poisson bracket of the classical supercharges gives the classical Hamiltonian:

$$H_W = H_0 + \frac{1}{2}\partial_M W\,\partial^M W + \psi^M\bar{\psi}^N\,\partial_M\partial_N W, \quad (8.49)$$

What is the Witten index of the new system? The point is that it is exactly the same as the index of the old one! Indeed, a *smooth* deformation cannot change the value of the index. It is possible that a zero-energy eigenstate of the Hamiltonian $\hat{H}_0$ acquires a nonzero energy after deformation or the other way round. However, it cannot do so *alone*, but only together with its superpartner — all the states with nonzero energies are grouped into supersymmetric doublets. This does not change the *difference* (6.55) of the bosonic and fermionic zero-energy states.

The index is still given by the integral (6.56) with $H_W$ substituted for $H_0$. *This* integral, in contrast to the integral (8.41), can be evaluated in a simple algebraic way. As is seen from Eq. (8.49), the Hamiltonian acquires after the deformation the potential term $\propto (\partial_M W)^2$. This potential turns to zero at the *critical points* of the function $W$. For a compact manifold and for a smooth function $W$, such critical points are always there.[9] With our smooth deformation, we can scale up $W$ so that the integrand in (6.56) would be exponentially suppressed everywhere except the neighbourhood of the critical points. We can thus evaluate the contribution of each critical point separately and then add them up.

Substituting (8.49) in (6.56) and integrating over the momenta, we obtain

$$I_W = \frac{1}{(2\pi\beta)^{D/2}}\int d^D x\sqrt{g}\prod_A d\psi_A d\bar{\psi}_A \exp\left[-\beta\left(\frac{1}{2}\partial_M W\partial_N W g^{MN} + \psi^M\bar{\psi}^N\,\partial_M\partial_N W + \frac{1}{4}R_{ABCD}\,\bar{\psi}_A\bar{\psi}_B\psi_C\psi_D\right)\right]. \quad (8.50)$$

At the vicinity of each critical point, one can expand

$$W(x) = W_0^{(P)} + \frac{1}{2}\,a_{MN}^{(P)}x^M x^N + \cdots \quad (8.51)$$

[9]For a Greek-style proof of this assertion, we invite the reader to *look* at Fig. 8.1.

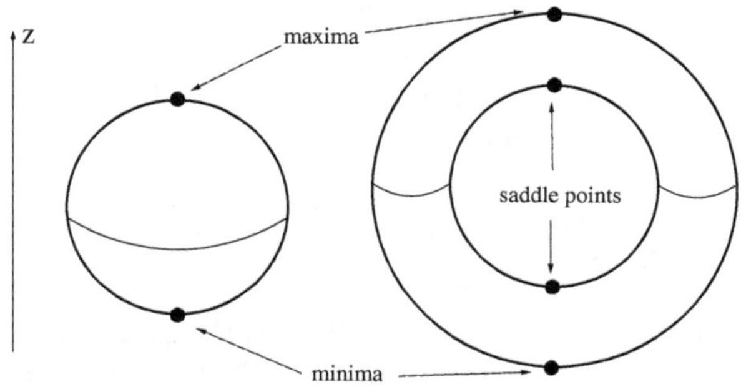


(a) $S^2$ (b) $T^2$

**Fig. 8.1.** Critical points of the Morse function $z(\mathcal{M})$. The Morse indices are $f_P = 0$ at the minima, $f_P = 2$ at the maxima, and $f_P = 1$ at the saddle points.

If $a^{(P)}_{MN}$ are chosen to be large enough, one can neglect in this expansion the higher-order terms and neglect also the term $\propto R_{ABCD}$ in the exponential. The fermion integral can be done using (4.50), and we can write

$$I_W^{(P)} = \left(\frac{\beta}{2\pi}\right)^{D/2} \int d^D x \sqrt{g} \det(e_A^M e_B^N a^{(P)}_{MN}) \exp\left\{-\frac{\beta}{2} a^{(P)}_{MQ} a^{(P)}_{NR} x^Q x^R g^{MN}\right\}$$
$$= \frac{\det(a^{(P)})}{|\det(a^{(P)})|} = (-1)^{f_P}, \tag{8.52}$$

where $f_P$ is the number of negative eigenvalues in the matrix $a^{(P)}_{MN}$. We have derived the formula representing the Euler characteristic of a manifold as a sum over the critical points of a generic smooth function[10] $\mathcal{M} \xrightarrow{W} \mathbb{R}$:

$$\chi(\mathcal{M}) = \sum_P (-1)^{f_P}. \tag{8.53}$$

This result is illustrated in Fig. 8.1. We embed $S^2$ and $T^2$ in $\mathbb{R}^3$. Let $W$ be the coordinate $z$ of a manifold point. For the sphere, $W$ has two critical points: a maximum and a minimum. The contribution of each point in the sum (8.53) is $+1$ and we derive $\chi(S^2) = 2$. For the torus, $W$ has 4 critical points: a maximum, a minimum and two saddle points. The contribution of each extremum is $+1$ and the contribution of each saddle point is $-1$. We derive that $\chi(T^2) = 0$ in agreement with (8.38).

[10] Mathematicians call it the *Morse function* and the integer $f_P$ is called the *Morse index* of the critical point $P$.

The Morse index $f_P$ also has a physical interpretation. Choose locally flat coordinates that diagonalize the matrix $a$ in the vicinity of the critical point $P$ and are rescaled so that $g_{MN}(P) = \delta_{MN}$. The dominant contribution to the quantum Hamiltonian is

$$\hat{H}^{(P)} = \frac{1}{2}\sum_M \left[\hat{P}_M^2 + \omega_M^2 (x^M)^2 + \omega_M(\psi^M \hat{\bar{\psi}}^M - \hat{\bar{\psi}}^M \psi^M)\right], \tag{8.54}$$

where $\omega_M$ are the eigenvalues of $a_{MN}$. This is the sum of $D$ Hamiltonians (5.29) describing supersymmetric oscillators. We have seen that the ground state of each term in the sum (5.29) has zero energy and is bosonic if $\omega_M > 0$ and fermionic if $\omega_M < 0$. If the matrix $a$ has $f_P$ negative eigenvalues, the ground state involves $f_P$ Grassmann factors and has fermion charge $f_P$.

We have constructed, however, only *approximate* ground states for the original Hamiltonian (8.48). The properties of the *true* ground states may not be the same. For example, for $S^2$, two supersymmetric bosonic states of (8.54) go over to two supersymmetric ground states of the full Hamiltonian because $I_W = 2$ in this case. On the other hand, $I_W = 0$ for the torus and the exact Hamiltonian may have no supersymmetric ground states. In fact, the Hamiltonian (8.31) (without the superpotential) *has* two bosonic and two fermionic zero-energy ground states (the harmonic forms in the geometric language), but for a generic nonzero superpotential, the states are shifted from zero and form two supersymmetric doublets.

The supersymmetric derivation of (8.53) and of other results of Morse theory is due to Witten [2].

Introducing the potential term (8.44) is not the only possible supersymmetric deformation of (8.6). We can also add the term [100, 101]

$$\boxed{L_{\mathcal{B}} = \frac{1}{2}\int d\bar{\theta}d\theta\, \mathcal{B}_{MN}(X^P)\, DX^M DX^N + \text{complex conjugate}} \tag{8.55}$$

with an antisymmetric $\mathcal{B}_{MN}(X^P)$. Substituting the component expansion of $X^M$, integrating over $d\bar{\theta}d\theta$ and excluding the auxiliary variables, we derive the following contribution to the component Lagrangian [97]:

$$L_{\mathcal{B}} = \frac{1}{2}\bigg[-3i\dot{x}^M(A_{MNP}\psi^N\psi^P + \bar{A}_{MNP}\bar{\psi}^N\bar{\psi}^P) + \nabla_M A_{NPQ}\,\psi^N\psi^P\psi^Q\bar{\psi}^M$$
$$+ \nabla_M \bar{A}_{NPQ}\,\bar{\psi}^N\bar{\psi}^P\bar{\psi}^Q\psi^M - \frac{1}{4}C^M C_M\bigg], \tag{8.56}$$

where $A_{NPQ} = \partial_{[N}\mathcal{B}_{PQ]}$, $\nabla_M A_{NPQ}$ and $\nabla_M \bar{A}_{NPQ}$ are the Levi-Civita covariant derivatives of the corresponding tensors (do not confuse $\nabla_M$ with the covariant time derivative $\boldsymbol{\nabla}$ in (8.5)) and

$$C_M = 3(A_{MNP}\psi^N\psi^P - \bar{A}_{MNP}\bar{\psi}^N\bar{\psi}^P). \tag{8.57}$$

The contribution (8.56) depends on an antisymmetric third-rank tensor $A_{NPQ}$, which makes one think about torsion, but it is not quite a torsion: the covariant derivative $\nabla_M$ keeps its torsionless form. In addition, ordinary torsions are all real, whereas $A_{MNP}$ can be complex. We will call the latter *quasitorsions*. We see that, besides the terms $\propto \psi\bar{\psi}$ and $\propto (\psi\bar{\psi})^2$, the Lagrangian (and the corresponding Hamiltonian) contain the terms $\propto \psi^3\bar{\psi}$ and $\propto \bar{\psi}^3\psi$. This means that the fermion charge is not conserved anymore.

The deformed covariant quantum supercharges can be obtained by our recipe consisting, as we recall, in three steps: (i) find classical supercharges by the Noether theorem; (ii) write the corresponding quantum operators using the Weyl ordering prescription; (iii) perform the similarity transformation (8.28). The deformed supercharges differ from the undeformed ones by the extra terms $\propto \psi^3$ and $\propto \hat{\bar{\psi}}^3$:

$$\hat{Q}_{\mathcal{B}} = \hat{Q}_0 + \frac{i}{\sqrt{2}} A_{MNP}\,\psi^M\psi^N\psi^P,$$

$$\hat{\bar{Q}}_{\mathcal{B}} = \hat{\bar{Q}}_0 + \frac{i}{\sqrt{2}} \bar{A}_{MNP}\hat{\bar{\psi}}^M\hat{\bar{\psi}}^N\hat{\bar{\psi}}^P. \tag{8.58}$$

By construction, these supercharges satisfy the $\mathcal{N} = 2$ supersymmetry algebra and hence are nilpotent. A dedicated reader who would like to check the latter fact explicity should take into account the action of the derivatives in $\hat{Q}_0$ and $\hat{\bar{Q}}_0$ not only on $A_{MNP}$ but also on the vielbeins in $\psi^M = e^M_A\psi_A$. The latter contributions exactly cancel the contributions coming from the anticommutators of the structures $\propto \omega\psi^2\hat{\bar{\psi}}$ and $\propto \omega\hat{\bar{\psi}}^2\psi$ in $\hat{Q}_0$ and $\hat{\bar{Q}}_0$ with the structures $\propto C\psi^3$ and $\bar{C}\hat{\bar{\psi}}^3$.

The supercharges (8.58) can be represented in such a way that their nilpotency becomes obvious:

$$\boxed{\begin{aligned}\hat{Q}_{\mathcal{B}} &= e^{\mathcal{B}_{MN}\psi^M\psi^N}\,\hat{Q}_0 e^{-\mathcal{B}_{MN}\psi^M\psi^N},\\ \hat{\bar{Q}}_{\mathcal{B}} &= e^{\mathcal{B}_{MN}\hat{\bar{\psi}}^M\hat{\bar{\psi}}^N}\,\hat{\bar{Q}}_0 e^{-\mathcal{B}_{MN}\hat{\bar{\psi}}^M\hat{\bar{\psi}}^N}.\end{aligned}} \tag{8.59}$$

This representation can be translated into the language of differential forms:

$$d_{\mathcal{B}}\alpha = d\alpha - d\mathcal{B} \wedge \alpha \tag{8.60}$$

($\mathcal{B} = \mathcal{B}_{MN}\, dx^M \wedge dx^N$) and

$$d^\dagger_{\mathcal{B}}\alpha^p = d^\dagger \alpha^p - \frac{p!}{(p-3)!} \bar{A}^{MNP} \alpha^p_{MNPR_4\cdots R_{p-3}} dx^{R_4} \wedge \cdots \wedge dx^{R_{p-3}} \tag{8.61}$$

if $p > 3$,

$$d^\dagger_{\mathcal{B}}\alpha^3 = d^\dagger \alpha^3 - 6\bar{A}^{MNP}\alpha^3_{MNP} \quad \text{if} \quad p = 3,$$

and $d^\dagger_{\mathcal{B}}\alpha^p = d^\dagger\alpha^p$ if $p < 3$.

The second term in (8.61) may be viewed of as a generalization of the interior product (1.30) when a 1-form $V$ is replaced by the 3-form $d\mathcal{B}$.

One can also make a more complicated deformation by adding to the Lagrangian the term

$$L_{\mathcal{E}} = \frac{1}{2} \int d\bar{\theta} d\theta\, \mathcal{E}_{MNPQ}\, DX^M DX^N DX^P DX^Q + \text{c.c.} \tag{8.62}$$

This gives the deformation like (8.60) with the 2-form $\mathcal{B}$ replaced by the 4-form $\mathcal{E}$. One can also write a term with 6 covariant derivatives, etc. But for a manifold of given dimension $D$, the number of different such deformations is finite. For 2-forms $\mathcal{B}$, the deformation (8.60) is non-trivial when $D \geq 3$, for 4-forms $\mathcal{E}$, the deformation $d_{\mathcal{E}}\alpha = d\alpha - d\mathcal{E} \wedge \alpha$ is non-trivial when $D \geq 5$, etc.

The Witten index of all the deformed models described above coincides with the Witten index of the undeformed model and is equal to the Euler characteristic of the manifold.

In this chapter, we only discussed the de Rham complex and its deformations that are known to mathematicians and have been studied by them previously. But one can also consider many other deformations that have not attracted the attention of mathematicians yet. These deformations and, in particular, the so-called *quasicomplex* deformation will be discussed in Chapters 10 and 12.

# Chapter 9

# Supersymmetric Description of the Dolbeault Complex

As we know from Chapter 2, there are two ways to describe complex manifolds. One way is to use complex coordinates. Another way is to use real coordinates but take into account the presence of the complex structure tensor satisfying the integrability conditions (Theorem 2.6). Similarly, there are two possible supersymmetric descriptions.

## 9.1 $\mathcal{N} = 2$ superfield description

Consider a set of $d$ holomorphic chiral $(\mathbf{2}, \mathbf{2}, \mathbf{0})$ fields $Z^m$ ($\bar{D}Z^m = 0$) and their antiholomorphic chiral conjugates $\bar{Z}^{\bar{m}}$. We generalize the superfield action (7.31) by writing

$$\boxed{S = \int d\bar{\theta} d\theta dt \left[ \frac{1}{4} h_{m\bar{n}}(Z, \bar{Z}) \bar{D}\bar{Z}^{\bar{n}} DZ^m + \mathcal{W}(Z, \bar{Z}) \right]} \tag{9.1}$$

with Hermitian $h_{m\bar{n}}$ and real $\mathcal{W}$. We see soon that the term $\propto \mathcal{W}$ describes a gauge field, like it does in the Landau system (7.31). But consider first the systems with $\mathcal{W} = 0$. The component Lagrangian takes the form [9]

$$\begin{aligned} L = {} & h_{m\bar{n}} \left[ \dot{z}^m \dot{\bar{z}}^{\bar{n}} + \frac{i}{2} \left( \psi^m \dot{\bar{\psi}}^{\bar{n}} - \dot{\psi}^m \bar{\psi}^{\bar{n}} \right) \right] \\ & + \frac{i}{2} \left[ (2\partial_{\bar{n}} h_{m\bar{k}} - \partial_{\bar{k}} h_{m\bar{n}}) \dot{\bar{z}}^{\bar{k}} - (2\partial_m h_{k\bar{n}} - \partial_k h_{m\bar{n}}) \dot{z}^k \right] \psi^m \bar{\psi}^{\bar{n}} \\ & + (\partial_k \partial_{\bar{q}} h_{m\bar{n}}) \psi^k \psi^m \bar{\psi}^{\bar{q}} \bar{\psi}^{\bar{n}}. \end{aligned} \tag{9.2}$$

We see that $h_{m\bar{n}}$ is the complex Hermitian metric.

By going to the quasireal notation (2.14) [with also $P_M = (p_m, p_{\bar{m}})$ and $\Psi^M = (\psi^m, \bar{\psi}^{\bar{m}})$], the Lagrangian (9.2) can be brought to the following nice form:

$$L = \frac{1}{2}\left[g_{MN}\,\dot{x}^M \dot{x}^N + i g_{MN}\,\Psi^M \nabla \Psi^N - \frac{1}{6}\,\partial_P C_{KLM}\,\Psi^P \Psi^K \Psi^L \Psi^M\right] \tag{9.3}$$

with real $\Psi^{M=1,\ldots,2d}$. The totally antisymmetric tensor $C_{KLM}$ has the meaning of torsion. It has the following nonvanishing components:

$$C_{kl\bar{m}} = (\partial_l h_{k\bar{m}} - \partial_k h_{l\bar{m}})\,, \quad C_{\bar{k}\bar{l}m} = \overline{C_{kl\bar{m}}} = (\partial_{\bar{l}} h_{m\bar{k}} - \partial_{\bar{k}} h_{m\bar{l}}). \tag{9.4}$$

This tensor was already written[1] in Chapter 2 [Eq. (2.18)]. It was mentioned there that the torsion (9.4) enters the Bismut affine connection

$$G^{(B)}_{N,SP} = \Gamma_{N,SP} + \frac{1}{2}\,C_{NSP}. \tag{9.5}$$

The covariant derivative

$$\nabla \psi^N = \dot{\psi}^N + G^N_{SP}\,\dot{x}^S \psi^P$$

entering (9.3) involves *this* connection. Note that the torsion (9.4) vanishes for Kähler manifolds.

The classical supercharges can be calculated using Noether's theorem in a standard way, similar to what we did in the preceding chapter for real supersymmetric sigma models. They read

$$\begin{aligned} Q &= \psi^m \left[h_{m\bar{n}}\dot{\bar{z}}^{\bar{n}} - \frac{i}{2} C_{mp\bar{q}} \psi^p \bar{\psi}^{\bar{q}}\right], \\ \bar{Q} &= \bar{\psi}^m \left[h_{n\bar{m}} z^n - \frac{i}{2} C_{\bar{m}\bar{p}q} \bar{\psi}^{\bar{p}} \psi^q\right] \end{aligned} \tag{9.6}$$

or [98]

$$\begin{aligned} Q &= \psi^m \left[\pi_m + \frac{i}{2}\Gamma_{m,n\bar{p}}\,\psi^n \bar{\psi}^{\bar{p}}\right] = \psi^m \left[\pi_m - \frac{i}{2}\partial_m h_{n\bar{p}}\,\psi^n \bar{\psi}^{\bar{p}}\right], \\ \bar{Q} &= \bar{\psi}^{\bar{m}} \left[\pi_{\bar{m}} - \frac{i}{2}\Gamma_{\bar{m},n\bar{p}}\,\psi^n \bar{\psi}^{\bar{p}}\right] = \psi^{\bar{m}} \left[\pi_{\bar{m}} + \frac{i}{2}\partial_{\bar{m}} h_{n\bar{p}}\,\psi^n \bar{\psi}^{\bar{p}}\right], \end{aligned} \tag{9.7}$$

[1] It is like a gun which an experienced playwright hangs on the wall in the study of the main character in the first act and which fires in the middle of the play. There are many such guns in our book.

where the canonical momenta $\pi_m$ and $\pi_{\bar{m}}$ are obtained by the differentiating of the Lagrangian over $\dot{z}^m, \dot{\bar{z}}^{\bar{m}}$ with keeping $\psi^m, \bar{\psi}^{\bar{m}}$ fixed.

Note that, for Kähler manifolds when the torsions vanish, the supercharges (9.6) reduce to linear combinations of the de Rham supercharges (8.14). This conforms to the fact that the action (9.1) can be expressed in this case in terms of real $\mathcal{N} = 2$ superfields $X^M$ — see Chapter 10.

Introducing the tangent space holomorphic and antiholomorphic Grassmann variables[2] $\psi^c = e^c_m \psi^m$ and $\bar{\psi}^{\bar{c}} = e^{\bar{c}}_{\bar{m}} \bar{\psi}^{\bar{m}}$ and canonical momenta $p_m$ and $p_{\bar{m}}$ given by the derivatives of the Lagrangian over the canonical velocities with fixed $\psi^c$ and $\bar{\psi}^{\bar{c}}$, we derive

$$Q = e^m_c \psi^c \left[ p_m - i\omega_{a\bar{b},m}\, \psi^a \bar{\psi}^{\bar{b}} \right],$$

$$\bar{Q} = e^{\bar{m}}_{\bar{c}} \bar{\psi}^{c} \left[ p_{\bar{m}} - i\omega_{a\bar{b},\bar{m}}\, \psi^a \bar{\psi}^{\bar{b}} \right], \tag{9.8}$$

where $\omega_{a\bar{b},m}$ and $\omega_{a\bar{b},\bar{m}}$ are the spin connections corresponding to the *standard* Levi-Civita affine connections $\Gamma^M_{NP}$ and presented in Eq. (2.25).[3]

Note, however, that the spin connections $\omega_{ab,\bar{m}}$ and $\omega_{\bar{a}\bar{b},m}$, which do not vanish for a generic complex manifold (see the discussion on p. 28–30), do not enter the expressions (9.8) for the supercharges.

The canonical classical Hamiltonian can be represented in the following compact form:

$$H = h^{\bar{n}m} \mathcal{P}_m \bar{\mathcal{P}}_{\bar{n}} - (\partial_k \partial_{\bar{l}}\, h_{m\bar{n}})\, \psi^k \psi^m \bar{\psi}^{\bar{l}} \bar{\psi}^{\bar{n}}, \tag{9.9}$$

where

$$\mathcal{P}_m = p_m - i\omega^{(B)}_{a\bar{b},m}\, \psi^a \psi^{\bar{b}}, \quad \bar{\mathcal{P}}_{\bar{n}} = \bar{p}_{\bar{n}} - i\omega^{(B)}_{a\bar{b},\bar{n}}\, \psi^a \psi^{\bar{b}}. \tag{9.10}$$

In contrast to the supercharges (9.8), the Hamiltonian (9.9) involves the Bismut spin connection

$$\omega^{(B)}_{AB,M} = e_{AN}(\partial_M e^N_B + G^N_{MK} e^K_B) = \omega_{AB,M} + \frac{1}{2}\, e^N_A e^K_B\, C_{NMK}. \tag{9.11}$$

Let us now turn to quantum theory. As in the previous chapter, we resolve the ordering ambiguities as prescribed in Ref. [52], i.e. we use the symmetric Weyl ordering for the supercharges supplemented by a similarity

---

[2]We remind that, in contrast to the real case where we did not have to care about the position of the flat tangent space indices, we should do so for complex geometry — see the footnote on p. 28.

[3]To see that the supercharges (9.8) are mutually conjugate, note the property

$$\omega_{a\bar{b},\bar{m}} = \overline{\omega_{\bar{a}b,m}} = -\overline{\omega_{b\bar{a},m}}.$$

transformation (8.28). We obtain the following geometric expressions for the covariant quantum supercharges:[4]

$$\boxed{\begin{aligned} \hat{Q} &= e^m_c \psi^c \left[ \hat{p}_m - \frac{i}{2} \partial_m (\ln \det e) - i\, \omega_{a\bar{b},m}\, \psi^a \hat{\bar{\psi}}^{\bar{b}} \right], \\ \hat{\bar{Q}} &= e^{\bar{m}}_{\bar{c}} \hat{\bar{\psi}}^{\bar{c}} \left[ \hat{p}_{\bar{m}} - \frac{i}{2} \partial_{\bar{m}} (\ln \det e) - i\, \omega_{a\bar{b},\bar{m}}\, \psi^a \hat{\bar{\psi}}^{\bar{b}} \right], \end{aligned}} \tag{9.12}$$

where $\hat{p}_m, \hat{p}_{\bar{m}}$ and $\hat{\bar{\psi}}^{\bar{a}}$ are the operators $-i\partial/\partial z^m, -i\partial/\partial \bar{z}^{\bar{m}}$ and $\partial/\partial\psi^a$. These expressions almost coincide by form with the classical expressions (9.8), but note the presence of the extra terms $\propto \partial_M \ln \det e$ in the quantum supercharges.

This extra term resembles the term $\propto \partial_M \ln g$ in (8.25). The difference is, however, that Eq. (8.25) defined the "flat" supercharges of the (**1**, **2**, **1**) model, [flat in a sense that they were mutually conjugated with the flat measure (8.28)]. The extra term was absent in the covariant supercharges (8.29) related to (8.25) by a proper similarity transformation.

As for the expressions (9.12), they describe the *covariant* supercharges of the model (9.1) with $\mathcal{W} = 0$, with the similarity transformation $\hat{Q}^{\text{flat}} \to \det^{-1/2}(h)\hat{Q}^{\text{flat}} \det^{1/2}(h)$ being already implemented. To understand the origin and the meaning of the extra terms in the covariant supercharges (9.12), we turn our attention to the second term in the action (9.1). The component Lagrangian acquires the extra term

$$\Delta L_{\mathcal{W}} = i(\dot{\bar{z}}^{\bar{m}} \partial_{\bar{m}} \mathcal{W} - \dot{z}^m \partial_m \mathcal{W}) + 2\, \partial_m \partial_{\bar{n}} \mathcal{W}\, \psi^m \bar{\psi}^{\bar{n}}. \tag{9.13}$$

This brings about the extra terms in the supercharges:

$$\Delta \hat{Q}_{\mathcal{W}} = i\psi^m \partial_m \mathcal{W}, \quad \Delta \hat{\bar{Q}}_{\mathcal{W}} = -i\hat{\bar{\psi}}^{\bar{m}} \partial_{\bar{m}} \mathcal{W}. \tag{9.14}$$

The shift (9.14) amounts to a similarity transformation,

$$\hat{Q}_{\mathcal{W}} = e^{\mathcal{W}} \hat{Q} e^{-\mathcal{W}}. \tag{9.15}$$

This resembles the supercharge (8.45) of the previous chapter that described the de Rham complex deformed by the extra scalar potential,[5] but the dynamical meanings of $W$ and $\mathcal{W}$ are rather different. Adding (9.14) to (9.12), we see that the momenta $\hat{p}_m$ and $\hat{p}_{\bar{m}}$ are substituted by the

[4]To avoid unnecessary complications, we assumed that the complex vielbeins $e^a_m$ are chosen so that their determinant is real: $\det e = \det \bar{e} = \sqrt{\det h}$.

[5]More examples of similarity transformations are given and discussed in Section 12.1.

structures $\hat{p}_m + i\partial_m \mathcal{W}$ and $\hat{p}_{\bar{m}} - i\partial_{\bar{m}} \mathcal{W}$. These structures also enter the Hamiltonian (9.18). This tells us that the vector

$$\boxed{A_M = (i\partial_m \mathcal{W},\ -i\partial_{\bar{m}} \mathcal{W})} \tag{9.16}$$

has the meaning of the *gauge vector potential* [cf. the expression (5.14) for the Pauli Hamiltonian where one should replace $e/c \to -1$]. The corresponding field strength tensor $F_{MN} = \partial_M A_N - \partial_N A_M$ has the components

$$F_{\bar{n}m} = -F_{m\bar{n}} = 2i\partial_m \partial_{\bar{n}} \mathcal{W}, \quad F_{mn} = F_{\bar{m}\bar{n}} = 0. \tag{9.17}$$

The classical Hamiltonian of a generic model (9.1) reads

$$\begin{aligned} H_{\mathcal{W}} = {} & h^{\bar{n}m} \left(\mathcal{P}_m + i\partial_m \mathcal{W}\right) \left(\bar{\mathcal{P}}_{\bar{n}} - i\partial_{\bar{n}} \mathcal{W}\right) \\ & - 2\,\partial_m \partial_{\bar{n}} \mathcal{W}\, \psi^m \bar{\psi}^{\bar{n}} - \left(\partial_k \partial_{\bar{l}}\, h_{m\bar{n}}\right) \psi^k \psi^m \bar{\psi}^{\bar{l}} \bar{\psi}^{\bar{n}}. \end{aligned} \tag{9.18}$$

Besides the shift of momenta $\mathcal{P}_M \to \mathcal{P}_M + A_M$, we observe an extra bifermion term proportional to $F_{MN}$ and a four-fermion term, which vanishes, however, in the Kähler case.

**Remark:** The form (9.17) for $F_{MN}$ means that $F_{AB} = e^M_A e^N_B F_{MN}$ is not generic but belongs to the algebra $u(2n) \subset so(4n)$, in the same way as the Bismut spin connection $\omega^{(B)}_{AB,M}$ does. And this means that $F_{AB}$ commutes with the complex structure $I_{AB}$. The inverse is also true: if $[F, I] = 0$, the holomorphic and antiholomorphic components of $F_{MN}$ vanish. Both statements follow rather straightforwardly from (9.17) and (2.53).

We now note that the contribution (9.14) has a similar form to the extra term $\propto \partial_M \ln \det e$ in the quantum supercharges. More precisely, by choosing $\mathcal{W} = (1/2) \ln \det e = (1/4) \ln \det h$, one can get rid of the extra term in $\hat{Q}$, while choosing $\mathcal{W} = -(1/4) \ln \det h$, one can get rid of the extra term in $\hat{\bar{Q}}$.

**Definition 9.1.** The Abelian gauge field (9.16) derived from the prepotential

$$\boxed{\mathcal{W} = \frac{1}{2} \ln \det h} \tag{9.19}$$

is called by mathematicians the connection of the *canonical line bundle*. The gauge field derived from $\mathcal{W} = (1/4) \ln \det h$ (which interests us more) is called the connection of the *square root* of the canonical line bundle. And the field derived from $\mathcal{W} = -(1/4) \ln \det h$ is the connection of the square root of the anticanonical line bundle.

I gave these translations for the benefit of our readers-mathematicians, but in the following, we stick to the physical terminology.

We are now ready to prove the central theorem, the complex analog of Theorem 8.1 of the preceding chapter.

**Theorem 9.1.** *Consider the supersymmetric system* (9.1) *with* $\mathcal{W} = (1/4)\ln\det h$. *The action of its quantum supercharges*

$$\hat{Q}_{\mathcal{W}} = -ie_c^m \psi^c \left[\frac{\partial}{\partial z^m} + \omega_{a\bar{b},m}\,\psi^a \frac{\partial}{\partial \psi^b}\right],$$

$$\hat{\bar{Q}}_{\mathcal{W}} = -ie_{\bar{c}}^{\bar{m}} \frac{\partial}{\partial \psi^c}\left[\frac{\partial}{\partial \bar{z}^{\bar{m}}} + \frac{1}{2}\partial_{\bar{m}}(\ln\det h) + \omega_{a\bar{b},\bar{m}}\,\psi^a \frac{\partial}{\partial \psi^b}\right] \qquad (9.20)$$

*on the wave functions* $\Psi(z^m, \bar{z}^{\bar{m}}, \psi^a)$ *is isomorphic to the action of the holomorphic exterior derivative operator* $\partial$ *and its conjugate* $\partial^\dagger$ *in the Dolbeault complex.*

**Proof.** The proof is exactly the same as in the de Rham case. Consider a wave function of fermion charge $p$ presented as

$$\Psi_p = A_{m_1\dots m_p}(x)\,\psi^{m_1}\cdots\psi^{m_p} \qquad (9.21)$$

and consider $\hat{Q}_{\mathcal{W}}\Psi_p$. The action of the derivative term is

$$-i\psi^m \frac{\partial}{\partial z^m}\Psi_p = -i\left(\partial_m A_{m_1\dots m_p}\right)\psi^m\psi^{m_1}\cdots\psi^{m_p}$$
$$- ip\,A_{m_1\dots m_p}(\partial_m e_a^{m_1})e_n^a\,\psi^m\psi^n\psi^{m_2}\cdots\psi^{m_p}. \qquad (9.22)$$

The action of the term $\sim \omega_{a\bar{b},m}$ gives

$$-ip\,\omega_{a\bar{b},m}\,e_n^a e_b^{m_1}\,A_{m_1\dots m_p}\,\psi^m\psi^n\psi^{m_2}\cdots\psi^{m_p}.$$

Adding this to the second term in (9.22), we obtain an expression involving the factor

$$X^{m_1} = e_n^a(\partial_m e_a^{m_1} + \omega_{a\bar{b},m}e_b^{m_1})\psi^m\psi^n. \qquad (9.23)$$

Look now at the "full" covariant derivative (1.55) with the Levi-Civita affine connections $\Gamma_{NP}^M$ and the Levi-Civita spin connections. It vanishes according to (1.57). Choose in the LHS of (1.57) all the free indices holomorphic, $M \to m_1, N \to m, A \to a$. We derive

$$\partial_m e_a^{m_1} + \Gamma_{mP}^{m_1}e_a^P + \omega_{aB,m}e_B^{m_1} = 0. \qquad (9.24)$$

Note now that the index $P$ is necessarily holomorphic because $e_a^{\bar{p}}$ does not exist. Also, the second term in (9.24) can be written as[6] $\omega_{a\bar{b},m}e_b^{m_1}$. Bearing

[6]Recall that $Y_A Z_A \equiv Y_a Z_{\bar{a}} + Y_{\bar{a}} Z_a$.

all this in mind, we derive

$$X^{m_1} = -\Gamma^{m_1}_{mn}\, \psi^m \psi^n = 0. \tag{9.25}$$

We are left with the result

$$\hat{Q}_{\mathcal{W}}\, \Psi_p = -i(\partial_m A_{m_1 \ldots m_p})\, \psi^m \psi^{m_1} \cdots \psi^{m_p}, \tag{9.26}$$

which has the same form, up to the factor $-i$, as Eq. (2.41) defining the action of the exterior holomorphic derivative $\partial$ on the holomorphic $(p, 0)$ forms. One has only to map $\psi^m \longrightarrow dz^m$.

The operator $\partial^\dagger$ is adjoint to $\partial$ with the measure (2.39) including the factor $\det h = \sqrt{\det g}$. And the operator $\hat{\bar{Q}}_{\mathcal{W}}$ is adjoint to $\hat{Q}_{\mathcal{W}}$ with the same covariant measure. Thus, if $\hat{Q}_{\mathcal{W}}$ is isomorphic to $\partial$, $\hat{\bar{Q}}_{\mathcal{W}}$ must be isomorphic to $\partial^\dagger$. □

**Remark:** This proof "explains" the "experimental fact" that the quantum supercharges of this model involve the *conventional* spin connection $\omega_{a\bar{b},m}$ rather than the Bismut spin connection (9.11). In the latter case, we would have the torsionful $G^{m_1}_{mn}$ instead of $\Gamma^{m_1}_{mn}$ in Eq. (9.25), the factor $X^{m_1}$ would not vanish and $\hat{Q}_{\mathcal{W}}$ would not have a nice geometric interpretation.

Consider now the gauge field derived from $\mathcal{W} = -(1/4) \ln \det h$ and consider the action of the corresponding quantum supercharges on the *dual* wave functions $\Psi(z^m, \bar{z}^m; \bar{\psi}^a)$ rather than on $\Psi(z^m, \bar{z}^m; \psi^a)$ (it is like going from the coordinate to momentum representation in ordinary quantum mechanics). One can prove in exactly the same way as above that the action of $\hat{\bar{Q}}_{\mathcal{W}}$ is isomorphic in this case to the action of the operator $\bar{\partial}$ on the forms (2.43) of the anti-Dolbeault complex. And $\hat{Q}_{\mathcal{W}}$ is isomorphic to $\bar{\partial}^\dagger$.

For an arbitrary $\mathcal{W}$, we are dealing with the *twisted Dolbeault complex*. The following fact, well known to mathematicians,[7] is a rather straightforward corollary of (9.12) and (9.14).

**Theorem 9.2.** *The action of a generic $\hat{Q}_{\mathcal{W}}$ on the wave functions $\Psi(z^m, \bar{z}^m; \psi^a)$ maps to the action of the nilpotent operator*

$$\boxed{\partial_{\mathcal{W}} = \partial - \partial\left(\mathcal{W} - \frac{1}{4} \ln \det h\right) \wedge} \tag{9.27}$$

[7] See e.g. Propositions 1.4.23 and 1.4.25 in Ref. [102].

*in the Dolbeault complex. The action of a generic* $\hat{\bar{Q}}_\mathcal{W}$ *on the wave functions* $\Psi(z^m, \bar{z}^m; \bar{\psi}^a)$ *maps to the action of the operator*

$$\boxed{\bar{\partial}_\mathcal{W} = \bar{\partial} + \bar{\partial}\left(\mathcal{W} + \frac{1}{4}\ln\det h\right)\wedge} \tag{9.28}$$

*in the anti-Dolbeault complex.*

**Remark for the future:** If the manifold is Kähler, the sum $\hat{Q}_\mathcal{W} + \hat{\bar{Q}}_\mathcal{W}$ is isomorphic to the *Dirac operator* living on the manifold with the gauge field (9.16). For a generic complex manifold, this sum is isomorphic to a *deformed* Dirac operator including the terms proportional to torsions [see Eq. (14.22)]. The Dirac complex and its relationship to the Dolbeaux complex are discussed in detail in Chapters 14, 15.

### 9.1.1 *Holomorphic torsions*

In contrast to the superpotential term (8.44) for the de Rham supersymmetric sigma model, the gauge field term in (9.1) involving $\int d\bar{\theta}d\theta\,\mathcal{W}(Z,\bar{Z})$ cannot be considered as a deformation. We see in Chapter 14 that $\mathcal{W}(Z,\bar{Z})$ cannot be infinitesimal — certain topological invariants expressed through the integrals involving[8] $\mathcal{W}$ are equal either to zero (when $\mathcal{W} = 0$) or to nonzero integers.

But the model (9.1) can also be deformed by adding the term [97]

$$\boxed{S_\mathcal{B} = \frac{1}{4}\int d\bar{\theta}d\theta dt\left[\mathcal{B}_{mn}(Z,\bar{Z})\,DZ^mDZ^n - \bar{\mathcal{B}}_{\bar{m}\bar{n}}(Z,\bar{Z})\,\bar{D}\bar{Z}^{\bar{m}}\bar{D}\bar{Z}^{\bar{n}}\right]} \tag{9.29}$$

with an arbitrary antisymmetric complex $\mathcal{B}_{mn}$ and its conjugate $\bar{\mathcal{B}}_{\bar{m}\bar{n}}$. This deformation is similar to the deformation (8.55) in the de Rham case, but $\mathcal{B}_{mn}$ carries now holomorphic indices.

The component Lagrangian now acquires the contribution

$$\Delta L_\mathcal{B} = -\,3i\,\partial_{[k}\mathcal{B}_{mn]}\,\dot{z}^k\psi^m\psi^n \;-\; \partial_{\bar{p}}\partial_k\mathcal{B}_{mn}\,\bar{\psi}^{\bar{p}}\psi^k\psi^m\psi^n + \text{ c.c.} \tag{9.30}$$

Note that adding this extra term to the Lagrangian (9.3) gives the Lagrangian of the same form where the torsion tensor now includes besides (9.4) pure holomorphic components $C_{kmn} = 12\,\partial_{[k}\mathcal{B}_{mn]}$ and $C_{\bar{k}\bar{m}\bar{n}} = 12\,\partial_{[\bar{k}}\bar{\mathcal{B}}_{\bar{m}\bar{n}]}$ (see also Theorem 9.5). Thus, the deformation (9.30) brings

[8]The simplest such invariant is the magnetic flux (5.25).

about extra *torsions* and not "quasitorsions" as was the case for the deformation (8.55).

The corresponding contributions to the supercharges are

$$\Delta\hat{Q}_{\mathcal{B}} = i\,\partial_k \mathcal{B}_{mn}\,\psi^k\psi^m\psi^n, \quad \Delta\hat{\bar{Q}}_{\mathcal{B}} = i\,\partial_{\bar{k}}\bar{\mathcal{B}}_{\bar{m}\bar{n}}\,\hat{\bar{\psi}}^{\bar{k}}\hat{\bar{\psi}}^{\bar{m}}\hat{\bar{\psi}}^{\bar{n}}. \tag{9.31}$$

Consider the $(2,0)$-form $\mathcal{B} = \mathcal{B}_{mn}\,dz^m \wedge dz^n$ and the complex conjugate $(0,2)$-form $\bar{\mathcal{B}}$. It is easy to see the following [97, 103]:

**Theorem 9.3.** *The action of a generic* $\hat{Q}_{\mathcal{W},B}$ *on the wave functions* $\Psi(z^m, \bar{z}^m; \psi^a)$ *maps to the action of the nilpotent operator*

$$\partial_{W,B} = \partial_{\mathcal{W}} - \partial\mathcal{B}\wedge \tag{9.32}$$

*in the Dolbeault complex*[9] *with* $\partial_{\mathcal{W}}$ *defined in Eq. (9.27). The action of a generic* $\hat{\bar{Q}}_{\mathcal{W},B}$ *on the wave functions* $\Psi(z^m, \bar{z}^m; \bar{\psi}^a)$ *maps to the action of the operator*

$$\bar{\partial}_{\mathcal{W},B} = \bar{\partial}_{\mathcal{W}} - \bar{\partial}\bar{\mathcal{B}}\wedge \tag{9.33}$$

*in the anti-Dolbeault complex with* $\bar{\partial}_{\mathcal{W}}$ *defined in Eq. (9.28).*

The deformations (9.32) and (9.33) are possible when the complex dimension $d$ is 3 or more. When $d \geq 5$, one can introduce a more complicated deformation

$$S_{\mathcal{E}} \propto \int d\bar{\theta}d\theta dt\,\mathcal{E}_{mnpq}\,DZ^m DZ^n DZ^p DZ^q + \text{c.c.} \tag{9.34}$$

Starting from $d = 7$, one can write a contribution with six covariant derivatives, etc.

## 9.2 $\mathcal{N} = 1$ superfield description

Take $D = 2d$ $\mathcal{N} = 1$ real superfields $\mathcal{X}^M = x^M + i\Theta\Psi^M$ [see Eq. (7.7)] and write the action in the following form [89]:

$$\boxed{\begin{aligned} S = \frac{i}{2}\int d\Theta dt\, g_{MN}(\mathcal{X})\dot{\mathcal{X}}^M\mathcal{D}\mathcal{X}^N \\ -\frac{1}{12}\int d\Theta dt\, C_{KLM}(\mathcal{X})\mathcal{D}\mathcal{X}^K\mathcal{D}\mathcal{X}^L\mathcal{D}\mathcal{X}^M, \end{aligned}} \tag{9.35}$$

with $\mathcal{D}$ given in Eq. (7.13). We added to the action corresponding to (8.3) the extra term involving a totally antisymmetric tensor $C_{KLM}$ playing the

[9]More precisely, in the *twisted deformed* Dolbeault complex, where "twisted" means the presence of a gauge field and "deformed" means the presence of an extra (2,0)-form $\mathcal{B}$.

role of the torsion. Doing the integral $\int d\Theta$, we reproduce the Lagrangian (9.3).

The $\mathcal{N} = 1$ supersymmetry of the action (9.35) is manifest. The integral $\int d\Theta dt$ is invariant under the change of variables (7.1) that generate the shift of the superfields

$$\delta_0 \mathcal{X}^M = \mathcal{X}^M(\Theta + E_0, t + iE_0\Theta) - \mathcal{X}^M(\Theta, t). \tag{9.36}$$

The components of $\mathcal{X}^M$ transform as in (7.9):

$$\delta_0 x^M = iE_0 \Psi^M, \quad \delta_0 \Psi^M = -E_0 \dot{x}^M. \tag{9.37}$$

We now require the action to be invariant under the following extra supersymmetry transformation:

$$\delta \mathcal{X}^M = E I_N{}^M \mathcal{D}\mathcal{X}^N, \tag{9.38}$$

where $I_N{}^M(\mathcal{X}^P)$ is a tensor to be specified below. The components in $\mathcal{X}^M$ transform as

$$\begin{aligned} \delta x^M &= iEI_N{}^M \Psi^N, \\ \delta \Psi^M &= E\left(I_N{}^M \dot{x}^N - i\,\partial_S I_N{}^M \Psi^S \Psi^N\right). \end{aligned} \tag{9.39}$$

The invariances (9.37) and (9.39) give us two real conserved supercharges $\mathcal{Q}_0$ and $\mathcal{Q}$.

It is rather evident that the transformations (9.36) and (9.38) commute. Indeed, $\mathcal{Y}^M = \delta\mathcal{X}^M$ is a superfield, and hence $\delta_0(\delta\mathcal{X}^M)$ and $\delta(\delta_0\mathcal{X}^M)$ coincide, having both the form (9.36) with $\mathcal{X}^M$ replaced by $\mathcal{Y}^M$. A corollary of this is the vanishing of the anticommutator $\{\hat{\mathcal{Q}}, \hat{\mathcal{Q}}_0\}$ of the quantum supercharges.

A nontrivial requirement that we impose is that the commutator of two transformations (9.38) boils down to the time translation. Then, the square of the supercharge associated with (9.39) coincides with the Hamiltonian, and we obtain the minimal $\mathcal{N} = 2$ supersymmetry algebra

$$\hat{\mathcal{Q}}_0^2 = \hat{\mathcal{Q}}^2 = \hat{H}, \quad \{\hat{\mathcal{Q}}_0, \hat{\mathcal{Q}}\} = 0. \tag{9.40}$$

This requirement and the requirement of invariance of the action under (9.38) imposes nontrivial constraints on the *complex structure* $I_N{}^M$ and the torsion tensor.[10] These constraints are specified by the following theorem:

**Theorem 9.4.** *The action* (9.35) *is invariant under* (9.38) *and the algebra* (9.40) *holds iff the following set of conditions is satisfied:*

[10]We see soon that $I_N{}^M$ has all the properties of an integrable complex structure and should be interpreted as such. And the tensor $C_{KLM}$ represents a sum of the torsion entering the relevant Bismut affine connection (9.5) and a possible pure holomorphic part coming from the deformation (9.30).

(1) $I^2 = -\mathbb{1}$.
(2) *The Nijenhuis tensor* (2.60) *vanishes.*
(3) *The matrix* $I_{MN} = g_{NK} I_M{}^K$ *is skew-symmetric.*

(4)
$$\nabla_L^{(B)} I_N{}^M + \nabla_N^{(B)} I_L{}^M = 0, \tag{9.41}$$

*where* $\nabla_L^{(B)}$ *is the Bismut covariant derivative with the connection* (9.5).

(5)
$$(\partial_L C_{[MSN})I_{R]}{}^L + I_{[N}{}^L(\partial_R C_{MS]L}) - 2C_{L[MS}(\partial_R I_{N]}{}^L) = 0. \tag{9.42}$$

This theorem was proved first in Ref. [35] by explicit component calculations. We give here, following Ref. [104], an alternative proof based on the language of $\mathcal{N} = 1$ superfields.

**Proof.** The conditions 1 and 2 follow from the algebra (5.1). To see that, note first that the law (9.38) implies

$$\begin{aligned}\delta(\mathcal{D}\mathcal{X}^N) &= \mathcal{D}(\delta\mathcal{X}^N) = -E\mathcal{D}(I_L{}^N \mathcal{D}\mathcal{X}^L) \\ &= -E(\partial_K I_L{}^N)\mathcal{D}\mathcal{X}^K\mathcal{D}\mathcal{X}^L + iEI_K{}^N\dot{\mathcal{X}}^K.\end{aligned} \tag{9.43}$$

The commutator of two supersymmetry transformations (9.38) is then derived to be

$$\begin{aligned}(\delta_2\delta_1 - \delta_1\delta_2)\,\mathcal{X}^M &= 2iE_1E_2(I^2)_K{}^M\dot{\mathcal{X}}^K + 2E_1E_2\Big[I_K{}^L\,(\partial_L I_N{}^M) \\ &\quad + (\partial_N I_K{}^L)\,I_L{}^M\Big]\,\mathcal{D}\mathcal{X}^K\mathcal{D}\mathcal{X}^N.\end{aligned} \tag{9.44}$$

If we want it to coincide with $-2iE_1E_2\,\partial_t\mathcal{X}^M$ [as is dictated by $\hat{\mathcal{Q}}^2 = \hat{H}$], the condition $I^2 = -\mathbb{1}$, as well as the identity

$$(\partial_L I_{[N}{}^M)\,I_{K]}{}^L + (\partial_{[N} I_{K]}{}^L)\,I_L{}^M = 0 \tag{9.45}$$

follow. Multiplying that by $I_M{}^S$, using $I^2 = -\mathbb{1}$ and flipping the derivative in the first term, the condition (9.45) can be brought to the form $\mathcal{N}_{[KN]}{}^S = 0$, with $\mathcal{N}_{[KN]}{}^S$ defined in (2.60).

The conditions 3 and 4 follow from the vanishing of the variation of the action under (9.38). The calculation gives

$$\begin{aligned}\delta S = E\int d\Theta dt\, I_{MN}\,\dot{\mathcal{X}}^M\dot{\mathcal{X}}^N - \frac{iE}{4}\int d\Theta dt\, P_{MSN}\,\dot{\mathcal{X}}^M\mathcal{D}\mathcal{X}^S\mathcal{D}\mathcal{X}^N \\ +\frac{E}{12}\int d\Theta dt\, T_{RMSN}\,\mathcal{D}\mathcal{X}^R\mathcal{D}\mathcal{X}^M\mathcal{D}\mathcal{X}^S\mathcal{D}\mathcal{X}^N,\end{aligned} \tag{9.46}$$

where

$$P_{MSN} = 2\nabla_M I_{SN} + 4\nabla_N I_{(MS)} - C^L{}_{SN} I_{ML},$$
$$T_{RMSN} = (\partial_L C_{MSN}) I_R{}^L - 3C_{LSN}\left(\partial_R I_M{}^L\right). \tag{9.47}$$

Note that $\nabla_M$ entering (9.47) are the ordinary Levi-Civita covariant derivatives.

Let us first see what happens if the torsion vanishes. In this case, the second line in (9.46) vanishes identically and the first line vanishes iff $I_{(MN)} = 0$ (which means together with the conditions 1 and 2, which we already derived, that $I_{MN}$ is an integrable complex structure) and $\nabla_M I_{SN} = 0$. We arrive at the *Kähler* geometry.

If $C_{MSN}$ does not vanish, the situation is somewhat more complicated. One of the complications is that the second and third integrals in (9.46) are not completely independent: a certain part of the second integral gives the same structure as the third one after flipping the derivative $\mathcal{D}$.

Indeed, look at the second term in the RHS of Eq. (9.46). Only the antisymmetric in $S \leftrightarrow N$ part of $P_{MSN}$ contributes. We represent

$$P_{M[SN]} = P_{[MSN]} + \frac{1}{3}\left(2P_{M[SN]} + P_{S[MN]} - P_{N[MS]}\right), \tag{9.48}$$

so that the structure in the parentheses gives zero after complete antisymmetrization over $[MSN]$. After trading $\dot{\mathcal{X}}^M$ for $i\mathcal{D}^2\mathcal{X}^M$ and integrating by parts, the integral involving $P_{[MSN]}$ gives

$$\int d\Theta dt\, P_{[MSN]} \dot{\mathcal{X}}^M \mathcal{D}\mathcal{X}^S \mathcal{D}\mathcal{X}^N$$
$$= -\frac{i}{3}\int d\Theta dt\, \partial_R P_{[MSN]} \mathcal{D}\mathcal{X}^R \mathcal{D}\mathcal{X}^M \mathcal{D}\mathcal{X}^S \mathcal{D}\mathcal{X}^N. \tag{9.49}$$

We thus present the variation as a sum of three independent structures

$$\delta S = E\int d\Theta dt\, I_{MN}\, \dot{\mathcal{X}}^M \dot{\mathcal{X}}^N$$
$$-\frac{iE}{12}\int d\Theta dt\, \left(2P_{M[SN]} + P_{S[MN]} - P_{N[MS]}\right) \dot{\mathcal{X}}^M \mathcal{D}\mathcal{X}^S \mathcal{D}\mathcal{X}^N$$
$$+\frac{E}{12}\int d\Theta dt\, \left(T_{RMSN} - \partial_{[R} P_{MSN]}\right) \mathcal{D}\mathcal{X}^R \mathcal{D}\mathcal{X}^M \mathcal{D}\mathcal{X}^S \mathcal{D}\mathcal{X}^N. \tag{9.50}$$

It vanishes iff

$$I_{(MN)} = 0, \tag{9.51}$$

$$2P_{M[SN]} + P_{S[MN]} - P_{N[MS]} = 0, \tag{9.52}$$

and

$$T_{[RMSN]} = \partial_{[R}P_{MSN]}. \tag{9.53}$$

Equation (9.51) gives the requirement 3 in the list above. The condition (9.52), after taking account of (9.51) and after substituting $P_{M[SN]}$ from Eq. (9.47), yields

$$2\nabla_M^{(B)} I_{SN} + \nabla_S^{(B)} I_{MN} - \nabla_N^{(B)} I_{MS} = 0 \tag{9.54}$$

with *Bismut* covariant derivatives. Symmetrizing this expression over $M \leftrightarrow S$ and lifting the index $N$ (we are allowed to do so due to the property $\nabla_M^{(B)} g^{QN} = 0$), we arrive at (9.41). Inversely, the LHS of Eq. (9.54) can be rewritten as $A_{(MS)N} - A_{(MN)S}$, where

$$A_{(MS)N} = \nabla_M^{(B)} I_{SN} + \nabla_S^{(B)} I_{MN}.$$

If $A_{(MS)N} = 0$, the validity of the identities (9.54) and hence (9.52) follows.

Finally, substituting into (9.53) the expression for $T_{RMSN}$ from (9.47) and

$$P_{[MSN]} = \left(\frac{2}{3}\partial_M I_{SN} - \frac{1}{3}C^L{}_{SN} I_{ML}\right) + \text{cycle}\,(M, S, N) \tag{9.55}$$

(the terms $\propto \Gamma^P_{MN}$ cancel), we derive the identity (9.42). □

**Remark:** If the dimension of the manifold is large enough, we can also consider an action including besides (9.35) the term $\sim \int dt d\Theta\, D_{KLMQR}$ $\mathcal{D}\mathcal{X}^K \mathcal{D}\mathcal{X}^L \mathcal{D}\mathcal{X}^M \mathcal{D}\mathcal{X}^Q \mathcal{D}\mathcal{X}^R$ and the terms including 7, 9 and more $\mathcal{D}\mathcal{X}$ factors. And one could then study the conditions to be imposed on the tensors $D_{KLMQR}$, etc., under which the model enjoys $\mathcal{N} = 2$ supersymmetry. Being expressed in terms of $\mathcal{N} = 2$ superfields, the action of such model should include the contributions (9.34), etc. That was never accurately done, however. Without doubt, this problem sounds a little academic, but a scholar with a flair for math may like it.

The conditions 1–3 above concern only the tensor $I$. According to Theorem 2.6 (the Newlander–Nirenberg theorem), they are necessary and sufficient for $I_{MN}$ to be an integrable complex structure. Actually, we have just given an alternative proof of the *necessity* of these conditions using supersymmetric methods.[11] Indeed, suppose the complex coordinates can be

[11] In fact, our reasoning was simply a translation into the supersymmetric language of the traditional proof of this fact given on p. 38. In particular, the commutator $[\delta_1, \delta_2]$ in the LHS of Eq. (9.44) has the same meaning as the commutator $[\mathcal{D}_M, \mathcal{D}_N]$ in the LHS of Eq. (2.62).

introduced. Then, the complex supercharges (9.12) can be written. Their real and imaginary parts satisfy the algebra (9.40). As we have just proved, the conditions 1–3 including the integrability condition 2 follow.

But the supersymmetric description also allows one to understand why these conditions are *sufficient* to introduce the complex coordinates $z^n$. The heuristic argument is the following. From Theorem 9.4, we have learned that, if $I^2 = -\mathbb{1}$ and the Nijenhuis tensor vanishes, the generators of the transformations (9.37) and (9.39) satisfy the $\mathcal{N} = 2$ supersymmetry algebra (9.40). We have altogether $2d$ real dynamical bosonic and $2d$ real dynamical fermionic variables. These variables provide an infinite-dimensional representation for the $\mathcal{N} = 2$ algebra, but this representation must be reducible, as the only known irreducible $\mathcal{N} = 2$ representation including the equal number of bosonic and fermionic dynamical variables is (**2**, **2**, **0**). If so, the representation (**2d**, **2d**, **0**) must be decomposable into a direct sum of $d$ representations (**2**, **2**, **0**) which transform as in (7.26). But this is tantamount to saying that the solution to the conditions (2.52) for $z^n$ exists.

In Chapter 2, we did not prove the NN theorem rigorously as we did not prove the convergence of the infinite series (2.73). But a rigorous mathematical proof of this theorem exists. That is why it is probably better to look at the problem from the opposite direction: *The NN theorem can be interpreted as a statement that a superfield representation of the $\mathcal{N} = 2$ superalgebra including $2d$ real dynamical bosonic and $2d$ real dynamical fermionic variables is necessarily reducible and can be decomposed into a direct sum of $d$ (**2**, **2**, **0**) representations* [12].

Look now at the conditions (9.41) and (9.42). Equation (9.41) reminds the condition $\nabla_N^{(B)} I_L{}^M = 0$, which uniquely defines the Bismut torsion (2.76) for an integrable complex structure $I_{MN}$. However, in our case where the covariant derivative of $I$ vanishes only after symmetrization over $N \leftrightarrow L$ and may not vanish otherwise, the torsion is not rigidly fixed.

The freedom of choice for the torsion is, however, limited: the condition (9.42) should be imposed.

To understand better its meaning, express it in the language of differential forms. Introduce the operator $\iota$ that acts on a generic $P$-form $\alpha = \alpha_{M_1 \ldots M_P} dx^{M_1} \wedge \cdots \wedge dx^{M_P}$ in a following way:

$$\iota\alpha = P\alpha_{N[M_2 \ldots M_P} I^N{}_{M_1]} dx^{M_1} \wedge \cdots \wedge dx^{M_P}.$$

For a form $\alpha_{p,q}$ with $p$ holomorphic and $q$ antiholomorphic indices with respect to the complex structure $I$ (so that $P = p + q$), the action of $\iota$ boils down to multiplication by $i(p - q)$.

We can prove the following theorem.

**Theorem 9.5.** [35,104] *Let $(g, I, C)$ satisfy conditions 1–5 of the preceding theorem. Consider the holomorphic decomposition of the torsion form with respect to the complex structure $I$:*

$$C = \sum_{q=0}^{3} C_{3-q,q} = C_{3,0} + C_{2,1} + C_{1,2} + C_{0,3}. \tag{9.56}$$

*Then,* (i) *the mixed part $C_{2,1} + C_{1,2}$ is the Bismut torsion* (2.18) *and* (ii) *the holomorphic and antiholomorphic parts are closed,*

$$\boxed{\partial C_{3,0} = \bar{\partial} C_{0,3} = 0.} \tag{9.57}$$

**Proof.** We express (9.41) in complex coordinates. The complex structure $I$ acquires a simple form (2.53). Choose $L = l, N = \bar{n}, M = \bar{m}$. Look at the first term in (9.41). We have $\partial_l \delta_{\bar{n}}^{\bar{m}} = 0$ and the two other contributions cancel out. For the second term, they add up, and we obtain the condition

$$G_{\bar{n}l}^{\bar{m}} = 0 \Longrightarrow G_{m,\bar{n}l} = 0, \tag{9.58}$$

as in (2.21). The components $G_{\bar{m},n\bar{l}}$ are the complex conjugates of $G_{m,\bar{n}l}$ and also vanish; this can also be derived directly from (9.41) if one chooses $L = l, N = \bar{n}, M = m$ there. From this, from the definition (9.5), and the expressions (2.17) for the Christoffel symbols, we may derive that the part in the torsion tensor corresponding to $C_{2,1} + C_{1,2}$ has the Bismut form (9.4). When all three indices in (9.41) are holomorphic or all of them are antiholomorphic, the equality is satisfied automatically. The choice $L = l, N = n, M = \bar{m}$ does not bring about new information.

One can also note that the forms $C_{2,1}$ and $C_{1,2}$ are exact:

$$C_{2,1} = 3C_{ml\bar{n}}\, dz^m \wedge dz^l \wedge d\bar{z}^{\bar{n}} = -6\partial\omega,$$

where $\omega = h_{l\bar{n}} dz^l \wedge d\bar{z}^{\bar{n}}$. Similarly, $C_{1,2} = 6\bar{\partial}\omega$.

Consider now the condition (9.42) and choose all the free indices holomorphic: $M = m, N = n, S = s, R = r$. Bearing in mind (2.53), we immediately derive $\partial_{[r} C_{mns]} = 0$, which means that $C_{3,0}$ is closed. Choosing all the indices antiholomorphic, we derive the closedness of $C_{0,3}$.

Note finally that the reality of $C$ in (9.56) implies that $C_{\bar{m}\bar{n}\bar{s}} = \overline{C_{mns}}$.

□

We can now make contact with the $\mathcal{N} = 2$ description of the preceding section. If $C_{3,0} = 0$, the torsion is Bismut and the system (9.35) is equivalent to the system (9.1) with $\mathcal{W} = 0$. As we learned in the preceding section, the latter maps to the twisted Dolbeault complex with the effective gauge potential [see (9.27) and (9.16)]

$$A_M^{\rm eff} = \left( -\frac{i}{4}\partial_m \ln \det h, \frac{i}{4}\partial_{\bar m} \ln \det h \right). \tag{9.59}$$

And the torsion with nonzero $C_{3,0}$ gives us the deformed twisted Dolbeault complex involving the extra contribution (9.29) in the action. Then, $C_{3,0} = 12\partial\mathcal{B}$ [see a remark after (9.30)]. And it is closed, indeed.

Bearing in mind the established mapping between the two approaches, it is easy to write the expressions for the real supercharges $\mathcal{Q}_0$ and $\mathcal{Q}$. They can be found as real and imaginary parts of the complex classical supercharge, taken with the factor 2. [With the chosen conventions, the complex supercharge is expressed as $(\mathcal{Q}_0 + i\mathcal{Q})/2$ — see Eq. (5.2).] When there is no deformation and $\mathcal{W} = 0$, we derive, starting from (9.8), the following expressions for the classical supercharges [7, 105, 106]:

$$\begin{aligned} \mathcal{Q}_0 &= \Psi^M \left( P_M - \frac{i}{2}\omega_{AB,M}\Psi^A\Psi^B + \frac{i}{12}C_{MKP}\Psi^K\Psi^P \right), \\ \mathcal{Q} &= \Psi^N I_N{}^M \left( P_M - \frac{i}{2}\omega_{AB,M}\Psi^A\Psi^B - \frac{i}{4}C_{MKP}\Psi^K\Psi^P \right), \end{aligned} \tag{9.60}$$

where $\omega_{AB,M}$ are the Levi-Civita spin connections and $P_M$ is obtained from the variation of the Lagrangian with fixed $\Psi^A$. The terms $\propto C_{MKP}$ cancel the holomorphic and antiholomorphic connections $\omega_{ab,\bar m}$ and $\omega_{\bar a\bar b,m}$, which are present in $\omega_{AB,M}$ [see (2.25) and (2.26)] but absent in (9.8). Bearing in mind the normalization $\{\Psi^A, \Psi^B\}_P = i\delta^{AB}$, the supercharges (9.60) obey the classical supersymmetry algebra.

The corresponding covariant quantum supercharges that are obtained by the ordering procedure outlined on p. 179 have the form

$$\begin{aligned} \hat{\mathcal{Q}}_0 &= \hat\Psi^M \left( \hat P_M - \frac{i}{8}\partial_M \ln\det g - \frac{i}{2}\omega_{AB,M}\hat\Psi^A\hat\Psi^B + \frac{i}{12}C_{MKP}\hat\Psi^K\hat\Psi^P \right), \\ \hat{\mathcal{Q}} &= \hat\Psi^N I_N{}^M \left( \hat P_M - \frac{i}{8}\partial_M \ln\det g - \frac{i}{2}\omega_{AB,M}\hat\Psi^A\hat\Psi^B - \frac{i}{4}C_{MKP}\hat\Psi^K\hat\Psi^P \right), \end{aligned} \tag{9.61}$$

where $\hat P_M = -i\partial_M$ and $\hat\Psi_A$ are the operators obeying the Clifford algebra

$$\{\hat\Psi_A, \hat\Psi_B\} = \delta_{AB} \quad \Rightarrow \quad \{\hat\Psi^M, \hat\Psi^N\} = g^{MN}. \tag{9.62}$$

Naturally, the supercharges (9.61) coincide up to the factor 2 with the Hermitian and anti-Hermitian parts of the complex quantum supercharge $\hat{Q}$ in Eq. (9.12).

Adding to the $\mathcal{N} = 1$ action (9.35) the term with five factors $\mathcal{D}\mathcal{X}^M$ in the integrand, one can reproduce the term (9.34) in the $\mathcal{N} = 2$ action, etc.

To describe in the $\mathcal{N} = 1$ language the system (9.1) with nonzero $\mathcal{W}$, one has to add to (9.35) an extra term with only one $\mathcal{D}\mathcal{X}^M$ factor:

$$\boxed{S_A = -i \int d\Theta dt \, A_M(\mathcal{X}) \, \mathcal{D}\mathcal{X}^M.} \tag{9.63}$$

In components, this gives

$$S_A = \int dt \left[ -A_M(x) \, \dot{x}^M + \frac{i}{2} F_{MN}(x) \, \Psi^M \Psi^N \right]. \tag{9.64}$$

**Theorem 9.6.** *The action* (9.63) *is invariant under the transformation* (9.38) *provided*

$$F_K{}^M I_M{}^L - I_K{}^M F_M{}^L = 0. \tag{9.65}$$

**Proof.** Bearing in mind (9.43), the variation of $S_A$ is

$$\delta S_A = -iE \int d\Theta dt \, [(\partial_K A_M) I_L{}^K \mathcal{D}\mathcal{X}^M \mathcal{D}\mathcal{X}^L + A_M I_L{}^M \mathcal{D}\mathcal{D}\mathcal{X}^L$$
$$+ A_M (\partial_K I_L{}^M) \mathcal{D}\mathcal{X}^K \mathcal{D}\mathcal{X}^L], \tag{9.66}$$

where we replaced $\partial/\partial t \to i\mathcal{D}\mathcal{D}$. Next, we integrate by parts the middle term. The part coming from differentiating $I_L{}^M(\mathcal{X})$ cancels the third term and we are left with

$$\delta S_A = i\epsilon \int d\Theta dt \, (\partial_K A_M) \, [I_L{}^M \mathcal{D}\mathcal{X}^K - I_L{}^K \mathcal{D}\mathcal{X}^M] \mathcal{D}\mathcal{X}^L. \tag{9.67}$$

After some massaging, this can be brought to the form

$$\delta S_A = \frac{i\epsilon}{2} \int d\Theta dt \, (I_M{}^K F_{KL} - F_M{}^K I_{KL}) \, \mathcal{D}\mathcal{X}^M \mathcal{D}\mathcal{X}^L, \tag{9.68}$$

which vanishes provided (9.65) holds. □

In fact, we obtained nothing new but simply reproduced, using the $\mathcal{N} = 1$ language, the result obtained earlier: to keep $\mathcal{N} = 2$ supersymmetry, the commutator $[F, I]$ must vanish. (See the *Remark* after Eq. (9.18).)

## 9.3 Dolbeault complex on $S^4\backslash\{\cdot\}$

We saw in the previous sections that both the classical Dolbeault complex and the Dolbeault complex twisted by the gauge field and/or deformed by the holomorphic torsions can be described in the language of supersymmetry. However, as was repeatedly announced before, the latter is more powerful than traditional mathematical methods. It allows one to describe systems that have not yet attracted much attention from mathematicians.

Consider $S^4$. It is not a complex manifold. Still, one can try to proceed along the same lines as for $S^2$: to trace at its south pole the four-dimensional tangent hyperplane and to make the stereographic projection. The points of $S^4$ are then parametrized by the Cartesian coordinates ($x^{M=1,2,3,4}$) of this hyperplane. We introduce the following complex coordinates:

$$z^1 = \frac{x^1 + ix^2}{\sqrt{2}}, \quad z^2 = \frac{x^3 + ix^4}{\sqrt{2}}. \tag{9.69}$$

The following expression for the metric can be written as follows [cf. Eq. (2.6); the radius of such sphere is equal to $1/\sqrt{2}$][12]:

$$ds^2 = \frac{2dz_k d\bar{z}_k}{(1 + z_j\bar{z}_j)^2}. \tag{9.70}$$

This metric is perfectly well defined everywhere except the north pole of $S^4$ where one of $z_j$ hits infinity. In contrast to $S^2$ where one can make a similar map for the northern hemisphere in such a way that the transition functions between the two sets of complex coordinates are analytic, one cannot do so for $S^4$ (see p. 26). But never mind. Just consider $S^4$ *without* the north pole and use the expression (9.70) for the metric.

Topologically, our manifold is now $\mathbb{R}^4$. In contrast to $S^4$, $\mathbb{R}^4$ *is* a complex manifold and one can well introduce there complex coordinates and define the Dolbeault complex. However, $\mathbb{R}^4$ is not compact, and one can formulate there essentially different spectral problems for different Hamiltonians associated with different Hilbert spaces distinguished by different measures. Thus, for the simplest spectral problem on $\mathbb{R}^4$, the Hamiltonian is the flat Laplacian. Its spectrum is continuous.

---

[12]The metric (9.70) of $S^4$ is conformally flat. In this situation, it is not convenient anymore to distinguish between the covariant and contravariant indices. Neither will we distinguish in this section the indices $j$ and $\bar{j}$. Further, we choose the vielbein in the conformally flat form, $e^j_a = f\,\delta^j_a$ and will not distinguish between the world and tangent space indices, displaying explicitly the conformal factor $f(z,\bar{z}) = 1 + z_j\bar{z}_j$ when it is necessary. The summation over the repeated indices is assumed, as usual.

What we now want to do, however, is to assume the spheric metric (9.70) and to include the factor

$$\det h \propto \frac{1}{(1+z\bar{z})^4} \tag{9.71}$$

in the measure (with the shorthand $z\bar{z} \equiv z_j\bar{z}_j$). Then, the volume of our manifold $V = \int d^2z d^2\bar{z} \, \det h$ is finite and the spectral problem is rather similar to spectral problems on compact complex manifolds.

Proceeding in a similar way to what we did for compact manifolds, we consider [107] the SQM action (9.1) with the metric (9.70). We choose $\mathcal{W} = \frac{1}{4}\ln\det h$ so that the covariant quantum supercharges of our model, which are given by (9.20), map to the operators $\partial$ and $\partial^\dagger$ of the standard (untwisted) Dolbeault complex. Their explicit expressions are

$$\begin{aligned} \hat{Q} &= -i\psi_j[(1+z\bar{z})\partial_j + \bar{z}_k\psi_k\hat{\bar{\psi}}_j], \\ \hat{\bar{Q}} &= -i\hat{\bar{\psi}}_j[(1+z\bar{z})\bar{\partial}_j - 2z_j + z_k\hat{\bar{\psi}}_k\psi_j]. \end{aligned} \tag{9.72}$$

Here, $\psi_j$ are *flat* variables: $\{\psi_j, \hat{\bar{\psi}}_k\} = \delta_{jk}$.

The operators (9.72) and the Hamiltonian $\hat{H} = \{\hat{Q}, \hat{\bar{Q}}\}$ act on the Hilbert space involving *square-integrable* wave functions with the measure

$$d\mu = \frac{d^2zd^2\bar{z}}{(1+z\bar{z})^4}\, d^2\psi d^2\bar{\psi}\, e^{-\psi_j\bar{\psi}_j} \tag{9.73}$$

[cf. Eq. (6.28)].

**Theorem 9.7.** *There are only three normalizable zero modes in the spectrum of the Hamiltonian* $\hat{H} = \{\hat{Q}, \hat{\bar{Q}}\}$*, annihilated by both* $\hat{Q}$ *and* $\hat{\bar{Q}}$*. They dwell in the sector* $F = 0$ *and read*

$$\Psi^{(0)} = 1, \bar{z}_1, \bar{z}_2. \tag{9.74}$$

**Proof.** (i) $F = 0$. The condition $\hat{\bar{Q}}\Psi^{F=0} = 0$ holds automatically. The condition $\hat{Q}\Psi = 0$ is satisfied iff $\Psi$ is either a constant or an antiholomorphic function. The requirement of normalizability leaves only three possibilities displayed in Eq. (9.74).

(ii) $F = 2$. The wave function has the form $\Psi^{F=2} = B(z,\bar{z})\psi^1\psi^2$. The condition $\hat{Q}\Psi = 0$ holds automatically. The condition $\hat{\bar{Q}}\Psi = 0$ yields

$$\varepsilon_{jk}\psi_k\big[(1+z\bar{z})\bar{\partial}_j - 2z_j\big]B(z,\bar{z}) = 0$$

or

$$\bar{\partial}_j \left[ \frac{B}{(1+z\bar{z})^2} \right] = 0. \tag{9.75}$$

The solutions to (9.75) have the form

$$B(z,\bar{z}) = (1+z\bar{z})^2 R(z_j).$$

They are not normalizable.

(ii) $F = 1$. The wave function has the form $\Psi^{F=1} = C_j(z,\bar{z})\psi_j$. The conditions $\hat{Q}\Psi = \hat{\bar{Q}}\Psi = 0$ imply

$$\begin{cases} \varepsilon_{jk}\left[(1+z\bar{z})\partial_j C_k - \bar{z}_j C_k\right] = 0, \\ (1+z\bar{z})\bar{\partial}_k C_k - 3z_k C_k = 0. \end{cases} \tag{9.76}$$

The first equation can be rewritten as

$$\varepsilon_{jk}(1+z\bar{z})^2 \partial_j \left[ \frac{C_k}{1+z\bar{z}} \right] = 0. \tag{9.77}$$

Its general solution is

$$C_k = (1+z\bar{z})\,\partial_k \Phi(z,\bar{z}). \tag{9.78}$$

Substituting this in the second equation, we derive

$$\left[-(1+z\bar{z})^2 \bar{\partial}_k \partial_k + 2(1+z\bar{z}) z_k \partial_k\right] \Phi = 0. \tag{9.79}$$

The operator acting on $\Phi$ happens to coincide with the Hamiltonian of the system in the sector $F = 0$. As we have just learned, $\Phi$ should be either a constant or an antiholomorphic function. But then $C_k$ in Eq. (9.78) vanishes. □

Thereby, the Witten index is equal to 3. The whole spectrum of the Hamiltonian was found in Ref. [107]. It is noteworthy that all the eigenstates in the spectrum are square-integrable and that the supercharges $\hat{Q}$ and $\hat{\bar{Q}}$ are *bound* in this Hilbert space. In other words, for any eigenstate $\Psi$ of the Hamiltonian, the functions $\hat{Q}\Psi$ and $\hat{\bar{Q}}\Psi$ either vanish or are also square-integrable eigenstates.[13] As a result, the spectrum of all excited states is (at least) double degenerate, as it should be in a supersymmetric system.

[13]That is a nontrivial fact. A glance at (9.72) tells one that the operators $\hat{Q}$ and $\hat{\bar{Q}}$ bring about an extra power of $z$ and could in principle make a non integrable at infinity function out of an integrable one. But this does not happen.

One also can solve the same problem for the twisted Dolbeault complex involving

$$\mathcal{W}_q = \frac{q}{4} \ln \det h \tag{9.80}$$

with an arbitrary integer $q$ (only integer $q$'s are admissible — see Chapter 14). In that case, the supercharges acquire the form[14]

$$\hat{Q} = -i\psi_j \left[ (1 + z\bar{z})\partial_j + (q-1)\bar{z}_j + \bar{z}_k \psi_k \hat{\bar{\psi}}_j \right],$$
$$\hat{\bar{Q}} = -i\hat{\bar{\psi}}_j \left[ (1 + z\bar{z})\bar{\partial}_j - (q+1)z_j + z_k \hat{\bar{\psi}}_k \psi_j \right]. \tag{9.81}$$

If $q$ is positive, the zero modes dwell in the sector $F = 0$. A general normalizable solution to the equation $\hat{Q}^{(0)}\Psi = 0$ (the condition $\hat{\bar{Q}}^{(0)}\Psi = 0$ is satisfied automatically) is

$$\Psi^{(0)}_{q>0} = \frac{P_{2q-1}(\bar{z}_k)}{(1 + z\bar{z})^{q-1}}. \tag{9.82}$$

A polynomial of degree $2q-1$ depending on two arguments has $2q^2 + q$ different coefficients, i.e. there are $2q^2 + q$ different bosonic vacuum states.

If $q$ is negative, the vacuum states are still bosonic, but they now dwell in the sector $F = 2$. A general normalizable solution to the equation $\hat{\bar{Q}}^{(0)}\Psi = 0$ (the condition $\hat{Q}^{(0)}\Psi = 0$ is satisfied in that case automatically) is

$$\Psi^{(0)}_{q<0} = \frac{P_{-2q-1}(\bar{z}_k)}{(1 + z\bar{z})^{-q-1}} \psi^1 \psi^2, \tag{9.83}$$

which gives $2q^2 - q$ states. In other words, the Witten index in the system with arbitrary $q$ is equal to

$$\boxed{I_W(q) = 2q^2 + |q|.} \tag{9.84}$$

The value $q = 0$ is special. The supercharges acquire the form

$$\hat{Q} = -i\psi_j[(1 + z\bar{z})\partial_j - \bar{z}_j + \bar{z}_k \psi_k \hat{\bar{\psi}}_j],$$
$$\hat{\bar{Q}} = -i\hat{\bar{\psi}}_j[(1 + z\bar{z})\bar{\partial}_j - z_j + z_k \hat{\bar{\psi}}_k \psi_j]. \tag{9.85}$$

In this case, one can write, however, *another* pair of supercharges:

$$\hat{R} = -i\varepsilon_{jk}\psi_k[(1 + z\bar{z})\bar{\partial}_j - z_j - z_l \psi_l \hat{\bar{\psi}}_j].$$
$$\hat{\bar{R}} = -i\varepsilon_{jk}\hat{\bar{\psi}}_k[(1 + z\bar{z})\partial_j - \bar{z}_j - \bar{z}_l \hat{\bar{\psi}}_l \psi_j]. \tag{9.86}$$

[14]Note a misprint in the corresponding expression for $Q$ after Eq. (22) in Ref. [107].

A direct calculation shows that the four supercharges (9.85) and (9.86) obey the extended $\mathcal{N} = 4$ supersymmetry algebra (5.36).

And this is not a miracle, but a corollary of the fact that $S^4 \backslash \{\cdot\}$ is in fact an *HKT manifold*! Indeed, its metric is conformaly flat and belongs to the class (3.88). And we see in the following chapter and then in Chapter 13 that the HKT manifolds admit the $\mathcal{N} = 4$ supersymmetric description.[15]

The index (9.84) vanishes in this case so that there are no zero-energy states in the spectrum. The states carrying nonzero energy split into degenerate quartets. One of such quartets with the supercharge actions is represented schematically in Fig. 9.1. Note that all the states in this quartet (and in all other such quartets) are normalizable with the measure (9.73) — similar to what we had for the Dolbeault complex discussed at the beginning of this section.

In Chapter 14, we go back to the supersymmetric sigma model on $S^4$ with[16] $q = 0$ and show that it can be extended to include *non-Abelian self-dual* gauge fields, like the *Yang monopole* and *instantons*, in such a way that the $\mathcal{N} = 4$ supersymmetry is left intact.

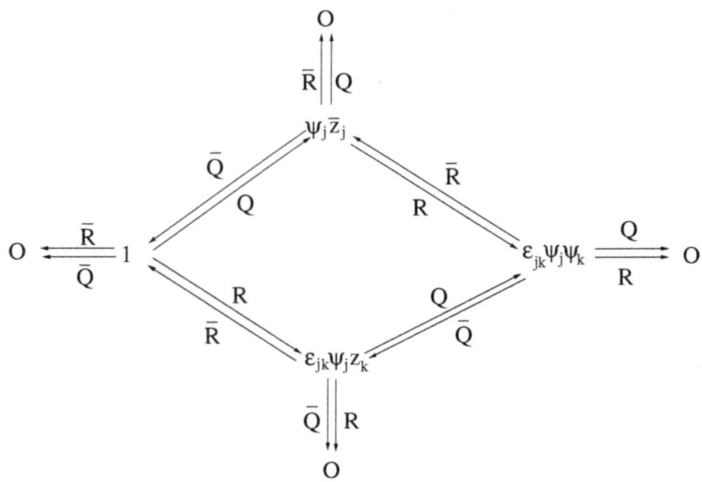


**Fig. 9.1.** A supersymmetric quartet of degenerate eigenstates of the Dirac Hamiltonian on $S^4$. Not wishing to clutter the figure, we have deprived the operators of their hats.

---

[15] Actually, the expressions (9.85) and (9.86) are nothing but the quantum counterparts of the classical supercharges (10.85) written for $S^4$.

[16] We learn there that it describes a "purely gravitational" *Dirac* complex involving extra torsions.

One can also consider $S^6$ and still higher even-dimensional spheres. It is an open question whether $S^6$ is a complex manifold or not. But even if it is, the true complex structure $I_{MN}$ and the correspondning complex coordinates are not known. One can make, however, for $S^{2d}$ the same stereographic projection as for $S^2$ and $S^4$, introduce the coordinates $z^1 = (x^1 + ix^2)/\sqrt{2}$, etc. and treat the problem for $S^{2d}\backslash\{\cdot\}$ in the same way as we did it for $S^4\backslash\{\cdot\}$. For the untwisted Dolbeault complex, one finds $C_{2d-1}^{d-1}$ bosonic zero modes. If $d = 3$, $I_W = 10$.

# Chapter 10

# Sigma Models with Extended Supersymmetries: $\mathcal{N} = 4$

In Chapters 2, 3, we defined some special classes of complex manifolds: the Kähler, hyperkähler and HKT manifolds. And now we are going to describe these and some other related geometries in the supersymmetric language.

The corresponding sigma models enjoy extended $\mathcal{N} = 4$ and $\mathcal{N} = 8$ supersymmetries. This chapter is devoted to $\mathcal{N} = 4$ models.

## 10.1 Kähler manifolds and around

The Kähler geometries admit several different supersymmetric descriptions. To begin with, consider the model (8.6) involving an even number of real $\mathcal{N} = 2$ superfields

$$X^M = x^M + \theta\psi^M + \bar{\psi}^M\bar{\theta} + \theta\bar{\theta}F^M. \tag{10.1}$$

By construction, the action of this model is invariant under the $\mathcal{N} = 2$ supersymmetry transformations (7.26):

$$\begin{aligned} \delta_0 x^M &= \epsilon_0\psi^M - \bar{\epsilon}_0\bar{\psi}^M, \\ \delta_0\psi^M &= \bar{\epsilon}_0(F^M - i\dot{x}^M), \\ \delta_0\bar{\psi}^M &= \epsilon_0(F^M + i\dot{x}^M), \\ \delta_0 F^M &= -i(\epsilon_0\dot{\psi}^M + \bar{\epsilon}\dot{\bar{\psi}}^M). \end{aligned} \tag{10.2}$$

We now impose a constraint and require the action to be invariant under the *extra* transformations

$$\boxed{\delta X^M = I_N{}^M(X)(\epsilon D X^N - \bar{\epsilon}\bar{D}X^N).} \tag{10.3}$$

For the lowest component, this means

$$\delta x^M = I_N{}^M(x)(\epsilon\psi^N - \bar\epsilon\bar\psi^N). \tag{10.4}$$

The expressions for $\delta\psi^M$ and $\delta F^M$ are a little longer.

By the same reasoning as in the previous chapter [see a comment after Eq. (9.39)], one can deduce that the transformations (10.2) and (10.3) commute and hence the anticommutators $\{\hat Q, \hat Q_0\}$, $\{\hat{\bar Q}, \hat Q_0\}$, $\{\hat Q, \hat{\bar Q}_0\}$, $\{\hat{\bar Q}, \hat{\bar Q}_0\}$ of the corresponding supercharges vanish.

One can now formulate a theorem analogous to Theorem 9.4 of the previous chapter[1]:

**Theorem 10.1.** *The action* (8.6) *is invariant under* (10.3) *and the generators of the transformations* (10.2) *and* (10.3) *satisfy the standard* $\mathcal{N} = 4$ *supersymmetry algebra iff* $I_N{}^M$ *is an integrable complex structure that is covariantly constant with respect to the Levi-Civita connection,* $\nabla_P I_M{}^N = 0$, *i.e. iff the manifold is Kähler.*

**Proof.** The proof follows the same lines as the proof of Theorem 9.4. Note first that the conditions $I^2 = -\mathbb{1}$ and $\mathcal{N}_{MN}{}^K = 0$ are necessary and sufficient to close the supersymmetry algebra — to ensure that the Lie bracket of two supertransformations (10.3) boils down to the time derivative:

$$(\delta_2\delta_1 - \delta_1\delta_2)X^M = 2i(\epsilon_2\bar\epsilon_1 - \epsilon_1\bar\epsilon_2)\dot X^M. \tag{10.5}$$

This can be checked by a direct calculation.

Consider now the supersymmetric variation of (8.6). Let us concentrate on the terms involving $\epsilon$. Bearing in mind $D^2 = 0$, we obtain

$$\begin{aligned}
\delta X^M &= \epsilon\, I_N{}^M(X) DX^N,\\
\delta(DX^M) &= -\epsilon(\partial_P I_N{}^M) DX^P DX^N,\\
\delta(\bar D X^M) &= -\epsilon\left[(\partial_P I_N{}^M)\bar D X^P DX^N + I_N{}^M \bar D D X^N\right].
\end{aligned} \tag{10.6}$$

The variation of the action (8.6) reads

$$\begin{aligned}
\delta S = \frac{\epsilon}{2}\int d\bar\theta d\theta dt\big[&(\partial_P g_{MN}) I_Q{}^P DX^Q \bar D X^M DX^N\\
&- g_{MN}(\partial_P I_Q{}^M)\bar D X^P DX^Q DX^N - I_{QN}(\bar D D X^Q) DX^N\\
&+ g_{MN}(\partial_P I_Q{}^N)\bar D X^M DX^P DX^Q\big].
\end{aligned} \tag{10.7}$$

[1] A similar theorem, formulated in the language of components for a field theory sigma model system, was proven in Ref. [108].

The first, second and fourth terms in this expression have the same structure $\sim \bar{D}XDXDX$. For the third term, we decompose $I_{QN} = I_{(QN)} + I_{[QN]}$. The part involving $I_{[QN]}$ can also be brought in the form $\sim \bar{D}XDXDX$ by flipping the Grassmann covariant derivative $\bar{D}$:

$$- \int d\bar{\theta}d\theta dt\, I_{[QN]}(\bar{D}DX^Q)DX^N$$
$$= \frac{1}{2}\int d\bar{\theta}d\theta dt\, (\partial_P I_{[QN]})\bar{D}X^P DX^Q DX^N. \tag{10.8}$$

The part involving $I_{(QN)}$ is a linearly independent structure. For the variation (10.7) to vanish, the condition $I_{(QN)} = 0$ should thus hold, which we assume in the following.

The structure $\sim \bar{D}XDXDX$ in the variation vanishes provided

$$I_Q{}^P \partial_P g_{MN} + g_{PN}\partial_M I_Q{}^P - \frac{1}{2}\partial_M I_{QN} + g_{PM}\partial_N I_Q{}^P - [Q \leftrightarrow N] = 0. \tag{10.9}$$

After some manipulations, this can be brought in the form

$$A_{M[QN]} = \nabla_M I_{QN} + \nabla_N I_{QM} - \nabla_Q I_{NM} = 0. \tag{10.10}$$

If $\nabla_Q I_{NM} = 0$, it obviosly vanishes. To show that (10.10) *implies* the covariant constancy of the complex structure, we symmetrize this relation over $M \leftrightarrow Q$ to derive $\nabla_M I_{QN} + \nabla_Q I_{MN} = 0$. By renaming indices, $Q \to M \to N \to Q$, we can also write $\nabla_M I_{QN} + \nabla_N I_{QM} = 0$. And from that and from (10.10), the condition $\nabla_Q I_{NM} = 0$ follows. □

Now, we proceed in the same way as in Chapters 8 and 9. At the first step, we go down to components and, using Noether's theorem, derive the expressions for the conserved classical complex supercharges corresponding to the invariances (10.2) and (10.3). These expressions are especially simple if written in terms of the fermion fields $\psi^M$ with the world indices and the canonical momenta $\Pi_M$ obtained by the variation of $L$ over $\dot{x}_M$ with fixed $\psi^M$ [109][2]:

$$Q_0 = \frac{1}{\sqrt{2}}\psi^M \left(\Pi_M - \frac{i}{2}\partial_M g_{PK}\, \psi^P \bar{\psi}^K\right),$$
$$\bar{Q}_0 = \frac{1}{\sqrt{2}}\bar{\psi}^M \left(\Pi_M + \frac{i}{2}\partial_M g_{PK}\, \psi^P \bar{\psi}^K\right),$$

[2]The expressions for $Q_0$ and $\bar{Q}_0$ were already written in (8.16).

$$Q_1 = \frac{1}{\sqrt{2}} \psi^N I_N{}^M \left( \Pi_M - \frac{i}{2} \partial_M g_{PK}\, \psi^P \bar{\psi}^K \right),$$
$$\bar{Q}_1 = \frac{1}{\sqrt{2}} \bar{\psi}^N I_N{}^M \left( \Pi_M + \frac{i}{2} \partial_M g_{PK}\, \psi^P \bar{\psi}^K \right). \tag{10.11}$$

At the second step, we quantize it using the recipe of Ref. [52]. For the quantum supercharges, the representation via the flat fermion variables $\psi^A$ is more convenient. We derive the following expressions for the covariant quantum supercharges [110]:

$$\boxed{\begin{aligned} \hat{Q}_0 &= -\frac{i}{\sqrt{2}}\, \psi^M D_M, \quad \hat{Q}_1 = -\frac{i}{\sqrt{2}}\, \psi^N I_N{}^M D_M, \\ \hat{Q}_0^\dagger &= -\frac{i}{\sqrt{2}}\, e_C^M \frac{\partial}{\partial \psi_C} D_M, \quad \hat{Q}_1^\dagger = -\frac{i}{\sqrt{2}}\, e_C^N I_N{}^M \frac{\partial}{\partial \psi_C} D_M, \end{aligned}} \tag{10.12}$$

where the operator of "spinor" covariant derivative,

$$D_M = \frac{\partial}{\partial x^M} + \omega_{AB,M}\, \psi_A \frac{\partial}{\partial \psi_B}, \tag{10.13}$$

was defined in (8.30) and the conjugated supercharges are marked by a dagger rather than a bar.

We now prove the following theorem representing an analog of Theorems 8.1 and 9.1 and clarifying the geometric meaning of the supercharges (10.12).

**Theorem 10.2.** *For Kähler manifolds, the action of the supercharges $\hat{Q}_0$ and $\hat{Q}_1$ on the wave functions $\Psi(x^M, \psi^M)$ is isomorphic to the action of the certain linear combinations of the operators $\partial$ and $\bar{\partial}$ in the Kähler–de Rham complex of the $(p, q)$ forms.*

**Proof.** Choose the complex coordinates $x^M \to (z^m, \bar{z}^{\bar{m}})$ such that the complex structure tensor $I_N{}^M$ acquires the form (2.53). Then, $\hat{Q}_0$ and $\hat{Q}_1$ can be represented as[3]

$$\hat{Q}_0 = -\frac{i}{\sqrt{2}} (\psi^m D_m + \psi^{\bar{m}} D_{\bar{m}}),$$
$$\hat{Q}_1 = \frac{1}{\sqrt{2}} (\psi^{\bar{m}} D_{\bar{m}} - \psi^m D_m). \tag{10.14}$$

[3]We do not write here a bar over $\psi^{\bar{m}}$ not to confuse it with the operators $\hat{\bar{\psi}} \equiv \hat{\psi}^\dagger = \partial/\partial\psi$.

Consider the action of the operator $\psi^m D_m$ on a wave function

$$\Psi(x^M, \psi^M) = A_{m_1 \ldots m_p \bar{n}_1 \ldots \bar{n}_q}(z, \bar{z})\, \psi^{m_1} \cdots \psi^{m_p}\, \psi^{\bar{n}_1} \cdots \psi^{\bar{n}_q}. \tag{10.15}$$

The same calculation as in the proof of Theorem 8.1 yields

$$\psi^s D_s \Psi = (\partial_s A_{m_1 \ldots m_p \bar{n}_1 \ldots \bar{n}_q}) \psi^s \psi^{m_1} \cdots \psi^{m_p}\, \psi^{\bar{n}_1} \cdots \psi^{\bar{n}_q} \tag{10.16}$$

plus some extra terms involving the structures $\Gamma^P_{sN} \psi^s \psi^N$.

Generically, these structures do not vanish (and that is why this theorem is not valid for generic complex manifolds), but in the Kähler case, the only nonvanishing Christoffel symbols have all holomorphic or all antiholomorphic indices [see Eq. (2.30)]. Hence, $\Gamma^{\bar{p}}_{sN} = 0$ and

$$\Gamma^p_{sN} \psi^s \psi^N = \Gamma^p_{sn} \psi^s \psi^n = 0$$

due to the symmetry $\Gamma^p_{sn} = \Gamma^p_{ns}$. Thus, the equality (10.16) is exact and the isomorphism $\psi^m D_m \leftrightarrow \partial$ is established.

The same reasoning allows one to establish the isomorphism $\psi^{\bar{m}} D_{\bar{m}} \leftrightarrow \bar{\partial}$. And the supercharges (10.12) map to the operators

$$\hat{Q}_0 \to -i(\partial + \bar{\partial})/\sqrt{2}, \quad \hat{Q}_1 \to (\bar{\partial} - \partial)/\sqrt{2} \tag{10.17}$$

and their Hermitian conjugates. □

We are able now to pay our old debt and prove Theorem 2.3. It affirmed that, for Kähler manifolds, the algebra

$$\partial^2 = \bar{\partial}^2 = \{\partial, \bar{\partial}\} = (\partial^\dagger)^2 = (\bar{\partial}^\dagger)^2 = \{\partial^\dagger, \bar{\partial}^\dagger\} = \{\partial, \bar{\partial}^\dagger\} = \{\bar{\partial}, \partial^\dagger\} = 0,$$
$$\{\partial, \partial^\dagger\} = \{\bar{\partial}, \bar{\partial}^\dagger\} \tag{10.18}$$

holds.

**Proof.** The isomorphism established above makes this statement trivial. The supercharges $\hat{Q}_{0,1}$ and their conjugates satisfy the extended $\mathcal{N} = 4$ supersymmetry algebra. The same concerns the operators in the RHS of Eq. (10.17). This algebra is equivalent to (10.18). □

To make contact with geometry, the description of this $\mathcal{N} = 4$ model in terms of $\mathcal{N} = 2$ superfields (10.1) is quite convenient. But if one sticks to their supersymmetric glasses and does not want or does not need to remove them, the most simple and natural description of the Kähler manifolds is based on $\mathcal{N} = 4$ chiral superfields (7.47).

Take $d$ such superfields,

$$\mathcal{Z}^{m=1,\dots,d} = z^m(t_L) + \sqrt{2}\theta_j \psi^{jm}(t_L) + 2\theta_1\theta_2 F^m(t_L). \tag{10.19}$$

Their lowest components will be interpreted as complex coordinates. Now, $\psi^{jm}$ are their fermion superpartners (in this case, there are *two* complex fermion fields for each complex coordinate $z^m$) and $F^m$ are complex auxiliary variables. Each superfield (10.19) represents a (**2**, **4**, **2**) multiplet, which can be decomposed into a pair of (**1**, **2**, **1**) multiplets (10.1).

Let us write the action in the form [111]

$$\boxed{S^{\text{kin}} = \frac{1}{4}\int d^2\bar{\theta} d^2\theta dt\, K[\mathcal{Z}^m(\theta_j, t_L), \bar{\mathcal{Z}}^{\bar{m}}(\bar{\theta}^j, t_R)],} \tag{10.20}$$

with real $K$. This is an obvious generalization of the first (kinetic) term in (7.50).

Integration over $d^2\bar{\theta}d^2\theta$ gives the following component Lagrangian:

$$\begin{aligned} L = \kappa_{m\bar{n}} &\left[\dot{z}^m \dot{\bar{z}}^{\bar{n}} + F^m \bar{F}^{\bar{n}} + \frac{i}{2}(\bar{\psi}^{\bar{n}} \boldsymbol{\nabla} \psi^m - \boldsymbol{\nabla}\bar{\psi}^{\bar{n}} \psi^m)\right] \\ &+ \frac{1}{2}\left[\partial_m \kappa_{p\bar{n}} \bar{F}^{\bar{n}} (\psi^m \psi^p) + \partial_{\bar{n}} \kappa_{m\bar{p}} F^m (\bar{\psi}^{\bar{p}} \bar{\psi}^{\bar{n}})\right. \\ &\left. + \partial_m \partial_{\bar{n}} \kappa_{p\bar{q}} (\bar{\psi}^{\bar{n}} \psi^m)(\bar{\psi}^{\bar{q}} \psi^p)\right], \end{aligned} \tag{10.21}$$

where

$$\kappa_{m\bar{n}} = \partial_m \partial_{\bar{n}} K, \tag{10.22}$$

we used the shorthand in (7.60), and the operator $\boldsymbol{\nabla}$ of the covariant time derivative was defined in Eq. (8.5). In the Kähler case,

$$\boldsymbol{\nabla}\psi^m = \dot{\psi}^m + \Gamma^m_{st} \dot{z}^s \psi^t, \qquad \boldsymbol{\nabla}\bar{\psi}^{\bar{n}} = \dot{\bar{\psi}}^{\bar{n}} + \Gamma^{\bar{n}}_{\bar{s}\bar{t}} \dot{\bar{z}}^{\bar{s}} \bar{\psi}^{\bar{t}}. \tag{10.23}$$

Excluding $F^m$ and $\bar{F}^{\bar{n}}$, we derive

$$\boxed{L = \kappa_{m\bar{n}} \left[\dot{z}^m \dot{\bar{z}}^{\bar{n}} + \frac{i}{2}(\bar{\psi}^{\bar{n}} \boldsymbol{\nabla}\psi^m - \boldsymbol{\nabla}\bar{\psi}^{\bar{n}} \psi^m)\right] + \frac{1}{2} R_{m\bar{n}p\bar{q}} (\bar{\psi}^{\bar{n}} \psi^m)(\bar{\psi}^{\bar{n}} \psi^m),} \tag{10.24}$$

where $R_{m\bar{n}p\bar{q}} = \partial_m \partial_{\bar{n}} \kappa_{p\bar{q}} - \kappa^{\bar{s}t}(\partial_m \kappa_{p\bar{s}})(\partial_{\bar{n}} \kappa_{t\bar{q}})$.

In other words, $K$ is nothing but the Kähler potential, $\kappa_{m\bar{n}}$ is the Hermitian Kähler metric and $R_{m\bar{n}p\bar{q}}$ is the Kähler Riemann tensor (2.32).

The complex structure has the canonical form (2.53) in complex and (2.51) in real terms.

One can add to (10.20) a superpotential term as in (7.50):

$$S^{\text{pot.}} = \frac{1}{2}\int d^2\theta dt_L\, W(\mathcal{Z}^m) + \frac{1}{2}\int d^2\bar{\theta} dt_R\, \bar{W}(\bar{\mathcal{Z}}^{\bar{m}}). \tag{10.25}$$

This brings about an extra contribution $-F^m \partial_m W - \bar{F}^{\bar{m}} \partial_{\bar{m}} \bar{W}$ in the component Lagrangian.

In this case, the equations of motion for $F^m$ are

$$F^m = -\frac{1}{2}\Gamma^m_{pq}(\psi^p \psi^q) + \kappa^{\bar{n}m}\partial_{\bar{n}}\bar{W}. \tag{10.26}$$

When the auxiliary variables are excluded, the component Lagrangian acquires new terms including the potential term

$$L^{\text{pot}} = -\kappa^{\bar{n}m}\partial_m W \partial_{\bar{n}}\bar{W}. \tag{10.27}$$

For completeness, we would also like to mention here that still another description of $\mathcal{N} = 4$ Kähler models is possible [9]. It is a description in terms of $\mathcal{N} = 2$ superfields of different nature: instead of taking $D = 2d$ real $(\mathbf{1},\, \mathbf{2},\, \mathbf{1})$ superfields, we may take half as much chiral superfields $(\mathbf{2},\, \mathbf{2},\, \mathbf{0})$ and the same number of Grassmann superfields $(\mathbf{0},\, \mathbf{2},\, \mathbf{2})$ including each the lowest fermion and the bosonic auxiliary component:

$$\Phi^n = \chi^n + \sqrt{2}\,\theta F^n - i\theta\bar{\theta}\dot{\chi}^n. \tag{10.28}$$

This description is similar to the description of the de Rham $\mathcal{N} = 2$ model in terms of the bosonic and Grassmann $\mathcal{N} = 1$ superfields given in Eq. (8.10). The Lagrangian of the model reads

$$\boxed{L = \int d\bar{\theta} d\theta\, \partial_m \partial_{\bar{n}} \mathcal{K}(Z, \bar{Z}) \left(\frac{1}{4}\bar{D}\bar{Z}^{\bar{n}} D Z^m + \frac{1}{2}\Phi^m \bar{\Phi}^{\bar{n}}\right).} \tag{10.29}$$

This Lagrangian is invariant under the extra $\mathcal{N} = 2$ supersymmetry transformations

$$\delta Z^n = -\eta \Phi^n, \quad \delta \bar{Z}^{\bar{n}} = \bar{\eta}\bar{\Phi}^{\bar{n}}, \quad \delta\Phi^n = i\bar{\eta}\dot{Z}^n, \quad \delta\bar{\Phi}^{\bar{n}} = -i\eta\dot{\bar{Z}}^{\bar{n}}. \tag{10.30}$$

Its component expansion coincides with (10.21).

### 10.1.1 *Quasicomplex Kähler models*

By construction, the deformation (10.25) preserves $\mathcal{N} = 4$ supersymmetry. But it is not the only possible such deformation. Besides the superpotential term (10.25), one can add in the action the following $\mathcal{N} = 4$ supersymmetric term involving time derivatives of the superfields $\mathcal{Z}^m$:

$$S^{\mathcal{A}} = \frac{1}{2}\int d^2\theta dt_L\, \mathcal{A}_m(\mathcal{Z})\,\dot{\mathcal{Z}}^m + \text{c.c.} \tag{10.31}$$

with an arbitrary vector holomorphic function $\mathcal{A}_m$. This structure, introduced in Ref. [109], is specific for quantum mechanics and does not have a field theory analog. That is probably the reason why it has not attracted an attention before.

Consider the component Lagrangian derived from the action

$$S = S^{\text{kin}} + S^{\mathcal{A}}. \tag{10.32}$$

The bosonic terms (see Refs. [82, 109] for the full expression) read

$$L_{\text{bos}} = \kappa_{m\bar{n}}(\dot{z}^m \dot{\bar{z}}^{\bar{n}} + F^m \bar{F}^{\bar{n}}) + \mathcal{F}_{nm}\dot{z}^n F^m + \bar{\mathcal{F}}_{\bar{n}\bar{m}}\dot{\bar{z}}^{\bar{n}}\bar{F}^{\bar{m}}, \tag{10.33}$$

where $\mathcal{F}_{nm} = \partial_n \mathcal{A}_m - \partial_m \mathcal{A}_n$. Excluding the auxiliary variables, we derive

$$L_{\text{bos}} = h_{m\bar{n}}\dot{z}^m \dot{\bar{z}}^n \tag{10.34}$$

with

$$\boxed{h_{m\bar{n}} = \kappa_{m\bar{n}} + \mathcal{F}_{ml}\kappa^{\bar{p}l}\bar{\mathcal{F}}_{\bar{p}\bar{n}}.} \tag{10.35}$$

The Lagrangian (10.34) describes the motion along a complex manifold with the metric (10.35). In contrast to $\kappa_{m\bar{n}}$, this metric is not Kähler![4]

Note that, after excluding the auxiliary variables, the Lagrangian does not include the terms linear in $\dot{z}^m$. The object $\mathcal{A}_m$ *looks* as an Abelian gauge field and $\mathcal{F}_{mn}$ as a field strength tensor, but their nature is in fact different.[5]

One can ask at this point: how is it possible? Our model enjoys the extended $\mathcal{N} = 4$ supersymmetry, and have we not just proven Theorem 10.1

[4] Nontrivialities start when the complex dimension of the manifold is 2 or more.

[5] We remind that, being complex, Kähler manifolds also admit the presence of gauge fields. The latter elongate the momenta and bring about an extra bifermionic term in the Hamiltonian [see Eq. (9.18)] but do not affect the metric.

affirming that the extra $\mathcal{N} = 2$ invariance (10.3) of the action (8.6) entails Kähler geometry?

The answer is the following. Being expressed in terms of the $\mathcal{N} = 2$ superfields $X^M$, the action (10.32) acquires the form (8.6), indeed, but the tensor $g_{MN}(X)$ is not real and symmetric anymore! It has an imaginary antisymmetric part:

$$g_{MN} = g_{(MN)} + ib_{[MN]}. \tag{10.36}$$

The second term does not spoil the reality of the action — the only essential and necessary requirement. The model (10.32) belongs to the class of *quasicomplex* supersymmetric sigma models [98]. The bosonic part of the action of a generic such model describes the motion over a manifold with the metric

$$G_{MN} = g_{(MN)} + b_{[MP]}g^{(PS)}b_{[SN]} \tag{10.37}$$

[cf. Eq. (10.35)]. We come back to the discussion of the quasicomplex models in Chapter 12. A generic quasicomplex model is $\mathcal{N} = 2$ supersymmetric, but the model (10.32) enjoys the extended $\mathcal{N} = 4$ supersymmetry.

Using a sloppy language, we call the model (10.32) the *quasicomplex Kähler model*. The manifold that it describes is not Kähler, however (you might recall at this point that neither the "hyperkähler with torsion" manifolds are hyperkähler or Kähler), and a more exact (but also more long) name for (10.32) would be "a quasicomplex deformation of a Kähler–de Rham supersymmetric sigma model".

### 10.1.2 *Bi-Kähler models*

The superfields (10.19) are ordinary chiral $\mathcal{N} = 4$ superfields satisfying $\bar{D}_j \mathcal{Z}^m = 0$. But as was mentioned at the end of Section 7.3.1, one can introduce also mirror (**2**, **4**, **2**) superfields satisfying the constraints[6]

$$\boxed{\bar{D}_1 \mathcal{U} = D^2 \mathcal{U} = 0.} \tag{10.38}$$

If all the fields entering the action are mirror, this amounts to redefining $\theta_2 \leftrightarrow \bar{\theta}^2$, which makes no difference. However, the situation changes when the action involves *both* ordinary and mirror multiplets.

---

[6] $D^2 \equiv D^{j=2}$, not a square of anything.

Consider a system involving $d_1$ ordinary $\mathcal{N} = 4$ chiral multiplets $\mathcal{Z}^m$ and $d_2$ mirror multiplets $\mathcal{U}^\mu$. Then, $d = d_1 + d_2$ is the complex dimension of the corresponding manifold. Consider the action

$$S = \frac{1}{4} \int d^2\bar{\theta} d^2\theta dt \, \mathcal{L}(\mathcal{Z}^m, \bar{\mathcal{Z}}^{\bar{m}}; \mathcal{U}^\mu, \bar{\mathcal{U}}^{\bar{\mu}}). \tag{10.39}$$

When going down to components, the bosonic part of the Lagrangian boils down to

$$L_{\text{bos}} = h_{m\bar{n}} \dot{z}^m \dot{\bar{z}}^{\bar{n}} + h_{\mu\bar{\nu}} \dot{u}^\mu \dot{\bar{u}}^{\bar{\nu}}, \tag{10.40}$$

where

$$h_{m\bar{n}} = \partial_m \partial_{\bar{n}} \mathcal{L}, \quad h_{\mu\bar{\nu}} = -\partial_\mu \partial_{\bar{\nu}} \mathcal{L}. \tag{10.41}$$

We see two distinct sectors. In both sectors, the metric is obtained as a double derivative of the prepotential $\mathcal{L}$, but this double derivative enters with a negative sign for the mirror sector.[7] The mixed components of the metric $h_{m\bar{\nu}}$ and $h_{\mu\bar{n}}$ vanish. This does not mean, however, that the corresponding manifold is a direct product of two Kähler manifolds: $h_{m\bar{n}}$ may depend not only on $z, \bar{z}$ but also on $u, \bar{u}$, and $h_{\mu\bar{\nu}}$ may depend not only on $u, \bar{u}$ but also on $z, \bar{z}$.

Both the ordinary chiral superfields $\mathcal{Z}^{m=1,\ldots,d_1}$ and the mirror superfields $\mathcal{U}^{\mu=1,\ldots,d_2}$ can be expressed via the real $(\mathbf{1},\ \mathbf{2},\ \mathbf{1})$ superfields $X^{M=1,\ldots,2d_1}$ and $Y^{M=2d_1+1,\ldots,2d}$, respectively. The action (10.39) expressed in these terms acquires the form (8.6) with a *real* tensor $g_{MN}(X, Y)$. One can ask again the following: how come the geometry of our model is not Kähler (which Theorem 10.1 seems to dictate) but has a more complicated bi-Kähler nature?

The answer is that the law of extra $\mathcal{N} = 2$ supersymmetry transformations is now more complicated than (10.3) and is different in the ordinary and mirror sectors. Generic such transformations can be brought to the following canonical form [81, 82]:

$$\begin{aligned} M, N = 1, \ldots, 2d_1 &\Rightarrow \delta X^M = I_N{}^M (\epsilon D - \bar{\epsilon}\bar{D}) X^N, \\ M, N = 2d_1 + 1, \ldots, 2d &\Rightarrow \delta Y^M = J_N{}^M (\bar{\epsilon} D - \epsilon \bar{D}) Y^N, \end{aligned} \tag{10.42}$$

where $I$ and $J$ are the canonical matrices[8] (2.51) of dimensions $2d_1$ and $2d_2$.

[7]This can be best understood by noting that the measure $d^2\bar{\theta}d^2\theta = d\bar{\theta}^2 d\bar{\theta}^1 d\theta_1 d\theta_2$ changes sign when $\theta_2$ and $\bar{\theta}^2$ are interchanged.

[8]They are *not* the same matrices as $I$ and $J$ in Eq. (3.2). Alas, the Latin alphabet has a finite length.

Introducing now

$$\epsilon = \epsilon_+ + i\epsilon_-, \quad D = D_+ + iD_- \tag{10.43}$$

with real $\epsilon_\pm$ and Hermitian $D_\pm$, one can rewrite (10.42) in the form

$$\boxed{\begin{aligned} \delta \begin{pmatrix} X \\ Y \end{pmatrix}^M &= 2i\epsilon_+ \begin{pmatrix} I & 0 \\ 0 & J \end{pmatrix}^M{}_N D_- \begin{pmatrix} X \\ Y \end{pmatrix}^N \\ &\quad + 2i\epsilon_- \begin{pmatrix} I & 0 \\ 0 & -J \end{pmatrix}^M{}_N D_+ \begin{pmatrix} X \\ Y \end{pmatrix}^N . \end{aligned}} \tag{10.44}$$

Both diag$(I, J)$ and diag$(I, -J)$ can be chosen for a complex structure of a bi-Kähler manifold.

The basic action (10.39) can be deformed by adding a potential or a quasicomplex term. Let us concentrate on the latter. Obviously, we can add the terms like in Eq. (10.31) independently in each sector and write

$$S^{\mathcal{A}}_{\text{bi-K}} = \frac{1}{2}\int d\theta_1 d\theta_2 dt_L\, \mathcal{A}_m(\mathcal{Z})\, \dot{\mathcal{Z}}^m + \frac{1}{2}\int d\theta_1 d\bar{\theta}^2 d\tilde{t}_L\, \mathcal{B}_\mu(\mathcal{U})\, \dot{\mathcal{U}}^\mu + \text{c.c.}, \tag{10.45}$$

where

$$t_L = t - i(\theta_1\bar{\theta}^1 + \theta_2\bar{\theta}^2), \quad \tilde{t}_L = t - i(\theta_1\bar{\theta}^1 - \theta_2\bar{\theta}^2) \tag{10.46}$$

($t_L$ vanishes under the action of $\bar{D}_{1,2}$ and $\tilde{t}_L$ vanishes under the action of $\bar{D}_1$ and $D^2$).

But one can also write $\mathcal{N} = 4$ invariants of more complicated structure. Consider the superfield

$$R = \bar{D}_2[\mathcal{A}_m(\mathcal{Z}, \mathcal{U}) D^2 \mathcal{Z}^m].$$

Bearing in mind that $\bar{D}_1\mathcal{Z}^m = \bar{D}_1\mathcal{U}^\mu = 0$ and $(\bar{D}_2)^2 = 0$, we deduce that $\bar{D}_j R = 0$ and hence $\int dt_L d\theta_1 d\theta_2\, R$ represents a supersymmetric invariant. Note that in the case when $\mathcal{A}_m$ does not depend on $\mathcal{U}^\mu$, one can pull $\bar{D}_2$ through to obtain $R = 2i\mathcal{A}_m \dot{\mathcal{Z}}^m$ so that the structure in (10.31) is reproduced.

The most general quasicomplex deformation reads

$$\begin{aligned} S^{\mathcal{A},\mathcal{B}}_{\text{bi-K}} &= -\frac{i}{4}\int d\theta_1 d\theta_2 dt_L \left\{ \bar{D}_2[\mathcal{A}^{(1)}_m(\mathcal{Z}, \mathcal{U}) D^2 \mathcal{Z}^m] + \bar{D}_1[\mathcal{A}^{(2)}_m(\mathcal{Z}, \bar{\mathcal{U}}) D^1 \mathcal{Z}^m] \right\} \\ &\quad - \frac{i}{4}\int d\theta_1 d\bar{\theta}^2 d\tilde{t}_L \left\{ D^2[\mathcal{B}^{(1)}_\mu(\mathcal{U}, \mathcal{Z}) \bar{D}_2 \mathcal{U}^\mu] + \bar{D}_1[\mathcal{B}^{(2)}_\mu(\mathcal{U}, \bar{\mathcal{Z}}) D^1 \mathcal{U}^\mu] \right\} \\ &\quad + \text{c.c.} \end{aligned} \tag{10.47}$$

## 10.2 Symplectic sigma models

The Kähler and related models of the preceding section were formulated for complex manifolds, whose real dimension is even. But for certain special manifolds of dimension $D = 3n$ or $D = 5n$ with integer $n$, one also can formulate supersymmetric sigma models that enjoy the extended $\mathcal{N} = 4$ and even $\mathcal{N} = 8$ (in the case $D = 5n$) supersymmetry. The $\mathcal{N} = 8$ models are discussed in the following chapter. Here, we concentrate on $\mathcal{N} = 4$ models describing $3n$-dimensional manifolds.

Consider the (**3**, **4**, **1**) superfield (7.61) with the component expansion (7.65). We can generalize the action (7.66), the bosonic part of which described free three-dimensional motion, to write[9] [85, 86]

$$\boxed{S = -\int d^2\bar{\theta}d^2\theta dt\, F(\Phi_M)} \tag{10.48}$$

with arbitrary $F(\Phi_M)$, $M = 1, 2, 3$. The bosonic part of this action reads

$$S_{\text{bos}} = \frac{1}{2}h(\mathbf{A})\dot{\mathbf{A}}^2, \tag{10.49}$$

where

$$h(\mathbf{A}) = 2\triangle F(\mathbf{A}). \tag{10.50}$$

It describes the motion along a conformally flat three-dimensional manifold.[10]

The component expressions for the quantum supercharges and the Hamiltonian may be found following the standard procedure described above. The supercharges read [84]

$$\begin{aligned}
\hat{Q}_j &= \frac{1}{\sqrt{2}}f(\mathbf{A})\hat{P}_M\,(\sigma_M)_j{}^k\psi_k + \frac{i}{\sqrt{2}}\,(\partial_M f)\,\hat{\bar{\psi}}\sigma_M\psi\,\psi_j, \\
\hat{\bar{Q}}^j &= \frac{1}{\sqrt{2}}\hat{P}_M f(\mathbf{A})\hat{\bar{\psi}}^k(\sigma_M)_k{}^j - \frac{i}{\sqrt{2}}(\partial_M f)\,\hat{\bar{\psi}}^j\hat{\bar{\psi}}\sigma_M\psi,
\end{aligned} \tag{10.51}$$

where $f(\mathbf{A})$ is an arbitrary function giving the metric, $h(\mathbf{A}) = f^{-2}(\mathbf{A})$, $\sigma_M$ are the Pauli matrices, and $\hat{\bar{Q}}^j = (\hat{Q}_j)^\dagger$.

---

[9]The opposite sign compared to Ref. [85] is due to the different sign convention in the fermion measure.

[10]Here $\triangle$ is not the covariant Laplace–Beltrami operator on this manifold but simply $\triangle = \partial_M\partial_M$.

We see that the supercharges carry spinorial indices belonging to the fundamental representation of the group $Spin(3) \simeq Sp(1)$. The same is true for the $\mathcal{N} = 8$ (**5**, **8**, **3**) sigma model to be considered in Chapter 11. Four complex supercharges of that model belong to the fundamental representation of $Spin(5) \simeq Sp(2)$. That is why we called the models of this type *symplectic* [83, 112]. These models describe the motion over a certain class of manifolds that we cannot call symplectic (this term is already taken) and, having not found a better word, we call these manifolds *spinorial.*

The anticommutator of the supercharges (10.51) gives the Hamiltonian

$$\boxed{\hat{H} = \frac{1}{2} f (\hat{P}_M)^2 f + \varepsilon_{MNP}\, \hat{\bar{\psi}} \sigma_M \psi\, f(\partial_N f) \hat{P}_P + \frac{1}{6} f(\triangle f)\, \hat{\bar{\psi}} \sigma_M \psi\, \hat{\bar{\psi}} \sigma_M \psi.} \tag{10.52}$$

An obvious generalization of the action (10.48) is the action

$$S = -\int d^2\bar{\theta} d^2\theta dt\, F(\mathbf{\Phi}_k) \tag{10.53}$$

including several (**3**, **4**, **1**) superfields $\mathbf{\Phi}_{k=1,\ldots,n}$. Its bosonic part describes the motion along a $3n$-dimensional manifold with the metric

$$g_{Mk,Nl} = 2\delta_{MN} \frac{\partial^2 F}{\partial A_{Pk}\, \partial A_{Pl}}. \tag{10.54}$$

It is difficult for us to say whether this new geometry is interesting or not from the mathematical point of view, but the system (10.53) has interesting applications in physics:

- For some special set of $\mathbf{\Phi}_k$ and for some special function $F(\mathbf{\Phi}_k)$, it gives the effective action for a 4-dimensional supersymmetric gauge theory placed in a small spatial box [84, 113].
- Surprisingly, the same model describes the dynamics of extremal Reissner-Nordström black holes [114].

Having started to talk about physics, I cannot resist mentioning another interesting class of (**3**, **4**, **1**) models that describes the effective Lagrangian of 4-dimensional *chiral* (i.e. having a left-right asymmetric fermion content) supersymmetric gauge theories squeezed in a small box. The simplest such SQM system was found in [115]. It has a pair of complex supercharges:

$$\hat{Q}_j = \frac{1}{\sqrt{2}}\left[(\sigma_M)_j{}^k\psi_k(\hat{P}_M + \mathcal{A}_M) + i\psi_j\mathcal{D}\right],$$
$$\hat{\bar{Q}}^j = \frac{1}{\sqrt{2}}\left[\hat{\bar{\psi}}^k(\sigma_M)_k{}^j(\hat{P}_M + \mathcal{A}_M) - i\hat{\bar{\psi}}^j\mathcal{D}\right], \tag{10.55}$$

where $\mathcal{A}$ and $\mathcal{D}$ are the functions of $\mathbf{A}$ such that

$$\boldsymbol{\nabla}\times\boldsymbol{\mathcal{A}} = \boldsymbol{\mathcal{B}} = \boldsymbol{\nabla}\mathcal{D}. \tag{10.56}$$

The standard $\mathcal{N} = 4$ supersymmetry algebra (5.36) holds, giving the Hamiltonian

$$\boxed{\hat{H} = \frac{1}{2}(\hat{\mathbf{P}} + \boldsymbol{\mathcal{A}})^2 + \frac{1}{2}\mathcal{D}^2 + \boldsymbol{\mathcal{B}}\,\hat{\bar{\psi}}\boldsymbol{\sigma}\psi.} \tag{10.57}$$

A physically interesting case is $\mathcal{D} = -1/(2|\mathbf{A}|) \equiv -1/(2A)$, where the Hamiltonian (10.57) describes the three-dimensional motion of a spin-0 particle (in the sectors of fermion charge $F = 0, 2$) or a spin-$\frac{1}{2}$ particle (in the sector $F = 1$) in the field of a Dirac magnetic monopole, $B_M = A_M/(2A^3)$, supplemented by the scalar potential $V(A) = 1/8A^2$. Remarkably, this Hamiltonian describes also the dynamics of four-dimensional *chiral supersymmetric electrodynamics* (the theory involving the photon, a right-handed fermion of charge 2, eight left-handed fermions[11] of charge 1, and their superpartners) placed in a small spatial box.[12]

Well, the statement above was not quite correct. In fact, the effective Hamiltonian for such a theory describes the motion in an infinite *cubic lattice* of monopoles carrying the magnetic charges +1 and −7 in such a way that the average magnetic charge is zero. But we do not want to harass our reader-mathematician more. A reader-physicist can consult the original paper [116] or Chapter 8 of Ref. [75].

It is possible to construct generalizations of this model. First of all, we can endow the three-dimensional manifold, along which the particle

---

[11] A remark addressed to the experts in particle physics: one has to take eight of them to fulfil the condition

$$\left(\sum_f q_f^3\right)_{\text{left}} = \left(\sum_f q_f^3\right)_{\text{right}},$$

which in necessary to cancel the chiral anomaly.

[12] To avoid confusion, note that the dynamical variables $\mathbf{A}$ have nothing to do in this case with the physical spatial coordinates. They have the meaning of the zero Fourier components of the physical vector potential [having again nothing to do with $\boldsymbol{\mathcal{A}}(\mathbf{A})$ entering Eqs. (10.55) and (10.57)].

moves, with a nontrivial conformally flat metric, $ds^2 = hd\boldsymbol{A}^2$. The quantum supercharges and the Hamiltonian acquire the form [84]

$$\begin{aligned}
\hat{Q}_j &= \frac{f}{\sqrt{2}} \left[ (\sigma_M)_j{}^k \psi_k (\hat{P}_M + \mathcal{A}_M) + i\psi_j \mathcal{D} \right] + \frac{i\partial_M f}{\sqrt{2}} \hat{\bar{\psi}} \sigma_M \psi \, \psi_j, \\
\hat{\bar{Q}}^j &= \left[ \hat{\bar{\psi}}^k (\sigma_M)_k{}^j (\hat{P}_M + \mathcal{A}_M) - i\hat{\bar{\psi}}^j \mathcal{D} \right] \frac{f}{\sqrt{2}} - \frac{i\partial_M f}{\sqrt{2}} \hat{\bar{\psi}}^j \hat{\bar{\psi}} \sigma_M \psi,
\end{aligned} \tag{10.58}$$

$$\begin{aligned}
\hat{H} &= f \left[ \frac{1}{2} (\hat{\boldsymbol{P}} + \boldsymbol{\mathcal{A}})^2 + \frac{1}{2} \mathcal{D}^2 + \mathcal{B}_M \hat{\bar{\psi}} \sigma_M \psi \right] f \\
&\quad + \varepsilon_{MNP} \, \hat{\bar{\psi}} \sigma_M \psi \, f(\partial_N f)(\hat{P}_P + \mathcal{A}_P) + f(\partial_M f) \mathcal{D} \, \hat{\bar{\psi}} \sigma_M \psi \\
&\quad + \frac{1}{6} f(\triangle f) \, \hat{\bar{\psi}} \sigma_M \psi \, \hat{\bar{\psi}} \sigma_M \psi,
\end{aligned} \tag{10.59}$$

where $f = 1/\sqrt{h}$. One can also take several $(\mathbf{3}, \mathbf{4}, \mathbf{1})$ superfields and describe in this way the effective Hamiltonians of *non-Abelian* chiral gauge theories in small boxes [117]. We present here the simplest nontrivial such Hamiltonian for a $SU(3)$ chiral anomaly-free theory which includes seven left chiral matter multiplets in the fundamental group representation and an additional left multiplet in the representation $\bar{\mathbf{6}}$. The effective Hamiltonian includes two symplectic superfields. It reads

$$\boxed{\begin{aligned}
\hat{H} &= \frac{2}{3} [(\hat{\mathbf{P}}^a + \boldsymbol{\mathcal{A}}^a)^2 + (\hat{\mathbf{P}}^b + \boldsymbol{\mathcal{A}}^b)^2 + (\hat{\mathbf{P}}^a + \boldsymbol{\mathcal{A}}^a) \cdot (\hat{\mathbf{P}}^b + \boldsymbol{\mathcal{A}}^b) \\
&\quad + (\mathcal{D}^a)^2 + (\mathcal{D}^b)^2 + \mathcal{D}^a \mathcal{D}^b] + \mathcal{B}^a_M \hat{\bar{\psi}}^a \sigma_M \psi^a \\
&\quad + \mathcal{B}^b_M \hat{\bar{\psi}}^b \sigma_M \psi^b + \mathcal{B}^{ab}_M (\hat{\bar{\psi}}^a \sigma_M \psi^b + \hat{\bar{\psi}}^b \sigma_M \psi^a),
\end{aligned}} \tag{10.60}$$

where $\mathcal{D}^{a,b}$, $\boldsymbol{\mathcal{A}}^{a,b}$ and $\boldsymbol{\mathcal{B}}^{a,b}$ are certain functions of the dynamical variables[13] $\mathbf{A}^{a,b}$. The former satisfy the constraints

$$\begin{aligned}
\mathcal{B}^a_M &= \varepsilon_{MNP} \, \partial^a_N \mathcal{A}^a_P = \partial^a_M \mathcal{D}^a, \\
\mathcal{B}^b_M &= \varepsilon_{MNP} \, \partial^b_N \mathcal{A}^b_P = \partial^b_M \mathcal{D}^b, \\
\mathcal{B}^{ab}_M &= \varepsilon_{MNP} \, \partial^a_N \mathcal{A}^b_P = \varepsilon_{MNP} \, \partial^b_N \mathcal{A}^a_P = \partial^a_M \mathcal{D}^b = \partial^b_M \mathcal{D}^a.
\end{aligned} \tag{10.61}$$

[13]The Born–Oppenheimer approximation, in which framework the effective Hamiltonian is calculated, looses its validity near the vertices of the dual lattice where the variables $\mathbf{A}^{a,b}$ live [cf. the monopole singularity at $\mathbf{A} = 0$ in the Hamiltonian (10.57)]. The expression (10.60) for the effective Hamiltonian is valid when we are far enough from the singularities.

The physical meaning of $\mathbf{A}^a$ and $\mathbf{A}^b$ are the zero Fourier modes of the components of gauge potential belonging to the Cartan subalgebra of $su(3)$ — their projections on the simple roots $a = \text{diag}(1,-1,0)$ and $b = \text{diag}(0,1,-1)$.

The superfield description of the symplectic models including gauge fields is only possible in the framework of the harmonic superspace approach [91]. One has to introduce a set of Grassmann-analytic double-charged superfields $V_k^{++}$ subject to the constraints $D^{++}V_k^{++} = 0$ [see Eq. (7.104) for their component expansion] and write the action as the integral over the full harmonic superspace:

$$S^{\text{kin}} = \int d^2\theta^+ \, d^2\theta^- \, dt \, du \, \mathcal{L}(V_k^{++}, V_k^{+-}, V_k^{--}, u), \tag{10.62}$$

where

$$V_k^{+-} = \frac{1}{2} D^{--} V_k^{++}, \quad V_k^{--} = \frac{1}{2}(D^{--})^2 V_k^{++} \tag{10.63}$$

and $\mathcal{L}$ carries zero harmonic charge. After integration over harmonics, the action (10.62) coincides with (10.53) and gives the metric (10.54). The contribution including the gauge fields $\mathcal{A}$ is reproduced, if one adds the structure

$$S^{\text{gauge}} = \int d^2\theta^+ dt_A \, du \, \mathcal{L}^{++}(V_k^{++}), \tag{10.64}$$

with an appropriate function[14] $\mathcal{L}^{++}$ of harmonic charge $+2$ [$t_A$ is the harmonic analytic time (7.91)].

We refer the reader to Ref. [91] for further details.

### 10.2.1 *Bi-symplectic models*

We have seen in the previous section that the supermultiplet (**2**, **4**, **2**) exists in two modifications — the ordinary and the mirror one. Both multiplets satisfy the constraint $\bar{D}_1 \mathcal{Z} = \bar{D}_1 \mathcal{U} = 0$, but the second constraint is different: $\bar{D}_2 \mathcal{Z} = 0$ for the ordinary multiplet and $D^2 \mathcal{U} = 0$ for the mirror one. This amounts to the interchange $\theta_2 \leftrightarrow \bar{\theta}^2$ in the $\theta$-expansion.

By the same token, on top of the ordinary symplectic multiplet $\boldsymbol{\Phi}$ whose expansion is given in Eq. (7.65), one can consider the mirror multiplet $\tilde{\boldsymbol{\Phi}}$

[14]To obtain the effective small-box effective Hamiltonian for a particular chiral theory, one should take some particular $\mathcal{L}$ and $\mathcal{L}^{++}$. But the system is evidently $\mathcal{N} = 4$ supersymmetric with arbitrary $\mathcal{L}$ and $\mathcal{L}^{++}$.

with the same expansion up to the interchange $\theta_2 \leftrightarrow \bar{\theta}^2$. If the model includes only mirror multiplets, the component action has the same form as with the ordinary multiplets. But if we take some number of ordinary *and* some number of mirror multiplets, we arrive at the sigma model of new type, which is natural to call a *bi-symplectic* sigma model.

Consider the simplest such model involving a single ordinary and a single mirror multiplet. Write the action as

$$S = -\int d^2\bar{\theta} d^2\theta dt\, F(\boldsymbol{\Phi}, \tilde{\boldsymbol{\Phi}}). \tag{10.65}$$

The bosonic part of the component Lagrangian reads in this case [cf. (10.40), (10.41)]

$$L_b = \frac{\partial^2 F(\mathbf{A}, \tilde{\mathbf{A}})}{(\partial \mathbf{A})^2} \dot{\mathbf{A}}^2 - \frac{\partial^2 F(\mathbf{A}, \tilde{\mathbf{A}})}{(\partial \tilde{\mathbf{A}})^2} \dot{\tilde{\mathbf{A}}}^2. \tag{10.66}$$

The corresponding six-dimensional metric represents a composition of two conformally flat blocks,

$$g_{MN} = 2\,\mathrm{diag}\left(\frac{\partial^2 F}{(\partial \mathbf{A})^2}, -\frac{\partial^2 F}{(\partial \tilde{\mathbf{A}})^2}\right). \tag{10.67}$$

A generalization to the case when one has several ordinary and mirror multiplets is obvious. The metric of a generic bi-spinorial manifold is block diagonal with the structure (10.54), taken with the positive or negative sign, depending on whether the corresponding multiplet is ordinary or mirror, in each block.

## 10.3 HKT and bi-HKT models

Interesting geometries arise from sigma models involving (**4**, **4**, **0**) superfields. As was the case for the (**2**, **4**, **2**) and (**3**, **4**, **1**) multiplets, the multiplets (**4**, **4**, **0**) exist in two modifications: the ordinary and the mirror ones [14, 35, 128]. A mirror multiplet is distinguished from an ordinary one by the interchange $\theta_2 \leftrightarrow \bar{\theta}^2$ in the $\theta$-expansion.

The models involving such multiplets *of the same type* describe HKT geometry. To obtain a full description of HKT metrics, one should use the harmonic superspace approach and take several *nonlinear* (**4**, **4**, **0**) multiplets satisfying the constraint (7.99). We treat such models in Chapter 13. In this chapter, we restrict ourselves by the case $\mathcal{L}^{+3a} = 0$. This defines linear multiplets, which can be alternatively described in the ordinary $\mathcal{N} = 4$ superspace.

As was shown in Chapter 7, an ordinary linear multiplet can be described by a pair of (**2**, **4**, **2**) superfields $\mathcal{Z}^A$ related by the constraint (7.68):

$$D_j \mathcal{Z}^A - \varepsilon^{AB} \bar{D}_j \bar{\mathcal{Z}}^B = 0. \tag{10.68}$$

An obvious generalization of the free action (7.70) reads

$$\boxed{S = \frac{1}{8} \int d^2\bar{\theta} d^2\theta dt \, \mathcal{L}(\mathcal{Z}_k^A, \bar{\mathcal{Z}}_k^A),} \tag{10.69}$$

where $k$ labels the multiplets and $\mathcal{L}$ is an arbitrary real function.

For the linear multiplets discussed in this chapter, the solution to the constraints (10.68) is easy to find. It is given by Eq. (7.69), where the multiplet $\mathcal{Z}^A$ is expressed in terms of a pair of unconstrained chiral (**2**, **2**, **0**) superfields $Z^A$. Now, we can substitute (7.69) into (10.69) and integrate over $d\bar{\theta}^2 d\theta_2$ to obtain

$$S = \frac{1}{4} \int d\bar{\theta}^1 d\theta_1 dt \, \left(\triangle^{kl}_{AB}\mathcal{L}\right) \bar{D}_1 \bar{Z}_l^B \, D^1 Z_k^A, \tag{10.70}$$

where

$$\triangle^{kl}_{AB}\mathcal{L} = \frac{1}{2}\left(\frac{\partial^2 \mathcal{L}}{\partial Z_k^A \partial \bar{Z}_l^B} + \varepsilon^{AC}\varepsilon^{BD} \frac{\partial^2 \mathcal{L}}{\partial \bar{Z}_k^C \partial Z_l^D}\right). \tag{10.71}$$

Integrating further over $d\bar{\theta}^1 d\theta_1$, we obtain the component Lagrangian. Its bosonic part reads[15]

$$L_{\text{bos}} = \left(\triangle^{kl}_{AB}\mathcal{L}\right) \dot{z}_k^A \dot{\bar{z}}_l^B. \tag{10.72}$$

In other words, the expression (10.71) is nothing but the Hermitian metric $h_{kA,\overline{lB}}$ on our complex manifold. If we have only one multiplet, the Lagrangian (10.72) reduces to

$$L_{\text{bos}} = \frac{1}{2} \frac{\partial^2 \mathcal{L}}{\partial z^A \partial \bar{z}^A} \, \dot{z}^B \dot{\bar{z}}^B. \tag{10.73}$$

This Lagrangian describes the motion along a conformally flat manifold of real dimension 4.

In this way of description, a half of the supersymmetries of the action (10.70) associated with the coordinate transformations in the superspace

[15] See Ref. [82] for the full expression.

$(t, \theta_1, \bar{\theta}^1)$ is manifest and another half is "hidden", being realized as

$$\delta Z_k^A = \eta \varepsilon^{AB} \bar{D}_1 \bar{Z}_k^B, \quad \delta \bar{Z}_k^A = -\bar{\eta} \varepsilon^{AB} D^1 Z_k^B, \tag{10.74}$$

where $\eta$ is a complex Grassmann parameter associated with the shift of $\theta_2$. Substituting the component expansion (7.27) and choosing the convention $\varepsilon^{12} = 1$, as in (7.68), we may rewrite it as

$$\boxed{\delta z_k^1 = \sqrt{2}\eta \bar{\psi}_k^2, \quad \delta z_k^2 = -\sqrt{2}\eta \bar{\psi}_k^1, \quad \delta \psi_k^1 = i\frac{\eta}{\sqrt{2}} \dot{\bar{z}}_k^2, \quad \delta \psi_k^2 = -i\frac{\eta}{\sqrt{2}} \dot{\bar{z}}_k^1.} \tag{10.75}$$

**Theorem 10.3.** *The metric* (10.71) *describes an Obata-flat HKT manifold.*

**Proof. Lemma 1.** *The manifold with the metric* (10.71) *is hypercomplex.*

**Proof.** We go one more step down and express each chiral multiplet $Z^m = Z_k^A$ as a pair of $\mathcal{N} = 1$ multiplets, as what was done in Section 9.2. The actions (10.69) and (10.70) can be then rewritten in the form (9.35).

By construction, this action enjoys four real supersymmetries. One of them is manifest, it is given by Eq. (9.37). And three others are hidden. One of the latter has the following component form:

$$\begin{aligned} \delta_1 x^M &= iE_1 I_N{}^M \Psi^N, \\ \delta_1 \Psi^M &= E_1 I_N{}^M \dot{x}^N, \end{aligned} \tag{10.76}$$

where the complex structure $I_N{}^M$ has the canonical form (2.51) with an even number of blocs. Note that (2.51) is a constant matrix and we could suppress by that reason the second term in (9.39).

Two remaining real supersymmetries follow from the complex supersymmetry (10.74). By posing $\eta = i(E_2 + iE_3)/\sqrt{2}$ with real $E_{2,3}$ and expressing the complex bosonic and fermionic coordinates via the real ones as[16]

$$\begin{aligned} z_1^1 &= \frac{x^1 + ix^2}{\sqrt{2}}, \quad z_1^2 = \frac{x^3 + ix^4}{\sqrt{2}}, \\ \psi_1^1 &= \frac{\Psi^1 + i\Psi^2}{\sqrt{2}}, \quad \psi_1^2 = \frac{\Psi^3 + i\Psi^4}{\sqrt{2}} \end{aligned} \tag{10.77}$$

(then $z_2^A$ are expressed via $x^{5,6,7,8}$, etc.), we can bring (10.75) into the same form as (10.76)

$$\begin{aligned} \delta_2 x^M &= iE_2 J_N{}^M \Psi^N, \quad \delta_3 x^M = iE_3 K_N{}^M \Psi^N, \\ \delta_2 \Psi^M &= E_2 J_N{}^M \dot{x}^N, \quad \delta_3 \Psi^M = E_3 K_N{}^M \dot{x}^N \end{aligned} \tag{10.78}$$

[16] Cf. Eq. (7.32).

with the constant matrices $J, K$ having the canonical form (3.2). The triple $(I, J, K)$ is quaternionic and hence the manifold is hypercomplex. □

**Lemma 2.** *The form* $\omega_+ = \omega_J + i\omega_K = (J_{MN} + iK_{MN})\, dx^M \wedge dx^N$ *is closed with respect to the exterior holomorphic derivative associated with the complex structure* $I$*:* $\partial_I \omega_+ = 0$.

**Proof.** According to Theorem 3.5, the only nonzero components of $\mathcal{I}^+ = (J + iK)_M{}^N/2$ are $(\mathcal{I}^+)_m{}^{\bar{n}}$. In our case, they are very simple:

$$(\mathcal{I}^+)_m{}^{\bar{n}} = \text{diag}(-\varepsilon, \dots, -\varepsilon). \tag{10.79}$$

The matrix $\mathcal{I}^+$ and the related matrix $\mathcal{I}^- = (J - iK)/2$, which in our case has the same form (10.79) modulo the interchange $m \leftrightarrow \bar{n}$, are the *hypercomplex structures* defined on p. 65. Note that the extra supersymmetry transformations (10.74) may be represented in the form

$$\begin{aligned} \delta Z_m &= -\eta(\mathcal{I}^+)_m{}^{\bar{n}} \bar{D}_1 \bar{Z}_{\bar{n}}, \\ \delta \bar{Z}_{\bar{n}} &= \bar{\eta}(\mathcal{I}^-)_{\bar{n}}{}^m D^1 Z_m, \end{aligned} \tag{10.80}$$

where the hypercomplex structures enter roughly in the same way as the ordinary complex structure enters the laws of the extra supersymmetry transformations in (9.38) and (10.3).

The form $\omega_+$ reads

$$\omega_+ = 2(\mathcal{I}^+)_m{}^{\bar{n}} h_{p\bar{n}} dz^m \wedge dz^p = 2\varepsilon_{AC} \frac{\partial^2 \mathcal{L}}{\partial \bar{z}_l^C \partial z_k^B}\, dz_k^B \wedge dz_l^A \tag{10.81}$$

(the two terms in (10.71) give the same contribution). When acting upon this with the exterior holomorphic derivative $\partial_{pD} dz_p^D \wedge$, we obtain zero. □

Two lemmas above and Theorem 3.6 allow us to conclude that the manifold is HKT. As the complex structures $I_N{}^M$, $J_N{}^M$ and $K_N{}^M$ are constant matrices in this case, the Obata connections (3.81) vanish. □

**Remark:** We have proven our theorem using certain specific features [like the constancy of the complex structures $(I, J, K)_N{}^M$] of the action (10.69) including the linear (**4, 4, 0**) multiplets. In Chapter 13, we prove a more general Theorem 13.4 asserting that any hypercomplex manifold described by a generic $\mathcal{N} = 4$ supersymmetric model (9.35) is HKT.

The invariances (9.37), (10.76) and (10.78) bring about four conserved real supercharges. They have the form [106]

$$\boxed{\begin{aligned}
\mathcal{Q}_0 &= \Psi^M \left( P_M - \frac{i}{2}\omega_{AB,M}\Psi^A\Psi^B + \frac{i}{12}C_{MKP}\Psi^K\Psi^P \right), \\
\mathcal{Q}_1 &= \Psi^N I_N{}^M \left( P_M - \frac{i}{2}\omega_{AB,M}\Psi^A\Psi^B - \frac{i}{4}C_{MKP}\Psi^K\Psi^P \right), \\
\mathcal{Q}_2 &= \Psi^N J_N{}^M \left( P_M - \frac{i}{2}\omega_{AB,M}\Psi^A\Psi^B - \frac{i}{4}C_{MKP}\Psi^K\Psi^P \right), \\
\mathcal{Q}_3 &= \Psi^N K_N{}^M \left( P_M - \frac{i}{2}\omega_{AB,M}\Psi^A\Psi^B - \frac{i}{4}C_{MKP}\Psi^K\Psi^P \right).
\end{aligned}} \tag{10.82}$$

In other words, two extra supercharges to be added to the supercharges (9.60) have the same form as $\mathcal{Q}_1$ where the complex structures $J, K$ are substituted for $I$. All the supercharges involve the Levi-Civita spin connection $\omega_{AB,M}$ and the universal Bismut torsion $C_{MKP}$. It is also instructive to write the explicit expressions for the complex supercharges [109]

$$S, \bar{S} = \frac{\mathcal{Q}_0 \pm i\mathcal{Q}_1}{2}, \quad R, \bar{R} = \frac{\mathcal{Q}_2 \pm i\mathcal{Q}_3}{2}. \tag{10.83}$$

We derive:

$$\begin{aligned}
S &= \psi^m \left( p_m - i\omega_{a\bar{b},m}\psi^a\bar{\psi}^{\bar{b}} \right), \\
\bar{S} &= \bar{\psi}^{\bar{m}} \left( p_{\bar{m}} - i\omega_{a\bar{b},\bar{m}}\psi^a\bar{\psi}^{\bar{b}} \right), \\
R &= \psi^k (\mathcal{I}^+)_k{}^{\bar{m}} \left( p_{\bar{m}} - i\omega_{a\bar{b},\bar{m}}\psi^a\bar{\psi}^{\bar{b}} - \frac{i}{2}C_{\bar{m}k\bar{l}}\psi^k\bar{\psi}^{\bar{l}} \right), \\
\bar{R} &= \bar{\psi}^{\bar{k}} (\mathcal{I}^-)_{\bar{k}}{}^{m} \left( p_m - i\omega_{a\bar{b},m}\psi^a\bar{\psi}^{\bar{b}} - \frac{i}{2}C_{mk\bar{l}}\psi^k\bar{\psi}^{\bar{l}} \right),
\end{aligned} \tag{10.84}$$

where $\mathcal{I}^\pm = (J \pm iK)/2$ are the matrices of hypercomplex structure.

The expressions (10.84) for the supercharges become somewhat more elegant if written in terms of the momenta $\pi_m$ and $\pi_{\bar{m}}$ as in (9.7):

$$\begin{aligned}
S &= \psi^m \left( \pi_m - \frac{i}{2}\partial_m h_{n\bar{p}}\psi^n\bar{\psi}^{\bar{p}} \right), \\
\bar{S} &= \bar{\psi}^{\bar{m}} \left( \pi_{\bar{m}} + \frac{i}{2}\partial_{\bar{m}} h_{n\bar{p}}\psi^n\bar{\psi}^{\bar{p}} \right),
\end{aligned}$$

$$R = \psi^k (\mathcal{I}^+)_k{}^{\bar m} \left( \pi_{\bar m} - \frac{i}{2} \partial_{\bar m} h_{n\bar p}\, \psi^n \bar\psi^{\bar p} \right),$$

$$\bar R = \bar\psi^{\bar k} (\mathcal{I}^-)_{\bar k}{}^{m} \left( \pi_{m} + \frac{i}{2} \partial_{m} h_{n\bar p}\, \psi^n \bar\psi^{\bar p} \right). \tag{10.85}$$

The first pair of the supercharges in (10.85) coincides with (9.7), which is not surprising: HKT manifolds are complex and HKT models belong to the broad class of $\mathcal{N} = 2$ Dolbeault models (9.1). In the second pair of the supercharges, the hypercomplex structures appear in exactly the same fashion as the complex structure appears in the expression (10.11) for the classical Kähler–de Rham supercharges. This is also not so surprising: as was mentioned in the *Geometry* part, HKT manifolds relate to generic hypercomplex manifolds roughly in the same way as Kähler manifolds relate to generic complex manifolds. The hypercomplex structure plays for the HKT manifolds the same role as the complex structure for the Kähler manifolds.

The reader may be confused at this point, having not observed in (10.84) a purely holomorphic term

$$\propto X = \psi^k (\mathcal{I}^+)_k{}^{\bar m}\, C_{\bar m pl}\, \psi^p \psi^l = 2 (\mathcal{I}^+)_k{}^{\bar m} (\partial_p h_{l\bar m}) \psi^k \psi^p \psi^l \tag{10.86}$$

in the expression for $R$ and and the similar structure $\bar X \propto \bar\psi\bar\psi\bar\psi$ in the expression for $\bar R$ that seem to pop out. Well, we had a reason for this omission.

**Theorem 10.4.** *For an HKT manifold, the expression $X$ vanishes identically.*

**Proof.** At the first step, we flip the derivative, capitalizing on the fact that the form (10.81) is closed with respect to $\partial_I = \partial_p dz^p \wedge$. This gives

$$X = -2\, \partial_p (\mathcal{I}^+)_k{}^{\bar m}\, h_{l\bar m}\, \psi^k \psi^p \psi^l. \tag{10.87}$$

The complex structures $I$, $J$ and $K$ are covariantly constant with respect to one and the same Bismut connection. It follows that

$$\begin{aligned} \partial_p (\mathcal{I}^+)_k{}^{\bar m} &= h^{\bar q s} G^{(B)}_{\bar q, pk} (\mathcal{I}^+)_s{}^{\bar m} - h^{\bar m s} G^{(B)}_{s, p\bar q} (\mathcal{I}^+)_k{}^{\bar q} \\ &= h^{\bar q s} (\partial_k h_{p\bar q}) (\mathcal{I}^+)_s{}^{\bar m} + h^{\bar m s} (\partial_s h_{p\bar q} - \partial_p h_{s\bar q}) (\mathcal{I}^+)_k{}^{\bar q}, \end{aligned} \tag{10.88}$$

where we used the explicit expressions (2.20) for the Bismut connection and (2.18) for the torsion. Then,[17]

$$X = 2\psi^k \psi^p \psi^l \left[\partial_k h_{p\bar q} (\mathcal{I}^+)_l{}^{\bar q} + (\partial_p h_{l\bar q} - \partial_l h_{p\bar q})(\mathcal{I}^+)_k{}^{\bar q}\right]$$
$$= 6\psi^k \psi^p \psi^l \, \partial_k h_{p\bar q} (\mathcal{I}^+)_l{}^{\bar q}. \tag{10.89}$$

Comparing this with (10.86), we deduce $X = 3X$ and hence $X = 0$. □

Note that the proof of Theorem 10.4 and hence the expressions (10.85) for the supercharges do not depend on the assumption of Obata-flatness. They hold for a generic HKT manifold.

The quantum supercharges that correspond to (10.82) have a simple interpretation in the language of differential forms [cf. Eq. (10.14)]:

$$\hat{\mathcal{Q}}_0 = -i(\tilde\partial_I + \tilde\partial_I^\dagger) = -i(\tilde\partial_J + \tilde\partial_J^\dagger) = -i(\tilde\partial_K + \tilde\partial_K^\dagger),$$
$$\hat{\mathcal{Q}}_1 = \tilde\partial_I^\dagger - \tilde\partial_I, \quad \hat{\mathcal{Q}}_2 = \tilde\partial_J^\dagger - \tilde\partial_J, \quad \hat{\mathcal{Q}}_3 = \tilde\partial_K^\dagger - \tilde\partial_K, \tag{10.90}$$

where $\tilde\partial_{I,J,K}$ are the twisted operators of holomorphic exterior derivatives with respect to the complex structures $I, J, K$. These operators were defined in Eq. (9.27), where one should set $\mathcal{W} = 0$. The operators $\tilde\partial^\dagger_{I,J,K}$ are their Hermitian conjugates.

Note finally that HKT models can be deformed by adding a gauge field. If we want to keep $\mathcal{N} = 4$ supersymmetry, this gauge field should have a special form. Generic description of such fields requires using the harmonic superspace formalism, and we postpone it till Chapter 13.

### 10.3.1 *Bi-HKT models*

Consider a model [82] involving some number of the ordinary linear $(\mathbf{4}, \mathbf{4}, \mathbf{0})$ multiplets $\mathcal{Z}^{A=1,2}_{k=1,\ldots,d_1/2}$ satisfying $\bar D_j \mathcal{Z}^A_k = 0$ and the constraints (10.68), and some number of mirror multiplets $\mathcal{U}^A_{\alpha=1,\ldots,d_2/2}$. (Naturally, $d_{1,2}$ should be even.) The mirror multiplets satisfy the constraints obtained from the constraints for $\mathcal{Z}^A_k$ by the interchange $D^2 = D_1 \leftrightarrow -\bar D_2$: $\bar D_1 \mathcal{U}^A_\alpha = D^2 \mathcal{U}^A_\alpha = 0$ and

$$\boxed{\bar D_2 \mathcal{U}^A_\alpha + \varepsilon^{AB} \bar D_1 \bar{\mathcal{U}}^B_\alpha = 0, \quad D^1 \mathcal{U}^A_\alpha - \varepsilon^{AB} D^2 \bar{\mathcal{U}}^B_\alpha = 0.} \tag{10.91}$$

The constraints (10.68) and (10.91) can be resolved to express each field $\mathcal{Z}^A_k$ via a pair of chiral multiplets $Z^A_k$ as in (7.69), and each field $\mathcal{U}^A_\alpha$ via a

[17]To handle the contribution of the first term in the second line in (10.88), we write $-\psi^k \psi^p \psi^l h_{l\bar m} h^{\bar q s} \partial_k h_{p\bar q} (\mathcal{I}^+)_s{}^{\bar m} = -\psi^k \psi^p \psi^l h_{l\bar m} \partial_k h_{p\bar q} (\mathcal{I}^+)^{\bar q \bar m} = \psi^k \psi^p \psi^l \partial_k h_{p\bar q} (\mathcal{I}^+)_l{}^{\bar q}$.

pair of chiral multiplets $U_\alpha^A$ as

$$\mathfrak{U}_\alpha^A = U_\alpha^A + \bar\theta^2 \varepsilon^{AB} \bar D_1 \bar U_\alpha^B + i\theta_2\bar\theta^2 \dot U_\alpha^A\,,$$
$$\bar{\mathfrak{U}}_\alpha^A = \bar U_\alpha^A - \theta_2 \varepsilon^{AB} D^1 U_\alpha^B - i\theta_2\bar\theta^2 \dot{\bar U}_\alpha^A\,. \tag{10.92}$$

A generic action reads

$$S = \frac{1}{8}\int d^2\bar\theta d^2\theta dt\, \mathcal{L}(\mathcal{Z}_k^A, \bar{\mathcal{Z}}_k^A; \mathfrak{U}_\alpha^A, \bar{\mathfrak{U}}_\alpha^A). \tag{10.93}$$

After integrating over $d\bar\theta^2 d\theta_2$, it acquires the form

$$S = \frac{1}{4}\int d\bar\theta^1 d\theta_1 dt \mathcal{L}' \tag{10.94}$$

with

$$\mathcal{L}' = (\triangle_{AB}^{kl}\mathcal{L})\, \bar D_1 \bar Z_l^B\, D^1 Z_k^A \;-\; (\triangle_{AB}^{\alpha\beta}\mathcal{L})\, \bar D_1 \bar U_\beta^B\, D^1 U_\alpha^A$$
$$+\frac{1}{2}\varepsilon_{AB}\varepsilon_{CD}\big[(\partial_{\overline{Bk}}\partial_{\overline{D\alpha}}\mathcal{L})\, D^1 Z_k^A D^1 U_\alpha^C - (\partial_{Bk}\partial_{D\alpha}\mathcal{L})\, \bar D_1 \bar Z_k^A \bar D_1 \bar U_\alpha^C\big], \tag{10.95}$$

where $\triangle_{AB}^{kl}\mathcal{L}$ was defined in (10.71), $\triangle_{AB}^{\alpha\beta}\mathcal{L}$ is defined analogously, $\partial_{Bk} = \partial/\partial Z_k^B$, etc. The Lagrangian (10.95) involves the terms $\sim DZ\bar D\bar Z$, $\sim DU\bar D\bar U$ and also the terms $\sim DZDU$ and $\sim \bar D\bar Z\bar D\bar U$. It belongs to the class of the Dolbeault SQM models twisted by holomorphic torsions that we discussed in Section 9.1.1. A half of the supersymmetries of the action (10.94) is realized in this formulation manifestly, while another half is "hidden". The latter is implemented as the invariance with respect to the following transformations of the $\mathcal{N}=2$ superfields:

$$\delta Z_k^A = \eta\, \varepsilon^{AB} \bar D_1 \bar Z_k^B\,, \quad \delta \bar Z_k^A = -\bar\eta\, \varepsilon^{AB} D^1 Z_k^B\,,$$
$$\delta U_\alpha^A = \bar\eta\, \varepsilon^{AB} \bar D_1 \bar U_\alpha^B\,, \quad \delta \bar U_\alpha^A = -\eta\, \varepsilon^{AB} D^1 U_\alpha^B\,. \tag{10.96}$$

With the Lagrangian (10.95) at hand, we can go down to components. The second derivatives of the prepotential $\mathcal{L}$ in the first line in (10.95) give the metric so that the bosonic part of the component Lagrangian reads

$$L_b = h_{Ak,\overline{Bl}}\, \dot z_k^A \dot{\bar z}_l^{\bar B} + h_{A\alpha,\overline{B\beta}}\, \dot u_\alpha^A \dot{\bar u}_\beta^{\bar B} \tag{10.97}$$

with

$$h_{Ak,\overline{Bl}} = \triangle_{AB}^{kl}\mathcal{L}, \quad h_{A\alpha,\overline{B\beta}} = -\triangle_{AB}^{\alpha\beta}\mathcal{L}. \tag{10.98}$$

The derivatives are taken now with respect to the bosonic fields $z_k^A$, $u_\alpha^A$. The second line in (10.95) involving holomorphic torsions does not contribute in the bosonic action but contributes in the terms involving fermions, like in (9.30). The full expression for the component Lagrangian is written in Ref. [82].

**Remark:** The reader has noted, of course, that the whole construction here is quite parallel to the construction of bi-Kähler and bi-symplectic models discussed above. In particular, formula (10.96) is an analog of (10.42) and the expression (10.98) for the metric is an analog of expressions (10.41) and (10.67). This is by no means accidental. In fact, the models (**4**, **4**, **0**), (**3**, **4**, **1**) and (**2**, **4**, **2**) are interrelated: two last models can be derived from the first one by the procedure of *Hamiltonian reduction*.[18] We discuss this in detail in Chapter 12.

Our next task is to determine the complex structures associated with the supersymmetry transformations in our model. We proceed in the same way as we did for the HKT models. One of the complex structures is associated with the manifest $\mathcal{N} = 2$ invariance of the action (10.94). It has the standard canonical form

$$I = \mathrm{diag}(\underbrace{\mathfrak{I}, \ldots, \mathfrak{I}}_{d/2}) \tag{10.99}$$

with the matrix $\mathfrak{I}$ given in (3.3). Here, $d$ is the complex dimension of the corresponding manifold. It is even. To find two other complex structures, we pose $\eta = i(E_2 + iE_3)/\sqrt{2}$, substitute it in (10.96) and go down to components: to the real coordinates $x^M$ and their superpartners $\Psi^M$. Then, the law of the supersymmetry transformations has the same form as in Eq. (10.78) with the following complex structures:

$$\begin{aligned} J &= \mathrm{diag}(\underbrace{\mathcal{J}, \ldots, \mathcal{J}}_{d_1/2}, \underbrace{-\mathcal{J}, \ldots, -\mathcal{J}}_{d_2/2}), \\ K &= \mathrm{diag}(\underbrace{\mathcal{K}, \ldots\ldots, \mathcal{K}}_{d/2=(d_1+d_2)/2}). \end{aligned} \tag{10.100}$$

In contrast to the HKT case, the triple of complex structures (10.99) and (10.100) is not quaternionic. On the other hand, the matrices $I, J, K$ still satisfy the Clifford algebra:

$$\boxed{I^p I^q + I^q I^p = -2\delta^{pq}\mathbb{1}.} \tag{10.101}$$

That is why the models of this type were called "Clifford Kähler with torsion" in Ref. [14]. We find, however, the name bi-HKT more adequate.

[18] That is why the multiplet (**4**, **4**, **0**) is sometimes called a *root* multiplet.

The point is that the algebra of the matrices (10.99) and (10.100) is not closed under multiplication, and it is a known mathematical fact [118] that its closure represents a direct sum of two quaternion algebras $\mathcal{H}_+ \oplus \mathcal{H}_-$. The subalgebras $\mathcal{H}_+$ and $\mathcal{H}_-$ correspond in our case to the ordinary and the mirror HKT sectors. Indeed, the whole set of matrices obtained by multiplying the matrices (10.99) and (10.100) is

$$\begin{aligned}
\mathbb{1}_{4d} &= \text{diag}(\underbrace{\mathbb{1}_4, \dots, \mathbb{1}_4}_{d/2}), \Delta = \text{diag}(\underbrace{\mathbb{1}_4, \dots, \mathbb{1}_4}_{d_1/2}, \underbrace{-\mathbb{1}_4, \dots, -\mathbb{1}_4}_{d_2/2}),\\
I_1 &= \text{diag}(\underbrace{\mathcal{I}, \dots, \mathcal{I}}_{d/2}), \quad I_2 = \text{diag}(\underbrace{\mathcal{I}, \dots, \mathcal{I}}_{d_1/2}, \underbrace{-\mathcal{I}, \dots, -\mathcal{I}}_{d_2/2}),\\
J_1 &= \text{diag}(\underbrace{\mathcal{J}, \dots, \mathcal{J}}_{d/2}), \quad J_2 = \text{diag}(\underbrace{\mathcal{J}, \dots, \mathcal{J}}_{d_1}, \underbrace{-\mathcal{J}, \dots, -\mathcal{J}}_{d_2}),\\
K_1 &= \text{diag}(\underbrace{\mathcal{K}, \dots, \mathcal{K}}_{d/2}), K_2 = \text{diag}(\underbrace{\mathcal{K}, \dots, \mathcal{K}}_{d_1}, \underbrace{-\mathcal{K}, \dots, -\mathcal{K}}_{d_2}).
\end{aligned} \tag{10.102}$$

Consider the linear combinations

$$\mathbb{1}^{\pm}_{4d} = \frac{\mathbb{1}_{4d} \pm \Delta}{2}, I^{\pm} = \frac{I_1 \pm I_2}{2}, J^{\pm} = \frac{J_1 \pm J_2}{2}, K^{\pm} = \frac{K_1 \pm K_2}{2}. \tag{10.103}$$

It is clear that the members of the "plus" quartet and the "minus" quartet satisfy separately the quaternion algebras and that the product of any "plus" matrix to any "minus" matrix vanishes. The diagonal subalgebra of $\mathcal{H}_+ \oplus \mathcal{H}_-$ involves the standard quaternionic matrices (3.2), which can be chosen for the complex structures in not only the HKT, but also the bi-HKT case.[19] This means that all bi-HKT manifolds are hypercomplex. However:

- This quaternion triple cannot be associated with the supersymmetry transformations as in (10.78).
- The exterior derivative $\partial_I(\omega_J + i\omega_K)$ does not vanish in this case.

[19]If we want to be precise, we must remind here that, for the HK and generic HKT manifolds, only the tangent space projections of the complex structures $(I, J, K)_{AB}$ have the form (3.2), while the tensors $(I, J, K)_M{}^N$ look more complicated. However, for the Obata-flat HKT manifolds, which we only study in this chapter, also $(I, J, K)_M{}^N$ are the simple constant matrices (3.2). Generic HKT models with nonvanishing Obata curvature will be studied in Chapter 13. We do not know whether generic not Obata-flat bi-HKT models exist.

In view of Theorem 3.6, the latter assertion means that, though bi-HKT manifolds are hypercomplex, they are not HKT. Examples of hypercomplex but not HKT manifolds were found earlier in Ref. [38], but they were essentially more complicated.

In Ref. [82], the explicit expressions for the conserved supercharges in the bi-HKT models were given. They are not so simple, however, and we have chosen not to present them here.

## Chapter 11

# Sigma Models with Extended Supersymmetries: $\mathcal{N} = 8$

### 11.1 Hyperkähler models

The most simple way to describe these models is to use the $\mathcal{N} = 2$ superspace language. Consider the model (8.6) involving $4n$ real $\mathcal{N} = 2$ superfields $X^M$ and require the action to be invariant under three different supersymmetry transformations

$$\boxed{\begin{aligned} \delta_1 X^M &= I_N{}^M(\epsilon_1 DX^N - \bar{\epsilon}_1 \bar{D} X^N), \\ \delta_2 X^M &= J_N{}^M(\epsilon_2 DX^N - \bar{\epsilon}_2 \bar{D} X^N), \\ \delta_3 X^M &= K_N{}^M(\epsilon_3 DX^N - \bar{\epsilon}_3 \bar{D} X^N) \end{aligned}} \tag{11.1}$$

with three different complex Grassmann parameters $\epsilon_1, \epsilon_2, \epsilon_3$. We know from Theorem 10.1 that the action is invariant under each of these transformations iff $I, J, K$ represent integrable complex structures that are covariantly constant with the ordinary Levi-Civita connection, i.e. iff the manifold is Kähler with respect to any one of these structures. The model involves eight supercharges:

$$\begin{aligned} \hat{Q}_0 &= -\frac{i}{\sqrt{2}} \psi^M D_M, & \hat{Q}_0^\dagger &= -\frac{i}{\sqrt{2}} e^M_C \frac{\partial}{\partial \psi_C} D_M, \\ \hat{Q}_I &= -\frac{i}{\sqrt{2}} \psi^N I_N{}^M D_M, & \hat{Q}_I^\dagger &= -\frac{i}{\sqrt{2}} e^N_C I_N{}^M \frac{\partial}{\partial \psi_C} D_M, \\ \hat{Q}_J &= -\frac{i}{\sqrt{2}} \psi^N J_N{}^M D, & \hat{Q}_J^\dagger &= -\frac{i}{\sqrt{2}} e^N_C J_N{}^M \frac{\partial}{\partial \psi_C} D_M, \\ \hat{Q}_K &= -\frac{i}{\sqrt{2}} \psi^N K_N{}^M D_M, & \hat{Q}_K^\dagger &= -\frac{i}{\sqrt{2}} e^N_C K_N{}^M \frac{\partial}{\partial \psi_C} D_M \end{aligned} \tag{11.2}$$

with $D_M$ defined in (8.30) and (10.13). Being translated into the language of differential forms, the supercharges $\hat{Q}_0, \hat{Q}_{I,J,K}$ look like

$$\hat{Q}_0 \to -\frac{i}{\sqrt{2}}(\partial_I + \bar{\partial}_I) = -\frac{i}{\sqrt{2}}(\partial_J + \bar{\partial}_J) = -\frac{i}{\sqrt{2}}(\partial_K + \bar{\partial}_K),$$

$$\hat{Q}_I \to \frac{\bar{\partial}_I - \partial_I}{\sqrt{2}}, \quad \hat{Q}_J \to \frac{\bar{\partial}_J - \partial_J}{\sqrt{2}}, \quad \hat{Q}_K \to \frac{\bar{\partial}_K - \partial_K}{\sqrt{2}}, \tag{11.3}$$

where $\partial_{I,J,K}$ and $\bar{\partial}_{I,J,K}$ are the exterior derivatives that are holomorphic and antiholomorhic with respect to the corresponding complex structures.[1] It follows from the analysis of Section 10.1 that

$$\boxed{\{\hat{Q}_0^\dagger, \hat{Q}_0\} = \{\hat{Q}_I^\dagger, \hat{Q}_I\} = \{\hat{Q}_J^\dagger, \hat{Q}_J\} = \{\hat{Q}_K^\dagger, \hat{Q}_K\} = \hat{H}} \tag{11.4}$$

and the anticommutators

$$\{\hat{Q}_0, \hat{Q}_{I,J,K}\}, \quad \{\hat{Q}_0^\dagger, \hat{Q}_{I,J,K}^\dagger\}, \quad \{\hat{Q}_0^\dagger, \hat{Q}_{I,J,K}\}, \quad \{\hat{Q}_0, \hat{Q}_{I,J,K}^\dagger\} \tag{11.5}$$

vanish.

Now, we prove the theorem:

**Theorem 11.1.** *For the anticommutators of the supercharges involving different complex structures, $\{\hat{Q}_I, \hat{Q}_J\}, \{\hat{Q}_I, \hat{Q}_J^\dagger\}$, etc., to vanish and the standard $\mathcal{N} = 8$ supersymmetry algebra with the only nontrivial anticommutators given in* (11.4) *to hold, it is necessary and sufficient for the triple $(I, J, K)$ of the covariantly constant complex structures to be quaternionic, so that the manifold is hyperkähler.*

**Proof. Sufficiency.** Introduce the operators

$$\hat{F}_I = I_M{}^N \psi^M \frac{\partial}{\partial \psi^N} = I_{AB}\, \psi_A \frac{\partial}{\partial \psi_B},$$

$$\hat{F}_J = J_M{}^N \psi^M \frac{\partial}{\partial \psi^N} = J_{AB}\, \psi_A \frac{\partial}{\partial \psi_B},$$

$$\hat{F}_K = K_M{}^N \psi^M \frac{\partial}{\partial \psi^N} = K_{AB}\, \psi_A \frac{\partial}{\partial \psi_B}. \tag{11.6}$$

When acting on a wave function including $p$ holomorphic and $q$ antiholomorphic (with respect to $I$) Grassmann factors, the operator $\hat{F}_I$ multiplies it by $i(q-p)$. Using the physical slang, $\hat{F}_I$ may be called the operator of fermion

[1]Note that, in constrast to the HKT supercharges (10.90), the supercharges (11.3) *are* not Hermitian. For example, $\sqrt{2}\hat{Q}_0^\dagger = i(\partial_I^\dagger + \bar{\partial}_I^\dagger) \neq \hat{Q}_0$.

charge (and there are three different fermion charges associated with three different complex structures).

**Lemma 1.** *The following identities hold:*

$$[\hat{F}_{I,J,K}, \hat{Q}_0] = \hat{Q}_{I,J,K}, \quad [\hat{F}_I, \hat{Q}_I] = [\hat{F}_J, \hat{Q}_J] = [\hat{F}_K, \hat{Q}_K] = -\hat{Q}_0,$$
$$[\hat{F}_I, \hat{Q}_J] = -[\hat{F}_J, \hat{Q}_I] = \hat{Q}_K, \quad [\hat{F}_J, \hat{Q}_K] = -[\hat{F}_K, \hat{Q}_J] = \hat{Q}_I,$$
$$[\hat{F}_K, \hat{Q}_I] = -[\hat{F}_I, \hat{Q}_K] = \hat{Q}_J, \tag{11.7}$$

*and similarly for* $\hat{Q}^\dagger_{I,J,K}$.

**Proof.** Consider the commutator

$$[\hat{F}_I, \hat{Q}_0] = \left[ I_M{}^N \psi^M \frac{\partial}{\partial \psi^N}, -\frac{i}{\sqrt{2}} \psi^P D_P \right]. \tag{11.8}$$

It involves the term $\propto \partial\psi^P/\partial\psi^N = \delta^P_N$, and this gives $\hat{Q}_I$. Besides, we see the contribution

$$\propto \psi^P I_{AB}\, \omega_{CD,P} \left[ \psi_A \frac{\partial}{\partial \psi_B}, \psi_C \frac{\partial}{\partial \psi^D} \right]$$
$$= \psi^P (I_{AB}\omega_{BD,P} - \omega_{AB,P} I_{BD}) \psi_A \frac{\partial}{\partial \psi^D}. \tag{11.9}$$

But the commutator $[I, \omega_P]$ vanishes for hyperkähler manifolds due to Theorem 3.2 and the relation (3.21).

The commutator

$$[\hat{F}_I, \hat{Q}_J] = \left[ I_M{}^N \psi^M \frac{\partial}{\partial \psi^N}, -\frac{i}{\sqrt{2}} \psi^P J_P{}^Q D_Q \right] \tag{11.10}$$

is treated in a similar way. The contribution involving $\partial\psi^P/\partial\psi^N$ gives[2] $\hat{Q}_K$ and the contribution $\propto [D_Q, \hat{F}_I]$ vanishes. All other commutators in (11.7) are derived analogously. □

The vanishing of, say, $\{\hat{Q}_I, \hat{Q}_J\}$ follows then from the vanishing of the anticommutators in (11.5) and from the Jacobi identity:

$$\{\hat{Q}_I, \hat{Q}_J\} = \{\hat{Q}_I, [\hat{F}_J, \hat{Q}_0]\} = [\hat{F}_J, \{\hat{Q}_I, \hat{Q}_0\}] - \{\hat{Q}_0, [\hat{F}_J, \hat{Q}_I]\}$$
$$= 0 + \{\hat{Q}_0, \hat{Q}_K\} = 0. \tag{11.11}$$

We have proven that the existence of a quaternion triple is sufficient for the manifold to be HK. To prove that it is also **necessary**, consider the

[2] The quaternionic nature of the triple $(I, J, K)$ is used at *this* step.

Lie bracket $(\delta_2\delta_1 - \delta_1\delta_2)X^M$ of the supersymmetry transformations (11.1). The bracket vanishes iff the anticommutator $\{\hat{Q}_I, \hat{Q}_J\}$ vanishes.

A calculation gives [cf. Eq. (9.44)]:

$$(\delta_2\delta_1 - \delta_1\delta_2)X^M = \epsilon_1\bar{\epsilon}_2[-i(IJ + JI)_P{}^M\dot{X}^P + \text{lin. independent structures}]$$
$$+ \epsilon_2\bar{\epsilon}_1[\cdots] + \epsilon_1\epsilon_2[\cdots] + \bar{\epsilon}_1\bar{\epsilon}_2[\cdots]$$

If the LHS vanishes, all four terms in the RHS should vanish, in particular, the term $\propto \epsilon_1\bar{\epsilon}_2$. If so, the contribution $\propto \dot{X}^P$ should vanish too. This means that $IJ + JI = 0$.

But a manifold possessing two covariantly constant anticommuting complex structures is hyperkählerian: the structure $K = IJ$ is also covariantly constant, and the triple $(I, J, K)$ is quaternionic (see the Remark on p. 51). □

The assertion that the hyperkähler geometry implies the $\mathcal{N} = 8$ supersymmetry of the action (8.6) was first made in Ref. [108]. The simple proof of the emergence of the $\mathcal{N} = 8$ algebra for hyperkähler manifolds that we outlined above represents an adaptation to the $\mathcal{N} = 8$ case of the similar proof given in Ref. [106] for the fact that the Obata-flat HKT geometry implies the $\mathcal{N} = 4$ supersymmetry algebra for the supercharges (10.82). A similar but not identical proof of the emergence of $\mathcal{N} = 8$ supersymmetry, which used, instead of the fermion charge operators (11.6) and the commutators (11.7), the Lefschetz operators $\Lambda, L$ and the Kodaira identities, was given in Ref. [37].

We have proven that the model (8.6) with a hyperkähler metric $g_{MN}$ enjoys $\mathcal{N} = 8$ supersymmetry. But the second part of Theorem 11.1 — that $\mathcal{N} = 8$ supersymmetry dictates hyperkähler nature of the manifold — was only proven in the assumption of invariance of the action under the transformations (11.1) associated with the covariantly constant complex structures $I, J, K$. This assertion is not valid for a more general class of the de Rham models — the bi-Kähler models discussed in the preceding chapter (Section 10.1.2). We have seen that the laws of supersymmetry transformations have the form (10.42), which is more complicated than (11.1).

A generic bi-Kähler model enjoys $\mathcal{N} = 4$ supersymmetry, but for a certain specific class of these models, which we are set to discuss in this section, the symmetry extends to $\mathcal{N} = 8$. Their metric is not hyperkähler and not even Kähler.

Bearing in mind the latter observation, we deduce from the second part of Theorem 11.1 the following simple **Corollary**:

*A Kähler sigma model enjoying $\mathcal{N} = 8$ supersymmetry is necessarily hyperkähler.*

## 11.2 $\mathcal{N} = 8$ supersymmetric bi-Kähler models

Consider a system (10.39) with an equal number of ordinary and mirror chiral $\mathcal{N} = 4$ multiplets:

$$S = \frac{1}{4} \int dt d^2\bar{\theta} d^2\theta \, \mathcal{L}(\mathcal{Z}_\alpha, \bar{\mathcal{Z}}_\alpha; \mathcal{U}_\alpha, \bar{\mathcal{U}}_\alpha). \tag{11.12}$$

Consider the following extra supersymmetry transformations mixing $(\mathcal{Z}_\alpha, \bar{\mathcal{Z}}_\alpha)$ and $(\mathcal{U}_\alpha, \bar{\mathcal{U}}_\alpha)$:

$$\boxed{\begin{aligned} \delta\mathcal{Z}_\alpha &= \epsilon\bar{D}_1\bar{\mathcal{U}}_\alpha + \eta\bar{D}_2\mathcal{U}_\alpha \Rightarrow \delta\bar{\mathcal{Z}}_\alpha = -\bar{\epsilon}D^1\mathcal{U}_\alpha - \bar{\eta}D^2\bar{\mathcal{U}}_\alpha, \\ \delta\mathcal{U}_\alpha &= -\epsilon\bar{D}_1\bar{\mathcal{Z}}_\alpha + \bar{\eta}D^2\mathcal{Z}_\alpha \Rightarrow \delta\bar{\mathcal{U}}_\alpha = \bar{\epsilon}D^1\mathcal{Z}_\alpha - \eta\bar{D}_2\bar{\mathcal{Z}}_\alpha. \end{aligned}} \tag{11.13}$$

These transformations respect the condition $\bar{D}_j\delta\mathcal{Z}_\alpha = 0$ and the conditions (10.38) for $\delta\mathcal{U}_\alpha$. Now, $\epsilon$ and $\eta$ are two different complex Grassmann numbers giving four extra supersymmetries. The transformations (11.13) play exactly the same role as the transformation (9.38) for the $\mathcal{N} = 2$ Dolbeaux model when it was expressed in terms of $\mathcal{N} = 1$ superfields or the transformation (10.3) for the $\mathcal{N} = 4$ model when it was expressed in terms of real $\mathcal{N} = 2$ superfields.

**Theorem 11.2.** *The action* (11.12) *is invariant under the transformations* (11.13) *iff the conditions*

$$\left( \frac{\partial^2}{\partial\mathcal{Z}_\alpha\partial\bar{\mathcal{Z}}_\beta} + \frac{\partial^2}{\partial\mathcal{U}_\beta\partial\bar{\mathcal{U}}_\alpha} \right) \mathcal{L} = 0 \tag{11.14}$$

*and*

$$\frac{\partial^2}{\partial\mathcal{Z}_{[\alpha}\partial\bar{\mathcal{U}}_{\beta]}}\mathcal{L} = 0 \tag{11.15}$$

*are fulfilled for all $\alpha, \beta$.*

**Proof. Necessity**. Consider the terms $\propto \epsilon$ in the variation $\delta\mathcal{L}$,

$$\delta_\epsilon\mathcal{L} = \epsilon \left( \frac{\partial\mathcal{L}}{\partial\mathcal{Z}_\alpha} \bar{D}_1\bar{\mathcal{U}}_\alpha - \frac{\partial\mathcal{L}}{\partial\mathcal{U}_\alpha} \bar{D}_1\bar{\mathcal{Z}}_\alpha \right). \tag{11.16}$$

For the action to stay intact, this variation should represent a total supersymmetric covariant derivative of some complex function $R$:

$$\delta_\epsilon \mathcal{L} = \epsilon \bar{D}_1 R = \epsilon \left( \frac{\partial R}{\partial \bar{\mathcal{Z}}_\alpha} \bar{D}_1 \bar{\mathcal{Z}}_\alpha + \frac{\partial R}{\partial \bar{\mathcal{U}}_\alpha} \bar{D}_1 \bar{\mathcal{U}}_\alpha \right). \tag{11.17}$$

Comparing this with (11.16), we obtain the system of differential equations for $R(\mathcal{Z}_\alpha, \bar{\mathcal{Z}}_\alpha; \mathcal{U}_\alpha, \bar{\mathcal{U}}_\alpha)$ representing a kind of generalized Cauchy–Riemann conditions:

$$\frac{\partial R}{\partial \bar{\mathcal{U}}_\alpha} = \frac{\partial \mathcal{L}}{\partial \mathcal{Z}_\alpha}, \quad \frac{\partial R}{\partial \bar{\mathcal{Z}}_\alpha} = -\frac{\partial \mathcal{L}}{\partial \mathcal{U}_\alpha}. \tag{11.18}$$

Differentiating the first equation over $\bar{\mathcal{Z}}_\beta$ and expressing $\partial R/\partial \bar{\mathcal{Z}}_\beta$ from the second equation, we arrive at (11.14) as an integrability condition for the system (11.18). Differentiating the first equation over $\bar{\mathcal{U}}_\beta$ or the second one over $\bar{\mathcal{Z}}_\beta$, we derive (11.15). The same integrability conditions follow if one considers the terms $\propto \bar{\epsilon}$ or the terms $\propto \eta$ or $\propto \bar{\eta}$ in $\delta \mathcal{L}$.

**Sufficiency.** The conditions (11.14) and (11.15) mean that the antiholomorphic form

$$\omega = \frac{\partial \mathcal{L}}{\partial \mathcal{Z}_\alpha} d\bar{\mathcal{U}}_\alpha - \frac{\partial \mathcal{L}}{\partial \mathcal{U}_\alpha} d\bar{\mathcal{Z}}_\alpha \tag{11.19}$$

is closed. But the topology of the space $(\bar{\mathcal{Z}}_\alpha, \bar{\mathcal{U}}_\alpha)$ is trivial and hence, according to Theorem 1.2, this form is also exact. And this gives us (11.18) and (11.17) implying invariance of the action. □

The pair $(\mathcal{Z}_\alpha, \mathcal{U}_\alpha)$ represents a large (**4**, **8**, **4**) $\mathcal{N} = 8$ supermultiplet. If the system has only one such multiplet, the condition (11.15) trivializes and (11.14) dictates that $\mathcal{L}$ is a $4D$ harmonic function. The simplest nontrivial such function is

$$\mathcal{L} = \frac{1}{\mathcal{Z}\bar{\mathcal{Z}} + \mathcal{U}\bar{\mathcal{U}}}. \tag{11.20}$$

The metric of the corresponding bi-Kähler manifold is[3]

$$ds^2 = 2\frac{z\bar{z} - u\bar{u}}{(z\bar{z} + u\bar{u})^3}(dzd\bar{z} + dud\bar{u}). \tag{11.21}$$

It is not positive definite, and one may not like it, but choosing a little more complicated harmonic superpotential

$$\boxed{\mathcal{L} = \ln \mathcal{Z} \ln \bar{\mathcal{Z}} - F\left(\frac{\mathcal{U}\bar{\mathcal{U}}}{\mathcal{Z}\bar{\mathcal{Z}}}\right),} \tag{11.22}$$

[3]Recall the convention (2.2).

where

$$F(x) = \int_1^x \frac{\ln(1+y)}{y} dy \tag{11.23}$$

is the Spence function, one arrives [119] at a nice positive definite harmonic metric

$$ds^2 = 2\frac{dzd\bar{z} + dud\bar{u}}{z\bar{z} + u\bar{u}}. \tag{11.24}$$

## 11.3 Legendre transformation

The $\mathcal{N} = 8$ bi-Kähler models analyzed above included the multiplets of type (**4**, **8**, **4**), which are also building blocks for the hyperkähler models. A remarkable fact is [29, 120] that the bi-Kähler models of a special kind, the models involving $U(1)$ isometries, are related to certain hyperkähler models by *Legendre transformations.*

Consider the model (11.12) with only one pair $(\mathcal{Z}, \mathcal{U})$ and harmonic prepotential $\mathcal{L}$. We also impose an additional requirements that $\mathcal{L}$ does not depend on $\mathcal{U}$ and $\bar{\mathcal{U}}$ separately, but only on the sum $X = \mathcal{U} + \bar{\mathcal{U}}$. Then, the harmonicity condition reduces to

$$\boxed{\frac{\partial^2 \mathcal{L}}{\partial \mathcal{Z} \partial \bar{\mathcal{Z}}} + \frac{\partial^2 \mathcal{L}}{\partial X^2} = 0.} \tag{11.25}$$

Introduce the function

$$\mathcal{K}(\mathcal{Z}, \bar{\mathcal{Z}}; V) = \mathcal{L}(\mathcal{Z}, \bar{\mathcal{Z}}; X) - XV, \tag{11.26}$$

where

$$V = \frac{\partial \mathcal{L}}{\partial X} \tag{11.27}$$

and consider the following action:

$$\boxed{S = \int dtd^2\bar{\theta}d^2\theta \, \mathcal{K}(\mathcal{Z}, \bar{\mathcal{Z}}; \Phi + \bar{\Phi}),} \tag{11.28}$$

where $\Phi$ and $\bar{\Phi}$ are *ordinary* left and right $\mathcal{N} = 4$ chiral multiplets. Equation (11.28) describes a conventional Kähler sigma model that enjoys by construction $\mathcal{N} = 4$ supersymmetry.

**Theorem 11.3.** *The metric following from* (11.28) *is hyperkähler.*

**Proof. Lemma 1.** *The Kähler potential $\mathcal{K}(\mathcal{Z}, \bar{\mathcal{Z}}, \Phi, \bar{\Phi})$ of Eq.* (11.26) *satisfies the equation*[4]

$$\begin{vmatrix} \frac{\partial^2 \mathcal{K}}{\partial\Phi\partial\bar{\Phi}} & \frac{\partial^2 \mathcal{K}}{\partial\mathcal{Z}\partial\bar{\Phi}} \\ \frac{\partial^2 \mathcal{K}}{\partial\Phi\partial\bar{\mathcal{Z}}} & \frac{\partial^2 \mathcal{K}}{\partial\mathcal{Z}\partial\bar{\mathcal{Z}}} \end{vmatrix} = 1. \tag{11.29}$$

**Proof.** The idea is to express the second partial derivatives of $\mathcal{K}$ in (11.29) via the second partial derivatives of $\mathcal{L}$ and use the harmonicity (11.25). Note first that

$$\frac{\partial \mathcal{K}}{\partial \Phi} = \frac{\partial \mathcal{K}}{\partial \bar{\Phi}} = \frac{\partial \mathcal{K}}{\partial V} = \frac{\partial \mathcal{L}}{\partial X}\frac{\partial X}{\partial V} - X - V\frac{\partial X}{\partial V} = -X.$$

It follows:

$$\frac{\partial^2 \mathcal{K}}{\partial\Phi\partial\bar{\Phi}} = \frac{\partial^2 \mathcal{K}}{\partial V^2} = -\frac{\partial X}{\partial V} = -\frac{1}{\partial V/\partial X} = -\frac{1}{\partial^2\mathcal{L}/\partial X^2}, \tag{11.30}$$

the derivatives being done assuming fixed $\mathcal{Z}$.

Now,

$$\frac{\partial^2 \mathcal{K}}{\partial\bar{\Phi}\partial\mathcal{Z}} = \frac{\partial^2 \mathcal{K}}{\partial V\partial\mathcal{Z}} = -\left.\frac{\partial X}{\partial \mathcal{Z}}\right|_V.$$

On the other hand,

$$0 = \left.\frac{\partial V}{\partial \mathcal{Z}}\right|_V = \left.\frac{\partial^2 \mathcal{L}}{\partial\mathcal{Z}\partial X}\right|_V = \frac{\partial^2 \mathcal{L}}{\partial X\partial\mathcal{Z}} + \frac{\partial^2 \mathcal{L}}{\partial X^2}\left(\frac{\partial X}{\partial \mathcal{Z}}\right)_V.$$

It follows:

$$\frac{\partial^2 \mathcal{K}}{\partial\bar{\Phi}\partial\mathcal{Z}} = \frac{\partial^2 \mathcal{L}}{\partial X\partial\mathcal{Z}}\left(\frac{\partial^2 \mathcal{L}}{\partial X^2}\right)^{-1}. \tag{11.31}$$

Similarly,

$$\frac{\partial^2 \mathcal{K}}{\partial\Phi\partial\bar{\mathcal{Z}}} = \frac{\partial^2 \mathcal{L}}{\partial X\partial\bar{\mathcal{Z}}}\left(\frac{\partial^2 \mathcal{L}}{\partial X^2}\right)^{-1}. \tag{11.32}$$

To evaluate $\partial^2\mathcal{K}/(\partial\mathcal{Z}\partial\bar{\mathcal{Z}})$, we start with

$$\frac{\partial \mathcal{K}}{\partial \mathcal{Z}} = \frac{\partial \mathcal{L}}{\partial \mathcal{Z}} + \frac{\partial \mathcal{L}}{\partial X}\left.\frac{\partial X}{\partial \mathcal{Z}}\right|_V - V\left.\frac{\partial X}{\partial \mathcal{Z}}\right|_V = \frac{\partial \mathcal{L}}{\partial \mathcal{Z}}.$$

[4]This is the so-called *heavenly* equation [121], a member of the *Monge–Ampère* family.

Then,

$$\frac{\partial^2 \mathcal{K}}{\partial \mathcal{Z} \partial \bar{\mathcal{Z}}} = \frac{\partial^2 \mathcal{L}}{\partial \mathcal{Z} \partial \bar{\mathcal{Z}}} + \frac{\partial^2 \mathcal{L}}{\partial \mathcal{Z} \partial X} \frac{\partial X}{\partial \bar{\mathcal{Z}}}\bigg|_V$$

$$= \frac{\partial^2 \mathcal{L}}{\partial \mathcal{Z} \partial \bar{\mathcal{Z}}} - \left(\frac{\partial^2 \mathcal{L}}{\partial \mathcal{Z} \partial X}\right)\left(\frac{\partial^2 \mathcal{L}}{\partial \bar{\mathcal{Z}} \partial X}\right)\left(\frac{\partial^2 \mathcal{L}}{\partial X^2}\right)^{-1}. \qquad (11.33)$$

The calculation of the determinant (11.29) gives

$$\det(\cdots) = -\frac{\partial^2 \mathcal{L}/(\partial \mathcal{Z} \partial \bar{\mathcal{Z}})}{\partial^2 \mathcal{L}/\partial X^2} = 1. \qquad (11.34)$$

(The condition (11.25) was used!) □

The metric following from the Kähler potential (11.26) is

$$\frac{ds^2}{2} = h_{\phi\bar{\phi}} d\phi d\bar{\phi} + h_{z\bar{z}} dz d\bar{z} + h_{\phi\bar{z}} d\phi d\bar{z} + h_{z\bar{\phi}} dz d\bar{\phi} \qquad (11.35)$$

with $h_{\phi\bar{\phi}} = \partial^2 \mathcal{K}/d\phi d\bar{\phi}$, etc. The corresponding closed Kähler form,

$$\omega^0 = 2i(h_{\phi\bar{\phi}} d\phi \wedge d\bar{\phi} + h_{z\bar{z}} dz \wedge d\bar{z} + h_{\phi\bar{z}} d\phi \wedge d\bar{z} + h_{z\bar{\phi}} dz \wedge d\bar{\phi}), \qquad (11.36)$$

is associated with the complex structure $I_M{}^N$ with nonzero elements

$$I_m{}^n = -i\delta^n_m, \quad I_{\bar{m}}{}^{\bar{n}} = i\delta^{\bar{n}}_{\bar{m}},$$

[see Eq. (2.53)]. Consider now the forms

$$\omega^+ = 2\, d\phi \wedge dz, \quad \omega^- = \overline{\omega^+} = 2\, d\bar{z} \wedge d\bar{\phi}. \qquad (11.37)$$

Obviously, they are closed.

**Lemma 2.** *The complex structures associated with the forms* (11.37) *are covariantly constant.*

**Proof.** Change the notation: $\phi \to w^1$, $z \to w^2$. The forms $\omega^\pm$ can be represented as $\pm \mathcal{I}^\pm_{MN}\, dx^M \wedge dx^N$, where the only nonzero components of $\mathcal{I}^+$ are $\varepsilon_{mn}$ and the only nonzero components of $\mathcal{I}^-$ are $\varepsilon_{\bar{m}\bar{n}}$. They are constant. For Kähler manifolds, the Christoffel symbols have only pure holomorphic and pure antiholomorphic components. Thus, we immediately see that $\nabla_{\bar{p}} \mathcal{I}^+ = \nabla_p \mathcal{I}^- = 0$. The derivative $\nabla_p \mathcal{I}^+$ is

$$\nabla_p \mathcal{I}^+_{mn} = \varepsilon_{qm} \Gamma^q_{pn} - \varepsilon_{qn} \Gamma^q_{pm}$$

with $\Gamma^q_{pn} = h^{\bar{s}q}\partial_p h_{n\bar{s}}$ [see Eq. (2.30)]. This expression is antisymmetric under $m \leftrightarrow n$. Take $m = 1, n = 2$. We derive

$$\nabla_p \mathcal{I}^+_{12} = -\Gamma^2_{p2} - \Gamma^1_{p1} = -h^{\bar{s}q}\partial_p h_{q\bar{s}} = -\partial_p[\ln\det(h_{s\bar{r}})], \tag{11.38}$$

which vanishes due to (11.29).

Similarly, $\nabla_P \mathcal{I}^-_{MN} = 0$. □

Define[5] $J = \mathcal{I}^+ + \mathcal{I}^-$ and $K = i(\mathcal{I}^- - \mathcal{I}^+)$.

**Lemma 3.** *The triple $(I, J, K)$ is quaternionic.*

**Proof.** We have

$$\begin{aligned} J_{12} &= -J_{21} = J_{\bar{1}\bar{2}} = -J_{\bar{2}\bar{1}} = 1, \\ K_{12} &= -K_{21} = K_{\bar{2}\bar{1}} = -K_{\bar{1}\bar{2}} = -i \end{aligned} \tag{11.39}$$

and hence

$$\begin{aligned} &J_1{}^{\bar{n}} = h^{\bar{n}2}, \quad J_2{}^{\bar{n}} = -h^{\bar{n}1}, \quad J_{\bar{1}}{}^m = h^{\bar{2}m}, \quad J_{\bar{2}}{}^m = -h^{\bar{1}m}, \\ &K_1{}^{\bar{n}} = -ih^{\bar{n}2}, \quad K_2{}^{\bar{n}} = ih^{\bar{n}1}, \quad K_{\bar{1}}{}^m = ih^{\bar{2}m}, \quad K_{\bar{2}}{}^m = -ih^{\bar{1}m}. \end{aligned} \tag{11.40}$$

Using this and the relation (11.29), one can derive $JJ = KK = -\mathbb{1}$ and also $IJ = -JI = K, JK = -KJ = I, KI = -IK = J$. □

A metric admitting three quaternionic covariantly constant complex structures is by definition a hyperkähler metric. □

**Remark:** The proofs of Lemmas 2 and 3 did not depend on the assumption that the Kähler potential of our manifold was derived as a result of the Legendre transformation as in (11.26). Only the fact $\det(h_{j\bar{k}}) = const$ was used. Thus, we have actually proved that *a Kähler four-dimensional manifold with a constant determinant of the metric is hyperkähler.*

And this can also be understood as follows. A Hermitian $2 \times 2$ matrix with unit determinant can be presented as

$$h_{j\bar{k}} = [\exp(v_a\sigma_a)]_{j\bar{k}} \tag{11.41}$$

with real $v_{1,2,3}$.

[5]Then, $\mathcal{I}^\pm = (J \pm iK)/2$. In fact, the tensors $\mathcal{I}^\pm$ are nothing but *hypercomplex structures*, which we already met in Chapter 3 (see the remark on p. 65) and in Chapter 10 (see the proof of Theorem 10.3). We discussed there the hypercomplex structures in relation to HKT manifolds. Generically, the hypercomplex structures are nontrivial functions, but for four-dimensional hyperkähler manifolds, they have the same form as in the flat space.

Indeed, the Pauli matrices $\sigma_a$ are proportional to the generators of $SU(2)$. Were $v_a$ purely imaginary, the RHS of (11.41) would be an element of $SU(2)$. As $v_a$ are real, $h_{j\bar{k}}$ belongs not to $SU(2)$, but to a coset: the complexified version of $SU(2)$ [i.e. $SL(2,\mathbb{C})$] factorized over its compact form. It is natural then that the holonomy group of such a manifold is $SU(2)$ rather than $U(2)$.

A direct proof of the last assertion is easy. If $\det h = 1$, one can choose the vielbeins such as $\det e = \det \bar{e} = 1$. Consider the Kähler spin connection (2.31). We see that

$$\omega_{a\bar{a},\bar{m}} = -\omega_{\bar{a}a,\bar{m}} = e_a^n \partial_{\bar{m}} e_n^a = \partial_{\bar{m}}(\ln \det e) = 0.$$

Similarly, $\omega_{a\bar{a},m} = 0$. Then, $\omega_{AB,M}dx^M \in su(2)$, the same concerns the curvature form $R_{AB}$, and the holonomy group is $SU(2)$.

Hence, according to Theorem 3.2, the metric is hyperkähler.

The action (11.28) can be expressed in terms of $\mathcal{N} = 2$ superfields. Having assured that the metric is hyperkähler, we can use the results of Theorem 11.1 and conclude that (11.28) enjoys $\mathcal{N} = 8$ supersymmetry.

Well, this could be expected in advance: the action (11.28) was obtained as a result of Legendre transformation of the bi-Kähler action (11.12) that *had* $\mathcal{N} = 8$ supersymmetry. It is only natural that its transform also has it. It is instructive, however, to find the laws of extra $\mathcal{N} = 4$ transformations for $\mathcal{N} = 4$ chiral superfields $\Phi, \mathcal{Z}$.

**Theorem 11.4.** *The action* (11.28) *is invariant under the following hidden $\mathcal{N} = 4$ supersymmetry transformations* [29, 120, 122]:

$$\begin{aligned}
\delta\Phi &= (\bar{D})^2 \left(\bar{\epsilon}_j \bar{\theta}^j \frac{\partial \mathcal{K}}{\partial \mathcal{Z}}\right) = -2\bar{\epsilon}^k \bar{D}_k \frac{\partial \mathcal{K}}{\partial \mathcal{Z}} + (\bar{\epsilon}_j \bar{\theta}^j)(\bar{D})^2 \frac{\partial \mathcal{K}}{\partial \mathcal{Z}}, \\
\delta\mathcal{Z} &= -(\bar{D})^2 \left(\bar{\epsilon}_j \bar{\theta}^j \frac{\partial \mathcal{K}}{\partial \Phi}\right) = 2\bar{\epsilon}^k \bar{D}_k \frac{\partial \mathcal{K}}{\partial \Phi} - (\bar{\epsilon}_j \bar{\theta}^j)(\bar{D})^2 \frac{\partial \mathcal{K}}{\partial \Phi}, \\
\delta\bar{\Phi} &= (D)^2 \left(\epsilon^j \theta_j \frac{\partial \mathcal{K}}{\partial \bar{\mathcal{Z}}}\right) = -2\epsilon_k D^k \frac{\partial \mathcal{K}}{\partial \bar{\mathcal{Z}}} + (\epsilon^j \theta_j)(D)^2 \frac{\partial \mathcal{K}}{\partial \bar{\mathcal{Z}}}, \\
\delta\bar{\mathcal{Z}} &= -(D)^2 \left(\epsilon^j \theta_j \frac{\partial \mathcal{K}}{\partial \bar{\Phi}}\right) = 2\epsilon_k D^k \frac{\partial \mathcal{K}}{\partial \bar{\Phi}} - (\epsilon^j \theta_j)(D)^2 \frac{\partial \mathcal{K}}{\partial \bar{\Phi}}.
\end{aligned} \tag{11.42}$$

*with*

$$(D)^2 = D^j D_j = 2D^1 D^2, \quad (\bar{D})^2 = \bar{D}_j \bar{D}^j = -2\bar{D}_1 \bar{D}_2.$$

**Proof.** Consider the terms $\propto \bar{\epsilon}^1 = -\bar{\epsilon}_2$ in the variation of the Kähler potential. We have

$$\delta_1 K = \frac{\partial \mathcal{K}}{\partial \Phi}\delta_1\Phi + \frac{\partial \mathcal{K}}{\partial \mathcal{Z}}\delta_1\mathcal{Z} = \bar{\epsilon}^1\left[2\left(\frac{\partial \mathcal{K}}{\partial \mathcal{Z}}\bar{D}_1\frac{\partial \mathcal{K}}{\partial \Phi} - \frac{\partial \mathcal{K}}{\partial \Phi}\bar{D}_1\frac{\partial \mathcal{K}}{\partial \mathcal{Z}}\right)\right.$$
$$\left. + \bar{\theta}^2\left(\frac{\partial \mathcal{K}}{\partial \mathcal{Z}}(\bar{D})^2\frac{\partial \mathcal{K}}{\partial \Phi} - \frac{\partial \mathcal{K}}{\partial \Phi}(\bar{D})^2\frac{\partial \mathcal{K}}{\partial \mathcal{Z}}\right)\right] \stackrel{\text{def}}{=} \bar{\epsilon}^1(A_1 + B_1). \qquad (11.43)$$

Look at $B_1$. Flipping the derivatives, we can bring it in the form

$$B_1 \equiv 2\left(\frac{\partial \mathcal{K}}{\partial \Phi}\bar{D}_1\frac{\partial \mathcal{K}}{\partial \mathcal{Z}} - \frac{\partial \mathcal{K}}{\partial \mathcal{Z}}\bar{D}_1\frac{\partial \mathcal{K}}{\partial \Phi}\right) + \bar{\theta}^2\left(\frac{\partial \mathcal{K}}{\partial \Phi}(\bar{D})^2\frac{\partial \mathcal{K}}{\partial \mathcal{Z}} - \frac{\partial \mathcal{K}}{\partial \mathcal{Z}}(\bar{D})^2\frac{\partial \mathcal{K}}{\partial \Phi}\right)$$
$$= -A_1 - B_1. \qquad (11.44)$$

Hence, $B_1 \equiv -A_1/2$. We are left with the structure

$$\delta S \propto \int dt d^2\theta d^2\bar{\theta} A_1 \propto \int dt d^2\theta d^2\bar{\theta}\left(\frac{\partial \mathcal{K}}{\partial \mathcal{Z}}\bar{D}_1\frac{\partial \mathcal{K}}{\partial \Phi} - \frac{\partial \mathcal{K}}{\partial \Phi}\bar{D}_1\frac{\partial \mathcal{K}}{\partial \mathcal{Z}}\right)$$
$$= 2\int dt d^2\theta d^2\bar{\theta}\,\frac{\partial \mathcal{K}}{\partial \mathcal{Z}}\bar{D}_1\frac{\partial \mathcal{K}}{\partial \Phi}. \qquad (11.45)$$

In the flat case, this reduces to $\int dt d^2\theta d^2\bar{\theta}\,\bar{\mathcal{Z}}\bar{D}_1\bar{\Phi}$. Consider it in the right chiral basis. Then, $\bar{\Phi}$ and $\bar{Z}$ do not depend on $\theta$, while $\bar{D}_1$ includes only one factor of $\theta$, which is not enough — the integral $\int d^2\theta$ gives zero.

On the other hand,

$$0 = \int dt d^2\theta d^2\bar{\theta}\,\bar{\mathcal{Z}}\bar{D}_1\bar{\Phi} = \frac{1}{2}\int dt d^2\theta(\bar{D})^2[(\bar{\mathcal{Z}}\bar{D}_1\bar{\Phi})]$$
$$= \int d^2\theta\bar{D}_2[(\bar{D}_1\bar{\mathcal{Z}})(\bar{D}_1\bar{\Phi})]. \qquad (11.46)$$

Now, consider

$$\int d^2\bar{\theta}\,\frac{\partial \mathcal{K}}{\partial \mathcal{Z}}\bar{D}_1\frac{\partial \mathcal{K}}{\partial \Phi} = \frac{1}{2}(\bar{D})^2\left(\frac{\partial \mathcal{K}}{\partial \mathcal{Z}}\bar{D}_1\frac{\partial \mathcal{K}}{\partial \Phi}\right) = \bar{D}_2\left[\left(\bar{D}_1\frac{\partial \mathcal{K}}{\partial \mathcal{Z}}\right)\left(\bar{D}_1\frac{\partial \mathcal{K}}{\partial \Phi}\right)\right]$$
$$= \bar{D}_2\left[\left(\frac{\partial^2 \mathcal{K}}{\partial \mathcal{Z}\partial\bar{\mathcal{Z}}}\frac{\partial^2 \mathcal{K}}{\partial \Phi\partial\bar{\Phi}} - \frac{\partial^2 \mathcal{K}}{d\mathcal{Z}d\bar{\Phi}}\frac{\partial^2 \mathcal{K}}{d\Phi d\bar{\mathcal{Z}}}\right)(\bar{D}_1\bar{\mathcal{Z}})(\bar{D}_1\bar{\Phi})\right]. \qquad (11.47)$$

Taking into consideration (11.29), we see that it also gives the structure (11.46) in the variation (11.45) and vanishes.

The vanishing of the contributions $\propto \bar{\epsilon}^2$ and $\propto \epsilon_{1,2}$ in the variation of the action can be proved in a similar way. □

### 11.3.1 *Example*

Consider the bi-Kähler model (11.12) with one hypermultiplet $(\mathcal{Z}, U)$ characterized by the prepotential [29, 122]

$$\boxed{\mathcal{L}(\mathcal{Z}, \bar{\mathcal{Z}}, X) = \lambda(\mathcal{Z}\bar{\mathcal{Z}} - X^2/2) + R - X\ln(X+R)} \tag{11.48}$$

with $R = \sqrt{X^2 + 4\mathcal{Z}\bar{\mathcal{Z}}}$. Here, $\lambda$ is a real positive number. The prepotential (11.48) is harmonic, the condition (11.25) holds. Introduce a real phase coordinate $\Psi$ and a complex coordinate $\phi$ such that

$$d\phi = \frac{1}{2}(d\mathcal{L}_x + id\Psi) = \frac{1}{2}(\mathcal{L}_{xx}dx + \mathcal{L}_{xz}dz + \mathcal{L}_{x\bar{z}}d\bar{z} + id\Psi). \tag{11.49}$$

$(\mathcal{L}_{xx} = \partial^2\mathcal{L}/dx^2$, etc). According to (11.30), (11.31), (11.32) and (11.33), the metric of the Kähler manifold that we obtain after the Legendre transformation is

$$\begin{aligned}
\frac{1}{2}ds^2 &= \mathcal{K}_{\phi\bar{\phi}}\, d\phi d\bar{\phi} + \mathcal{K}_{\phi\bar{z}}\, d\phi d\bar{z} + \mathcal{K}_{z\bar{\phi}}\, dz d\bar{\phi} + \mathcal{K}_{z\bar{z}}\, dz d\bar{z} \\
&= -\frac{1}{4\mathcal{L}_{xx}}(d\Psi + A)^2 - \frac{\mathcal{L}_{xx}}{4}dx^2 + \frac{1}{4\mathcal{L}_{xx}}B^2 \\
&\quad + \frac{1}{4\mathcal{L}_{xx}}A^2 - \left(\mathcal{L}_{x\bar{x}} + \frac{\mathcal{L}_{xz}\mathcal{L}_{x\bar{z}}}{\mathcal{L}_{xx}}\right) dz d\bar{z},
\end{aligned} \tag{11.50}$$

where $A = i(\mathcal{L}_{xz}dz - \mathcal{L}_{x\bar{z}}d\bar{z})$ and $B = \mathcal{L}_{xz}dz + \mathcal{L}_{x\bar{z}}d\bar{z}$. This gives

$$\frac{1}{2}ds^2 = -\frac{1}{4\mathcal{L}_{xx}}(d\Psi + A)^2 - \frac{\mathcal{L}_{xx}}{4}(dx^2 + 4dzd\bar{z}). \tag{11.51}$$

In our case, $\mathcal{L}_{xx} = -\lambda - 1/r$, $dx^2 + 4dzd\bar{z}$ may be presented as $(d\boldsymbol{x})^2$, and the metric (11.51) coincides up to a factor with the conventional form (3.39) of the Taub–NUT metric!

Note that $\mathcal{L}_{xx}$ is harmonic as $\mathcal{L}$ was. Choosing a more complicated prepotential,

$$\mathcal{L}(\mathcal{Z}, \bar{\mathcal{Z}}, X) = \lambda(\mathcal{Z}\bar{\mathcal{Z}} - X^2/2) + \sum_{i=1}^{n}\left[R_i - (X - a_i)\ln(X - a_i + R_i)\right] \tag{11.52}$$

with $R_i = \sqrt{(X-a_i)^2 + 4(\mathcal{Z} - z_i)(\bar{\mathcal{Z}} - \bar{z}_i)}$, one can derive *multicenter* Taub–NUT metrics. Actually, any metric (11.51) with harmonic $\mathcal{L}_{xx}$ is hyperkähler.

Note also that the metric (11.51) with $\lambda = 0$ is in fact a *flat* $\mathbb{R}^4$ metric expressed in unusual variables.[6] Choose the gauge (3.41):

$$A = \boldsymbol{A} \cdot d\boldsymbol{x} \; = \frac{1}{r(r+x_3)}(x_2 dx_1 - x_1 dx_2). \tag{11.53}$$

Then, the variable change [124]

$$\begin{aligned} y_1 &= -\sqrt{\frac{r+x_3}{2}} \sin\frac{\Psi}{2}, \quad y_2 = \sqrt{\frac{r+x_3}{2}} \cos\frac{\Psi}{2}, \\ y_3 &= \frac{1}{\sqrt{2(r+x_3)}} \left( x_1 \cos\frac{\Psi}{2} + x_2 \sin\frac{\Psi}{2} \right), \\ y_4 &= \frac{1}{\sqrt{2(r+x_3)}} \left( x_2 \cos\frac{\Psi}{2} - x_1 \sin\frac{\Psi}{2} \right), \end{aligned} \tag{11.54}$$

which implies

$$\boxed{\begin{aligned} x_1 &= 2(y_1 y_4 + y_2 y_3), \quad x_2 = 2(y_2 y_4 - y_1 y_3), \\ x_3 &= y_1^2 + y_2^2 - y_3^2 - y_4^2, \quad \Psi = -2 \arctan\frac{y_1}{y_2}, \end{aligned}} \tag{11.55}$$

brings (11.51) in the form

$$ds^2 = 2(dy_1^2 + dy_2^2 + dy_3^2 + dy_4^2). \tag{11.56}$$

This nontrivial fact can be deduced using a quaternionic representation of the metric — see Eq. (11) and around in Ref. [30] — but a reader of my dream can follow the author's example and check it explicitly.

### 11.3.2 *Many dimensions*

The analysis presented above can be generalized to construct multidimensional hyperkähler manifolds [29, 120]. We start with a $\mathcal{N} = 8$ bi-Kähler model (11.12) where the prepotential $\mathcal{L}$ has isometries depending only on the *sums* $X_\alpha = \mathfrak{U}_\alpha + \bar{\mathfrak{U}}_\alpha$ and satisfies the conditions (11.14) and (11.15). The latter acquire the form

$$\boxed{\frac{\partial^2 \mathcal{L}}{\partial \mathfrak{Z}_\alpha \partial \bar{\mathfrak{Z}}_\beta} + \frac{\partial^2 \mathcal{L}}{\partial X_\alpha \partial X_\beta} = 0} \tag{11.57}$$

[6] That was first mentioned in Ref. [123].

and

$$\boxed{\frac{\partial^2 \mathcal{L}}{\partial \mathcal{Z}_{[\alpha} \partial X_{\beta]}} = 0.} \tag{11.58}$$

We perform the Legendre transformation and construct the Kähler potential

$$\mathcal{K}(\mathcal{Z}_\alpha, \bar{\mathcal{Z}}_\alpha, V_\alpha) = \mathcal{L}(\mathcal{Z}_\alpha, \bar{\mathcal{Z}}_\alpha, X_\alpha) - X_\alpha V_\alpha, \tag{11.59}$$

where

$$V_\alpha = \Phi_\alpha + \bar{\Phi}_\alpha = \frac{\partial \mathcal{L}}{\partial X_\alpha}. \tag{11.60}$$

One can generalize Theorem 11.4 and show that the action

$$S = \int d^2\theta d^2\bar{\theta}\, \mathcal{K}(\Phi_\alpha, \bar{\Phi}_\alpha, \mathcal{Z}_\alpha, \bar{\mathcal{Z}}_\alpha) \tag{11.61}$$

is invariant under the following nonmanifest supersymmetry transformations:

$$\begin{aligned} \delta\Phi_\alpha &= (\bar{D})^2 \left[ (\bar{\epsilon}_j \bar{\theta}^j) \left( \frac{\partial \mathcal{K}}{\partial \mathcal{Z}_\alpha} + \frac{\partial \mathcal{K}}{\partial \Phi_\beta} R_{\beta\alpha} \right) \right], \\ \delta\mathcal{Z}_\alpha &= -(\bar{D})^2 \left[ (\bar{\epsilon}_j \bar{\theta}^j) \frac{\partial \mathcal{K}}{\partial \Phi_\alpha} \right] \end{aligned} \tag{11.62}$$

with

$$R_{\beta\alpha} = \frac{\partial^2 \mathcal{K}}{\partial \mathcal{Z}_\beta \partial \bar{\Phi}_\gamma} \left( \frac{\partial^2 \mathcal{K}}{\partial \Phi_\alpha \partial \bar{\Phi}_\gamma} \right)^{-1} - \frac{\partial^2 \mathcal{K}}{\partial \mathcal{Z}_\alpha \partial \bar{\Phi}_\gamma} \left( \frac{\partial^2 \mathcal{K}}{\partial \Phi_\beta \partial \bar{\Phi}_\gamma} \right)^{-1}. \tag{11.63}$$

The model (11.61) is Kähler and it enjoys $\mathcal{N} = 8$ supersymmetry. Based on the remark at the very end of Section 11.1, we can conclude that this model is also hyperkähler. It would be also nice, however, to see it *directly* as we did so in the four-dimensional case: to show in an explicit way that the Kähler potential (11.59) admits a quaternionic triple of the covariantly constant complex structures.

I want to mention now the following nice theorem [125, 126]:

**Theorem 11.5.** *A Kähler manifold with the metric $h_{i\bar{k}}$ is hyperkähler iff the following relation holds*:

$$\boxed{h_{i\bar{k}} h_{j\bar{l}} \Omega^{\bar{k}\bar{l}} = C \Omega_{ij},} \tag{11.64}$$

*where $C > 0$, while $\Omega_{ij}$ and $\Omega^{\bar{k}\bar{l}}$ are the symplectic matrices:* $\Omega = \mathrm{diag}(\varepsilon, \ldots, \varepsilon)$.

The relation (11.64) is a natural generalization of the heavenly equation (11.29) for the multidimensional case. It has the same form as the condition $U\Omega U^T = \Omega$, which defines an $Sp(n)$ matrix $U$, but the Hermitian metric $h_{j\bar{k}}$ is not an element of $Sp(n)$, but rather, in a full analogy with (11.41), has the form

$$h_{j\bar{k}} = [\exp(v_a t_a)]_{j\bar{k}}\,, \tag{11.65}$$

where $t_a$ are the Hermitian generators of $Sp(n)$ satisfying the condition $(t_a)^T\Omega + \Omega t_a = 0$ and $v_a$ are real. Thus, $h_{j\bar{k}}$ is an element of the coset: the complexified version of $Sp(n)$ factorized over its compact form.

It is natural then that the holonomy group of the manifold is $Sp(n)$ and hence the manifold is hyperkähler.

**Proof.** For the formal proof, a math reader can consult Proposition 2.3 in Ref. [125] but we also give here an elementary explicit proof [127]. We choose the complex structures with the following nonzero components:

$$\begin{aligned} &I_{m\bar{n}} = -ih_{m\bar{n}} \implies I_m{}^n = -i\delta^n_m,\ I_{\bar{m}}{}^{\bar{n}} = i\delta^{\bar{n}}_{\bar{m}}, \\ &J_{mn} = \Omega_{mn},\ J_{\bar{m}\bar{n}} = \Omega_{\bar{m}\bar{n}}, \quad K_{mn} = -i\Omega_{mn},\ K_{\bar{m}\bar{n}} = i\Omega_{\bar{m}\bar{n}}. \end{aligned} \tag{11.66}$$

In the flat case, they reduce to (3.2), bearing in mind our convention $z^1 = (x^1 + ix^2)/\sqrt{2}$, etc.

Using the condition (11.64), it is easy to show that they are quaternionic:

$$I^2 = J^2 = K^2 = -\mathbb{1}, \quad IJ = K,\ JK = I,\ KI = J. \tag{11.67}$$

To prove the theorem, one should also show that they are covariantly constant. This is obvious for the conventional Kähler complex structure $I$.

It is convenient to consider the hypercomplex structures $\mathcal{I}^\pm = (J \pm iK)/2$. $J + iK$ has only holomorphic and $J - iK$ only antiholomorphic components.

The ordinary derivatives of $J \pm iK$ vanish, and we have to prove that

$$\nabla_P(J \pm iK)_{MN} = \Gamma^S_{PM}(J \pm iK)_{SN} + \Gamma^S_{PN}(J \pm iK)_{MS} = 0. \tag{11.68}$$

According to Eq. (2.30), the only nonzero Christoffels for Kähler manifolds, are holomorphic, $\Gamma^s_{mn} = h^{\bar{q}s}\partial_m h_{n\bar{q}}$, or antiholomorphic. Hence $\nabla_{\bar{p}}(J + iK)_{mn} \propto \partial_{\bar{p}}\Omega_{mn} = 0$. Now,

$$\nabla_p(J + iK)_{mn} = 2\,h^{\bar{q}s}\partial_m h_{n\bar{q}}\Omega_{sn} - (m \leftrightarrow n). \tag{11.69}$$

The matrix $(X_p)^s_m = h^{\bar{q}s} \partial_m h_{n\bar{q}}$ is an element of $sp(n)$, as follows from (11.65). Then,

$$\nabla_p (J + iK)_{mn} \propto (X_p)^s_m \Omega_{sn} - (X_p)^s_n \Omega_{sm} = [X_p \Omega + \Omega (X_p)^T]_{mn} = 0. \quad (11.70)$$

The property $\nabla_P (J - iK) = 0$ is proved in the same way. □

The metric derived from (11.59) must satisfy (11.64). It would be nice to show it explicitly.

Nontrivial hyperkähler metrics of dimension 4 and more can also be derived in a pure geometric way using the procedure of *hyperkähler reduction* from flat metrics [29, 30]. In particular, the Taub–NUT manifold may be represented as a quotient of $R^8$ that keeps its hyperkähler structure (we will explain how it works in the following chapter). In addition, multidimensional metrics of Taub–NUT type may be derived from the analysis of certain three-dimensional $\mathcal{N} = 4$ supersymmetric gauge theories [27, 31]. But supersymmetric field theories are beyond the scope of our book, and we do not explain here this method, addressing the reader to the original papers.

## 11.4 $\mathcal{N} = 8$ bi-HKT models

The results of the previous section imply that Kähler sigma models (11.61) that enjoy $\mathcal{N} = 8$ supersymmetry describe hyperkähler geometry. As we mentioned above, it is not necessarily so for generic sigma models (8.6). Certain bi-Kähler models have $\mathcal{N} = 8$ symmetry, but the corresponding metrics are not Kähler and are not hyperkähler.

The same concerns bi-HKT, bi-symplectic and some other models. In this section, we dwell on the bi-HKT case [82, 128, 129]. Consider a bi-HKT model including an equal number of ordinary multiplets $\mathcal{Z}^A_{\alpha=1,\ldots,n}$ subject to the constraints (10.68) and the mirror multiplets $\mathcal{U}^A_{\alpha=1,\ldots,n}$ subject to the constraints (10.91). The action (10.93) acquires the form

$$\boxed{S = \frac{1}{8} \int d^2\bar{\theta} d^2\theta dt \, \mathcal{L}(\mathcal{Z}^A_\alpha, \bar{\mathcal{Z}}^A_\alpha; \mathcal{U}^A_\alpha, \bar{\mathcal{U}}^A_\alpha).} \quad (11.71)$$

Generically, it describes an $\mathcal{N} = 4$ supersymmetric sigma model on a certain $8n$-dimensional bi-HKT manifold. But for some special Lagrangians $\mathcal{L}$, the symmetry extends up to $\mathcal{N} = 8$. In such a case, the set $(\mathcal{Z}^A_\alpha, \mathcal{U}^A_\alpha)$

constitutes for a given $\alpha$ one large $\mathcal{N}=8$ multiplet $(\mathbf{8},\ \mathbf{8},\ \mathbf{0})$. The four extra supersymmetry transformation can be represented as

$$\begin{aligned}
\delta\mathcal{Z}^A_\alpha &= i\bar{D}_1(\eta_1\,\bar{\mathcal{U}}^A_\alpha + \eta_2\,\varepsilon^{AB}\,\bar{\mathcal{U}}^B_\alpha) + i\bar{D}_2(\eta_3\,\mathcal{U}^A_\alpha + \eta_4\,\varepsilon^{AB}\,\mathcal{U}^B_\alpha),\\
\delta\bar{\mathcal{Z}}^A_\alpha &= iD^1(\eta_1\,\mathcal{U}^A_\alpha + \eta_2\,\varepsilon^{AB}\,\mathcal{U}^B_\alpha) + iD^2(\eta_3\,\bar{\mathcal{U}}^A_\alpha + \eta_4\,\varepsilon^{AB}\,\bar{\mathcal{U}}^B_\alpha),\\
\delta\mathcal{U}^A_\alpha &= -i\bar{D}_1(\eta_1\,\bar{\mathcal{Z}}^A_\alpha + \eta_2\,\varepsilon^{AB}\,\bar{\mathcal{Z}}^B_\alpha) - iD^2(\eta_3\,\mathcal{Z}^A_\alpha + \eta_4\,\varepsilon^{AB}\,\mathcal{Z}^B_\alpha),\\
\delta\bar{\mathcal{U}}^A_\alpha &= -iD^1(\eta_1\,\mathcal{Z}^A_\alpha + \eta_2\,\varepsilon^{AB}\,\mathcal{Z}^B_\alpha) - i\bar{D}_2(\eta_3\,\bar{\mathcal{Z}}^A_\alpha + \eta_4\,\varepsilon^{AB}\,\bar{\mathcal{Z}}^B_\alpha)
\end{aligned} \tag{11.72}$$

with real Grassmann $\eta_{1,2,3,4}$. One can verify by a direct calculation that the superfield variations (11.72) are consistent with the constraints that we imposed on $\mathcal{Z}^A_\alpha$ and $\mathcal{U}^A_\alpha$.

**Conjecture:** *The action* (11.71) *is invariant under the transformations* (11.72) *iff the conditions*

$$\left(\frac{\partial^2}{\partial\mathcal{Z}^A_\alpha\partial\bar{\mathcal{Z}}^A_\beta} + \frac{\partial^2}{\partial\mathcal{U}^A_\beta\partial\bar{\mathcal{U}}^A_\alpha}\right)\mathcal{L} = 0 \tag{11.73}$$

*and*

$$\left(\frac{\partial^2}{\partial\mathcal{Z}^A_\alpha\partial\bar{\mathcal{U}}^A_\beta} - \frac{\partial^2}{\partial\mathcal{Z}^A_\beta\partial\bar{\mathcal{U}}^A_\alpha}\right)\mathcal{L} = 0 \tag{11.74}$$

*are fulfilled for all* $\alpha, \beta$.

This assertion is similar in spirit to Theorem 11.2 for the bi-Kähler $\mathcal{N}=8$ models, but its proof meets in this case certain difficulties. The main obstacle is that, in contrast to the independent chiral $\mathcal{N}=8$ superfields $\mathcal{Z}_\alpha$ and $\mathcal{U}_\alpha$, the definition of the HKT superfields $\mathcal{Z}^A_\alpha$ and $\mathcal{U}^A_\alpha$ involves nontrivial constraints (10.68) and (10.91).

The problem simplifies if we have only one pair of multiplets $(\mathcal{Z}^A, \mathcal{U}^A)$. Then, the condition (11.74) trivializes and (11.73) is reduced to the harmonicity condition $\triangle^{D=8}\mathcal{L} = 0$. This system was analyzed at the component level in Ref. [129]. The simplest nontrivial solution to this condition is

$$\boxed{\mathcal{L} = \frac{1}{\mathcal{Z}^A\bar{\mathcal{Z}}^A + \mathcal{U}^A\bar{\mathcal{U}}^A}\left(\frac{1}{\mathcal{Z}^B\bar{\mathcal{Z}}^B} - \frac{1}{\mathcal{U}^B\bar{\mathcal{U}}^B}\right).} \tag{11.75}$$

This prepotential gives a nice positive definite harmonic metric

$$ds^2 = 2\frac{dz^A d\bar{z}^A + du^A d\bar{u}^A}{(z^B\bar{z}^B + u^B\bar{u}^B)^3}. \tag{11.76}$$

## 11.5 Symplectic model of the second kind

The whole zoo of $\mathcal{N} = 8$ models was systematically studied in [130]. And not all the animals there belong to the "bi" genus. As the last example, consider the model involving two different $\mathcal{N} = 4$ superfields: the (**3**, **4**, **1**) superfield $\Phi_M$ defined in Eq. (7.65) and the chiral (**2**, **4**, **2**) superfield $\mathcal{Z}$ in Eq. (7.47). Consider the action

$$S = \int d^2\bar{\theta} d^2\theta dt\, \mathcal{L}(\Phi_M, \bar{\mathcal{Z}}, \mathcal{Z}) \tag{11.77}$$

with a real prepotential $\mathcal{L}$.

We require this action to be invariant under the following extra $\mathcal{N} = 4$ supersymmetry transformations:

$$\boxed{\begin{aligned} \delta\mathcal{Z} &= \frac{2i}{3}\,\bar{\epsilon}^j (\sigma_M)_j{}^k \bar{D}_k \Phi_M, \\ \delta\bar{\mathcal{Z}} &= \frac{2i}{3} (\sigma_M)_j{}^k \epsilon_k D^j \Phi_M, \\ \delta\Phi_M &= -\,i\bar{\epsilon}^j (\sigma_M)_j{}^k \bar{D}_k \bar{\mathcal{Z}} - i(\sigma_M)_j{}^k \epsilon_k D^j \mathcal{Z}. \end{aligned}} \tag{11.78}$$

One can show that, up to coeffients, these are the only transformations that mix $\Phi_M \leftrightarrow (\mathcal{Z}, \bar{\mathcal{Z}})$ and respect the constraints $\bar{D}_j \mathcal{Z} = 0$ and the constraints $\Phi_M = \overline{\Phi_M}$ and (7.62) for $\Phi_M$.

If the invariance under (11.78) holds, we are dealing with not just two separate $\mathcal{N} = 4$ supermultiplets but with a single $\mathcal{N} = 8$ multiplet of type (**5**, **8**, **3**).

**Theorem 11.6.** [112, 131] *The action* (11.77) *is invariant under* (11.78) *iff the condition*

$$\frac{\partial^2 \mathcal{L}}{\partial \Phi_M \partial \Phi_M} + 2\frac{\partial^2 \mathcal{L}}{\partial \bar{\mathcal{Z}} \partial \mathcal{Z}} = 0 \tag{11.79}$$

*is fulfilled.*

**Proof.** The variation of the Lagrangian under (11.78) reads

$$\delta\mathcal{L} = i\epsilon_k (\sigma_M)_j{}^k \left[ \frac{2}{3}\frac{\partial \mathcal{L}}{\partial \bar{\mathcal{Z}}} D^j \Phi_M - \frac{\partial \mathcal{L}}{\partial \Phi_M} D^j \mathcal{Z} \right] + \text{c.c.} \tag{11.80}$$

We now proceed in the full analogy with what we did to prove Theorem 11.2. The variation of the action vanishes iff the variation (11.80) represents a sum of total derivatives:

$$(\sigma_M)_j{}^k \left[\frac{2}{3}\frac{\partial \mathcal{L}}{\partial \bar{\mathcal{Z}}} D^j \Phi_M - \frac{\partial \mathcal{L}}{\partial \Phi_M} D^j \mathcal{Z}\right] = D^k R + (\sigma_M)_j{}^k D^j L_M, \tag{11.81}$$

where $R$ and $L_M$ are some complex functions of $\Phi_M, \mathcal{Z}$ and $\bar{\mathcal{Z}}$. Bearing in mind that $\bar{D}_j \mathcal{Z} = D^j \bar{\mathcal{Z}} = 0$ and that $D^j \Phi_M = (\sigma_M)_k{}^j \,\bar{\Xi}^k$ with some $\bar{\Xi}^k$ [see Eq. (7.63)], we can write

$$\begin{aligned} D^k R &= \frac{\partial R}{\partial \mathcal{Z}} D^k \mathcal{Z} + \frac{\partial R}{\partial \Phi_M} (\sigma_M)_p{}^k \,\bar{\Xi}^p, \\ D^j L_M &= \frac{\partial L_M}{\partial \mathcal{Z}} D^j \mathcal{Z} + \frac{\partial L_M}{\partial \Phi_N} (\sigma_N)_p{}^j \,\bar{\Xi}^p. \end{aligned} \tag{11.82}$$

Then, comparing the coefficients of the independent structures in both sides of (11.81), we end up with the following set of linear partial differential equations:

$$\begin{aligned} 2\frac{\partial \mathcal{L}}{\partial \bar{\mathcal{Z}}} - \frac{\partial L_M}{\partial \Phi_M} &= 0, \\ \frac{\partial \mathcal{L}}{\partial \Phi_M} + \frac{\partial L_M}{\partial \mathcal{Z}} &= 0, \\ i\varepsilon_{MNP} \frac{\partial L_M}{\partial \Phi_N} - \frac{\partial R}{\partial \Phi_P} &= 0, \\ \frac{\partial R}{\partial \mathcal{Z}} &= 0. \end{aligned} \tag{11.83}$$

Differentiating the first equation over $\mathcal{Z}$, the second over $\Phi_M$ and adding the results, we arrive at (11.79) as an integrability condition for the system (11.83). This proves the necessity of (11.79). Two other equations impose some constraints on the function $R$ and $L_M$, which present for us no particular interest.

Now, let $\mathcal{L}$ satisfies the constraint (11.79). This is nothing but a five-dimensional harmonicity condition

$$\boxed{\frac{\partial^2 \mathcal{L}}{\partial \tilde{\Phi}_M \partial \tilde{\Phi}_M} = 0} \tag{11.84}$$

with

$$\tilde{\Phi}_{M=1,2,3,4,5} = \left(\Phi_{M=1,2,3}, \sqrt{2}\,\mathrm{Re}(\mathcal{Z}), \sqrt{2}\,\mathrm{Im}(\mathcal{Z})\right). \tag{11.85}$$

A generic real solution to (11.84) is a linear function of $\tilde{\Phi}_M$ plus a sum of singular "Coulomb" functions, like

$$\frac{1}{(\tilde{\Phi}_M \tilde{\Phi}_M)^{3/2}}, \tag{11.86}$$

and their derivatives. It is then not difficult to write for any such $\mathcal{L}$ an expression for the harmonic function $Q$ satisfying the additional requirement $-\partial Q/\partial \mathcal{Z} = \mathcal{L}$. Then, one of many possible solutions for the system (11.83) that guarantees the invariance of the action under (11.78) reads

$$R = 0, \quad L_M = \frac{\partial Q}{\partial \Phi_M}. \tag{11.87}$$

This proves the sufficiency of (11.79). □

The system (11.77) with harmonic $\mathcal{L}$ has four complex supercharges. In a complete analogy with the ordinary symplectic (**3**, **4**, **1**) system where the supercharges lie in the fundamental representation of $Spin(3) \equiv Sp(1)$, in this case, the supercharge lies in the fundamental representation of $Spin(5) \equiv Sp(2)$. One may thus call the model (11.77) a symplectic model of the second kind.

We have presented here its description in terms of $\mathcal{N} = 4$ superfields. But one can also describe the whole (**5**, **8**, **3**) multiplet as a single $\mathcal{N} = 8$ superfield living in the corresponding harmonic superspace [112].

As was the case with all other models considered in this section, multidimensional versions of the symplectic models of the second kind exist. Take some number $n$ of (**5**, **8**, **3**) multiplets and write the action

$$\boxed{S = \int d^2\bar{\theta} d^2\theta dt\, \mathcal{L}(\tilde{\Phi}^\alpha_M),} \tag{11.88}$$

$\alpha = 1, \ldots, n$. Proceeding in the same way as before, we can prove the theorem [112]:

**Theorem 11.7.** *For the action* (11.88) *to stay invariant under the transformations* (11.78) *for each supermultiplet* (11.85), *the conditions*

$$\frac{\partial^2 \mathcal{L}}{\partial \tilde{\Phi}^\alpha_M \partial \tilde{\Phi}^\beta_M} = 0 \tag{11.89}$$

*must be fulfilled for all $\alpha, \beta$.*

**Proof.** The invariance of the action implies in particular that

$$2\frac{\partial \mathcal{L}}{\partial \bar{\mathcal{Z}}^\alpha} - \frac{\partial L_M}{\partial \Phi_M^\alpha} = 0, \qquad \frac{\partial \mathcal{L}}{\partial \Phi_M^\beta} + \frac{\partial L_M}{\partial \mathcal{Z}^\beta} = 0 \tag{11.90}$$

with some $L_M$. Differentiating the first equation over $\mathcal{Z}^\beta$, the second one over $\Phi_M^\alpha$ and adding, we arrive at

$$2\frac{\partial^2 \mathcal{L}}{\partial \bar{\mathcal{Z}}^\alpha \partial \mathcal{Z}^\beta} + \frac{\partial L_M}{\partial \Phi_M^\alpha \partial \Phi_M^\beta} = 0,$$

which coincides with (11.89). □

One can conjecture that these conditions are also sufficient.

# Chapter 12

# Taming the Zoo of Models

A reader who has just finished reading the previous chapter could be somehow confused. In Chapters 10 and 11, a large hip of different models were described, and it might be not immediately clear whether there is a relationship between them, whether this hip has a *structure*. In this chapter, we answer positively to this rhetoric question.

## 12.1 Similarity transformations

Suppose we have a $\mathcal{N} = 2$ SQM model involving the algebra (5.1). Consider the operator

$$\hat{Q}' = e^{\hat{R}} \hat{Q} e^{-\hat{R}} \tag{12.1}$$

with an arbitrary $\hat{R}$. As the supercharge $\hat{Q}$ is nilpotent, the same obviously concerns the new supercharge $\hat{Q}'$. The Hermitially conjugated supercharge $\hat{Q}'^{\,\dagger} \equiv \hat{\bar{Q}}'$ is

$$\hat{Q}'^{\,\dagger} = e^{-\hat{R}^\dagger} \hat{Q}^\dagger e^{\hat{R}^\dagger}. \tag{12.2}$$

The new Hamiltonian is

$$\hat{H}' = \{\hat{Q}', \hat{Q}'^{\,\dagger}\}. \tag{12.3}$$

If $\hat{R}$ is anti-Hermitian, $\hat{Q}$ and $\hat{Q}^\dagger$ are rotated with the same unitary matrix and the same concerns the Hamiltonian. This is just a unitary transformation, which boils down to a redefinition of the basis. But if $\hat{R}^\dagger \neq -\hat{R}$, the transformations (12.1) and (12.2) produce a different Hamiltonian (12.3) with a different spectrum.

**Theorem 12.1.** [110] *The supercharge $\hat{Q}$ in* (8.29) *of a generic de Rham supersymmetric sigma model is obtained from the flat supercharge*

$$\hat{Q}^{\text{flat}} = \frac{1}{\sqrt{2}}\psi_A \hat{P}_A, \tag{12.4}$$

$\hat{P}_A = -i\partial/\partial x^A$, *by a similarity transformation* (12.1).

**Proof.** Choose

$$\hat{R} = \alpha_{AB}\,\psi_A \hat{\bar{\psi}}_B \tag{12.5}$$

with $\hat{\bar{\psi}}_B = \partial/\partial\psi_B$ and $\alpha_{AB}(x)$ being a real symmetric function of all the coordinates (an antisymmetric $\alpha_{AB}$ would lead to an uninteresting unitary transformation). Evaluate $\hat{Q} = e^{\hat{R}}\hat{Q}^{\text{flat}}e^{-\hat{R}}$. We use the Hadamard formula

$$e^{\hat{R}}\hat{X}e^{-\hat{R}} = \hat{X} + [\hat{R},\hat{X}] + \frac{1}{2}[\hat{R},[\hat{R},\hat{X}]] + \cdots. \tag{12.6}$$

In our case, this implies

$$\begin{aligned} e^{\hat{R}}\psi_A e^{-\hat{R}} &= \psi_B(e^{\alpha})_{BA}, \\ e^{\hat{R}}\partial_A e^{-\hat{R}} &= \partial_A + (e^{\alpha}\partial_A e^{-\alpha})_{CD}\,\psi_C\hat{\bar{\psi}}_D \end{aligned} \tag{12.7}$$

and

$$\boxed{\hat{Q} = \frac{1}{\sqrt{2}}\psi_B(e^{\alpha})_{BA}\left[\hat{P}_A - i(e^{\alpha}\partial_A e^{-\alpha})_{CD}\psi_C\hat{\bar{\psi}}_D\right].} \tag{12.8}$$

Introducing the vielbeins as

$$e_A^M = (e^{\alpha})_{AM}, \quad e_{MA} = (e^{-\alpha})_{MA}, \tag{12.9}$$

which give the metric

$$g_{MN} = (e^{-2\alpha})_{MN}, \quad g^{MN} = (e^{2\alpha})_{MN}, \tag{12.10}$$

one can show by a direct calculation that the result (12.8) can be represented as

$$\hat{Q} = \frac{1}{\sqrt{2}}\,\psi^M(\hat{P}_M - i\omega_{CD,M}\,\psi_C\hat{\bar{\psi}}_D), \tag{12.11}$$

where $\omega_{CD,M}$ is the standard Levi-Civita spin connection

$$\omega_{CD,M} = e_{CN}(\partial_M e_D^N + \Gamma^N_{MP}e_D^P). \tag{12.12}$$

The expression (12.11) for the supercharge coincides with that in (8.29). □

As we know from Theorem 8.1, the supercharge (12.11) maps to the exterior derivative operator $d$ of the de Rham complex.

In other words, the SQM sigma model describing the de Rham complex is "genetically" related to the free model living on $\mathbb{R}^D$ and involving $D$ real superfields (**1**, **2**, **1**). By adding to $\hat{R}$ a simple function $-W(x)$, we may also obtain the deformed supercharge $\hat{Q}_W$ in Eq. (8.45). An additional similarity transformation,

$$\hat{Q} \to e^{\mathcal{B}_{MN}\psi^M\psi^N}\,\hat{Q}\,e^{-\mathcal{B}_{MN}\psi^M\psi^N}, \tag{12.13}$$

generates the model with quasitorsions discussed at the end of Chapter 8.

Another family of SQM sigma models involves the chiral multiplets (**2**, **2**, **0**). These models discussed in Chapter 9 can also be obtained by a similarity transformation from flat models. Consider a flat model living on $\mathbb{C}^d$ with the supercharges[1]

$$\hat{Q}^{\text{flat}} = \psi^a \hat{p}_a, \quad \hat{\bar{Q}}^{\text{flat}} = \hat{\bar{\psi}}^{\bar{a}} \hat{p}_{\bar{a}} \tag{12.14}$$

($\hat{p}_a = -i\partial/\partial z^a, \hat{p}_{\bar{a}} = -i\partial/\partial \bar{z}^a, \hat{\bar{\psi}}^{\bar{a}} = \partial/\partial\psi^a$). Consider a similarity transformation with

$$\hat{R} = \beta_{a\bar{b}}\,\psi^a \hat{\bar{\psi}}^{\bar{b}} \tag{12.15}$$

such that the matrix $\beta_{a\bar{b}}(z,\bar{z})$ is Hermitian (an anti-Hermitian $\beta_{a\bar{b}}$ amounts to a unitary transformation).

**Theorem 12.2.** *The similarity transformation* (12.1) *of the supercharge* $\hat{Q}^{\text{flat}}$ *in* (12.14) *with* $\hat{R}$ *written in Eq.* (12.15) *gives the supercharge that maps to the exterior holomorphic derivative* $\partial$ *of the Dolbeault complex with the complex vielbeins*

$$e_a^m = (e^\beta)_a{}^m, \quad e_m^a = (e^{-\beta})_m{}^a. \tag{12.16}$$

**Proof.** The line of reasoning is the same as for the preceding theorem. Using the Hadamard formula, we derive for the transformed supercharge

$$\boxed{\hat{Q} = \psi^b (e^\beta)_b{}^a\,[\hat{p}_a - i(e^\beta \partial_a e^{-\beta})_c{}^d\,\psi^c \hat{\bar{\psi}}^{\bar{d}}].} \tag{12.17}$$

One can verify by a direct calculation that the supercharge (12.17) can be represented in the form

$$\hat{Q} = \psi^m(\hat{p}_m - i\omega_{a\bar{b},m}\,\psi^a \hat{\bar{\psi}}^{\bar{b}}), \tag{12.18}$$

[1]We are writing the indices at the two levels now — see the footnote on p. 28.

where

$$\omega_{a\bar{b},m} = -\omega_{\bar{b}a,m} = -e_p^b(\partial_m e_a^p + \Gamma^p_{mk} e_a^k) \tag{12.19}$$

are the relevant components of the Levi-Civita spin connection. The result (12.18) coincides with the sum of the contributions in (9.12) and in (9.14) with $\mathcal{W} = (1/4) \ln \det h$.

Theorem 9.1 tells us that this supercharge maps to the operator $\partial$ of the untwisted Dolbeault complex. □

One can also make an additional transformation

$$\hat{Q} \to e^{\tilde{\mathcal{W}}} \hat{Q} e^{-\tilde{\mathcal{W}}}.$$

This brings about an extra gauge field (9.16): $A_M = (i\partial_m \tilde{\mathcal{W}}, -i\partial_{\bar{m}} \tilde{\mathcal{W}})$. As was mentioned in Chapter 9 and will be discussed in more details in Chapter 14, the function $\tilde{\mathcal{W}}(z, \bar{z})$ is not arbitrary here, for the topological charge associated with this gauge field should better be integer.

And the deformation

$$\hat{Q} \to e^{\mathcal{B}_{mn}\psi^m\psi^n}\, \hat{Q}\, e^{-\mathcal{B}_{mn}\psi^m\psi^n} \tag{12.20}$$

amounts to the shift (9.31) and generates the Dolbeault complex with holomorphic torsions.

A similarity transformation makes a $\mathcal{N} = 2$ supersymmetric model out of a $\mathcal{N} = 2$ supersymmetric model. But flat models often enjoy extended supersymmetries. For example, any model with the supercharge (12.4), involving an even number of variables $(x_A, \psi_A)$, also admits another complex supercharge

$$\hat{S}^{\text{flat}} = \frac{1}{\sqrt{2}} \psi_A I_{AB} \hat{P}_B, \tag{12.21}$$

where $I_{AB}$ is a complex structure: $I = -I^T, I^2 = -\mathbb{1}$. In fact, many such supercharges with different complex structures can be written, but one cannot take all of them, they would not mutually anticommute. However, a single extra supercharge in the form (12.21), anticommuting with $\hat{Q}^{\text{flat}}$ and $\hat{\bar{Q}}^{\text{flat}}$, can always be chosen on $\mathbb{R}^{2d}$. This gives $\mathcal{N} = 4$ supersymmetry. By rotations, one can always bring $I$ in the canonical form $I = \text{diag}(-\varepsilon, \ldots, -\varepsilon)$.

We now apply a similarity transformation for the supercharges $\hat{Q}^{\text{flat}}$ and $\hat{S}^{\text{flat}}$ *with the same* Hermitian operator $\hat{R}$ in (12.5). By construction, the transformed complex supercharges

$$\hat{Q} = e^{\hat{R}} \hat{Q}^{\text{flat}} e^{-\hat{R}}, \quad \hat{S} = e^{\hat{R}} \hat{S}^{\text{flat}} e^{-\hat{R}} \tag{12.22}$$

given by the expressions in (10.12) are nilpotent and anticommute. But the anticommutators $\{\hat{\bar{Q}}, \hat{Q}\}$ and $\{\hat{\bar{S}}, \hat{S}\}$ do not necessarily coincide and the anticommutators $\{\hat{\bar{Q}}, \hat{S}\}$ and $\{\hat{\bar{S}}, \hat{Q}\}$ do not necessarily vanish.

However, there are some special deformations when the latter property holds. *If* the complex structure $I_M{}^N = e_{MA} e_B^N I_{AB}$ thus obtained turns out to be covariantly constant with respect to the Levi-Civita connections, we are led to the Kähler geometry, and in this case, the supercharges (10.12) satisfy the $\mathcal{N} = 4$ superalgebra.[2]

If the number of flat space variables $D$ is an integer multiple of 4, one can write not one but three additional complex supercharges

$$\hat{S}_I^{\text{flat}} = \frac{1}{\sqrt{2}} \psi_A I_{AB} \hat{P}_B, \quad \hat{S}_J^{\text{flat}} = \frac{1}{\sqrt{2}} \psi_A J_{AB} \hat{P}_B,$$
$$\hat{S}_K^{\text{flat}} = \frac{1}{\sqrt{2}} \psi_A K_{AB} \hat{P}_B, \tag{12.23}$$

where three quaternionic covariantly constant complex structures $I, J, K$ can be brought to a canonical form (3.2). The supercharges (12.4) and (12.23) and their conjugates obey the $\mathcal{N} = 8$ algebra. An arbitrary similarity transformation with the operator (12.5) keeps $\mathcal{N} = 2$ but not $\mathcal{N} = 8$ supersymmetry. However, if the metric (12.10) thus obtained turns out to be hyperkähler and all three complex structures are covariantly constant, the anticommutators $\{\hat{\bar{Q}}, \hat{S}_{I,J,K}\}$, $\{\hat{Q}, \hat{\bar{S}}_{I,J,K}\}$ and $\{\hat{\bar{S}}_I, \hat{S}_{J \neq I}\}$ all vanish, the anticommutators $\{\hat{\bar{Q}}, \hat{Q}\}$, $\{\hat{\bar{S}}_I, \hat{S}_I\}$ all coincide, and the deformed model is $\mathcal{N} = 8$ supersymmetric.

As the last example, consider the SQM model on $\mathbb{C}^{d=2k}$ involving $k$ $(\mathbf{4}, \mathbf{4}, \mathbf{0})$ supermultiplets. This model involves two pairs of complex supercharges satisfying the $\mathcal{N} = 4$ algebra:

$$\hat{Q}^{\text{flat}} = \psi^a \hat{p}_a, \quad \hat{\bar{Q}}^{\text{flat}} = \hat{\bar{\psi}}^{\bar{a}} \hat{p}_{\bar{a}},$$
$$\hat{S}^{\text{flat}} = \psi^a (\mathcal{I}^+)_a{}^{\bar{b}} \hat{p}_{\bar{b}}, \quad \hat{\bar{S}}^{\text{flat}} = \hat{\bar{\psi}}^{\bar{a}} (\mathcal{I}^-)_{\bar{a}}{}^b \hat{p}_b, \tag{12.24}$$

---

[2]This is clear from the considerations in Section 10.1, especially from Theorem 10.2 demonstrating the isomorphism between the supercharges (10.12) and the operators of the Kähler–de Rham complex, but the reader can also consult the Appendix in Ref. [110], where it was proven by a direct calculation of the anticommutators.

where $(\mathcal{I}^+)_a{}^{\bar b} = (J + iK)_a{}^{\bar b}/2$ and $(\mathcal{I}^-)_{\bar a}{}^b = (J - iK)_{\bar a}{}^b/2$ are the flat hypercomplex structures, which can be both chosen as $\mathrm{diag}(-\varepsilon, \ldots, -\varepsilon)$.

Perform now a *simultaneous* similarity transformation (12.1) of $\hat Q^{\text{flat}}$ and $\hat S^{\text{flat}}$ with $\hat R = \beta_{a\bar b}\,\psi^a \hat{\bar\psi}^{\bar b}$. For a generic Hermitian $\beta$, $\mathcal{N} = 4$ supersymmetry of the flat model will be lost. But for some special $\beta$ giving a hyperkähler or an HKT metric $h_{m\bar n} = (e^{-2\beta})_{m\bar n}$, it is left intact.

In the simplest four-dimensional case, it is easy to derive in this way the conformally flat HKT metric (3.88).

## 12.2 Hamiltonian reduction

The procedure of Hamiltonian reduction is well known to physicists. It is indispensible for an accurate formulation of gauge theories and was worked out by Dirac almost hundred years ago [132]. For the benefit of our reader-mathematician, we have chosen to outline here its general features in the simplest examples before going to the discussion of its geometric aspects representing main interest for us in this book.

### 12.2.1 *A toy model*

Consider the quantum Hamiltonian

$$\boxed{\hat H_0 = \frac{1}{2}(\hat P_x^2 + \hat P_y^2) + \frac{\omega^2}{2}(x^2 + y^2),} \tag{12.25}$$

which describes the two-dimensional oscillator. This Hamiltonian admits an integral of motion $\hat P_\phi = x\hat P_y - y\hat P_x : [\hat P_\phi, \hat H] = 0$. It has the meaning of the angular momentum [cf. Eq. (4.24)]. The spectrum of (12.25) may be expressed as

$$E_{n_r,m} = \omega(1 + 2n_r + |m|), \tag{12.26}$$

where an integer $m$ is the eigenvalue of $\hat P_\phi$ and an integer non negative $n_r$ is the radial quantum number.

Now, suppose that we are not interested in the full spectrum of (12.25) but only in its part including the states $\Psi$ having zero angular momentum: $\hat P_\phi \Psi = 0$. The spectrum

$$E = \omega(1 + 2n_r) \tag{12.27}$$

coincides up to an irrelevant shift with the spectrum of the one-dimensional oscillator describing radial motion on the half-line $0 \le r < \infty$.

The corresponding eigenfunctions do not depend on the angular variable $\phi$. Thus, the number of dynamical degrees of freedom is effectively reduced from two to one. The word "dynamical" is important here. In fact, the angular variable $\phi$ has not disappeared completely. It has become a *gauge degree of freedom.* To understand the latter statement and the meaning of the word "gauge", consider the classical Hamiltonian describing our reduced system. It reads

$$H = H_0 + \lambda P_\phi. \tag{12.28}$$

The equation

$$\frac{\partial H}{\partial \lambda} = P_\phi = 0 \tag{12.29}$$

is our constraint. The other Hamilton equations of motion (4.3) are

$$\begin{aligned} \dot{P}_i + \omega^2 x_i + \lambda \varepsilon_{ij} P_j &= 0, \\ \dot{x}_i - P_i + \lambda \varepsilon_{ij} x_j &= 0. \end{aligned} \tag{12.30}$$

Performing the standard Legendre transformation (4.8), we derive the canonical Lagrangian

$$\boxed{L = \frac{1}{2}(\dot{x}_i + \lambda \varepsilon_{ij} x_j)^2 - \frac{\omega^2}{2} x_i^2.} \tag{12.31}$$

The equations of motion following from (12.31) are equivalent to (12.30).

**Theorem 12.3.** *The Lagrangian* (12.31) *is invariant under the gauge transformation*

$$\begin{aligned} x_i &\to O_{ij} x_j \\ \lambda &\to \lambda + \dot{\chi} \end{aligned} \tag{12.32}$$

*with*

$$O = \begin{pmatrix} \cos\chi & -\sin\chi \\ \sin\chi & \cos\chi \end{pmatrix}. \tag{12.33}$$

**Proof.** The simplest way to see it is to consider an infinitesimal transformation

$$\delta x_i = -\chi \varepsilon_{ij} x_j, \quad \delta\lambda = \dot{\chi} \tag{12.34}$$

and show that the variation $\delta L$ vanishes. □

The variable $\lambda(t)$ may be interpreted as a time-dependent angular velocity with which the reference system rotates.

Note the difference with *rigid* symmetries like in (4.19) or in (5.49), where the parameters of the transformations were ordinary or Grassmann constants. The same concerns the rigid rotational symmetry of the original system (12.25). But in (12.32), the parameter of the transformation $\chi(t)$ is *time-dependent*.[3] Noether's theorem does not apply in this case. In fact, gauge symmetry is not quite a symmetry but rather a manifestation of the fact that the chosen description of the system involves redundant unphysical variables (in our case, $\lambda$ and $\phi$). In our trivial case, it is simple to get rid of $\lambda$ and $\phi$ altohether (in the Hamiltonian language, the constraint $\hat{P}_\phi = 0$ allows one to *fix the gauge*: to set $\phi = 0$) and describe the radial motion of the system by the Hamiltonian

$$\hat{H}^{\text{reduced}} = \frac{\hat{P}_r^2}{2} + \frac{\omega^2 r^2}{2} \tag{12.35}$$

with the boundary condition $\Psi(r = 0) = 0$ to be imposed on the wave function.

But in many cases (especially, in field theories), getting rid of the redundant gauge variables is not practically feasible, and one has to work with the Lagrangians of the type (12.31) where the redundant variables are present.

We do not need it in the following, but having started this discussion, I cannot resist giving a nontrivial example of such a system. Consider the following quantum Hamiltonian involving nine dynamical variables $A_{j=1,2,3}^{a=1,2,3}$ and the corresponding momenta $\hat{E}_j^a = -i\partial/\partial A_j^a$:

$$\boxed{\hat{H} = \frac{1}{2}\left[(\hat{E}_j^a)^2 + (B_j^a)^2\right]} \quad \text{with} \quad B_j^a = \tfrac{1}{2}\varepsilon^{abc}\varepsilon_{jkl}A_k^b A_l^c. \tag{12.36}$$

It is actually the Hamiltonian of Yang–Mills theory with the $SU(2)$ gauge group in the quantum mechanical limit (the vector potentials $A_j^a$ depend only on time, but not on spatial coordinates). Impose the constraints

$$\hat{G}^a\Psi = \varepsilon^{abc}A_j^b\hat{E}_j^c\,\Psi = 0. \tag{12.37}$$

[3]And that is what the word "gauge" means — it refers to the situation when the action is invariant under the transformations whose parameters depend on time. In field theories, also on spatial coordinates.

One can do so because

- the operators $\hat{G}^a$ commute with the Hamiltonian,
- the algebra of the constraints is closed:

$$[\hat{G}^a, \hat{G}^b] = i\varepsilon^{abc}\hat{G}^c \tag{12.38}$$

so that the requirement $\hat{G}^a\Psi = 0$ is self-consistent.

The constraints (12.37) kill 3 out of 9 dynamical variables (like the constraint $\hat{P}_\phi\Psi = 0$ killed the angular variable in our toy model). For the quantum mechanical system (12.36), one can in principle *resolve the constraints*, i.e. get rid of three gauge degrees of freedom and find a Hamiltonian, an analog of (12.35), that depends only on gauge-invariant variables [133] (for the "parent" field theory, it is not technically possible), but this Hamiltonian is complicated and not so convenient. It is better to introduce three Lagrange multipliers $\lambda^a \equiv A_0^a$ and write an analog of (12.31) — the Lagrangian that depends on $A_\mu^a = (A_0^a, A_j^a)$. It has a nice form

$$\boxed{L = -\frac{1}{4}F_{\mu\nu}^a F^{\mu\nu\, a},} \tag{12.39}$$

where the nonzero components of $F_{\mu\nu}^a$ are

$$F_{jk}^a = -F_{kj}^a = \varepsilon^{abc}A_j^b A_k^c, \quad F_{0j}^a = -F_{j0}^a = \dot{A}_j^a + \varepsilon^{abc}A_0^b A_j^c \tag{12.40}$$

and the indices are lifted and lowered with the Minkowski metric $\eta_{\mu\nu} = (1, -1, -1, -1)$ (a birthmark from the field-theory origin of this model). This Lagrangian is invariant under certain gauge transformations. Infinitesimally, under the variations

$$\delta A_0^a = \dot{\chi}^a + \varepsilon^{abc}A_0^b\chi^c, \quad \delta A_j^a = \varepsilon^{abc}A_j^b\chi^c. \tag{12.41}$$

### 12.2.2 (2, 2, 0) ⟶ (1, 2, 1)

We are going to show now that the two types of SQM models discussed above — the models involving the chiral supermultiplets (**2**, **2**, **0**) and the models involving the real supermultiplets (**1**, **2**, **1**) are genetically related: the latter are obtained from the former by Hamiltonian reduction. The main idea is very simple. Take the chiral supersymmetric sigma model (9.1) with $\mathcal{W} = 0$ and assume that the metric $h_{m\bar{n}}$ does not depend on the imaginary parts of the complex coordinates $z^m$. If nothing depends on $\text{Im}(z^m)$, we can get rid of it and obtain a model defined on the real manifold with the coordinates $\text{Re}(z^m)$.

What *is* this model? One way to derive it is to go down to components, to the Hamiltonian (9.9) (or rather to its quantum counterpart), to observe that, under the assumption that $h_{m\bar{n}}$ does not depend on $\mathrm{Im}(z^m)$, the Hamiltonian commutes with $\mathrm{Im}(\hat{p}_m)$ and impose the constraint $\mathrm{Im}(\hat{p}_m)\Psi = 0$. Fixing the gauge $\mathrm{Im}(z^m) = 0$, one can thus derive the Hamiltonian involving only $\mathrm{Re}(z^m)$ and the corresponding canonical momenta. Observing further that $\mathrm{Im}(\hat{p}_m)$ commutes not only with the Hamiltonian but also with the supercharges, we derive that the reduced model enjoys the same $\mathcal{N} = 2$ supersymmetry as the original one.

However, this procedure does not allow one to trace the superspace and superfield structure of the model thus derived. To do so, one should perform the *Hamiltonian* reduction in the *Lagrangian* superfield formalism. The appropriate technique was developed in Refs. [134].

Consider the supersymmetry transformations (7.30) of the components of a chiral superfield. We distinguish the real and imaginary parts of $z = (x + iy)/\sqrt{2}$ to obtain

$$\begin{aligned}
\delta x &= \epsilon\psi - \bar{\epsilon}\bar{\psi}, \\
\delta y &= -i(\epsilon\psi + \bar{\epsilon}\bar{\psi}), \\
\delta\psi &= \bar{\epsilon}(\dot{y} - i\dot{x}), \\
\delta\bar{\psi} &= \epsilon(\dot{y} + i\dot{x}).
\end{aligned} \tag{12.42}$$

One can now observe that the transformations (12.42) *coincide* with the transformations (7.26) for the real superfield if one identifies $\dot{y} \equiv F$.

Consider now the Lagrangian

$$L = \frac{1}{4}\int d\bar{\theta}d\theta\, h_{m\bar{n}}(Z,\bar{Z})\, \bar{D}\bar{Z}^{\bar{n}} DZ^m \tag{12.43}$$

and suppose that the metric does not depend on $y^m$. Also, the component Lagrangian (9.2) does not depend in this case on $y^m$ but only on $\dot{y}^m$. Suppose first that the metric $h_{m\bar{n}}$ is real and symmetric. One can be convinced that, by setting $h_{m\bar{n}} \to g_{MN}$ and trading $\dot{y}^m \to F^M$, we reproduce the component Lagrangian (8.7) of the de Rham supersymmetric sigma model. Heuristically, the superfield expression of this Lagrangian,

$$L = \frac{1}{2}\int d\bar{\theta}d\theta\, g_{MN}(X)\, \bar{D}X^M DX^N, \tag{12.44}$$

is obtained from (12.43) by setting $Z^m, \bar{Z}^{\bar{m}} \to \sqrt{2}\,X^M,\ h \to g$.

However, the Hermitian metric $h_{m\bar{n}}$ need not be necessarily real. It might contain an imaginary antisymmetric part. The Hamiltonian reduction of a generic model (12.43) gives the *quasicomplex* $\mathcal{N} = 2$ model mentioned in Section 10.1.1. The superfield form of its Lagrangian is[4]

$$\boxed{L = \frac{1}{2}\int d\bar{\theta}d\theta\,[g_{(MN)}(X) + ib_{[MN]}(X)]\,\bar{D}X^M DX^N,} \tag{12.45}$$

with real symmetric $g_{(MN)}$ and real skew-symmetric $b_{[MN]}$. The bosonic part of the component Lagrangian (the full expression is written in Ref. [98]) is

$$L = \frac{1}{2}\left(g_{(MN)} + b_{[MP]}g^{(PQ)}b_{[QN]}\right)\dot{x}^M\dot{x}^N. \tag{12.46}$$

In other words, the true metric of the manifold is not $g_{(MN)}$ but depends also on $b_{[MN]}$, being given by the combination in (12.46).

The conserved supercharges of this model read [cf. (8.16)]

$$\begin{aligned} Q &= \frac{1}{\sqrt{2}}\psi^M\left[\Pi_M - \frac{i}{2}\partial_M(g_{(PQ)} + ib_{[PQ]})\,\psi^P\bar{\psi}^Q\right], \\ \bar{Q} &= \frac{1}{\sqrt{2}}\bar{\psi}^M\left[\Pi_M + \frac{i}{2}\partial_M(g_{(PQ)} - ib_{[PQ]})\,\psi^P\bar{\psi}^Q\right]. \end{aligned} \tag{12.47}$$

It is remarkable that the same quasicomplex supercharges may be derived by not only Hamiltonian reduction from the Dolbeault model but also similarity transformation from flat quantum supercharges in Euclidean space $\mathbb{R}^D$. To this end, one should take the supercharge (12.4) and transform it as in (12.1) with the operator (12.5), not requiring anymore the reality for the matrix $\alpha_{AB}$. A generic Hermitian matrix $\alpha_{AB}$ brings about a nontrivial deformation, and one can see [110] that the deformed supercharges coincide with the quantum version of the quasicomplex supercharges (12.47)!

In other words, quasicomplex sigma model may be obtained from the parent flat complex model by a subsequent application of two operations performed in any order: a similarity transformation and Hamiltonian reduction (see Fig. 12.1).

In the discussion above, we assumed that the metric did not depend on $\mathrm{Im}(z^m)$. But reduction can be performed not only in this restricted case

[4]Note that if one also adds to this Lagrangian the quasitorsion term $\sim \int d\bar{\theta}d\theta(\mathcal{B}_{MN}\, DX^M DX^N + \text{c.c.})$, one arrives at the model mentioned in passing in Ref. [35] — see Eq. (3.10) there.

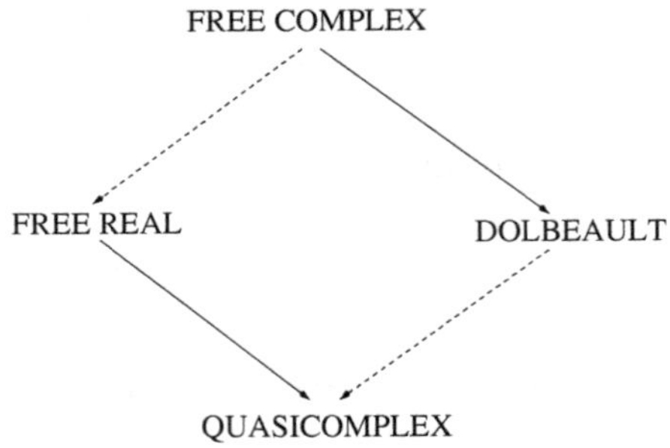


**Fig. 12.1.** The rhombus of sigma models. Solid arrows stand for similarity transformations and dashed arrows for Hamiltonian reduction.

but for *any metric with isometries*. To illustrate this, consider the model involving a single chiral superfield $Z$ with the free action:

$$S = \frac{1}{4} \int d\bar{\theta} d\theta dt \, \bar{D}\bar{Z}DZ. \tag{12.48}$$

One of the isometries of the flat metric are rotations $z \to ze^{i\phi}$. In the same way as for the toy model of the preceding section, we can impose the constraint $\hat{p}_\phi \Psi = 0$ on the wave functions.[5] The action of the reduced model is

$$S = \frac{1}{2} \int d\bar{\theta} d\theta dt \, \bar{D}R \, DR, \tag{12.49}$$

where $R$ is the real $(\mathbf{1}, \mathbf{2}, \mathbf{1})$ superfield whose lowest component is $r = |z|\sqrt{2}$. The motion goes in the case along the ray $[0, \infty)$ with the boundary condition $\Psi(0) = 0$.

### 12.2.3 (4, 4, 0) ⟶ (2, 4, 2)

Consider an Obata-flat HKT model given by the Lagrangian (10.70). Suppose that its metric (10.71) depends only on $z_k^{A=1}$ but not on $z_k^{A=2}$. Then, the quantum Hamiltonian and the supercharges commute with $\hat{p}_k^{A=2} = -i\partial/\partial z_k^{A=2}$ and the Hamiltonian reduction preserving the $\mathcal{N} = 4$ supersymmetry of the original model is possible.

[5]For sure, one can also introduce the variable $w = \ln z$. In this case, the angular variable $\phi$ is presented as $\text{Im}(w)$ and we come back to the situation analyzed above. But it is not always easy to choose the variables in such a way, and it is actually not necessary.

From the $\mathcal{N} = 2$ viewpoint, a (**4**, **4**, **0**) multiplet represents a pair of (**2**, **2**, **0**) multiplets. As we have learned above, the supersymmetry transformations in the multiplet (**2**, **2**, **0**) coincide with those in the multiplet (**1**, **2**, **1**) under the identification $\sqrt{2}\,\mathrm{Im}\,(\dot{z}) \equiv F$; the reduced model involves only the real parts of $z^m$, while their imaginary parts show up as auxiliary variables. The same holds for the large multiplets (**4**, **4**, **0**). A direct calculation shows that one can choose there any two real coordinates, identify their time derivatives with two real auxiliary variables of a (**2**, **4**, **2**) multiplet and observe that, under this identification, the supersymmetry transformations in the two multiplets coincide.

The reduced model must therefore be expressed in terms of (**2**, **4**, **2**) multiplets and represent a Kähler model. Generically — a quasicomplex Kähler model whose action (10.32) is given by the sum of standard kinetic term (10.20) and the extra term (10.31). The only question is how the Kähler potential $\mathcal{K}$ and the functions $\mathcal{A}_m$ determining the action of the reduced model depend on the HKT prepotential $\mathcal{L}$ of the original model.

Let us pose $\mathcal{W}^k = \mathcal{Z}^1_k$ and $\Xi^k = \bar{\mathcal{Z}}^2_k$, with $w^k, \xi^k$ being their bosonic components. In these terms, the bosonic part (10.72) of the component HKT Lagrangian can be represented as

$$\boxed{L^{\mathrm{HKT}}_{\mathrm{bos}} = \kappa_{k\bar{l}}\,(\dot{w}^k\dot{\bar{w}}^{\bar{l}} + \dot{\xi}^k\dot{\bar{\xi}}^{\bar{l}}) + \mathfrak{F}_{kl}\,\dot{w}^k\dot{\xi}^l + \bar{\mathfrak{F}}_{\bar{k}\bar{l}}\,\dot{\bar{w}}^{\bar{k}}\dot{\bar{\xi}}^{\bar{l}},} \tag{12.50}$$

where

$$\kappa_{k\bar{l}} = \frac{1}{2}\left(\frac{\partial^2}{\partial w^k \partial \bar{w}^{\bar{l}}} + \frac{\partial^2}{\partial \xi^k \partial \bar{\xi}^{\bar{l}}}\right)\mathcal{L} \tag{12.51}$$

and

$$\mathfrak{F}_{kl} = \frac{1}{2}\left(\frac{\partial^2}{\partial w^k \partial \xi^l} - \frac{\partial^2}{\partial \xi^k \partial w^l}\right)\mathcal{L},$$

$$\bar{\mathfrak{F}}_{\bar{k}\bar{l}} = \overline{\mathfrak{F}_{kl}} = \frac{1}{2}\left(\frac{\partial^2}{\partial \bar{w}^{\bar{k}} \partial \bar{\xi}^{\bar{l}}} - \frac{\partial^2}{\partial \bar{\xi}^{\bar{k}} \partial \bar{w}^{\bar{l}}}\right)\mathcal{L}. \tag{12.52}$$

For the model to be reducible, $\kappa_{k\bar{l}}$ and $\mathfrak{F}_{kl}$ should not depend on $\xi^p$ and $\bar{\xi}^{\bar{p}}$. What constraints does it impose on the prepotential?

**Theorem 12.4.** *In one of the charts with trivial topology, into which our complex manifold is subdivided, the prepotential $\mathcal{L}$, satisfying the requirements*

$$\frac{\partial}{\partial \xi^p}\kappa_{k\bar{l}} = \frac{\partial}{\partial \bar{\xi}^{\bar{p}}}\kappa_{k\bar{l}} = \frac{\partial}{\partial \xi^p}\mathfrak{F}_{kl} = \frac{\partial}{\partial \bar{\xi}^{\bar{p}}}\mathfrak{F}_{kl} = 0, \tag{12.53}$$

*can be chosen in the form*

$$\frac{\mathcal{L}}{2} = \mathcal{K}(\mathcal{W}, \bar{\mathcal{W}}) + \mathcal{A}_k(\mathcal{W})\, \Xi^k + \bar{\mathcal{A}}_{\bar{k}}(\bar{\mathcal{W}})\, \bar{\Xi}^{\bar{k}}, \tag{12.54}$$

*where* $\mathcal{K}$ *and* $\mathcal{A}_k$ *are such that*

$$\kappa_{k\bar{l}} = \partial_k \partial_{\bar{l}} \mathcal{K}, \quad \mathfrak{F}_{kl} = \partial_k \mathcal{A}_l - \partial_l \mathcal{A}_k \tag{12.55}$$

$(\partial_k \equiv \partial/\partial w^k)$.

**Proof.** We prove first the following lemma:

**Lemma.** The forms $\kappa_{k\bar{l}}\, dw^k \wedge d\bar{w}^{\bar{l}}$, $\mathfrak{F}_{kl}\, dw^k \wedge dw^l$ and $\bar{\mathfrak{F}}_{\bar{k}\bar{l}}\, d\bar{w}^{\bar{k}} \wedge d\bar{w}^{\bar{l}}$ are closed under the condition (12.53).

**Proof.** The closeness of the forms $\mathfrak{F}_{kl}\, dw^k \wedge dw^l$ and $\bar{\mathfrak{F}}_{\bar{k}\bar{l}}\, d\bar{w}^{\bar{k}} \wedge d\bar{w}^{\bar{l}}$ is an immediate corollary of the definitions (12.52). Consider now $\kappa_{k\bar{l}}$. The first term in (12.51) gives a closed form and the only problem is with the second term. Expand $\mathcal{L}$ in $\xi^p$ and $\bar{\xi}^{\bar{p}}$. A "dangerous" $(\xi, \bar{\xi})$-independent contribution to $\kappa_{k\bar{l}}$ in (12.51) may come from the structure

$$X = R_{m\bar{n}}(w, \bar{w})\, \xi^m \bar{\xi}^{\bar{n}} \tag{12.56}$$

in the prepotential. But the same structure gives the contribution $\propto$ $(\partial R_{l\bar{n}}/\partial w^k - \partial R_{k\bar{n}}/\partial w^l)\bar{\xi}^{\bar{n}}$ in $\mathfrak{F}_{kl}$. Bearing in mind (12.53), this contribution must vanish, and we derive $\partial_k R_{l\bar{n}} - \partial_l R_{k\bar{n}} = 0$. Considering the contribution of (12.56) into $\bar{\mathfrak{F}}_{\bar{k}\bar{l}}$, we derive $\partial_{\bar{k}} R_{m\bar{l}} - \partial_{\bar{l}} R_{m\bar{k}} = 0$. It follows that

$$\frac{\partial^2 X}{\partial \xi^k \partial \bar{\xi}^{\bar{l}}}\, dw^k \wedge d\bar{w}^{\bar{l}} = (1/2) R_{k\bar{l}}\, dw^k \wedge d\bar{w}^{\bar{l}}$$

is closed. □

If the forms are closed and the topology of a chart is trivial, they are also exact (see Theorem 1.2). And this means that $\kappa_{k\bar{l}}$ and $\mathfrak{F}_{kl}$ may be presented in the form (12.55) and the prepotential $\mathcal{L}$ may be chosen in the form (12.54). □

**Remark:** One *can* choose $\mathcal{L}$ in the form (12.54), but $\mathcal{L}$ does not *necessarily* have this form. One can always add to $\mathcal{L}$ the structures like

$$\Delta \mathcal{L} = \mathcal{B}_{\bar{k}}(\mathcal{W})\, \bar{\Xi}^{\bar{k}} \quad \text{or} \quad \Delta \mathcal{L} = Q(\mathcal{W}, \bar{\mathcal{W}}) - (\partial_k \partial_{\bar{l}} Q)\, \Xi^k \bar{\Xi}^{\bar{l}}, \tag{12.57}$$

which do not affect $\kappa_{k\bar{l}}$ and $\mathfrak{F}_{kl}$. And it is the latter objects that determine geometry and that we are finally interested in. The shifts (12.57) are analogous to gauge transformations of electromagnetic vector potential which do not affect the observable field density.

We are ready now to prove the following:

**Theorem 12.5.** *The HKT SQM model with the prepotential* (12.54) *gives after reduction the quasicomplex Kähler model with the action*

$$\boxed{S = \frac{1}{4}\int d^2\bar\theta d^2\theta dt\, \mathcal{K}(\mathcal{W},\bar{\mathcal{W}}) + \frac{1}{2}\left[\int d^2\theta dt_L\, \mathcal{A}_k(\mathcal{W})\,\dot{\mathcal{W}}^k + \text{c.c.}\right].} \tag{12.58}$$

**Proof.** One of the ways to prove it is to compare the bosonic terms (12.50) in the original HKT model with the bosonic terms (10.33) in the model (12.58) and observe that they coincide under the identification $\dot\xi \equiv F$. All other terms in the both models are reproduced by $\mathcal{N} = 4$ supersymmetry. They should also coincide, and one can check that they do. □

One can also formulate the following theorem:

**Theorem 12.6.** *In a topologically trivial chart, the prepotential* $\mathcal{L}(\mathcal{Z}_k^A, \bar{\mathcal{Z}}_k^A; \mathcal{U}_\alpha^A, \bar{\mathcal{U}}_\alpha^A)$ *of a generic reducible bi-HKT model* (10.93) *can be chosen, modulo gauge transformations that do not affect geometrical properties, in the form*

$$\begin{aligned}\frac{1}{2}\mathcal{L} &= \mathcal{K}(\mathcal{W},\bar{\mathcal{W}};\mathcal{V},\bar{\mathcal{V}}) + \left[\mathcal{A}_k^{(1)}(\mathcal{W},\mathcal{V}) + \mathcal{A}_k^{(2)}(\mathcal{W},\bar{\mathcal{V}})\right]\Xi^k \\ &\quad + \left[\bar{\mathcal{A}}_{\bar k}^{(1)}(\bar{\mathcal{W}},\bar{\mathcal{V}}) + \bar{\mathcal{A}}_{\bar k}^{(2)}(\bar{\mathcal{W}},\mathcal{V})\right]\bar\Xi^{\bar k} + \left[\mathcal{B}_\alpha^{(1)}(\mathcal{V},\mathcal{W}) + \mathcal{B}_\alpha^{(2)}(\mathcal{V},\bar{\mathcal{W}})\right]\Sigma^\alpha \\ &\quad + \left[\bar{\mathcal{B}}_{\bar\alpha}^{(1)}(\bar{\mathcal{V}},\bar{\mathcal{W}}) + \bar{\mathcal{B}}_{\bar\alpha}^{(2)}(\bar{\mathcal{V}},\mathcal{W})\right]\bar\Sigma^{\bar\alpha},\end{aligned} \tag{12.59}$$

*where* $\mathcal{W}^k = \mathcal{Z}_k^{A=1}$, $\Xi^k = \bar{\mathcal{Z}}_k^{A=2}$; $\mathcal{V}^\alpha = \mathcal{U}_\alpha^{A=1}$, $\Sigma^\alpha = \bar{\mathcal{U}}_\alpha^{A=2}$. *The reduced model is the generic quasicomplex bi-Kähler model with the action*

$$\begin{aligned}S &= \frac{1}{4}\int d^2\bar\theta d^2\theta dt\, \mathcal{K}(\mathcal{W},\bar{\mathcal{W}};\mathcal{V},\bar{\mathcal{V}}) \\ &\quad - \frac{i}{4}\left[\int d\theta_1 d\theta_2 dt_L \left\{\bar D_2[\mathcal{A}_k^{(1)}(\mathcal{W},\mathcal{V})D^2\mathcal{W}^k] + \bar D_1[\mathcal{A}_k^{(2)}(\mathcal{W},\bar{\mathcal{V}})D^1\mathcal{W}^k]\right\}\right. \\ &\quad \left. + \int d\theta_1 d\bar\theta^2 d\tilde t_L \left\{D^2[\mathcal{B}_\alpha^{(1)}(\mathcal{V},\mathcal{W})\bar D_2\mathcal{V}^\alpha] + \bar D_1[\mathcal{B}_\alpha^{(2)}(\mathcal{V},\bar{\mathcal{W}})D^1\mathcal{V}^\alpha]\right\} - \text{c.c.}\right]\end{aligned} \tag{12.60}$$

*with* $t_L$ *and* $\tilde t_L$ *defined in* (10.46).

**Proof.** The second part of this theorem [that the Hamiltonian reduction of (12.59) gives (12.60)] was proved in Ref. [82] in exactly the same way as Theorem 12.5: by comparison of the bosonic Lagrangians in the two models, bearing in mind the identification of the time derivatives of the coordinates to be reduced in the bi-HKT Lagrangian with the auxiliary variables in the bi-Kähler Lagrangian. We address the reader to that paper for further details.

The first statement [that the choice (12.59) is always possible] can be justified by noting that (12.60) *is* the most general form (10.47) of the quasicomplex bi-Kähler action. If the original bi-HKT Lagrangian involves the terms like $\mathcal{A}_k(\mathcal{W}, \mathcal{V}, \bar{\mathcal{V}})\, \Xi^k$ and is reducible (its metric does not depend on $\xi$ and $\sigma$), it should still give (12.60) after reduction. And this means that the bi-HKT action can be brought to the form (12.59) by a gauge transformation. □

A dedicated reader is welcome to construct a more rigorous proof of this fact in the spirit of Theorem 12.4.

### 12.2.4 *Symplectic and other models*

Up to now, we discussed models where the metric did not depend on a half of real coordinates. But one can also consider the situation when the metric involves an arbitrary number of isometries. Performing a reduction, we may obtain a vast number of different supersymmetric sigma models. We can start e.g. with an HKT model involving some number of *root* (**4**, **4**, **0**) multiplets and consider the case when the metric does not depend on a *quarter* of coordinates — of one real coordinate in each superfield. Quite naturally, we obtain in this way a symplectic model involving the multiplets (**3**, **4**, **1**), which was treated in Section 10.2.

Or one can consider a model where the reduction is performed with respect to three quarter of the coordinates and obtain a model involving the multiplets (**1**, **4**, **3**), where each dynamical coordinate has two different complex superpartners. A certain class of such models was studied in Ref. [135].

One can also start with the model involving $\mathcal{N} = 8$ (**8**, **8**, **0**) root multiplets and perform different reductions. The symplectic model of the second kind discussed in Section 11.5 can be obtained in this way.

### 12.2.5 *Kähler reduction and hyperkähler reduction*

We outlined above the procedure of Hamiltonian reduction, which allowed us to relate the SQM models of different dimension. We imposed constraints on the wave functions which commuted with the Hamiltonian and the supercharges. As a result, the reduced models enjoyed the same symmetry as the original ones.

There also exist related but different procedures, the *Kähler reduction* and *hyperkähler reduction*, which allow one to construct Kähler quotients of Kähler manifolds and hyperkähler quotients of hyperkähler manifolds. The main idea is to factorize the manifold under the action of a group leaving (hyper)Kähler forms intact.

To understand that it is not the same as Hamiltonian reduction of the type discussed in Sections 12.2.2 and 12.2.3, consider the SQM model on the plane $\mathbb{R}^2$. It has two complex supercharges, $\hat{Q}_0 = \hat{p}_1\psi_1 + \hat{p}_2\psi_2$ and $\hat{Q}_1 = \hat{p}_1\psi_2 - \hat{p}_2\psi_1$. Consider the $U(1)$ group rotating the plane. It keeps intact the Kähler form $dx_1 \wedge dx_2$, but its generator $\hat{G} = x_1\hat{p}_2 - x_2\hat{p}_1$ does not commute with the supercharges. We have instead $[\hat{G}, \hat{Q}_0] = -i\hat{Q}_1$ and $[\hat{G}, \hat{Q}_1] = i\hat{Q}_0$.

In constract to Hamiltonian reduction that is formulated in physical terms for dynamical supersymmetric models (and thereby is in line with the main ideology of our book), (hyper)kähler reduction is formulated on purely geometric language. Thus, speaking logically, this subject belongs to Chapter 3 rather than to this chapter. Still, we decided to be inconsistent and outline this method here.

#### 12.2.5.1 *Toy model*

As the simplest nontrivial example [127], consider the space $\mathbb{R}^2 \times (\mathbb{R} \times S^1)$ with the flat metric expressed in the form

$$ds^2 = dr^2 + r^2 d\phi^2 + dx^2 + a^2 d\theta^2, \tag{12.61}$$

where $\phi, \theta \in (0, 2\pi)$ and $a$ is the radius of the circle. One of the isometries of this manifold is a simultaneous shift of $\phi$ and $\theta$:

$$\phi \to \phi + \alpha, \quad \theta \to \theta + \alpha. \tag{12.62}$$

The difference

$$\chi = \phi - \theta \tag{12.63}$$

stays invariant under (12.62), and the same concerns the Kähler form

$$\omega = \omega_{MN} dx^M \wedge dx^N = r dr \wedge d\phi + a dx \wedge d\theta. \tag{12.64}$$

Consider the 1-form $\omega_\alpha$ obtained by a *contraction* of the form (12.64) with the vector $V^M = (0_r, 1_\phi, 0_x, 1_\theta)$ generating the transformation (12.62):

$$\omega_\alpha = \omega_{MN} dx^M V_N = r dr + a dx. \tag{12.65}$$

The form (12.65) is obviously closed and also exact, $\omega_\alpha = d\mu_\alpha$, with

$$\boxed{\mu_\alpha = \frac{r^2}{2} + ax.} \tag{12.66}$$

The function $\mu_\alpha$ is called a *moment map.*

Personally, I find this term a little awkward, but it is widely used in the literature, and we use it too. In fact, "moment map" is an abbreviation of "momentum map". Here, "momentum" refers to the generator of the transformation (12.62) (or a more complicated $U(1)$ isometry in more complicated cases) and "map" is a mathematical way to think of functions. Indeed, $\mu_\alpha$ describes a map from our manifold to $\mathbb{R}$.

To perform Kähler reduction, we set $\mu_\alpha = 0$ and substitute

$$x = -\frac{r^2}{2a} \tag{12.67}$$

into (12.61). The metric acquires the form

$$ds^2 = dr^2 \left(1 + \frac{r^2}{a^2}\right) + r^2 (d\theta + d\chi)^2 + a^2 d\theta^2. \tag{12.68}$$

Next, we factorize the three-dimensional manifold thus obtained with respect to the isometry (12.62). To this end, we represent the metric as

$$ds^2 = dr^2 \left(1 + \frac{r^2}{a^2}\right) + \frac{a^2 r^2}{r^2 + a^2} d\chi^2 + (r^2 + a^2)\left(d\theta + \frac{r^2}{r^2 + a^2} d\chi\right)^2 \tag{12.69}$$

and *suppress* the last term with complete square. We are left with

$$\boxed{ds^2 = dr^2 \left(1 + \frac{r^2}{a^2}\right) + \frac{a^2 r^2}{r^2 + a^2} d\chi^2.} \tag{12.70}$$

The last operation has a dynamical interpretation. Let us write the Lagangian (8.2) for the metric (12.68). It reads

$$L = \left(1 + \frac{r^2}{a^2}\right)\frac{\dot{r}^2}{2} + \frac{r^2}{2}\dot{\chi}^2 + \frac{r^2 + a^2}{2}\dot{\theta}^2 + r^2\dot{\chi}\dot{\theta}. \tag{12.71}$$

The corresponding canonical Hamiltonian is

$$H = \frac{p_r^2}{2(1 + r^2/a^2)} + \frac{p_\chi^2(1 + r^2/a^2)}{2r^2} + \frac{p_\theta^2}{2a^2} - \frac{p_\chi p_\theta}{a^2}. \tag{12.72}$$

The Poisson bracket $\{p_\theta, H\}_P$ vanishes, and we can impose the constraint $p_\theta = 0$. This gives the reduced Hamiltonian

$$H = \frac{p_r^2}{2(1 + r^2/a^2)} + \frac{p_\chi^2(1 + r^2/a^2)}{2r^2}. \tag{12.73}$$

Its Legendre transform,

$$L = \frac{\dot{r}^2}{2}\left(1 + \frac{r^2}{a^2}\right) + \frac{\dot{\chi}^2}{2}\frac{r^2}{1 + r^2/a^2}, \tag{12.74}$$

corresponds to the metric (12.70).

This part of the procedure looks like a genuine Hamiltonian reduction!

Topologically, the metric (12.70) describes a sphere. Indeed, the Gaussian curvature is equal in this case to curvature is

$$\mathbb{R} = R_{1212}g^{11}g^{22} = \frac{8a^4}{(r^2 + a^2)^3}, \tag{12.75}$$

and the Euler characteristic (8.40) is

$$\chi = 8a^4 \int_0^\infty \frac{rdr}{(r^2 + a^2)^3} = 2.$$

As any 2D manifold, this manifold is Kähler. Its Kähler form,

$$\tilde{\omega} = rdr \wedge d\chi, \tag{12.76}$$

*coincides* with (12.64), bearing in mind (12.63) and (12.67).

In this case, it is not difficult to see by a direct calculation that the complex structure $I_{r\chi} = -I_{\chi r} = r$ associated with $\tilde{\omega}$ is covariantly constant.[6] Moreover, mathematicians can prove (see Ref. [136] and Theorem 3.1 in Ref. [29]) that it is *always* so and that the complex structure on the quotient

---

[6]It is true for any metric

$$ds^2 = f(r)dr^2 + \frac{r^2}{f(r)}d\chi^2.$$

derived by applying the procedure described above to any Kähler manifold is covariantly constant so that the quotient is also Kähler.[7]

#### 12.2.5.2 *Taub–NUT*

Hyperkähler reduction is Kähler reduction with respect to all the three complex structures. It is possible to perform for hyperkähler manifolds possessing an isometry leaving intact three Kähler forms. The simplest nontrivial example is $\mathbb{R}^8$ or rather $\mathbb{R}^7 \times S^1 = \mathbb{R}^4 \times (\mathbb{R}^3 \times S^1)$ with the Taub–NUT manifold as a quotient [30].[8]

The flat $\mathbb{R}^4$ metric

$$ds^2 = \sum_M dy^2_{M=1,2,3,4} \tag{12.77}$$

admits three quaternionic complex structures (3.3). The corresponding Kähler forms are

$$\begin{aligned} \omega_I &= dy_1 \wedge dy_2 + dy_3 \wedge dy_4, \\ \omega_J &= dy_1 \wedge dy_3 - dy_2 \wedge dy_4, \\ \omega_K &= dy_1 \wedge dy_4 + dy_2 \wedge dy_3. \end{aligned} \tag{12.78}$$

These forms are invariant under the isometry

$$\begin{aligned} y_1' &= y_1 \cos\alpha - y_2 \sin\alpha, \quad & y_2' &= y_1 \sin\alpha + y_2 \cos\alpha, \\ y_3' &= y_3 \cos\alpha + y_4 \sin\alpha, \quad & y_4' &= -y_3 \sin\alpha + y_4 \cos\alpha. \end{aligned} \tag{12.79}$$

As was mentioned in Chapter 11, the metric[9] (12.77) can also be expressed in the coordinates $\{\boldsymbol{x}, \Psi\}$ in (11.55) (they play the same role as the polar coordinates on $\mathbb{R}^2$) as follows:

$$\boxed{ds^2 = \frac{r}{4}(d\Psi + \boldsymbol{A} \cdot d\boldsymbol{x})^2 + \frac{1}{4r} d\boldsymbol{x} \cdot d\boldsymbol{x},} \tag{12.80}$$

where $\mathbf{A}$ given in (3.41) is the Dirac monopole field. Then, the isometry (12.79) boils down to the shift $\Psi \to \Psi + 2\alpha$ while the variables $x_j$ do not change.

---

[7] I have to admit that I do not completely understand the proof of this theorem; a more detailed and explicit proof would not be out of place. But I do not have any doubt that this statement is correct.

[8] See also Ref. [124] and Section 2.5 of Ref. [137] for pedagogical expositions.

[9] Note a slight change of convention compared to (11.56).

Consider now the manifold $\mathbb{R}^4 \times (\mathbb{R}^3 \times S^1)$ choosing the metric (12.80) for the first factor and the Cartesian metric $ds^2 = d\boldsymbol{X}^2 + a^2 d\theta^2$ for $\mathbb{R}^3 \times S^1$, $a$ being the radius of the circle. Consider the isometry[10]

$$G: \quad \Psi \to \Psi + 2\alpha, \quad \theta \to \theta + \alpha. \tag{12.81}$$

We choose the hyperkähler triple in the form

$$\begin{aligned} \omega_I &= dy_1 \wedge dy_2 + dy_3 \wedge dy_4 + dX_1 \wedge dX_2 + dX_3 \wedge (a d\theta), \\ \omega_J &= dy_1 \wedge dy_3 - dy_2 \wedge dy_4 + dX_2 \wedge dX_3 + dX_1 \wedge (a d\theta), \\ \omega_K &= dy_1 \wedge dy_4 + dy_2 \wedge dy_3 - dX_1 \wedge dX_3 + dX_2 \wedge (a d\theta). \end{aligned} \tag{12.82}$$

The corresponding complex structures read

$$I = \mathrm{diag}(\mathcal{I}, \mathcal{I}); \quad J = \mathrm{diag}(\mathcal{J}, \mathcal{K}); \quad K = \mathrm{diag}(\mathcal{K}, -\mathcal{J}) \tag{12.83}$$

with $\mathcal{I}, J, K$ defined in (3.3). Such a "twisted" choice is convenient to simplify the following formulas. Note that the structures (12.83) are quaternionic in both sectors (though a choice where they would be quaternionic in one sector and antiquaternionic in the other sector would also be possible).

The forms (12.82) are invariant under $G$. The combination $\chi = \Psi - 2\theta$ also stays invariant. At the next step, we determine the contractions of these forms with the vector field

$$V^M = (-y_2, y_1, y_4, -y_3, \mathbf{0}, 1) \tag{12.84}$$

defining the infinitesimal rotation (12.79), (12.81). We obtain

$$\begin{aligned} \omega_I^\alpha &= y_1 dy_1 + y_2 dy_2 - y_3 dy_3 - y_4 dy_4 + a dX_3 = d\left(\frac{x_3}{2} + aX_3\right), \\ \omega_J^\alpha &= y_4 dy_1 + y_3 dy_2 + y_2 dy_3 + y_1 dy_4 + a dX_1 = d\left(\frac{x_1}{2} + aX_1\right), \\ \omega_K^\alpha &= -y_3 dy_1 + y_4 dy_2 - y_1 dy_3 + y_2 dy_4 + a dX_2 = d\left(\frac{x_2}{2} + aX_2\right). \end{aligned} \tag{12.85}$$

These forms represent exterior derivatives of the moment maps:

$$\boxed{\mu_I = \frac{x_3}{2} + aX_3, \quad \mu_J = \frac{x_1}{2} + aX_1, \quad \mu_K = \frac{x_2}{2} + aX_2.} \tag{12.86}$$

---

[10] $\theta \in (0, 2\pi)$ and $\Psi \in (0, 4\pi)$.

To perform the hyperkähler reduction, we impose on the flat $\mathbb{R}^8$ metric the constraints $\mu_{I,J,K} = 0$. This gives us a five-dimensional manifold with the metric

$$ds^2 = \frac{1}{4}\left(\frac{1}{r} + \frac{1}{a^2}\right) d\boldsymbol{x}^2 + \frac{r}{4}(d\chi + 2d\theta + \boldsymbol{A} \cdot d\boldsymbol{x})^2 + a^2 d\theta^2, \qquad (12.87)$$

where we also plugged $\Psi = \chi + 2\theta$. For a further factorization with respect to the action of (12.81), we rewrite (12.87) distinguishing a complete square,

$$\begin{aligned} ds^2 &= \frac{1}{4}\left(\frac{1}{r} + \frac{1}{a^2}\right) d\boldsymbol{x}^2 + \frac{(d\chi + \boldsymbol{A} \cdot d\boldsymbol{x})^2}{4\left(\frac{1}{r} + \frac{1}{a^2}\right)} \\ &\quad + (r + a^2)\left[d\theta + \frac{r}{2(r + a^2)}(d\chi + \boldsymbol{A} \cdot d\boldsymbol{x})\right]^2, \end{aligned} \qquad (12.88)$$

and cross the complete square out.

As promised, we have derived the Taub–NUT metric.[11]

The hyperkähler triple on the quotient may be derived starting from (12.82) in the same way as we derived (12.76) starting from (12.64) in the toy two-dimensional model. An explicit calculation was done in Ref. [124], which we follow. At the first step, one should express the flat $\mathbb{R}^4$ triple (12.78) in terms of $\boldsymbol{x}$ and $\Psi$, using (11.54). After some calculation, one arrives at rather simple expressions:

$$\begin{aligned} \omega_I^{\text{flat}} &= \frac{1}{4r} dx_1 \wedge dx_2 + \frac{1}{4} dx_3 \wedge (d\Psi + A_1 dx_1 + A_2 dx_2), \\ \omega_J^{\text{flat}} &= \frac{1}{4r} dx_2 \wedge dx_3 + \frac{1}{4} dx_1 \wedge (d\Psi + A_2 dx_2), \\ \omega_K^{\text{flat}} &= \frac{1}{4r} dx_3 \wedge dx_1 + \frac{1}{4} dx_2 \wedge (d\Psi + A_1 dx_1), \end{aligned} \qquad (12.89)$$

where $A_{1,2} \equiv A_{x,y}$ are the vector potentials in (3.41).

Now, we substitute it in (12.82). After setting there $X_j = -x_j/(2a)$ and $d\Psi = d\chi + 2d\theta$, one may observe that the terms $\propto dx_j \wedge d\theta$ cancel, and

[11] That was the third derivation of this metric in our book: (i) the original derivation of its discoverers [19], who obtained the metric in the form (3.33), (ii) the derivation in the preceding chapter that used the Legendre transform and now we have done it with the hyperkähler reduction. In the following chapter, our patient reader will learn the fourth derivation based on the technique of harmonic superspace.

we arrive at a surprisingly simple answer:

$$\boxed{\begin{aligned} \omega_I^{\rm TN} &= \frac{1}{4} dx_1 \wedge dx_2 \left(\frac{1}{r} + \frac{1}{a^2}\right) + \frac{1}{4} dx_3 \wedge (d\chi + \boldsymbol{A} \cdot d\boldsymbol{x}), \\ \omega_J^{\rm TN} &= \frac{1}{4} dx_2 \wedge dx_3 \left(\frac{1}{r} + \frac{1}{a^2}\right) + \frac{1}{4} dx_1 \wedge (d\chi + \boldsymbol{A} \cdot d\boldsymbol{x}), \\ \omega_K^{\rm TN} &= \frac{1}{4} dx_3 \wedge dx_1 \left(\frac{1}{r} + \frac{1}{a^2}\right) + \frac{1}{4} dx_2 \wedge (d\chi + \boldsymbol{A} \cdot d\boldsymbol{x}). \end{aligned}} \tag{12.90}$$

The only difference compared to the flat structures (12.89) is the change $1/r \to 1/r + 1/a^2$.

We mention here also Ref. [21] where the Taub–NUT complex structures expressed in terms of the variables $r, \sigma_{1,2,3}$ in Eq. (3.33) were derived. These expressions are rather bulky. Probably, they reduce to (12.90) after an appropriate variable change.

Starting from the flat metric in $\mathbb{R}^{4n}$ with $n \geq 3$, one can derive many nontrivial multidimensional hyperkähler metrics using the procedure described above [30]. Unfortunately, a similar derivation for the multidimensional hyperkähler metrics (3.44) describing multimonopole dynamics meets difficulties, and the only way that I know to derive it is to resort to $\mathcal{N} = 4$ supersymmetric $3D$ field theory.

But once this multimonopole metric is established, one can reduce it further to obtain the metric (3.49) and more complicated metrics for more complicated groups [31].

# Chapter 13

# HK and HKT through Harmonic Glasses

In Chapters 10–12, we discussed, among other extended supersymmetric models, the hyperkähler and HKT models, but the efficiency and generality of our discussion were limited:

- A hyperkähler model was defined as an $\mathcal{N} = 2$ model with three extra complex supersymmetries (11.1). We proved that the metric is hyperkähler iff the complex structures are quaternionic, but the extra supersymmetries had an external nature; they did not appear as superspace symmetries. As a result, we did not have "industrial" tools that would allow one to construct the models enjoying these extra symmetries automatically.
- We showed that the models with the superspace action (10.69) are HKT, and their metrics are given by the double derivatives of the prepotential, as in (10.72). But all these metrics belonged to a rather special class of HKT metrics, being *Obata-flat*. The complex structures $(I, J, K)_M{}^N$ are constant matrices in this case.
- We described two ingenious methods to construct hyperkähler models. One of them used Legendre transformation and the other was based on hyperkähler reduction. But these methods have a limited scope of applications, and I would not call them really *industrial*.

How to construct generic HK and HKT metrics in superspace approach? We have already briefly touched upon this question in Section 7.4. We wrote the harmonic superfield action (7.109) involving $2n$ $\mathcal{N} = 8$ harmonic G-analytic pseudoreal superfields $Q^{+a}$ and claimed that the corresponding geometry is hyperkähler. We also claimed that a model, involving $2n$ $\mathcal{N} = 4$ harmonic G-analytic pseudoreal superfields $q^{+a}$ subject to the nonlinear constraints (7.99), describes HKT geometry.

In this chapter, we justify these claims and show how, given a superspace action, to derive the corresponding HK or HKT metric. Generically, it is much more difficult than for e.g. Kähler manifolds, where the metric is always given by the double derivative of the Kähler potential, $h_{m\bar{n}} = \partial_m \partial_{\bar{n}} K$. For hyperkähler and not Obata-flat HKT manifolds, the metric can only be found as a result of solution of rather complicated harmonic differential equations. Only in rare cases it is possible to present the solutions in a closed analytical form.

## 13.1 Hyperkähler models

The harmonic $\mathcal{N} = 8$ superspace construction which describes hyperkähler geometry was worked out in Ref. [138] (see also Chapter 5 of Ref. [90]). For reader's convenience, we remind here some definitions that we gave at the end of Chapter 7.

Our starting point is the ordinary $\mathcal{N} = 8$ superspace that involves time, four complex odd coordinates $\theta_{j\alpha}$ $(\alpha, j = 1, 2)$ and their conjugates $\bar{\theta}^{j\alpha} = \overline{\theta_{j\alpha}}$. We harmonize it by introducing[1]

$$\theta^\pm_\alpha = \theta^j_\alpha u^\pm_j, \quad \bar{\theta}^{\pm\alpha} = \bar{\theta}^{j\alpha} u^\pm_j, \quad t_A = t + i(\theta^+_\alpha \bar{\theta}^{-\alpha} + \theta^-_\alpha \bar{\theta}^{+\alpha}), \tag{13.1}$$

where $u^{+j}$ is a complex doublet of unit length parametrizing $S^2$, $u^-_j = \overline{u^{+j}}$ and $t_A$ is the Grassmann analytic time. The analytic subspace $(\zeta, u) = (t_A, \theta^+_\alpha, \bar{\theta}^{+\alpha}; u^\pm_j)$ of the large superspace stays invariant under the supertransformations and under the so-called pseudoconjugation defined in Eq. (7.78).[2]

Now, we take $2n$ G-analytic superfields

$$\begin{aligned} Q^{+a}(\zeta, u) = {} & F^{+a}(t_A, u) + \theta^+_\alpha\, \chi^{\alpha a}(t_A, u) \\ & + \bar{\theta}^{+\alpha}\, \kappa^a_\alpha(t_A, u) + \theta^+_\alpha \bar{\theta}^{+\alpha}\, A^{-a}(t_A, u) + \cdots \\ & + (\theta^+_\alpha \bar{\theta}^{+\alpha})^2\, P^{-3a}(t_A, u), \quad a = 1, \ldots, 2n \end{aligned} \tag{13.3}$$

(only *some* terms relevant for us in what follows are displayed in the expansion). The superscripts like "+", "−" or "−3" refer to the harmonic

---

[1] Both Latin and Greek indices are lifted and lowered with the convention (5.37).

[2] The presence of the index $\alpha$ does not change much. The odd coordinates transform now as

$$\widetilde{\theta^\pm_\alpha} = \bar{\theta}^{\pm\alpha}, \quad \widetilde{\bar{\theta}^{\pm\alpha}} = -\theta^\pm_\alpha. \tag{13.2}$$

charges: each field entering Eq. (13.3) is an eigenstate of the operator $D^0$, defined in (7.83), with the corresponding eigenvalue. To make the representation (13.3) irreducible, we impose the pseudoreality constraint

$$\boxed{\widetilde{Q_a^+} = \Omega^{ab} Q_b^+ \equiv Q^{+a}} \qquad \Rightarrow \qquad \widetilde{Q^{+a}} = -Q_a^+, \tag{13.4}$$

where $\Omega^{ab} = -\Omega_{ab}$ is the symplectic matrix and $\widetilde{\ }$ is the pseudoconjugation.

We write the harmonic superspace action as an integral over the analytic superspace

$$\boxed{S = \int dt_A\, du\, d^2\bar\theta^+ d^2\theta^+ \left[\frac{1}{8}\, Q_a^+ D^{++} Q^{+a} + \frac{1}{4}\mathcal{L}^{+4}(Q^{+a}, u)\right],} \tag{13.5}$$

with[3]

$$D^{++} = \partial^{++} + 2i\theta_\alpha^+ \bar\theta^{+\alpha} \frac{\partial}{\partial t_A}, \qquad \partial^{++} = u^{+j} \frac{\partial}{\partial u_j^-}. \tag{13.6}$$

The equations of motion derived from it are

$$D^{++} Q^{+a} = \Omega^{ab} \frac{\partial \mathcal{L}^{+4}}{\partial Q^{+b}}. \tag{13.7}$$

Substituting here the expansion (13.3), we obtain a set of equations for the components. In particular, we derive

$$\partial^{++} F^{+a} = \Omega^{ab} \frac{\partial \mathcal{L}^{+4}}{\partial F^{+b}} \tag{13.8}$$

and

$$\partial^{++} A^{-a} - \Omega^{ab} \frac{\partial^2 \mathcal{L}^{+4}}{\partial F^{+b} \partial F^{+c}} A^{-c} = -2i\dot F^{+a} + \text{fermionic terms}. \tag{13.9}$$

Consider equation (13.8). If $\mathcal{L}^{+4}$ were absent, the linear equation $\partial^{++} F^{+a} = 0$ would have a simple solution,

$$F^{+a} \equiv F_0^{+a} = x^{ja} u_j^+, \tag{13.10}$$

with $x^{ja}$ obeying the pseudoreality condition (3.16) that follows from (13.4).

To be *quite* precise, $x^{ja}$ depends on $t_A$ which includes the odd coordinates, and one should rather say that $x^{ja}$ satisfies the condition $\widetilde{x^{ja}} = x_{aj}$.

[3]This is the first line in Eq. (7.97) where we added the summation over $\alpha$ and suppressed the terms $\propto \partial/\partial\theta_\alpha^-, \partial/\partial\bar\theta^{-\alpha}$ — they do not work due to analyticity of $Q^{+a}$.

However, bearing in mind that $t_A$ is invariant under pseudoconjugation (which plays the role of the ordinary complex conjugation in this approach), one can still write the condition (3.16) and treat $t_A$ as an ordinary real time $t$. In the following, the index "A" is suppressed.

When $\mathcal{L}^{+4} \neq 0$, it is a rather difficult task to find a solution to (13.8). To date, it was found in a closed form only for a few particular choices of $\mathcal{L}^{+4} \neq 0$ [90,138–140], including the choice corresponding to the Taub–NUT manifold to be discussed in detail in the following. In the general case, the solution to (13.8) can be found by iterations: redefine $\mathcal{L}^{+4} \to \lambda\mathcal{L}^{+4}$ and represent the solution as a formal series in $\lambda$:

$$F^{+a} = F_0^{+a} + \lambda F_1^{+a} + \lambda^2 F_2^{+a} + \cdots \tag{13.11}$$

We obtain the chain of equations

$$\begin{aligned} \partial^{++} F_1^{+a} &= \Omega^{ab}\, \frac{\partial \mathcal{L}^{+4}}{\partial F_0^{+b}}, \\ \partial^{++} F_2^{+a} &= \Omega^{ab}\, \frac{\partial^2 \mathcal{L}^{+4}}{\partial F_0^{+b} \partial F_0^{+c}} F_1^{+c}, \\ \ldots &= \ldots \end{aligned} \tag{13.12}$$

These are in fact algebraic equations, as becomes clear if one expands their left-hand and right-hand sides in a proper harmonic basis. For example, we represent

$$F_1^{+a} = A^{(ijk)a}\, u_i^+ u_j^+ u_k^- + B^{(ijklp)a}\, u_i^+ u_j^+ u_k^+ u_l^- u_p^- + \cdots \tag{13.13}$$

(the linear term $\propto u_j^+$ does not contribute in the left-hand sides of (13.12); it is attributed to $F_0^+$) and

$$\Omega^{ab}\, \frac{\partial \mathcal{L}^{+4}}{\partial F_0^{+b}} = C^{(ijk)a}\, u_i^+ u_j^+ u_k^+ + D^{(ijklp)a}\, u_i^+ u_j^+ u_k^+ u_l^+ u_p^- + \cdots \tag{13.14}$$

Then, the first equation in (13.12) implies

$$A = C, \quad B = \frac{1}{2} D, \quad \text{etc.}$$

We first solve the equation for $F_1^{+a}$, then we substitute its solution to the equation for $F_2^{+a}$, solve it, substitute into the equation for $F_3^{+a}$, etc. As a result, $F^{+a}(t,u)$ is expressed via the harmonic-independent coefficients $x^{ja}(t)$, which have the meaning of the coordinates on the hyperkähler manifolds that we are set to describe. Note that the pseudoreality condition (3.16) for $x^{ja}$ implies that the vectors

$$x^M = -\frac{i}{\sqrt{2}}(\Sigma^M)_{aj}x^{ja} \tag{13.15}$$

[with the constant matrices $\Sigma^M = \Sigma_M \equiv \Sigma_A$ defined in Eq. (3.9)] are real. Any other choice of $4n$ real coordinates $x^M$ is also possible.

Consider now Eq. (13.9) where the fermion terms are suppressed. The solution $\tilde{A}^{-a}(t,u)$ of this truncated equation is also expressed via $x^{ja}(t)$. It follows from (13.9) and (13.10) that

$$\tilde{A}^{-a} = -2i\dot{x}^{ja}u_j^- + \text{nonlinear in harmonics terms.} \tag{13.16}$$

To find the metric of the manifold that our action describes, we substitute the solutions thus obtained for $F^{+a}(t,u)$ and $\tilde{A}^{-a}(t,u)$ into the bosonic part of the action (13.5). Indeed, if one expresses the latter via the components, one obtains, using the equation of motion (13.9), a very simple expression

$$S_{\text{bos}} = \frac{i}{2}\int dt\, du\, \tilde{A}_a^- \dot{F}^{+a}, \tag{13.17}$$

and the other bosonic components of the superfield (13.3) do not contribute! After expressing $\tilde{A}_a^-$ and $F^{+a}$ through $x^{ja}(t)$, the action (13.17) takes the generic form

$$S_{\text{bos}} = \frac{1}{2}\int dt\, g_{aj,bk}\,\dot{x}^{ja}\dot{x}^{kb}, \quad g_{aj,bk} = \varepsilon_{jk}\Omega_{ab} + O(\lambda). \tag{13.18}$$

Note the following. The action (13.18) was derived using the equations of motion (13.8) and (13.9). As was mentioned, these equations have algebraic nature and allow one to express the multitude of auxiliary variables dwelling in $F^{+a}(t,u)$ and $A^{-a}(t,u)$ via the dynamical variables $x^{ja}(t)$. But the superfield equation of motion (13.7) contains also the *dynamical* equation for $x^{ja}(t)$. The latter is hidden in the component equation

$$\partial^{++}P^{-3a} - \Omega^{ab}\frac{\partial^2\mathcal{L}^{+4}}{\partial F^{+b}\partial F^{+c}}P^{-3c} = -2i\dot{A}^{-a} + \text{ferm. terms} \tag{13.19}$$

that we did not discuss before. Substituting there the solution (13.16), we could derive in the flat case $\mathcal{L}^{+4} = 0$ the equations $P^{-3a} = 0$ and $\ddot{x}^{ja} = 0$, giving the straight geodesic.

A similar program can be carried out for the fermionic components. In particular, comparing the terms $\propto \theta^+_\alpha$ on the both sides of Eq. (13.7), we derive

$$\boxed{\mathcal{D}^{++}\chi^{\alpha a} \stackrel{\text{def}}{=} \partial^{++}\chi^{\alpha a} - \Omega^{ab}\frac{\partial^2\mathcal{L}^{+4}}{\partial F^{+b}\partial F^{+c}}\,\chi^{\alpha c} = 0.} \tag{13.20}$$

The variables $\kappa^a_\alpha$ satisfy a similar equation, but they are not independent, being expressed via $\chi^{\alpha a}$ in virtue of (13.4):

$$\kappa_{\alpha a} = \widetilde{\chi^{\alpha a}}. \tag{13.21}$$

A general solution to (13.20) can be presented as[4]

$$\chi^{\alpha a}(t,u) = (M^{-1})^a{}_{\underline{b}}(u)\,\psi^{\alpha\underline{b}}(t), \tag{13.22}$$

where $(M^{-1})^a{}_{\underline{b}}$, which satisfies the equation

$$\mathcal{D}^{++}(M^{-1})^a{}_{\underline{b}} \equiv \partial^{++}(M^{-1})^a{}_{\underline{b}} - \Omega^{ab}\,\frac{\partial^2\mathcal{L}^{+4}}{\partial F^{+b}\partial F^{+c}}\,(M^{-1})^c{}_{\underline{b}} = 0, \tag{13.23}$$

is a very important object called the *bridge*. It relates the fields carrying the *world* symplectic index $a$ to those carrying the *tangent space* underlined symplectic index $\underline{a}$.[5] The inverse matrix $M^{\underline{b}}{}_a$ satisfies the equation

$$\partial^{++}M^{\underline{b}}{}_a + \Omega^{cb}\,\frac{\partial^2\mathcal{L}^{+4}}{\partial F^{+a}\partial F^{+b}}M^{\underline{b}}{}_c = 0. \tag{13.24}$$

Note that the field $\psi^{\alpha\underline{b}}(t)$ (through which the field $\chi^{\alpha a}$ is expressed) does not carry harmonic dependence. It is natural to associate it with the fermion field in tangent space. One can also define the harmonic-independent fermion field $\psi^{\alpha a}(t)$ carrying the world symplectic index (we do so a little bit later).

There are infinitely many solutions to Eqs. (13.23) and (13.24) interrelated by the transformation (a kind of gauge transformation)

$$M^{\underline{b}}{}_a \to R^{\underline{b}}{}_{\underline{c}}\,M^{\underline{c}}{}_a \tag{13.25}$$

with an arbitrary harmonic-independent $R^{\underline{b}}{}_{\underline{c}}$. We will make a particular convenient choice for $R$ and $M$ (pick up a particularly convenient gauge) in Sections 13.3 and 13.4.

---

[4]The notation $M^{-1} \to M$ would probably be more natural, but we follow the conventions of Refs. [92, 104].

[5]See Ref. [92] for more details.

One can show that all the other components of $Q^{+a}$ in the expansion (13.3) are auxiliary variables; they are algebraically expressed via $x^{ja}$ and $\psi^{\alpha\underline{a}}$.

We have altogether $4n$ complex dynamical fermionic variables — one complex fermionic variable for each real bosonic coordinate. Their contribution to the Lagrangian includes the bilinear structure[6] $\propto \bar{\psi}\dot{\psi}$. The variables can be chosen such that the coefficients in the fermion kinetic term and in the bosonic kinetic term are given by the same metric tensor. There is also a four-fermionic term coming from the expansion of $\mathcal{L}^{+4}$. One thus obtains the Lagrangian of a supersymmetric sigma model that can be expressed in terms of $4n$ $(\mathbf{1}, \mathbf{2}, \mathbf{1})$ multiplets, a particular case of the generic Lagrangian (8.6) giving in components (8.9).

Our model possesses by construction extended $\mathcal{N} = 8$ supersymmetry. This suggests that the corresponding manifold is hyperkähler but does not prove it yet — we know from the analysis in Chapter 11 that bi-Kähler or bi-HKT manifolds of special type may also enjoy $\mathcal{N} = 8$ supersymmetry.

**Theorem 13.1.** *The action* (13.5) *describes a hyperkähler supersymmetric sigma model.*

**Proof.** It is sufficient to study in this model the supersymmetry transformations of the coordinates $x^M$ defined in (13.15) and show that they coincide with the corollary of (11.1):

$$\begin{aligned} \delta_1 x^M &= I_N{}^M(\epsilon_1\psi^N - \bar{\epsilon}_1\bar{\psi}^N), \\ \delta_2 x^M &= J_N{}^M(\epsilon_2\psi^N - \bar{\epsilon}_2\bar{\psi}^N), \\ \delta_3 x^M &= K_N{}^M(\epsilon_3\psi^N - \bar{\epsilon}_3\bar{\psi}^N), \end{aligned} \tag{13.26}$$

with quaternionic complex structures and properly defined $\psi^M$. This will exclude the bi-HKT possibility (HKT manifolds are not Kähler) and also the bi-Kähler possibility — we have seen that in the latter case the law of supersymmetry transformations is more complicated [see Eq. (10.42)].

At the level of the harmonic superfield (13.3), the supersymmetry transformations are generated by the shifts $\delta\theta^+_\alpha = \epsilon^+_\alpha, \delta\bar{\theta}^{+\alpha} = \bar{\epsilon}^{+\alpha}$. This gives, bearing in mind (13.21),

$$\delta F^{+a} = \delta_\epsilon F^{+a} + \delta_{\bar{\epsilon}} F^{+a} = \epsilon^+_\alpha \chi^{\alpha a} - \bar{\epsilon}^{+\alpha}\widetilde{\chi^a_\alpha}. \tag{13.27}$$

[6]It comes from the term $\propto Q^+_a D^{++} Q^{+a}$ in (13.5), where one has to substitute $\theta^+_\alpha \chi^{\alpha a} + \bar{\theta}^{+\alpha}\kappa^a_\alpha$ for $Q^{+a}$.

Let us concentrate on $\delta_\epsilon F^{+a}$. Using (13.22) and multiplying both sides of (13.27) by $M^{\underline{b}}{}_a$, we may write

$$M^{\underline{b}}{}_a\,(\partial_{bk}F^{+a})\,\delta_\epsilon x^{kb} = \epsilon^+_\alpha \psi^{\alpha\underline{b}}(t). \tag{13.28}$$

The right-hand side of this relation is proportional to $u^+_{\underline{k}}$ (we define $\epsilon^+_\alpha = \epsilon_\alpha{}^{\underline{k}} u^+_{\underline{k}}$), while higher-order terms in the harmonic expansion are absent. The same should be true for the left-hand side.[7] We can thus define

$$M^{\underline{b}}{}_a\,(\partial_{bk}F^{+a}) = e^{\underline{kb}}_{bk}(t)\,u^+_{\underline{k}}. \tag{13.29}$$

Then, the transformation law (13.28) acquires the form

$$\delta_\epsilon x^{kb} = e^{kb}_{\underline{bk}}\,\epsilon_\alpha{}^{\underline{k}}\,\psi^{\alpha\underline{b}}, \tag{13.30}$$

where $e^{kb}_{\underline{bk}}$ is the inverse of $e^{\underline{kb}}_{bk}$. Later, we will see that these matrices are nothing but the ordinary vielbeins under a particular choice of the matrix $R$ in Eq. (13.25).

Next, we rename the dummy index $\alpha \to \underline{j}$ so that $\psi^{\alpha\underline{b}} \to \psi^{\underline{jb}}$ and define

$$\psi^{jc} = e^{jc}_{\underline{bj}}\,\psi^{\underline{jb}}, \qquad \epsilon_{\underline{j}}{}^{\underline{k}} = \epsilon_0\,\delta^{\underline{k}}_{\underline{j}} + i\epsilon_p(\sigma_p)_{\underline{j}}{}^{\underline{k}}, \tag{13.31}$$

where $\sigma_p$ are the Pauli matrices. We derive

$$\delta_\epsilon x^{kb} = e^{\underline{jb}}_{cj}\left[\epsilon_0\,\delta^{\underline{k}}_{\underline{j}} + i\epsilon_p(\sigma_p)_{\underline{j}}{}^{\underline{k}}\right] e^{kb}_{\underline{bk}}\,\psi^{jc}. \tag{13.32}$$

No harmonics anymore!

At this stage, we may go to the vector notation (13.15) and also

$$\psi^M = -\frac{i}{\sqrt{2}}\,(\Sigma^M)_{aj}\psi^{ja}, \tag{13.33}$$

restore the variation $\delta_{\bar\epsilon} x^M$ from the requirement for $x^M$ and $\delta x^M$ to be real and derive

$$\delta x^M = \epsilon_0\psi^M + \epsilon_p(I^p)_N{}^M\psi^N - \bar\epsilon_0\bar\psi^M - \bar\epsilon_p(I^p)_N{}^M\bar\psi^N \tag{13.34}$$

with

$$(I^p)_N{}^M = \frac{i}{2}(\Sigma^\dagger_N)^{jc}\,e^{\underline{jb}}_{cj}\,(\sigma_p)_{\underline{j}}{}^{\underline{k}}\,e^{kb}_{\underline{bk}}\,(\Sigma^M)_{bk} \tag{13.35}$$

(we remind that $\Sigma^M = \Sigma_M$ are flat and $\mathrm{Tr}\{\Sigma^\dagger_N\Sigma_M\} = 2\delta_{NM}$).

---

[7]And it is. See Ref. [104] for an independent derivation.

One can observe that the complex structures (13.35) are quaternionic. To see it, consider first the flat four-dimensional case. The expression (13.35) reduces to[8]

$$I^p_{\nu\mu} = \frac{i}{2}(\sigma^p)_j{}^k(\sigma^\dagger_\nu\sigma_\mu)_k{}^j = -\eta^p_{\nu\mu} \tag{13.36}$$

[see Eq. (3.10)]. These structures are quaternionic due to (3.7). In the multidimensional flat case, we derive

$$I^p_{MN} = -\mathrm{diag}(\eta^p_{\mu\nu}, \ldots, \eta^p_{\mu\nu}), \tag{13.37}$$

which corresponds to (3.19). For curved manifolds, the property $I^pI^q = -\delta^{pq} + \varepsilon^{pqr}I^r$ may be established dressing the flat complex structure with the symplectic vielbeins, as is dictated by (13.35).

The terms involving $\epsilon_0$ and $\bar\epsilon_0$ in (13.34) give the standard $\mathcal{N} = 2$ supertransformation of the lowest component of the superfield $X^M$ and the terms involving $\epsilon_p$ and $\bar\epsilon_p$ coincide with the lowest component of (11.1). □

We have shown that the model (13.5) describes hyperkähler geometry. Two remarks are in order:

(1) A legitimate question is whether *any* hyperkähler metric can be derived this way. The answer to this question is positive. In Ref. [141] (see also Chapter 11 in Ref. [90]), the problem was solved in a different way — not invoking supersymmetry but solving instead the constraint $R_A{}^B \in sp(n)$ for the curvature form (see Theorem 3.2). It was shown that a *general* solution to this constraint depends on an arbitrary harmonic function $\mathcal{L}^{+4}(Q^{+a}, u)$ and that this solution coincides with the solution following from (13.5).

(2) Our discussion was local — we described all hyperkähler *metrics* in a given chart with trivial topology. We said nothing about how different charts are glued together and thus we have not obtained the classification of hyperkähler *manifolds*. Similarly, the expression $h_{m\bar n} = \partial_m\partial_{\bar n}K$ gives a complete classification of Kähler metrics in a given chart but does not give a classification of Kähler manifolds.

[8]Note that $(\sigma^p)_j{}^k = (\sigma^p)^k{}_j$. It follows from the fact that $(\sigma^p)_{jk}$ are symmetric matrices.

## 13.2 HKT models

These models enjoy $\mathcal{N} = 4$ supersymmetry and, to give their superspace description, one should take $\mathcal{N} = 4$ superspace including time, a doublet $\theta_j$ and a doublet $\bar{\theta}^j$. However, as we mentioned in the beginning of this chapter, this formalism is not sufficient if one wants to describe generic HKT manifolds with a nonzero Obata curvature. To do so, one has to harmonize the $\mathcal{N} = 4$ superspace. The basic definitions were given in Chapter 7. They coincide with the definitions of the preceding section, where one only has to suppress the index $\alpha$.

We take $2n$ G-analytic superfields (7.94),

$$\boxed{q^{+a}(\zeta, u) = f^{+a}(t, u) + \theta^+ \chi^a(t, u) + \bar{\theta}^+ \kappa^a(t, u) + \theta^+ \bar{\theta}^+ A^{-a}(t, u)} \tag{13.38}$$

($t \equiv t_A$), and impose the nonlinear constraint (7.99),

$$D^{++} q^{+a} = \mathcal{L}^{+3a}(q^{+b}, u), \tag{13.39}$$

with arbitrary $\mathcal{L}^{+3a}$, supplemented by the pseudoreality constraint $\widetilde{q^{+a}} = -q^+_a$. Substituting (13.38) into (13.39), we derive for the components:

$$\partial^{++} f^{+a} = \mathcal{L}^{+3a}(f^{+b}, u), \tag{13.40}$$

$$\mathcal{D}^{++} \chi^a = \mathcal{D}^{++} \kappa^a = 0, \tag{13.41}$$

$$\mathcal{D}^{++} A^{-a} = -2i\dot{f}^{+a} + \frac{\partial^2 \mathcal{L}^{+3a}}{\partial f^{+b} \partial f^{+c}} \kappa^b \chi^c, \tag{13.42}$$

where the action of the covariant harmonic derivative $\mathcal{D}^{++}$ on any symplectic vector $X^a(t, u)$ is defined as[9]

$$\mathcal{D}^{++} X^a = \partial^{++} X^a - \frac{\partial \mathcal{L}^{+3a}}{\partial f^{+b}} X^b. \tag{13.44}$$

Note that the constraint (13.39) plays a completely different role compared to (13.8). The latter was the superfield *equations of motion* following from the action (13.5). And the constraint (13.39) is an *external* constraint

[9] We note right now that in the particular case

$$\mathcal{L}^{+3a} = \Omega^{ac} \frac{\partial \mathcal{L}^{+4}}{\partial f^{+c}} \tag{13.43}$$

$\mathcal{D}^{++}$ acts as in Eq. (13.20).

imposed to make the representation (13.38) irreducible. At this stage, we have not written yet the action describing the dynamics of $q^{+a}$. We do so soon.

The pseudoreality constraint implies that

$$f^{+a}(t,u) = x^{ja} u_j^+ + \text{higher harmonic terms} \tag{13.45}$$

with pseudoreal $x^{ja}$. The latter can be traded for $4n$ real coordinates $x^M$ as in (13.15).

In the fermion sector, we derive

$$\kappa_a(t, u_j^\pm) = \widetilde{\chi^a(t, u_j^\pm)}. \tag{13.46}$$

By the same (group theory) reasons that we defined $\overline{\theta_j} = \bar\theta^j$, it is convenient to define[10]

$$\overline{\chi_a} = \bar\chi^a \quad \Rightarrow \quad \overline{\chi^a} = -\bar\chi_a, \tag{13.47}$$

the fundamental representation of $Sp(n)$ becomes antifundamental after conjugation and the other way round. The fermion components in $q^{+a}$ acquire thus the form

$$(q^{+a})^{\text{ferm}} = \theta^+ \chi^a(t,u) - \bar\theta^+ \bar\chi^a(t,\tilde u). \tag{13.48}$$

We have $2n$ complex or $4n$ real fermion fields — a real fermion superpartner $\psi^M$ for each real coordinate. As for $A^{-a}$, they are expressed via $x^M$ and $\psi^M$ in virtue of (13.42).

The most general $\mathcal{N} = 4$ supersymmetric action involving the fields $q^{+a}$ reads

$$\boxed{\begin{aligned} S = S_0 + S_{\text{gauge}} &= \int dt du\, d^2\theta^+ d^2\theta^- \mathcal{L}(q^{+a}, q^{-b}, u^\pm) \\ &\quad + i \int dt_A du\, d^2\theta^+ \mathcal{L}^{++}(q^{+a}, u^\pm). \end{aligned}} \tag{13.49}$$

The term $S_0$ represents an integral over the whole superspace. The integrand $\mathcal{L}$ depends on the harmonics, on $q^{+a}$ and also on $q^{-a} = D^{--}q^{+a}$. To make the action real, $\mathcal{L}$ should be real. The second term represents an integral over the analytic superspace. To make $S_{\text{gauge}}$ real, the integrand $\mathcal{L}^{++}$ should be odd under the pseudoconjugation (7.78).[11]

We can now formulate the central statement to be proven in this section.

[10] Note that in Refs. [92, 104], the opposite sign convention $\overline{\chi^a} = \bar\chi_a$ was chosen.
[11] Thus, our model includes all together *three* prepotentials: $\mathcal{L}, \mathcal{L}^{++}$ and $\mathcal{L}^{+3a}$.

*The action $S_0$ in* (13.49), *where the pseudoreal superfields $q^{+a}$ are subject to the constraints* (13.39), *describes HKT geometry. The second term $S_{\text{gauge}}$ describes the gauge field of some special form, living on the manifold.*

The proof of this fact involves several steps.

To begin with, we have just seen that the dynamical degrees of freedom in our model include $4n$ real coordinates $x^M$ and $4n$ their real fermionic superpartners. It is exactly the right number of degrees of freedom that corresponds to the model (9.35) including $D\ \mathcal{N}=1$ superfields $\mathcal{X}^M$ with the possible extra term (9.63).

We want to emphasize that the action[12]

$$S = \int d\Theta dt \left[\frac{i}{2} g_{MN}(\mathcal{X}) \dot{\mathcal{X}}^M \mathcal{D}\mathcal{X}^N - \frac{1}{12} C_{KLM}(\mathcal{X}) \mathcal{D}\mathcal{X}^K \mathcal{D}\mathcal{X}^L \mathcal{D}\mathcal{X}^M\right]$$
$$- i \int d\Theta dt A_M(\mathcal{X})\, \mathcal{D}\mathcal{X}^M \qquad (13.50)$$

describes a *most general* $\mathcal{N}=1$ model including an equal number of bosonic and fermionic dynamic variables and not involving higher derivatives. To construct such a model, we have only the superfields $\mathcal{X}^M = x^M + i\Theta\Psi^M$ at our disposal. The fermionic superfields like $\Phi = \phi + \Theta F$ do not fit: if we want $\phi$ to enter the component Lagrangian with a derivative, the field $F$ is necessarily auxiliary, entering without derivatives (see the a remark on p. 175). That would lead to a mismatch of bosonic and fermionic dynamic variables.

Let us first concentrate on $S_0$ and prove the following:

**Theorem 13.2.** *The action $S_0$ in* (13.49) *can be expressed in terms of $\mathcal{X}^M$ as in the first line of* (13.50) *without an admixture of the second line.*

**Proof.** The component Lagrangian corresponding to (9.35) [the first line in (13.50)] was given in Eq. (9.3). In contrast to the Lagrangian in (9.64) that follows from (9.63), it involves only the terms $\propto \dot{x}^2$ but not the terms $\propto \dot{x}$. Thus, we only have to show that the action $S_0$ expressed in components involves only the quadratic in velocities term.

Using the expansion (13.38) and the expansion

$$q^{-a} = D^{--} q^{+a} = \partial^{--} f^{+a} + 2i\theta^- \bar{\theta}^- \dot{f}^{+a} + \theta^+ \bar{\theta}^+ \partial^{--} A^{-a}$$
$$+ (\theta^- \bar{\theta}^+ + \theta^+ \bar{\theta}^-) A^{-a} + 2i\theta^+ \bar{\theta}^+ \theta^- \bar{\theta}^- \dot{A}^{-a} + \text{fermion terms}, \qquad (13.51)$$

[12]Please do not confuse the gauge potential $A_M$ with the component $A^{-a}$ of the superfield $q^{+a}$.

we derive[13]

$$S_0^{\text{bos}} = 2i \int dudt \left[ (\dot{f}^{+b} \tilde{A}^{-a} - \dot{f}^{+a} \tilde{A}^{-b}) \frac{\partial^2 \mathcal{L}}{\partial f^{+a} \partial f^{-b}} \right.$$

$$\left. + \left( \dot{f}^{+a} \partial^{--} \tilde{A}^{-b} + \frac{i}{2} \tilde{A}^{-a} \tilde{A}^{-b} - \dot{f}^{-a} \tilde{A}^{-b} \right) \frac{\partial^2 \mathcal{L}}{\partial f^{-a} \partial f^{-b}} \right], \quad (13.52)$$

where we defined $f^{-a} = \partial^{--} f^{+a}$ and $\tilde{A}^{-a}$ is the solution to the *truncated* constraint (13.42) with suppressed fermionic term in the right-hand side.

This solution is proportional to $\dot{f}^{+a}$ and finally to $\dot{x}^{ja}$. This means that the component bosonic Lagrangian derived from $S_0$ involves only the terms $\propto \dot{x}^2$. □

After doing the harmonic integral in (13.52), we may derive the metric. We do so later.

We know from Theorem 9.4 and the remark at the end of Section 9.2 that, if the metric $g_{MN}$, the torsion $C_{KLM}$ and the gauge potential $A_M$ satisfy certain conditions, the model enjoys $\mathcal{N} = 2$ supersymmetry. In such models, one of the supersymmetries mixes the components of $\mathcal{X}^M$ according to (7.9) and another supersymmetry transforms the whole $\mathcal{N} = 1$ superfield $\mathcal{X}^M$ as in (9.38). The model that we are now interested in has, however, not one but three extra supersymmetries. According to Theorem 9.4, *each* of them should have a form (9.38) with its own complex structure $(I^p)_N{}^M$ that satisfies the standard conditions: $(I^p)^2 = -\mathbb{1}, I^p_{MN} = -I^p_{NM}$ and the vanishing of the Nijenhuis tensor. The action should thus be invariant under

$$\begin{aligned}
\delta_1 \mathcal{X}^M &= E_1 (I^1)_N{}^M \mathcal{D}\mathcal{X}^N, \\
\delta_2 \mathcal{X}^M &= E_2 (I^2)_N{}^M \mathcal{D}\mathcal{X}^N, \\
\delta_3 \mathcal{X}^M &= E_3 (I^3)_N{}^M \mathcal{D}\mathcal{X}^N
\end{aligned} \quad (13.53)$$

with real Grassmann $E_{1,2,3}$.

[13]Note that the expansion (13.38) was written in the *analytic* basis, and the expansion (13.51), which was derived by acting on (13.38) with the operator $D^{--}$ in the form (7.97), is written also in the analytic basis. Thus, when deriving (13.52) from the integral for $S_0$ in (13.49), we actually assumed that this integral was done over $dt_A$ rather than over $dt$. But this makes no difference because the super-Jacobian of the transformation $t \to t_A$ with other variables fixed is equal to 1. After the integral is done, we can suppress the subscript $A$ again.

**Theorem 13.3.** *For the $\mathcal{N} = 4$ supersymmetry algebra to hold, the complex structures $I^p$ should satisfy the Clifford algebra*

$$I^p I^q + I^q I^p = -2\delta^{pq}. \tag{13.54}$$

**Proof.** Consider the Lie bracket $(\delta_2\delta_1 - \delta_1\delta_2)\mathcal{X}^M$. The calculation gives

$$(\delta_2\delta_1 - \delta_1\delta_2)\mathcal{X}^M = -iE_1E_2(I^1I^2 + I^2I^1)_N{}^M\dot{\mathcal{X}}^N + \text{independent terms.}$$

For the supercharges $\hat{Q}_1$ and $\hat{Q}_2$ to anticommute, this Lie bracket must vanish. This gives $\{I^1, I^2\} = 0$. In the same way, we derive $\{I^1, I^3\} = \{I^2, I^3\} = 0$. □

In Chapter 10, we discussed $\mathcal{N} = 4$ supersymmetric models with this set of variables (a real fermion for each real coordinate) and found two types of such models: the HKT models and the bi-HKT models. The HKT models were distinguished compared to bi-HKT ones by *quaternionic* nature of their complex structures. In Chapter 10, only Obata-flat models were discussed, but that property holds also in the general case. Let us prove it.

**Theorem 13.4.** *A model* (9.35) *with three extra invariances* (13.53) *where the complex structures are quaternionic describes HKT geometry.*

**Proof.** The proof is purely geometric. The invariance under (13.53) and Theorem 9.4 imply in particular the condition (9.41) for each complex structure:

$$\nabla_L^{(B)}(I^p)_N{}^M + \nabla_N^{(B)}(I^p)_L{}^M = 0, \tag{13.55}$$

where $\nabla^{(B)}$ is a covariant derivative with the Bismut connection (2.19), one and the same for all three complex structures. We show that in the case when the complex structures are quaternionic, they satisfy the stronger condition

$$\nabla_N^{(B)}(I^p)_L{}^M = 0, \tag{13.56}$$

which means the covariant invariance of the triple with respect to the universal Bismut connection so that the manifold is HKT.

The integrability allows us to choose the complex coordinates associated with any of the complex structures. Let us do so for $I^1 \equiv I$. Then, the only nonzero components of the tensor ${I_N}^M$ are displayed in Eq. (2.53):

$$ {I_m}^n = -i\delta_m^n, {I_{\bar{m}}}^{\bar{n}} = i\delta_{\bar{m}}^{\bar{n}}, \quad {I_m}^{\bar{n}} = {I_{\bar{m}}}^n = 0. \tag{13.57} $$

Consider the tensors[14] $I^2 \pm iI^3 \equiv J \pm iK$. As follows from Theorem 3.5 [specifically, from the relations (3.66)], their only nonzero components are $(J+iK)_m{}^{\bar{n}} = 2J_m{}^{\bar{n}}$ and $(J-iK)_{\bar{m}}{}^n = 2J_{\bar{m}}{}^n$.

Note that the properties $J^2 = K^2 = -\mathbb{1}$ and $JK = -KJ = I$ [this *is* the step where the quaternionic nature of the triple $(I, J, K)$ is used] together with (13.57) imply

$$ \begin{aligned} (J-iK)_{\bar{m}}{}^n \, (J+iK)_n{}^{\bar{l}} &= -4\delta_{\bar{m}}^{\bar{l}}, \\ (J+iK)_n{}^{\bar{l}} \, (J-iK)_{\bar{l}}{}^m &= -4\delta_n^m. \end{aligned} \tag{13.58} $$

Considering the constraint (13.55) for the complex structure $I$, we derived earlier (see the proof of Theorem 9.5) that certain components of the Bismut affine connections vanish:

$$ G^n_{l\bar{s}} = G^{\bar{n}}_{\bar{l}s} = 0. \tag{13.59} $$

Consider now the constraint (13.55) for $J + iK$:

$$ \nabla_L^{(B)} (J+iK)_N{}^M + \nabla_N^{(B)} (J+iK)_L{}^M = 0. \tag{13.60} $$

We choose $L = l$ and $M = m$ but do not specify the holomorphicity of $N$. Using the fact that the only nonzero components of $J + iK$ have an antiholomorphic upper index, we see that in this case most of the terms in (13.60) vanish and we are left with

$$ G^m_{l\bar{s}} (J+iK)_N{}^{\bar{s}} + G^m_{N\bar{s}} (J+iK)_l{}^{\bar{s}} = 0. \tag{13.61} $$

However, the first term in the left-hand side vanishes due to (13.59), and the constraint boils down to $G^m_{N\bar{s}} (J+iK)_l{}^{\bar{s}} = 0$. Multiplying this by $(J-iK)_{\bar{r}}{}^l$ and using the first relation in (13.58), we arrive at the constraint

$$ G^m_{N\bar{r}} = 0 \tag{13.62} $$

for all $N$, holomorphic and antiholomorphic. The constraint $G^{\bar{m}}_{Ns} = 0$ can be derived in a similar way by exploring the condition (13.55) for the structure $J - iK$.

---

[14] $(J \pm iK)/2$ are the hypercomplex structures $\mathcal{I}^{\pm}$ defined on p. 65.

Consider now the covariant derivative of $I$ without symmetrization,

$$\nabla_L^{(B)} I_N{}^M = G^Q_{LN} I_Q{}^M - G^M_{LQ} I_N{}^Q \tag{13.63}$$

$[\partial_L I_N{}^M = 0$ due to (13.57)] and choose again $M = m$ (the case $M = \bar{m}$ is treated analogously). We note that if $N = n$, the right-hand side of (13.63) vanishes identically due to the cancellation of the two terms and if $N = \bar{n}$, it boils down to $-2iG^m_{L\bar{n}}$, which vanishes due to (13.62).

Thus, $\nabla_L^{(B)} I_M{}^N = 0$. In fact, we have proven that if the complex structures are quaternionic, the holomorphic and antiholomorphic parts $C_{3,0}$ and $C_{0,3}$ in the decomposition (9.56) are *absent* (Theorem 9.5 only said that the corresponding forms must be closed) and the torsion $C$ is the Bismut torsion for $I$. Choosing the complex coordinates associated with $J$ or with $K$, we can repeat all the arguments and prove that $\nabla_L^{(B)} J_M{}^N = \nabla_L^{(B)} K_M{}^N = 0$. The Bismut torsion is universal. □

**Remark:** The term "HKT" in the formulation of this theorem should be understood in a generic sense including also hyperkähler manifolds with zero torsion. In the latter case, the second term in the action (9.35) is absent. It is also in this sense that the term "HKT" is used in the formulation of the following theorem.

**Theorem 13.5.** *The model $S_0$ involving pseudoreal superfields $q^{+a}$ subject to the constraints* (13.39) *is HKT.*

**Proof.** We only need to show that the complex structures are quaternionic. The proof of that is similar to the proof of Theorem 13.1.

Supertransformations of the bosonic components $f^{+a}$ of the superfields $q^{+a}$ are generated by the shifts $\delta\theta^+ = \epsilon^+$ and $\delta\bar{\theta}^+ = \bar{\epsilon}^+$. They read

$$\delta_{\epsilon,\bar{\epsilon}} f^{+a}(t,u) = \epsilon^+ \chi^a(t,u) - \bar{\epsilon}^+ \bar{\chi}^a(t,\tilde{u}). \tag{13.64}$$

Similarly to (13.22), we express the solution of Eq. (13.41) as

$$\chi^a(t,u) = (M^{-1})^a{}_{\underline{b}} \psi^{\underline{b}}(t), \quad \bar{\chi}^a(t,\tilde{u}) = (M^{-1})^a{}_{\underline{b}} \bar{\psi}^{\underline{b}}(t), \tag{13.65}$$

where the bridge $M^{-1}$ satisfies the equation

$$\mathcal{D}^{++}(M^{-1})^a{}_{\underline{b}} \equiv \partial^{++}(M^{-1})^a{}_{\underline{b}} - \frac{\partial \mathcal{L}^{+3a}}{\partial f^{+b}} (M^{-1})^b{}_{\underline{b}} = 0. \tag{13.66}$$

Then, (13.64) may be rewritten in a way analogous to (13.28):

$$M^{\underline{b}}{}_a \, (\partial_{bk} f^{+a}) \, \delta_{\epsilon,\bar{\epsilon}} x^{kb}(t) = \epsilon^+ \psi^{\underline{b}}(t) - \bar{\epsilon}^+ \bar{\psi}^{\underline{b}}(t). \tag{13.67}$$

The right-hand side and hence the left-hand side of Eq. (13.67) involve only the linear term $\propto u^+_{\underline{k}}$ in the harmonic expansion. Defining the vielbeins as in (13.29),

$$M^{\underline{b}}{}_a\,(\partial_{bk} f^{+a}) = e^{\underline{kb}}_{bk}\, u^+_{\underline{k}}\,, \tag{13.68}$$

we obtain the transformation law for harmonic-independent fields:

$$\delta_{\epsilon,\bar\epsilon} x^{kb} = e^{kb}_{\underline{bk}}\,(\epsilon^{\underline{k}}\,\psi^{\underline{b}} - \bar\epsilon^{\underline{k}}\,\bar\psi^{\underline{b}}). \tag{13.69}$$

The fields $\psi^{\underline{b}}$ and $\bar\psi^{\underline{b}}$ can be joined into a quartet $\psi^{\underline{kb}}$ according to

$$\psi^{\underline{1b}} \equiv \psi^{\underline{b}}, \qquad \psi^{\underline{2b}} \equiv -\bar\psi^{\underline{b}}.$$

We also define the matrix $\epsilon_{\underline{l}}{}^{\underline{k}}$ as

$$\epsilon_{\underline{1}}{}^{\underline{k}} \equiv \epsilon^{\underline{k}}, \qquad \epsilon_{\underline{2}}{}^{\underline{k}} \equiv \bar\epsilon^{\underline{k}} \tag{13.70}$$

and represent it as

$$\epsilon_{\underline{l}}{}^{\underline{k}} = iE_0\delta^{\underline{k}}_{\underline{l}} + E_p(\sigma^p)_{\underline{l}}{}^{\underline{k}} \tag{13.71}$$

with real[15] $E_0, E_p$.

In this notation, the transformation law (13.69) is expressed as

$$\delta x^{kb} = e^{kb}_{\underline{bk}}\,\epsilon_{\underline{l}}{}^{\underline{k}}\,\psi^{\underline{lb}} = iE_0\, e^{kb}_{\underline{bk}}\,\psi^{\underline{kb}} + E_p\, e^{kb}_{\underline{bk}}\,(\sigma^p)_{\underline{l}}{}^{\underline{k}}\,\psi^{\underline{lb}}. \tag{13.72}$$

If we define

$$\psi^{kb} \equiv e^{kb}_{\underline{bk}}\,\psi^{\underline{kb}}, \tag{13.73}$$

the first term in (13.72) reads simply $\delta x^{kb} = iE_0\psi^{kb}$. This is a $\mathcal{N}=1$ supersymmetry transformation indicating that $x^{kb}$ and $\psi^{kb}$ are superpartners: they represent the components of the $\mathcal{N}=1$ superfield [cf. (9.37)].

The second term in (13.72) can be represented as

$$\delta x^{kb} = i\,E_p\, e^{\underline{cl}}_{lc}\,(I^p)_{\underline{cl}}{}^{\underline{kb}}\, e^{kb}_{\underline{bk}}\psi^{lc} \equiv i\,E_p\,(I^p)_{cl}{}^{kb}\,\psi^{lc}, \tag{13.74}$$

where

$$(I^p)_{\underline{cl}}{}^{\underline{kb}} = -i\,(\sigma^p)_{\underline{l}}{}^{\underline{k}}\,\delta^{\underline{b}}_{\underline{c}}. \tag{13.75}$$

Lifting (or lowering) the indices in (13.75), we reproduce the flat quaternionic complex structures (3.19). The complex structures $(I^p)_N{}^M = e_{NA}(I^p)_{AB}\, e^M_B$, or those in the symplectic notation,

$$(I^p)_{cl}{}^{kb} = -ie^{\underline{lc}}_{cl}\,(\sigma^p)_{\underline{l}}{}^{\underline{k}}\, e^{kb}_{\underline{ck}}, \tag{13.76}$$

[15]The reality of $E_0$ and $E_p$ follows from the convention (13.47), which also holds for $\epsilon_{\underline{a}}$.

which enter (13.74), are also quaternionic. Hence, according to Theorem 13.4, our manifold is HKT. □

**Remark:** The fact that our action can be expressed in the language of $\mathcal{N} = 1$ superfields and its invariance under the transformations (13.53) implies the existence of the conserved real supercharge $Q_0$ associated with the transformations within the $\mathcal{N} = 1$ multiplets and three extra conserved real supercharges $Q_p$. Each of the latter depends on the corresponding complex structure as in (9.60). We arrive at the expressions (10.82) for the supercharges, which are thus valid not only for Obata-flat models discussed in Chapter 10 but also in the general case. These expressions are especially nice and simple in the hyperkähler case when the torsion $C_{MKP}$ vanishes.

## 13.3 Metric □ Obata families □ HKT → HK.

The theorems proved in the previous sections would be useless if one could not use for practical purposes the harmonic technique presented there: could not derive explicit expressions for the metric of the HKT and HK manifolds. In this section, we describe a general method on how this can be achieved and we show in the following section how this method works in particular cases.

Consider the expression (13.52) for the bosonic action of a generic HKT model. To find the metric, we should do the harmonic integral and express it through the harmonic-independent coordinates $x^{ja}$. Bearing in mind (13.68), we derive

$$\dot{f}^{+a} = (\partial_{bk} f^{+a})\dot{x}^{kb} = (M^{-1})^a{}_{\underline{a}}\, e^{j\underline{a}}_{bk}\, \dot{x}^{kb}\, u^+_{\underline{j}}\,. \tag{13.77}$$

The solution to the truncated constraint (13.42) reads

$$\tilde{A}^{-a} = -2i(M^{-1})^a{}_{\underline{a}}\, e^{j\underline{a}}_{bk}\, \dot{x}^{kb}\, u^-_{\underline{j}}\,, \tag{13.78}$$

as one can be convinced using $\mathcal{D}^{++}M^{-1} = 0$ and $\partial^{++}u^- = u^+$.

The calculation is simple when $\mathcal{L}$ is quadratic,

$$\mathcal{L} = -\frac{1}{8}q^{+a}q^-_a = -\frac{1}{8}\Omega_{ab}\, q^{+a}q^{-b}. \tag{13.79}$$

Then, the term $\propto \partial_{-a}\partial_{-b}\mathcal{L}$ in (13.52) is absent, $\partial_{+a}\partial_{-b}\mathcal{L} = -\Omega_{ab}/8$, and we obtain

$$S_0^{\text{bos}} = \frac{i}{2}\Omega_{ab}\int dudt\, \dot{f}^{+a}\tilde{A}^{-b}. \tag{13.80}$$

Substituting there (13.77) and (13.78), we derive

$$L_0^{\rm bos} = \dot{x}^{jc}\dot{x}^{kd}\, e_{cj}^{\underline{ja}}\, e_{dk}^{\underline{kb}}\, \Omega_{ab} \int du\, u_{\underline{j}}^+ u_{\underline{k}}^- (M^{-1})^a{}_{\underline{a}} (M^{-1})^b{}_{\underline{b}}. \quad (13.81)$$

One can observe by symmetry considerations that the expression multiplying the structure $u_{\underline{j}}^+ u_{\underline{k}}^-$ is antisymmetric under $\underline{j} \leftrightarrow \underline{k}$. Then, we can replace using (7.76) $u_{\underline{j}}^+ u_{\underline{k}}^- \to \varepsilon_{\underline{jk}}/2$ to derive

$$L_0^{\rm bos} = \frac{1}{2} g_{cj,dk}\, \dot{x}^{jc}\dot{x}^{kd}, \quad (13.82)$$

where

$$g_{cj,bk} = e_{cj}^{\underline{ja}}\, e_{dk}^{\underline{kb}}\, \varepsilon_{\underline{jk}}\, \Omega_{ab} \int du\, (M^{-1})^a{}_{\underline{a}} (M^{-1})^b{}_{\underline{b}} \stackrel{\rm def}{=} e_{cj}^{\underline{ja}}\, e_{dk}^{\underline{kb}}\, \varepsilon_{\underline{jk}}\, G_{\underline{ab}}. \quad (13.83)$$

When $\mathcal{L}^{+3a}$ are absent, $(M^{-1})^a{}_{\underline{a}} = \delta^a_{\underline{a}}$, the vielbeins can be chosen as $e_{cj}^{\underline{ja}} = \delta_j^{\underline{j}}\, \delta_{\underline{c}}^{\underline{a}}$, and we obtain the flat metric (3.14).

For an arbitrary $\mathcal{L}$, one can obtain after some work [92] the result for the metric that has the same structure as in (13.83), but the harmonic integral for $G_{\underline{ab}}$ involves now an extra factor in the integrand:

$$\boxed{G_{\underline{ab}}^{\text{any}\ \mathcal{L}} = \int du\, \mathcal{F}_{ab}\, (M^{-1})^a{}_{\underline{a}}\, (M^{-1})^b{}_{\underline{b}},} \quad (13.84)$$

where

$$\mathcal{F}_{ab} = -8\left[\partial_{+[a}\partial_{-b]} - (M^{-1})^c{}_{\underline{c}}\, (\partial^{--} M^{\underline{c}}{}_{[a})\, \partial_{-b]}\partial_{-c}\right] \mathcal{L}. \quad (13.85)$$

We see that, generically, the metric is not just a convolution of the matrices $e_{ja}^{\underline{ja}}$ but involves an extra factor — the antisymmetric matrix $G_{\underline{ab}}$ carrying the symplectic indices. However, one can bring $G_{\underline{ab}}$ to the form $\Omega_{\underline{ab}}$ using the gauge freedom (13.25) with a harmonic-independent matrix $R$. Indeed, let $G_{\underline{ab}}^{(0)} = \Omega_{\underline{ab}}$ and $R^{\underline{a}}{}_{\underline{b}} = \delta_{\underline{b}}^{\underline{a}} + \lambda^{\underline{a}}{}_{\underline{b}}$ with $\lambda \ll 1$. We can then write

$$G_{\underline{ab}} = R^{\underline{c}}{}_{\underline{a}} R^{\underline{d}}{}_{\underline{b}}\, \Omega_{\underline{cd}} = \Omega_{\underline{ab}} + 2\lambda_{\underline{[ab]}} + o(\lambda). \quad (13.86)$$

It is clear that a finite version of this transformation can produce an arbitrary antisymmetric matrix $G_{\underline{ab}}$.

Note that the metric (13.83) is invariant under the gauge rotations with the matrix $R^{\underline{a}}{}_{\underline{b}}$ defined in (13.25). It is immediately seen from the definition

(13.29): the objects $e^{ja}_{\underline{aj}}$ are rotated by $R$ in the same way as $M^{\underline{a}}{}_a$, whereas the factors $M^{-1}$ in Eq. (13.84) are rotated by $R^{-1}$. If the gauge $G_{\underline{ab}} = \Omega_{\underline{ab}}$ is chosen, $e^{ja}_{\underline{aj}}$ acquire the meaning of the true vielbeins and the metric is given by the standard expression

$$g_{aj,bk} = e^{ja}_{\underline{aj}} e^{kb}_{\underline{bk}} \varepsilon_{\underline{jk}} \Omega_{\underline{ab}}. \tag{13.87}$$

It is simply the familiar identity $g_{MN} = e_{AM} e_{AN}$ expressed in symplectic notation. In the generic gauge where the expression (13.83) for the metric involves an additional structure $G_{\underline{ab}}$, one may call $e^{ja}_{\underline{aj}}$ *quasivielbeins.*

We have shown how, given two harmonic functions $\mathcal{L}^{+3a}$ and $\mathcal{L}$, one can derive the metric. Generically, this metric is HKT. We have proven this indirectly, and in Ref. [92], the reader can find the *direct* calculations of the complex structures, the vielbeins and the connections. With these expressions in hand, one can check explicitly that the Bismut connections for the three complex structures coincide and the metric is HKT. These explicit calculations are, however, not at all simple.

Again, one can ask at this point the following: is it true that *any* HKT metric can be derived this way? We can *conjecture* that yes, it can, but, in constrast to the hyperkähler case, an exact proof of this assertion has not been given yet.

Another remark is that our description has a high level of redundancy. One can perform a variable change $q^{+a} \to q_1^{+a}(q^{+a}, u)$. Infinitesimally,

$$q^{+a} \to q^{+a} + \Lambda^{+a}(q^{+a}, u), \quad \Lambda \ll q. \tag{13.88}$$

This transformation induces the transformation of the prepotentials $\mathcal{L}^{+3a}$ which does not affect geometry.

Let us prove this.

**Theorem 13.6.** *For a given* $\mathcal{L}$*, the metric* (13.87) *is invariant under the transformations* (13.88) *accompanied by a proper deformation of* $\mathcal{L}^{+3a}$.

**Proof.** The transformation (13.88) induces the transformation of the bosonic components of $q^{+a}$. Infinitesimally,

$$\delta_\lambda f^{+a} = \lambda^{+a}(f, u) \ll f^{+a}. \tag{13.89}$$

We have

$$\mathcal{L}^{+3a}(f) = \partial^{++} f^{+a}. \tag{13.90}$$

After the transformation, $\mathcal{L}^{+3a}$ is shifted[16]:

$$\delta_\lambda \mathcal{L}^{+3a} = \partial^{++}\lambda^{+a} = \hat{\partial}^{++}\lambda^{+a} + (\partial_{+b}\lambda^{+a})\,\mathcal{L}^{+3b}, \qquad (13.91)$$

where $\partial_{+a} = \partial/\partial f^{+a}$ and $\hat{\partial}^{++}$ is the part of the full harmonic derivative $\partial^{++}$ due to the explicit harmonic dependence:

$$\partial^{++}\Phi(f^{+a}, u) = \hat{\partial}^{++}\Phi + (\partial_{+b}\Phi)\,\mathcal{L}^{+3b}. \qquad (13.92)$$

Consider now equation (13.66) for the bridge $(M^{-1})^a{}_{\underline{b}}$ and determine how its solution transforms under (13.89). To this end, one has first to find out the law of transformation for the structure

$$E^{+2a}{}_b = \frac{\partial \mathcal{L}^{+3a}}{\partial f^{+b}}. \qquad (13.93)$$

Using (13.89), (13.91) and (13.92), we derive[17]

$$\delta_\lambda E^{+2a}{}_b = -E^{+2a}{}_c(\partial_{+b}\lambda^{+c}) + E^{+2c}{}_b(\partial_{+c}\lambda^{+a}) + \partial^{++}(\partial_{+b}\lambda^{+a}). \qquad (13.94)$$

For $\mathcal{D}^{++}M^{-1}$ to stay zero after the transformation, the bridge should transform, bearing in mind (13.94), as

$$\delta_\lambda (M^{-1})^a{}_{\underline{b}} = (\partial_{+b}\lambda^{+a})(M^{-1})^b{}_{\underline{b}}. \qquad (13.95)$$

The inverse matrix transforms as

$$\delta_\lambda M^{\underline{b}}{}_a = -(\partial_{+a}\lambda^{+b})M^{\underline{b}}{}_b. \qquad (13.96)$$

In other words, in contrast to $E^{+2a}{}_b$, the bridges $M^{-1}$ and $M$ transform as contravariant and covariant symplectic vectors under the shifts of analytic harmonic coordinates,[18]

The same concerns the structure $\partial_{bk} f^{+a}$ — it transforms as

$$\delta_\lambda(\partial_{bk} f^{+a}) = \partial_{bk}\lambda^{+a} = (\partial_{bk} f^{+c})(\partial_{+c}\lambda^{+a}) \qquad (13.97)$$

and represents a contravariant vector. Note that the index "b" in the spatial derivative $\partial_{bk}$ does not "feel" the analytic shift (13.89)!

---

[16]To write (13.91), we have used the identity (13.90). Note that the deformation (13.91) does not coincide with what one would obtain by simply shifting the argument of $\mathcal{L}^{+3a}(q)$ according to (13.89).

[17]The last term in (13.94) comes from the derivative of the first term $\hat{\partial}^{++}\lambda^{+a}$ in the variation (13.91) *and* from the derivative of the factor $\partial_{+b}\lambda^{+a}$ in the second term there. These contributions add up to produce the *total* harmonic derivative $\partial^{++}$.

[18]Speaking of $E^{+2a}{}_b$, it plays the role of the *connection* associated with the analytic diffeomorphism (13.89). Its law of transformation includes, besides the first two tensorial terms, the extra harmonic gradient term, in the full analogy with the transformation law $\delta A_M = \partial_M X$ for the (Abelian) gauge potential or with the second term in the transformation law (1.11) for the ordinary affine connection.

It follows that the LHS of Eq. (13.68) is a scalar under (13.89) and the same concerns the quasivielbeins in its RHS. The metric (13.83) is given by the convolution of the quasivielbeins multiplied by the factor $G_{\underline{ab}}$ that depends only on $\mathcal{L}$. It also stays invariant. □

Consider now a set of models characterized by the same potential $\mathcal{L}^{+3a}$ and hence the same harmonic constraints (13.39) but different actions $S_0$. Different $\mathcal{L}$'s result in different $\mathcal{F}_{ab}$'s in (13.85), which affect $G_{\underline{ab}}$. As we have seen, this modification may be compensated by the appropriate gauge rotations of the bridges and, correspondingly, of the vielbeins. We thus obtain a family of models whose vielbeins are interrelated by the transformations

$$\tilde{e}^{ja}_{\underline{aj}} = R^{\ \underline{a}}{}_{\underline{b}}\, e^{jb}_{\underline{aj}} \tag{13.98}$$

where all the dependence on $\mathcal{L}$ is encoded in the matrix $R$. To keep $e_{AM}$ real, the matrix $R^{\underline{a}}{}_{\underline{b}}$ should satisfy pseudoreality conditions, like in (3.16). Then, it involves $(2n) \times (2n)/2 = 2n^2$ complex or $4n^2$ real parameters. $2n^2 + n$ of them correspond to the action of $Sp(n)$ group, which does not affect the metric (13.87). So, we are left with $2n^2 - n$ essential parameters. For $n = 1$, only one such parameter is left, which corresponds to multiplying the metric by a conformal factor.

An important observation is that the complex structures in all such models coincide. Indeed, looking at the expression (13.76), we observe that it is invariant under $R$ transformations because the upper index $\underline{c}$ in (13.76) is rotated by the matrix $R$, while the contracted lower index $\underline{c}$ is rotated by the inverse matrix $R^{-1}$. So, we have

$$(\tilde{I}^p)_{cl}{}^{kb} = (I^p)_{cl}{}^{kb}. \tag{13.99}$$

In particular, in all the models expressed in terms of linear (**4**, **4**, **0**) multiplets with vanishing $\mathcal{L}^{+3a}$, the complex structures keep their flat form (3.17).

An important corollary of this observation is as follows:

**Theorem 13.7.** *The Obata curvature invariants for a family of the HKT metrics, characterized by a particular nonlinear constraint* (13.39) *but having different actions $S_0$ in* (13.49)*, coincide.*[19]

**Proof.** This statement follows from Theorem 3.8, which says that there exists a frame where the components of the Obata connection are expressed

[19] We will call such a family an *Obata family*.

via the complex structures, and the fact that the latter do not depend on $\mathcal{L}$. Then, this connection also cannot depend on $\mathcal{L}$ in this frame and the same is true in any frame for the Obata curvature invariants. □

It follows, in particular, that all the models based on linear $(\mathbf{4}, \mathbf{4}, \mathbf{0})$ multiplets are Obata-flat.

Up to now, we discussed in this and in the preceding section a generic HKT model characterized by arbitrary $\mathcal{L}^{+3a}$ and $\mathcal{L}$. But there is an important particular case.

**Theorem 13.8.** *Consider a limited class of models where the function* $\mathcal{L}^{+3a}$ *in* (13.39) *has the form*

$$\boxed{\mathcal{L}^{+3a} = \Omega^{ab} \frac{\partial \mathcal{L}^{+4}}{\partial q^{+b}},} \tag{13.100}$$

*as in* (13.43), *and the function* $\mathcal{L}$ *in* (13.49) *is quadratic,* $\mathcal{L} = A q^{+a} q_a^-$. *In this case, the metric is hyperkähler.*

**Proof.** The point is that the bosonic metric following from the quadratic $S_0$ in (13.49) with the constraints (13.39) involving (13.100) *exactly coincides* with the metric derived from the equation of motion (13.7) for the model (13.5) involving the "large" multiplet (13.3). And, as we know from Theorem 13.1, the latter model enjoys $\mathcal{N} = 8$ supersymmetry and gives rise to a hyperkähler metric.

To prove that the metrics of the two models are, indeed, the same, compare the bosonic action of the two models. We see that in both cases, it is given by the expression (13.17), which is equivalent to (13.80) (one only has to replace $F^{+a}$ by $f^{+a}$). If the condition (13.100) is fulfilled, $F^{+a}$ and $f^{+a}$ satisfy *the same* equations (13.8) $\equiv$ (13.40). Hence, they are expressed through the coordinates $x^{ja}$ in the same way. The same is true for $\tilde{A}^{-a}$: if (13.100) holds, equations (13.9) and (13.42) coincide. Thus, the bosonic actions and hence the metrics in the two models coincide. □

Note that Theorem 13.8 only gives *sufficient* conditions for the metric to be hyperkähler but not the necessary ones. As was discussed above, our description has a high level of redundancy. We can take the hyperkähler model with quadratic $\mathcal{L}$ and $\mathcal{L}^{+3a}$ that satisfies (13.100) and deform $\mathcal{L}$ in such a way that the matrix $R^{\underline{a}}{}_{\underline{b}}$, to a rotation by which this deformation boils down, belongs to $Sp(n)$. In this case, the metric stays hyperkähler, even though $\mathcal{L}$ is no longer quadratic.

On the other hand, for the quadratic $\mathcal{L}$, the condition (13.100) *is* necessary for the metric to be HK. Heuristically, this is more or less clear: indeed, let $\mathcal{L} = Aq^{+a}q^-_a$ and suppose that one cannot represent $\mathcal{L}^{+3a}$ as in (13.100), but the metric is still HK. We know, however, from the results of Ref. [141] that *any* HK metric is derived from some prepotential $\mathcal{L}^{+4}$ which enters the $\mathcal{N} = 8$ Lagrangian (7.109). As we have just seen, the same metric can be derived from the $\mathcal{N} = 4$ model (13.49) with the quadratic $\mathcal{L}$ and the constraint involving

$$\tilde{\mathcal{L}}^{+3a} = \Omega^{ab}\frac{\partial \mathcal{L}^{+4}}{\partial q^{+b}}. \tag{13.101}$$

But it is hardly possible that, given the quadratic $\mathcal{L}$ and hence the expression (13.80) for the bosonic action, one and the same metric follows from two different prepotentials $\tilde{\mathcal{L}}^{+3a} \neq \mathcal{L}^{+3a}$. It is clear from the derivation above that the bridge $M$ and hence the expressions of $f^{+a}$ and $A^{-a}$ [to be substituted into (13.80)] via the coordinates $x^{ja}$ depend on the prepotential in an essential way.

This heuristic argument is confirmed by the explicit calculations in Ref. [92]. Note first of all that the "analytic connection" $E^{+2a}{}_b$ that enters Eq. (13.66) belongs to the algebra $sp(n)$ iff the representation (13.100) holds.[20] Then and only then Eq. (13.66) admits the solutions for the bridge that belong to the group $Sp(n)$. The analysis of the explicit formulas of Ref. [92] exhibits that, given a quadratic $\mathcal{L}$, in this and only in this case the Levi-Civita spin connection $\omega_{AB,M}\, dx^M$ also belongs to $sp(n)$, which is a necessary and sufficient condition for the metric to be hyperkähler, in virtue of Theorem 3.2.

We keep now $\mathcal{L}$ quadratic and allow for an arbitrary $\mathcal{L}^{+3a}$. A noteworthy fact is as follows:

**Theorem 13.9.** *The torsion form for an HKT model with* $\mathcal{L} = Aq^{+a}q^-_a$ *is closed,* $dC = 0$.

In other words, we are dealing in this case with the so-called *strong* HKT geometry [34, 35].

[20]Indeed, a $2n$-dimensional matrix $E^{+2a}{}_b$ belongs to $sp(n)$ iff the identity $\Omega E^{+2} + (E^{+2})^T\Omega$ is fullfilled, i.e. iff the matrix $\Omega_{ac}E^{+2c}{}_b$ is symmetric. And this is true iff (13.100) holds and

$$E^{+2a}{}_b = \Omega^{ac}\frac{\partial^2 \mathcal{L}^{+4}}{\partial q^{+c}\partial q^{+b}}.$$

**Proof.** A generic $\mathcal{N} = 4$ component Lagrangian has the form (9.3) and includes a 4-fermion term. The latter vanishes iff the torsion form is closed. Thus, it is sufficient to prove that the full component action of the HKT model with quadratic $\mathcal{L}$ does not involve such term.

Let us substitute in $\mathcal{L} = -q^{+a}q_a^-/8$ the expansions (13.38) and (13.51), where now the fermion terms are also taken into account. The explicit form of the latter is

$$\begin{aligned}(q^{+a})_{\rm ferm} &= \theta^+\chi^a - \bar\theta^+\bar\chi^a,\\ (q_a^-)_{\rm ferm} &= \theta^-\chi_a - \bar\theta^-\bar\chi_a\\ &\quad + \theta^+\partial^{--}\chi_a - \bar\theta^+\partial^{--}\bar\chi_a + 2i\theta^-\bar\theta^-(\theta^+\dot\chi_a - \bar\theta^+\dot{\bar\chi}_a), \end{aligned}\tag{13.102}$$

but we do not need it so much; the only thing that we need is the fact that both $q^{+a}$ and $q^{-a}$ involve only the linear in fermion fields terms. Then, the quadratic $\mathcal{L}$ may generate only bi-fermion terms [note that they are also present in the term $\propto \dot f^{+a}A_a^-$ in $\int d^2\theta^+ d^2\theta^-(q^{+a}q_a^-)$ that includes the field $A_a^-(t,u)$ satisfying the untruncated constraint (13.42)], while 4-fermion terms are absent. □

When $\mathcal{L}$ is not quadratic, the torsion form is not closed in most cases. But in some cases, it can happen to be closed. For example, in the linear case $\mathcal{L}^{+3a} = 0$, the metric coincides for $n = 1$ with the four-dimensional flat metric multiplied by a certain conformal factor $\lambda(x)$ [see Eq. (10.73)]. In fact, we discussed this example in Chapter 3 on p. 71. According to (3.87), the torsion form is

$$C = \varepsilon_{\mu\nu\rho\sigma}\,\partial_\sigma\lambda(x)\,dx^\mu\wedge dx^\nu\wedge dx^\rho. \tag{13.103}$$

If $\partial_\sigma\partial_\sigma\lambda(x) = 0$, this form is closed.

The results outlined above are illustrated in Table 13.1.

**Table 13.1.** HK and HKT geometries.

| | $\mathcal{L} = Aq^{+a}q_a^-$ | $\mathcal{L} \neq Aq^{+a}q_a^-$ |
|---|---|---|
| $\mathcal{L}^{+3a} = \Omega^{ab}\dfrac{\partial\mathcal{L}^{+4}}{\partial q^{+b}}$ | HK<br>$C = 0$ | weak HKT<br>$dC \neq 0$ (generically) |
| $\mathcal{L}^{+3a} \neq \Omega^{ab}\dfrac{\partial\mathcal{L}^{+4}}{\partial q^{+b}}$ | strong HKT<br>$dC = 0$ | |

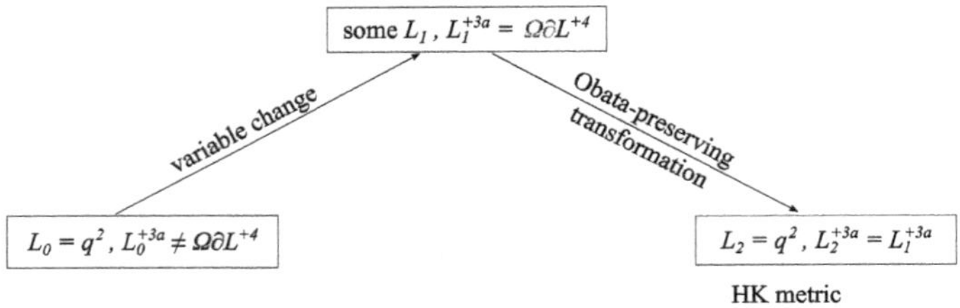


**Fig. 13.1.** A reducible HKT metric playing hide and seek with analytic frames.

The whole set of HKT metrics can be divided into two classes: *reducible* metrics, i.e. the metrics belonging to an Obata family (see the footnote on p. 310) including a hyperkähler member, and *irreducible* metrics that belong to families deprived of such a member and that cannot be transformed to hyperkähler form by a rotation of symplectic vielbeins as in (13.98). A nontrivial statement is the following:

**Theorem 13.10.** *Irreducible metrics exist.*

This statement is nontrivial due to the redundancy of description mentioned above. The metric is determined by a pair $\{\mathcal{L}^{+3a}, \mathcal{L}\}$, but many (an infinity) of such pairs are equivalent under the analytic variable change (13.88) accompanied by (13.91). As a result, one and the same metric can belong to an infinity of apparently different but in fact equivalent Obata families. Even though the "stem member" of a given Obata family with quadratic $\mathcal{L}$ may have a nontrivial $\mathcal{L}^{+3a}$ not given by (13.100) and the corresponding manifold is not hyperkähler, it is not obvious that the constraints $\mathcal{L}^{+3a}$ cannot be brought to a form which satisfy (13.100) by a transformation (13.88), after which $\mathcal{L}$ would not be quadratic any more. Choosing then quadratic $\mathcal{L}$, while not modifying $\mathcal{L}^{+3a}$, we would arrive to a hyperkähler metric (see Fig. 13.1). There are many such metrics which seem to be irreducible but are in fact reducible in disguise.

To prove the theorem, one has to give at least one example of a metric that is definitely irreducible. We do it in the following section.

## 13.4 Examples

### 13.4.1 *Taub–NUT metric*

To show how the procedure described in the preceding section really works, we present here a detailed harmonic derivation of this metric.

Consider the four-dimensional $n = 1$ model involving two analytic harmonic fields $q^{+(a=1,2)}$ with the hyperkähler prepotential $\mathcal{L}^{+4}$ chosen as

$$\boxed{\mathcal{L}^{+4} = -\frac{\lambda}{2}\,(q^{+1})^2(q^{+2})^2,} \tag{13.104}$$

from which $\mathcal{L}^{+3a} = \varepsilon^{ab}\partial_{+b}\mathcal{L}^{+4}$ is derived. In this case, the constraints (13.40)–(13.42) can be solved analytically. We start with Eq. (13.40) for $f^{+a}$. It reads[21]

$$\partial^{++}f^{+1} = \lambda(f^{+1})^2 f^{+2}, \quad \partial^{++}f^{+2} = -\lambda f^{+1}(f^{+2})^2. \tag{13.105}$$

The solution to this equation has the form

$$f^{+1} = \exp\{\lambda J\}x^{+1}, \quad f^{+2} = \exp\{-\lambda J\}x^{+2}, \tag{13.106}$$

where $x^{\pm a} = x^{ia}u_i^{\pm}$ with harmonic-independent $x^{ia}$ and

$$J = \frac{1}{2}\,(x^{+1}x^{-2} + x^{-1}x^{+2}). \tag{13.107}$$

When the nonlinearity is switched off, $f^{+a} \to x^{+a}$.

To find the metric, we need to know the bridge. The bridge $M$ satisfies the homogeneous equation (13.24). As was mentioned before, the analytic connection

$$E^{+2a}{}_b = \Omega^{ac}\frac{\partial^2\mathcal{L}^{+4}}{\partial f^{+b}\partial f^{+c}} \tag{13.108}$$

that enters this equation belongs to the algebra $sp(n)$ and hence there are solutions for the bridge that belong to the group $Sp(n)$. In our case, $Sp(1) \equiv SU(2)$.

Even after we impose the condition for the bridge $M$ to belong to $SU(2)$, there is a freedom associated with multiplication by constant $SU(2)$ matrices as in (13.25). We choose a solution that reduces to the unit matrix in the limit when the nonlinearity associated with $\mathcal{L}^{+4}$ is switched off. It reads

$$M^{\underline{b}}{}_a = \frac{1}{\sqrt{1+\lambda x^1\cdot x^2}}\begin{pmatrix} e^{-\lambda J}(1-\lambda x^{-1}x^{+2}) & -\lambda e^{\lambda J}x^{-1}x^{+1} \\ \lambda e^{-\lambda J}x^{-2}x^{+2} & e^{\lambda J}(1+\lambda x^{-2}x^{+1}) \end{pmatrix}, \tag{13.109}$$

[21] We remind that $\varepsilon^{12} = -\varepsilon_{12} = -1$.

where $x^1 \cdot x^2 = -x^2 \cdot x^1 = x^{j1} x^2_j$. The index $\underline{b}$ marks the lines and the index $a$ marks the columns. In the most following formulas, we set $\lambda = 1$.

The vielbeins can be found from the solutions (13.106) and (13.109) and from the definition (13.68). They are[22]

$$\sqrt{1 + x^1 \cdot x^2}\, e^{\underline{k1}}_{1i} = -\frac{1}{2} x^1_i x^{k2} + \delta^k_i \left(1 + \frac{1}{2} x^1 \cdot x^2\right),$$

$$\sqrt{1 + x^1 \cdot x^2}\, e^{\underline{k1}}_{2i} = -\frac{1}{2} x^1_i x^{k1}, \quad \sqrt{1 + x^1 \cdot x^2}\, e^{\underline{k2}}_{1i} = \frac{1}{2} x^2_i x^{k2},$$

$$\sqrt{1 + x^1 \cdot x^2}\, e^{\underline{k2}}_{2i} = \frac{1}{2} x^2_i x^{k1} + \delta^k_i \left(1 + \frac{1}{2} x^1 \cdot x^2\right), \tag{13.111}$$

where the world spinor index $k$ in the coordinate components on the right-hand sides has the same value as the tangent space index $\underline{k}$ on the left-hand sides.

A generic expression for the HKT metric given in Eq. (13.83) involves an extra factor $G_{\underline{ab}}$. Thus, generically, the objects $e^{\underline{kb}}_{ai}$ found through the definition (13.68) are not quite vielbeins; they are quasivielbeins, and one should rotate them by the matrix $R$ as in (13.98) to obtain the true vielbeins. But in the hyperkähler case when $M, M^{-1} \in Sp(n)$ and $\mathcal{L} = -q^{+a} q^-_a/8$, we derive $\mathcal{F}_{ab} = \Omega_{ab}$ and $G_{\underline{ab}} = \Omega_{\underline{ab}}$. Then, $e^{\underline{kb}}_{ai}$ are the true vielbeins from which the metric is constructed according to (13.87).

In our case, the components of the metric are

$$g_{1i,1j} = x^2_i x^2_j \frac{1 + \frac{1}{2} x^1 \cdot x^2}{1 + x^1 \cdot x^2}, \quad g_{2i,2j} = x^1_i x^1_j \frac{1 + \frac{1}{2} x^1 \cdot x^2}{1 + x^1 \cdot x^2},$$

$$g_{1i,2j} = g_{2j,1i} = \varepsilon_{ij}(1 + x^1 \cdot x^2) + x^2_i x^1_j \frac{1 + \frac{1}{2} x^1 \cdot x^2}{1 + x^1 \cdot x^2}. \tag{13.112}$$

The metric (13.112) is equivalent to the Taub–NUT metric written in a more familiar form (3.39), which we used in Chapters 11 and 12 and are rewriting here:

$$ds^2 = V^{-1}(r)\, (d\Psi + \boldsymbol{A} d\boldsymbol{X})^2 + V(r)\, d\boldsymbol{X} d\boldsymbol{X}, \tag{13.113}$$

---

[22] The higher terms in the harmonic expansion in the LHS of Eq. (13.68) cancel, as they should. The reader may be amused to check it explicitly using the identity

$$u^+_m u^+_n u^-_k = u^+_{(m} u^+_n u^-_{k)} + \frac{1}{3}(\varepsilon_{mk} u^+_n + \varepsilon_{nk} u^+_m). \tag{13.110}$$

where $\Psi \in (0, 4\pi)$,

$$V(r) = \frac{1}{r} + \lambda \tag{13.114}$$

$(r = |\mathbf{X}|)$ and

$$A^1 = \frac{X^2}{r\,(r + X^3)}, \quad A^2 = -\frac{X^1}{r\,(r + X^3)}, \quad A^3 = 0. \tag{13.115}$$

The correspondence is established by setting in (13.114) $\lambda = 1$ and changing the variables as[23]

$$\begin{aligned} X^1 &= x^{21}x^{22} - x^{11}x^{12}, \quad X^2 = i(x^{11}x^{12} + x^{21}x^{22}), \\ X^3 &= x^{11}x^{22} + x^{21}x^{12}, \quad \Psi = i\ \ln\frac{x^{22}}{x^{11}}. \end{aligned} \tag{13.116}$$

The relations

$$\begin{aligned} r = x^1 \cdot x^2, \quad d\Psi + \mathbf{A}d\mathbf{X} &= \frac{i}{x^1 \cdot x^2}\left(x^2 \cdot dx^1 + x^1 \cdot dx^2\right), \\ (d\mathbf{X})^2 = (x^1 \cdot dx^2)^2 &+ (x^2 \cdot dx^1)^2 + 2(x^1 \cdot dx^1)(x^2 \cdot dx^2) \\ &+ 2(x^1 \cdot x^2)(dx^1 \cdot dx^2) \end{aligned} \tag{13.117}$$

hold.[24] Inserting (13.114), (13.117) and (13.116) in (13.113), we obtain

$$ds^2 = \frac{2 + x^1 \cdot x^2}{1 + x^1 \cdot x^2}\,(x^2 \cdot dx^1 + x^1 \cdot dx^2)^2 + 4(1 + x^1 \cdot x^2)dx^1 \cdot dx^2, \tag{13.118}$$

which coincides up to the extra irrelevant factor 2 with what the metric (13.112) gives.

Another example where the constraints can be explicitly solved and the metric explicitly found is the model described by

$$\boxed{\mathcal{L}^{+4} = \frac{(\xi^{jk}u_j^+u_k^+)^2}{(q^{+a}u_a^-)^2}} \tag{13.119}$$

---

[23] A more compact form of the first line in (13.116) is $X^p = -(\sigma^p)_{ij}\, x^{i1}x^{j2}$ with $(\sigma^p)_{ij}$ defined in (3.18). It is the same variable change as in Eq. (11.55), which can be seen if trading $x^{ja}$ for a real 4-vector $x^M$ according to (3.13).

[24] Note that the variable $r$ as defined in (13.117) is real and nonnegative in virtue of pseudoreality of $x^{ia}$.

with an arbitrary symmetric $\xi^{jk}$. After performing the program outlined above and choosing the coordinates in an appropriate way, one arrives [90, 139] at the hyperkähler Eguchi–Hanson metric (3.61).

### 13.4.2 *Delduc–Valent metric*

Consider now a four-dimensional model ($n = 1$) with $\mathcal{L} = -q^{+a}q_a^-/8$ and the constraints [42]

$$\boxed{\begin{aligned} D^{++}q^{+1} &\equiv \mathcal{L}^{(+3)1} = (q^{+1})^2 q^{+2}\,(\lambda + i\rho) \\ D^{++}q^{+2} &\equiv \mathcal{L}^{(+3)2} = -q^{+1}(q^{+2})^2\,(\lambda - i\rho) \end{aligned}} \tag{13.120}$$

with real $\lambda, \rho$. When $\rho = 0$, we go back to the Taub–NUT system, but if $\rho \neq 0$, $\partial_a \mathcal{L}^{+3a} \neq 0$, the constraints (13.120) are not expressed as in (13.100) and we are dealing with a nontrivial HKT manifold. Using a general technique described above, one can derive its metric[25]:

$$4d\tau^2 = V^{-1}(s)(d\Psi + \omega)^2 + V(s)\Gamma, \tag{13.121}$$

where

$$\begin{aligned} V(s) &= \frac{1}{s} + \lambda, \quad \omega = \cos\theta\, d\phi - \frac{\rho(1+\lambda s)}{1+\rho^2 s^2} ds, \\ \Gamma &= \frac{ds^2}{(1+\rho^2 s^2)^2} + \frac{s^2}{1+\rho^2 s^2}(d\theta^2 + \sin^2\theta\, d\phi^2). \end{aligned} \tag{13.122}$$

In the limit $\rho \to 0$, this metric goes over to the Taub–NUT metric (13.113). For $\rho \neq 0$, the Bismut connection of this model involves a nontrivial torsion. The explicit expression for the torsion form is

$$C = -\frac{\rho s(1+\lambda s)}{2(1+\rho^2 s^2)^2}\, ds \wedge \sin\theta d\theta \wedge d\phi. \tag{13.123}$$

In accord with Theorem 13.9, this form is closed.

Let us show now that this metric is *irreducible* and prove thereby Theorem 13.10.

**Proof.** If the HKT metric (13.121) were reducible, a certain conformal transformation

$$d\tilde{\tau}^2 = G(s, \Psi, \theta, \phi) d\tau^2 \tag{13.124}$$

[25] We quote Eq. (24) of Ref. [42] where we set $\gamma_0 = 1$.

making the metric $d\tilde{\tau}^2$ hyperkähler would exist [see the remark after Eq. (13.98)]. Equivalently, the vielbeins would be conformal to the hyperkähler ones, with a real conformal factor $\sqrt{G}$. Then, the ordinary Levi-Civita covariant derivatives of the complex structures [the latter do not change under such conformal transformations — see Eq. (13.99)] with the Christoffel symbols following from (13.124) must vanish:

$$\nabla_M I_N^P = \nabla_M J_N^P = \nabla_M K_N^P = 0. \tag{13.125}$$

This condition can be solved to determine the conformal factor $G$ [142], but it turns out[26] that in the particular case of the Delduc–Valent metric (13.121), this procedure gives a *complex* function $G(s, \Psi, \theta, \phi)$ and a complex "hyperkähler metric". Real solutions to the constraints (13.125) do not exist. □

**Remark:** In Chapter 3, we discussed group manifolds and showed that many of them are HKT. The simplest such manifold (Hopf manifold) describes the group $SU(2) \times U(1)$. Its metric (3.88) is reducible. For more complicated groups from the list (3.114), the metrics are probably irreducible, but this interesting question has not been studied yet.

### 13.4.3 *Eight dimensions*

The examples considered so far were four-dimensional. We also know an example when harmonic formalism spelled out above allowed one to construct the metric of a eight-dimensional hyperkähler manifold [140]. Take four analytic fields $q^{+a}$ and choose a prepotential

$$\boxed{L^{+4} = -q^{+1}q^{+2}q^{+3}q^{+4}.} \tag{13.126}$$

To calculate the vielbeins and then the metric, we should first solve the equations (13.40) for $f^{+a}$ (the lowest components of $q^{+a}$) and then solve the equations (13.24) for the bridge. The expressions for the vielbeins (*achtbeins*) will then follow from the definition (13.68).

The equations for $f^{+a}$ have in this case a rather simple form:

$$\partial^{++}f^{+1} = f^{+1}f^{+3}f^{+4}, \quad \partial^{++}f^{+2} = -f^{+2}f^{+3}f^{+4},$$
$$\partial^{++}f^{+3} = f^{+3}f^{+1}f^{+2}, \quad \partial^{++}f^{+4} = -f^{+4}f^{+1}f^{+2}. \tag{13.127}$$

[26] We address the reader to Refs. [143, 144] for details.

Their solution is

$$f^{+1} = x^{+1} e^{\mathcal{K}}, \quad f^{+2} = x^{+2} e^{-\mathcal{K}},$$
$$f^{+3} = x^{+3} e^{\mathcal{J}}, \quad f^{+4} = x^{+4} e^{-\mathcal{J}}, \tag{13.128}$$

where

$$\mathcal{J} = \frac{1}{2}(x^{+1}x^{-2} + x^{-1}x^{+2}), \quad \mathcal{K} = \frac{1}{2}(x^{+3}x^{-4} + x^{-3}x^{+4}). \tag{13.129}$$

And here is the solution for the bridge (the most convenient solution which belongs to $Sp(2)$ and becomes a unit matrix when the interaction is switched off):

$$\Delta^{1/4} M^{\underline{b}}{}_a = \begin{pmatrix} e^{-\mathcal{K}} & 0 & -x^{-1}x^{+4}e^{-\mathcal{J}}, & -x^{-1}x^{+3}e^{\mathcal{J}} \\ 0 & e^{\mathcal{K}} & x^{-2}x^{+4}e^{-\mathcal{J}} & x^{-2}x^{+3}e^{\mathcal{J}} \\ -x^{-3}x^{+2}e^{-\mathcal{K}}, & -x^{-3}x^{+1}e^{\mathcal{K}} & e^{-\mathcal{J}} & 0 \\ x^{-4}x^{+2}e^{-\mathcal{K}} & x^{-4}x^{+1}e^{\mathcal{K}} & 0 & e^{\mathcal{J}} \end{pmatrix}, \tag{13.130}$$

where the index $\underline{b}$ marks the lines and the index $a$ marks the columns like in Eq. (13.109),

$$\Delta = 1 - (x^1 \cdot x^2)(x^3 \cdot x^4), \tag{13.131}$$

and $(a \cdot b) \equiv a^j b_j$.

We skip the expressions for the vielbeins (the reader can find them in Ref. [140]) but quote the expressions for the metric. Unfortunately, they are rather bulky:

$$\Delta^{1/2} g_{j1,k1} = \frac{1}{4}[(x^2 \cdot x^4)(23)_{jk} - \{3 \leftrightarrow 4\}],$$

$$\Delta^{1/2} g_{j2,k2} = \frac{1}{4}[(x^1 \cdot x^4)(13)_{jk} - \{3 \leftrightarrow 4\}],$$

$$\Delta^{1/2} g_{j3,k3} = \frac{1}{4}[(x^4 \cdot x^2)(14)_{jk} - \{1 \leftrightarrow 2\}],$$

$$\Delta^{1/2} g_{j4,k4} = \frac{1}{4}[(x^3 \cdot x^2)(13)_{jk} - \{1 \leftrightarrow 2\}],$$

$$\Delta^{1/2} g_{j1,k3} = [\varepsilon_{jk}(x^4 \cdot x^2) + x^4_j x^2_k], \quad \Delta^{1/2} g_{j1,k4} = [\varepsilon_{jk}(x^3 \cdot x^2) + x^3_j x^2_k],$$
$$\Delta^{1/2} g_{j2,k3} = [\varepsilon_{jk}(x^4 \cdot x^1) + x^4_j x^1_k], \quad \Delta^{1/2} g_{j2,k4} = [\varepsilon_{jk}(x^3 \cdot x^1) + x^3_j x^1_k],$$

$$\Delta^{1/2} g_{j1,k2} = \varepsilon_{jk}\left[1 + \frac{1}{4}(x^2 \cdot x^4)(x^1 \cdot x^3) - \frac{1}{4}(x^2 \cdot x^3)(x^1 \cdot x^4)\right]$$
$$+ \frac{1}{4}\left[(x^2 \cdot x^4)x_j^1 x_k^3 + (x^1 \cdot x^4)x_j^3 x_k^2\right.$$
$$\left. - (x^1 \cdot x^3)x_j^4 x_k^2 - (x^2 \cdot x^3)x_j^1 x_k^4\right] + \frac{1}{4}(x^1 \cdot x^2)(x_j^3 x_k^4 - x_j^4 x_k^3),$$
$$\Delta^{1/2} g_{j3,k4} = \varepsilon_{jk}\left[1 + \frac{1}{4}(x^2 \cdot x^4)(x^1 \cdot x^3) - \frac{1}{4}(x^2 \cdot x^3)(x^1 \cdot x^4)\right]$$
$$+ \frac{1}{4}\left[(x^1 \cdot x^4)x_j^3 x_k^2 + (x^1 \cdot x^3)x_j^2 x_k^4\right.$$
$$\left. - (x^2 \cdot x^4)x_j^3 x_k^1 - (x^2 \cdot x^3)x_j^1 x_k^4\right] + \frac{1}{4}(x^3 \cdot x^4)(x_j^1 x_k^2 - x_j^2 x_k^1),$$
$$(13.132)$$

where $(23)_{jk} = x_j^2 x_k^3 + x_j^3 x_k^2$, etc.

We *conjecture*, however, that the metric thus obtained is equivalent to the metric (3.49) — an eight-dimensional generalization of the Taub–NUT metric based on the group $SO(5) \simeq Sp(2)$ and describing a "hyperkähler orientifold" discussed in Chapter 3. One metric should be derivable from another by a coordinate change. This is suggested by the fact that the two metrics have similar symmetries. In particular, the prepotential (13.126) is clearly "on good terms" with the group $Sp(2)$, on which the metric (3.49) is also based. In addition, both metrics have similar discrete $\mathbb{Z}_2$ isometries: the interchange $\boldsymbol{r}_1 \leftrightarrow \boldsymbol{r}_2$ for the metric (3.49) and the simultaneous interchange $x^1 \leftrightarrow x^3, x^2 \leftrightarrow x^4$ for the metric (13.132).

For sure, it would be interesting to *prove* this conjecture.

## 13.5 Gauge fields

In the four preceding sections, we discussed and analyzed only the first term in the generic $\mathcal{N} = 4$ action (13.49). Let us now look at the second term:

$$S_{\text{gauge}} = i \int dt_A du\, d^2\theta^+ \mathcal{L}^{++}(q^{+a}, u^\pm). \qquad (13.133)$$

Substitute there the component expansion (13.38) and integrate over $d^2\theta^+ = d\bar{\theta}^+ d\theta^+$. The bosonic term in the action reads

$$S^{\text{bos}}_{\text{gauge}} = i \int dt du\, \frac{\partial \mathcal{L}^{++}(f^{+b}, u)}{\partial f^{+a}}\, \tilde{A}^{-a}, \qquad (13.134)$$

where $\tilde{A}^{-a}$ is expressed via $\dot{x}^{kb}$ as in (13.78). Thus, we can write

$$S^{\text{bos}}_{\text{gauge}} = -\int dt\, A_{bk}\, \dot{x}^{kb} \tag{13.135}$$

with the Abelian gauge potential (which has nothing to do with $\tilde{A}^{-a}$):

$$A_{bk}(t) = -2e^{ja}_{kb} \int du\, (\partial_{+a}\mathcal{L}^{++})(M^{-1})^{a}{}_{\underline{a}}\, u^{-}_{\underline{j}}\,. \tag{13.136}$$

The full component expression for $S_{\text{gauge}}$ is given by equation (9.64), which we also reproduce here:

$$S_A = \int dt \left[-A_M\, \dot{x}^M + \frac{i}{2} F_{MN}\, \Psi^M \Psi^N\right] \tag{13.137}$$

($x^M = -i(\Sigma^M)_{aj}x^{ja}/\sqrt{2}$, $A_M = i(\Sigma^\dagger_M)^{kb}A_{bk}/\sqrt{2}$ and $F_{MN} = \partial_M A_N - \partial_N A_M$).

**Theorem 13.11.** *For the action* (13.137) *to enjoy* $\mathcal{N} = 4$ *supersymmetry, the field density* $F_{AB} = e^M_A e^N_B F_{MN}$ *should belong to the algebra* $sp(n)$.

**Proof.** In Theorem 9.6, we have established that the action (13.137) enjoys the extra supersymmetry (9.38), provided the field strength tensor commutes with the complex structure. And to enjoy three extra supersymmetries (13.53), it should commute with all three complex structures:

$$F_{AC}I^p_{CB} - I^p_{AC}F_{CB} = 0. \tag{13.138}$$

This condition has exactly the same form as (3.21), and we can use the result of Theorem 3.2 where the fact that any matrix satisfying this condition has the form (3.26) and belongs to $sp(n)$ was proven. □

This fact is confirmed by the direct calculations [92]. Let us show how it works for the flat model with $\mathcal{L}^{+3a} = 0$ and $\mathcal{L} \propto q^+q^-$. In this case, (13.136) reduces to

$$A_{bk} = -2\int du\, (\partial_{+b}\mathcal{L}^{++})\, u^-_k\,. \tag{13.139}$$

We have

$$\begin{aligned} F_{aj,\,bk} &= \partial_{aj}A_{bk} - \partial_{bk}A_{aj} \\ &= -2\int du \left(\frac{\partial^2\mathcal{L}^{++}}{\partial f^{+a}\partial f^{+b}}\right) u^+_j u^-_k - (aj \leftrightarrow bk). \end{aligned} \tag{13.140}$$

Bearing in mind the symmetry of $\partial^2 \mathcal{L}^{++}/(\partial f^{+a} \partial f^{+b})$, the field density is proportional to $u_j^+ u_k^- - u_k^+ u_j^- = \varepsilon_{jk}$ and has, indeed, the form (3.26).

For $n = 1$, $F_{AB} \in su(2)$. This means that $F_{AB}$ is *anti-self-dual*,

$$F_{AB} = -\frac{1}{2}\varepsilon_{ABCD} F_{CD},$$

if the complex structures (with which $F_{AB}$ commutes) are chosen self-dual, as in (3.3). Naturally, a choice when the complex structures are anti-self-dual while $F_{AB}$ is self-dual is also possible. The fact that, to keep $\mathcal{N} = 4$ supersymmetry of a four-dimensional HKT model after the inclusion of gauge fields, the latter must be (anti-)self-dual was noted in Refs. [91, 145].

# Chapter 14

# Gauge Fields on the Manifolds

What physicists call *gauge fields,* mathematicians call *principal G-connections.* We gave the mathematical definition of the latter in Section 1.3. We discussed systems including gauge fields in the several places in this book — in particular, we did so in Chapters 5, 6, 9 and 13. We mentioned there an essential feature of such systems: gauge fields are not arbitrary; certain integrals that involve field densities, like the magnetic flux (6.68), may acquire only a particular set of discrete (quantized) values. However, we did not discuss this remarkable feature (quantization of topological charge) in detail. The time has come to do so.

Gauge fields may live both on noncompact and compact manifolds. For example, the Landau problem discussed in Chapter 5 was defined on $\mathbb{R}^2$ (and even on $\mathbb{R}^3$ if one also takes into account free motion along the direction of the magnetic field). However, in this and the following chapter, we mostly limit our discussion to compact manifolds, where both physics and mathematics is simpler and more interesting in our opinion.

## 14.1 Spinors □ Dirac operator

In Section 1.4, we introduced the object $\omega_{AB,M}$ and called it the *spin connection.* But the "etymology" of this name was left obscur there. We will elucidate it *now.*

Consider first flat $D$-dimensional space. The group $SO(D)$ [or, better to say, its double covering $Spin(D)$ — see the following] involves spinor representations. For odd $D$, there is a unique irreducible spinor representation, but, to make contact with supersymmetry, we are only interested with *even* $D$, where there are two irreducible representations $S_L$ and $S_R$ — left and right spinors. Consider the set of complex functions $\Phi_\alpha(x)$ that belong to the sum $S_L \oplus S_R$ (the *bispinor* representation[1]). The index $\alpha$ takes

$$N_D = 2^{D/2}$$

different values. Now, we introduce a convenient object:

**Definition 14.1.** Hermitian matrices $\gamma^{A=1,\ldots,D}$ of dimension $N_D \times N_D$ satisfying the Clifford algebra

$$\gamma^A\gamma^B + \gamma^B\gamma^A = 2\delta^{AB} \tag{14.1}$$

are called Euclidean $D$-dimensional *gamma matrices.*[2]

Explicit choices for $\gamma^A$ may be different, and we need not specify it.

In these terms, the law of infinitesimal global rotations for the bispinor $\Phi_\alpha$ acquires a nice form:

$$\delta\Phi_\alpha = \frac{1}{8}\theta_{AB}\,(\gamma^A\gamma^B - \gamma^B\gamma^A)_{\alpha\beta}\,\Phi_\beta, \tag{14.2}$$

where $\theta_{AB} = -\theta_{BA}$ is a small rotation angle in the plane $[AB]$. The reader can be convinced that the generators $t^{AB} = -i(\gamma^A\gamma^B - \gamma^B\gamma^A)/4$ commute as they should in $so(D)$,

$$[t^{AB}, t^{CD}] = i(\delta^{AC}t^{BD} - \delta^{BC}t^{AD} + \delta^{BD}t^{AC} - \delta^{AD}t^{BC}), \tag{14.3}$$

and that the norm $\Phi^*_\alpha\Phi_\alpha$ is invariant under the rotations (14.2). One can also note that the matrix

$$\exp\left\{\frac{\pi}{2}(\gamma^A\gamma^B - \gamma^B\gamma^A)\right\}$$

of the finite rotation by $2\pi$ in the plane $[AB]$ (or in any other plane) coincides with $-\mathbb{1}$. That is why mathematicians call the group involving spinor representations $Spin(D)$ rather than $SO(D)$, with $Spin(D)$ representing a two-fold covering of $SO(D)$. But we need not plunge further into these group theory details.

[1]For $D = 2$, physicists habitually call this representation "spinor" rather than "bispinor", which is not quite exact — the complex two-dimensional representation of $SO(2)$ is reducible. We also remind that what physicists call "representation" is "representation space" in the mathematical language.

[2]They were first introduced almost hundred years ago by Dirac in Minkowski four-dimensional space. In this case, Dirac matrices satisfy the relation $\{\gamma^\mu, \gamma^\nu\} = 2\eta^{\mu\nu} = 2\,\mathrm{diag}(1, -1, -1, -1)$.

Consider the expression

$$(\not{\mathcal{D}}\Phi)_\alpha = (\gamma^M)_{\alpha\beta}\,\partial_M \Phi_\beta. \tag{14.4}$$

Assuming that $\gamma^M$ transform under rotations as a vector, in the same way as the gradient operator $\partial_M$, it is easy to see that $(\not{\mathcal{D}}\Phi)_\alpha$ transforms in the same way as $\Phi_\alpha$, i.e. it represents a bispinor

The operator (14.4) is called the *Dirac operator* in flat space in the absence of the gauge fields.

The gauge field is described by the vector potential (alias, a principal $G$-connection) $A_M(x)$. It can be Abelian (mathematicians call the corresponding bundle a *line bundle*) or non-Abelian. In the latter case, it represents a Hermitian matrix, an element of a Lie algebra $\mathfrak{g}$, while the bispinor field should be placed now in the fundamental representation of $G$ and endowed with an extra group index, $\Phi_\alpha \to \Phi_\alpha^a$. The Dirac operator acquires the extra term:

$$\not{\mathcal{D}} = \gamma^M(\partial_M + iA_M). \tag{14.5}$$

As was explained in Section 1.3, $[(\partial_M + iA_M)\Phi]^a$ and hence $(\not{\mathcal{D}}\Phi)^a$ transform under gauge rotations in the same way as $\Phi^a$.

Finally, let us define the Dirac operator on a curved manifold in a nontrivial gauge background as

$$\boxed{\not{\mathcal{D}} = e_A^M \gamma^A \left(\partial_M + \frac{1}{4}\omega_{BC,M}\,\gamma^B\gamma^C + iA_M\right).} \tag{14.6}$$

With this definition, $(\not{\mathcal{D}}\Phi)_\alpha^a$ transforms in the same way as $\Phi_\alpha^a$ under both gauge rotations and orthogonal rotations in the tangent plane with $x$-dependent parameters:

$$\delta\Phi = \frac{1}{4}\theta_{BC}(x)\,\gamma^B\gamma^C\,\Phi \quad \Rightarrow \quad \delta(\not{\mathcal{D}}\Phi) = \frac{1}{4}\theta_{BC}(x)\,\gamma^B\gamma^C\,\not{\mathcal{D}}\Phi. \tag{14.7}$$

Indeed, one can be convinced that the term $\sim \partial_M\theta_{BC}$ in $\not{\mathcal{D}}\delta\Phi$ exactly cancels the term with the same structure coming from the variation $\delta\omega_{BC,M}$ in $\not{\mathcal{D}}$ [see Eq. (1.56)].

The notion of spin connection was first introduced by physisists to describe spinor fields in general relativity.

Equation (14.6) is a local expression. A nontrivial question is whether the Dirac operator can be defined *globally* on the whole manifold. To do so, one has to define the *spinor bundle*, i.e. divide the manifold in several charts and define the bispinor field in each chart in such a way that, in the intersection regions, the bispinor fields from two different charts $U_\alpha$ and $U_\beta$ are interrelated by a transformation $\Omega_{\alpha\beta} \in Spin(D)$. If nonempty overlap regions $U_\alpha \cap U_\beta \cap U_\gamma$ of three different charts exist, the group-valued functions $\Omega_{\alpha\beta}$ should satisfy there the consistency condition $\Omega_{\alpha\beta}\Omega_{\beta\gamma} = \Omega_{\alpha\gamma}$.

Consider the "purely gravitational" Dirac operator (14.6) with $A_M = 0$ and suppose that, in each chart, the bispinor field represents an eigenfucntion of $\not{\mathcal{D}}$. Suppose further that, in the intersection regions, the sets of vielbeins $e_A^M$ defined in each chart are related by the $SO(D)$ rotations that correspond to $\Omega_{\alpha\beta}$.

**Definition 14.2.** People say that the manifolds admitting spinor bundles satisfying the conditions above admit a *spin structure.*

Not all manifolds do. For example, $\mathbb{CP}^{2k}$ do not admit spin structure and the Dirac operator not including an extra gauge field $A_M$ is not well defined there. However, we see soon that the so-called *spin*$^{\mathbb{C}}$ *structure* on $\mathbb{CP}^{2k}$ is admissible: if an Abelian gauge field $A_M$ of a special form is present, the Dirac operator is globally well defined. The bispinor fields in the overlapping charts are then related by $\Omega_{\alpha\beta} \in Spin(D) \times U(1)$.

### 14.1.1 *Dirac operator and the supercharges*

We have already met the flat Abelian Dirac operator (14.5) in Chapters 5 and 6. We showed there that the problem of two-dimensional motion with the Hamiltonian proportional to the square of the operator (14.5) is in fact supersymmetric. We show now that it is also true in more general cases.

**Theorem 14.1.** *Consider the Dirac operator* (14.6) *defined on an even-dimensional manifold with an Abelian or non-Abelian gauge field* $A_M(x)$. *Then, the operators*

$$\boxed{\hat{Q} = -\frac{i}{2\sqrt{2}}\not{\mathcal{D}}\,(1-\gamma^{D+1}), \quad \hat{Q}^\dagger = -\frac{i}{2\sqrt{2}}\not{\mathcal{D}}\,(1+\gamma^{D+1}),} \tag{14.8}$$

*where*

$$\gamma^{D+1} = \frac{(-i)^{D/2}}{D!}\,\varepsilon_{A_1,\ldots,A_D}\gamma^{A_1}\cdots\gamma^{A_D}, \tag{14.9}$$

*obey the supersymmetry algebra* (5.1) *with the Hamiltonian*

$$\hat{H} = -\frac{\not{\mathcal{D}}^2}{2}. \tag{14.10}$$

**Proof.** This follows immediately from the fact that $\gamma^{D+1}$ anticommutes with all $\gamma^{A_j}$ and that $(\gamma^{D+1})^2 = \mathbb{1}$. □

**Remarks:**

- The requirement for $D$ to be even is essential. If $D$ is odd, the RHS of Eq. (14.9) would not give a nontrivial matrix anticommuting with $\gamma^{A_j}$. It would be proportional to $\mathbb{1}$.
- A collorary from $(\gamma^{D+1})^2 = \mathbb{1}$ is that $\gamma^{D+1}$ can be diagonalized with the eigenvalues $\pm 1$. In fact, a half of these eigevalues are $+1$ and a half $-1$.

**Definition 14.3.** A bispinor satisfying the condition $\gamma^{D+1}\Psi = \Psi$ is called *right-handed* and a bispinor satisfying the condition $\gamma^{D+1}\Psi = -\Psi$ is called *left-handed.*

As (14.9) commutes with the Hamiltonian, the eigenstates of the latter can be divided into the two classes: right-handed and left-handed ones. They belong to irreducible representations of $Spin(D)$.

In the rest of this and in the two following sections (14.2 and 14.3), we concentrate on Abelian gauge fields, postponing the discussion of the non-Abelian case to Sections 14.4 and 14.5.

The Hilbert space where the supercharges (14.8) and the Hamiltonian (14.10) act involves the bispinors $\Phi_\alpha(x)$. Let us now try to treat this system in the same way as we did for other supersymmetric systems considered in this book: to define a Hilbert space involving the functions depending on the coordinates and also on holomorphic Grassmann variables and to map the triple $\hat{Q}, \hat{Q}^\dagger, \hat{H}$ to the operators acting there. We also would like to derive the expressions for the *classical* supercharges and the Hamiltonian.

At first sight, this seems to be feasible. The $\gamma$-matrices that satisfy the Clifford algebra (14.1) can be mapped to the quantum operators $\hat{\Psi}^A\sqrt{2}$ satisfying the same algebra. Their classical counterparts are the nilpotent Grassmann numbers[3] $\Psi^A\sqrt{2}$.

Well, as far as the Dirac operator is concerned, there is no problem. It has indeed a classical counterpart — a classical real supercharge. For example, in the flat case and in the absence of the gauge field, the latter

---

[3]As was mentioned on p. 101, "Grassmann" becomes "Clifford" after quantization. And this means that "Clifford" goes over to "Grassmann" when we go down from quantum theory to classical one.

may be defined as the counterpart of $\hat{\mathcal{Q}}_1 = \hat{Q} + \hat{Q}^\dagger = -i\not{\mathcal{D}}/\sqrt{2}$, which gives

$$Q^{\rm cl} = \Psi^M P_M, \tag{14.11}$$

[cf. Eq. (9.60)]. With the natural definition of the Grassmann Poisson bracket, $\{\Psi^M, \Psi^N\}_P = i\delta^{MN}$ [cf. the first line in (4.58)], we derive the $\mathcal{N} = 1$ classical supersymmetry algebra $\{Q^{\rm cl}, Q^{\rm cl}\}_P = 2iH^{\rm cl}$ as in (5.44).

It is difficult, however, to find a classical counterpart of the nonlinear operator (14.9) and it is difficult to find a classical counterpart of the *second* Hermitian supercharge $\hat{\mathcal{Q}}_2 = i(\hat{Q}^\dagger - \hat{Q}) = \not{\mathcal{D}}\gamma^{D+1}/\sqrt{2}$. For a generic even-dimensional manifold, this is impossible [64]. We have a *classical supersymmetry anomaly* here [63]: $\mathcal{N} = 2$ supersymmetry that our quantum system enjoys cannot be preserved at the classical level!

The situation is better when our manifold is *Kähler*. We map

$$\gamma^A \to \sqrt{2}\,\hat{\Psi}^A$$

and represent the Dirac operator as

$$\not{\mathcal{D}} = \sqrt{2}\, e_A^M \hat{\Psi}^A \left(\partial_M + \frac{1}{2}\omega_{BC,M}\hat{\Psi}^B\hat{\Psi}^C + iA_M\right). \tag{14.12}$$

Now, we distinguish the holomorphic and antiholomorphic fermion components, $\hat{\Psi}^A \equiv (\psi^a, \hat{\bar{\psi}}^{\bar{a}})$ with $\hat{\bar{\psi}}^{\bar{a}} = \partial/\partial\psi^a$. Plugging this into (14.12), we obtain an operator which acts on the wave functions $\Phi(x^M, \psi^a)$. The latter have $2^{D/2}$ components — the same number as the number $N_D$ of the components of the bispinor $\Phi_\alpha(x)$, and there is a bijective correspondence between the two Hilbert spaces.

**Theorem 14.2.** *One can define this bijective mapping in such a way that the right-handed spinors go over to the bosonic states and the left-handed spinors to the fermionic states. Then,*

$$\gamma^{D+1} \leftrightarrow (-1)^{\hat{F}}. \tag{14.13}$$

**Proof.** This immediately follows from the mentioned fact that a half of the eigenvalues of $\gamma^{D+1}$ are $+1$ and another half are $-1$. □

**Remark:** Obviously, one can also map left-handed spinors to boson states and right-handed spinors to fermion states. In fact, we did it in Chapter 5 for the Landau problem of electron motion in magnetic field. We will see in the following chapter (see the footnote on p. 368 and the last paragraph in Section 15.3.1) that it more convenient to keep the convention (14.13) for the manifolds of dimension $D = 4k$ and reverse it for $D = 2(2k+1)$. In other words, we assume

$$\boxed{\gamma^{D+1} \leftrightarrow (-1)^{D/2}(-1)^{\hat{F}}.} \tag{14.14}$$

Capitalizing on the fact that, for the Kähler manifolds, the components $\omega_{bc,M}$ and $\omega_{\bar{b}\bar{c},M}$ vanish [see Eq. (2.31)], we can represent

$$\not{\mathcal{D}} = \not{\mathcal{D}}^{\rm Hol} - (\not{\mathcal{D}}^{\rm Hol})^\dagger, \tag{14.15}$$

where

$$\not{\mathcal{D}}^{\rm Hol} = \sqrt{2}\, e_a^m \psi^a \left[\partial_m + \frac{1}{2}\omega_{\bar{b}c,m}\,(\hat{\bar{\psi}}^{\bar{b}}\psi^c - \psi^c\hat{\bar{\psi}}^{\bar{b}}) + iA_m\right], \tag{14.16}$$

$$(\not{\mathcal{D}}^{\rm Hol})^\dagger = -\sqrt{2}\, e_{\bar{a}}^{\bar{m}} \hat{\bar{\psi}}^{\bar{a}} \left[\partial_{\bar{m}} + \frac{1}{2}\omega_{b\bar{c},\bar{m}}\,(\psi^b\hat{\bar{\psi}}^{\bar{c}} - \hat{\bar{\psi}}^{\bar{c}}\psi^b) + iA_{\bar{m}}\right]. \tag{14.17}$$

**Theorem 14.3.** *For the Kähler manifolds, the holomorphic and antiholomorphic parts of the Dirac operator* (14.16) *and* (14.17) *with* $A_m = A_{\bar{m}} = 0$ *coincide up to an extra factor* $i\sqrt{2}$ *with the covariant quantum supercharges* (9.12) *for the system* (9.1) *with* $\mathcal{W} = 0$.

**Proof.** To compare $\not{\mathcal{D}}^{\rm Hol}$ with $\hat{Q}$ in Eq. (9.12), we send the operator $\hat{\bar{\psi}}^{\bar{b}}$ in the expression (14.16) to the right, which gives us the extra term $\propto \omega_{\bar{b}b,m}$. For Kähler manifolds, when $\omega_{\bar{b}c,m} = e_{\bar{b}}^{\bar{n}}\partial_m e_{\bar{n}}^{\bar{c}}$, [see Eq. (2.31)], we derive

$$\omega_{\bar{b}b,m} = \partial_m \ln\det\bar{e}. \tag{14.18}$$

Performing a similar transformation for $(\not{\mathcal{D}}^{\rm Hol})^\dagger$ gives us a term $\propto \omega_{b\bar{b},\bar{m}} = \partial_{\bar{m}} \ln\det e$. These extra contributions exactly coincide with the structures displayed in (9.12) (we remind that the expression (9.12) was written under a simplifying assumption $\det e = \det\bar{e} = \sqrt{\det h}$). □

Suppose now that an *Abelian* gauge field $A_M$ is present. In the particular case when it has the form (9.16):

$$A_M = (i\partial_m\mathcal{W},\ -i\partial_{\bar{m}}\mathcal{W}), \tag{14.19}$$

the extra contributions in (14.16) and (14.17) coincide up to the extra factor $i\sqrt{2}$ with the shift (9.14) of the supercharges (9.12).

In Chapter 9 (see Theorem 9.2), we established that the supercharges (9.12) with the shift (9.14) map to the operators $\partial_{\mathcal{W}}$ and $\partial^\dagger_{\mathcal{W}}$ of the twisted Dolbeaux complex with the twisted exterior holomorphic derivative

$$\partial_{\mathcal{W}} = \partial - \partial\left(\mathcal{W} - \frac{1}{4}\ln\det h\right)\wedge. \tag{14.20}$$

Alternatively, they can be mapped to the operators $\bar\partial_{\mathcal{W}}$ and $\bar\partial^\dagger_{\mathcal{W}}$ of the twisted anti-Dolbeaux complex with

$$\bar\partial_{\mathcal{W}} = \bar\partial + \bar\partial\left(\mathcal{W} + \frac{1}{4}\ln\det h\right)\wedge. \tag{14.21}$$

Bearing all this in mind, we arrive at the theorem:

**Theorem 14.4.** *For the Kähler manifolds, the Dirac operator* (14.6) *with the gauge field* (14.19) *can be decomposed into the holomorphic and antiholomorphic parts as in* (14.15). *The holomorphic part maps to the exterior holomorphic derivative* (14.20) *of the twisted Dolbeault complex. The antiholomorphic part maps to the exterior antiholomorphic derivative* (14.21).

The fact that the Dolbeault complex maps to the Dirac complex and that the sum $\hat{Q}_{\mathcal{W}} + \hat{\bar{Q}}^\dagger_{\mathcal{W}}$ is isomorphic to the Dirac operator was mentioned before (see the pagagraph after Theorem 9.2). In Chapter 9, we did not prove this fact. We promised to do so in Chapter 14, and we have now fulfilled this promise. However, the formulation of Theorem 14.4 involves an important restriction: the equivalence has been proven only for the Kähler manifolds.

A natural question is what happens for generic complex, not necessarily Kähler manifolds. We now show that a more general version of such equivalence can also be established in this case. The proof above relied on the decomposition (14.15). It does not hold in a generic complex case for the ordinary Dirac operator because of the presence of the components $\omega_{ab,\bar m}$ and $\omega_{\bar a\bar b,m}$ in the spin connection. We can define, however, a *deformed* Dirac operator that involves not only the spin connection but also the torsion:

$$\boxed{\not{\mathcal{D}}^{\,\text{tors}} = e^M_A\gamma^A\left(\partial_M + \frac{1}{4}\omega_{BC,M}\gamma^B\gamma^C - \frac{1}{24}C_{MKP}\gamma^K\gamma^P + iA_M\right),} \tag{14.22}$$

where the torsion $C_{MKP}$ is given by (2.18). This is nothing but the quantum version of the supercharge $\mathcal{Q}_0$ in (9.60), where we also added the gauge field term. As was noted in the paragraph after Eq. (9.60), the role of the torsion term is to *cancel* the troublesome terms $\propto \omega_{ab,\bar{m}}$ and $\propto \omega_{\bar{a}\bar{b},m}$ in the expression for $\not{\mathcal{D}}^{\,\text{tors}}$. The operator $\not{\mathcal{D}}^{\,\text{tors}}$ admits the decomposition (14.15) and we can claim the following:

**Theorem 14.5.** *For a generic Dolbeault complex involving the twisted exterior holomorphic derivative* (14.20), *the operator* $\partial_{\mathcal{W}} + \partial^\dagger_{\mathcal{W}}$, *where* $\partial^\dagger_{\mathcal{W}}$ *is the Hermitian conjugate of* $\partial_{\mathcal{W}}$, *maps to the deformed torsionful Dirac operator* (14.22) *with the gauge field* (14.19).

Thus, for a generic complex manifold, one can define at least two different quantum supersymmetric problems:

(1) the Dirac problem including the Hermitian supercharges $-i\not{\mathcal{D}}/\sqrt{2}$ and $\not{\mathcal{D}}\gamma^{D+1}/\sqrt{2}$,
(2) the Dolbeault problem including the Hermitian supercharges, one of which is the deformed Dirac operator $-i\not{\mathcal{D}}^{\,\text{tors}}/\sqrt{2}$ and another one is the quantum version of the operator $\mathcal{Q}$ in (9.60) (translated if you will into the spinor language).

Generically, the Hamiltonians of these two quantum problems do not coincide, though for the Kähler manifolds (and, obviously, in flat space), they do.

We note finally that the Dirac problem can be defined for any even-dimensional manifold, while the Dolbeault problem is usually defined only for the complex manifolds.[4]

## 14.2 Magnetic field on $S^2$

We are going here to treat in much detail the simplest nontrivial case — an Abelian gauge field on $S^2$.

In the mathematical language, we are dealing with a particular case of the principal fiber bundle — a *line bundle.* In Section 1.3, we defined the principal fiber bundle locally, in a particular chart, and did not discuss what happens in the overlap region. Neither did we do so in the preceding section. But now we will.

---

[4]We say "usually" because, as we saw in Section 9.3, there are manifolds like $S^4$ which are not complex but can be made complex by removing a point. And the Dolbeault Hamiltonian problem on $S^4\backslash\{\}$ with square-integrable wave functions makes sense and is supersymmetric.

We are interested in the bundles equiped with principal $\mathcal{G}$-connections (gauge fields). In each chart, such a connection represents a vector function $A_M(x)$, whose components are elements of the corresponding Lie algebra represented by Hermitian matrices. In the overlap region between two charts, we can require $A'_M = A_M$, but this requirement is too rigid. We can relax it by requiring for $A'_M$ and $A_M$ to be related by a gauge transformation (1.45). For Abelian line bundles, when $\omega(x) = e^{i\chi(x)}$ and $A_M(x)$ are ordinary functions, this boils down to

$$A'_M = A_M - \partial_M \chi(x). \tag{14.23}$$

The Abelian field density $F_{MN} = \partial_M A_N - \partial_N A_M$ (the curvature of our line bundle) stays invariant under this transformation.

$S^2$ may be covered by two charts — the northern and the southern hemispheres overlapping at the equator, where the condition (14.23) is imposed. An important remark is that the symbol $\chi(x)$ entering (14.23) is not necessarily a uniquely defined function. The element of the $U(1)$ gauge group $\omega(x) = e^{i\chi(x)}$ *is*, but $\chi(x)$ is *not*. We can allow for the phase $\chi(x)$ to be shifted by $2\pi q$ with an integer $q$ when we go around the equator $S^1$. This integer is a topological invariant characterizing the mapping $U(1) \to S^1$.

Consider now the integral $\int_{S^2} F = \int_{S^2} dA$ with $A = A_M dx^M$. It is the integral of a total derivative and, by Stokes' theorem, it seems to vanish because $S^2$ has no borders. However, one should take into account the fact that $A_M$ in the northern and southern hemispheres may not be same, being related as in (14.23). Calculating separately the integrals for the northern and southern hemispheres, using Stokes' theorem and (14.23), we derive

$$\int_{S^2} F = \oint_{\text{equator}} \partial_M \chi(x)\, dx^M = 2\pi q. \tag{14.24}$$

We arrive at the following theorem:

**Theorem 14.6.** *For a line bundle on $S^2$, the flux of the magnetic field $\Phi = \int F$ is quantized as in* (14.24).

A very similar statement was derived in Chapter 6: the Witten index (6.68) for the supersymmetric system (6.66) that desribed the two-dimentional motion in a magnetic field (and, by definition, the Witten index is an integer) is equal to $\Phi/2\pi = q$.

However, there is a certain difference. In Chapters 5 and 6, we discussed the motion in flat two-dimensional space, while now we are dealing with the motion over a curved 2-sphere. If one looks for a physical interpretation, we are dealing with the motion in the field of a *magnetic monopole* placed at the center of the "round" sphere with a homogeneous field density $F_{MN}\,\varepsilon^{MN} = \text{const}$. In this case, the radial component of the three-dimensional vector potential vanishes, an electron moving in the radial direction does not feel the presence of the field, and only the angular motion, the motion over $S^2$, is nontrivial. The quantization of the magnetic flux boils down then to the quantization of the magnetic charge $m$:

$$em = \frac{q\hbar c}{2}, \tag{14.25}$$

where we restored the physical constants $e, c, \hbar$ present in the physical Pauli Hamiltonian (5.14). The quantization condition (14.25) was derived long time ago by Dirac [23] from the requirement that the angular momentum of the electromagnetic field created by an electron and the magnetic monopole is an integer multiple of $\hbar$. Its derivation from the mathematical fiber bundle construction is due to Wu and Yang [146].

The vector potential of the Dirac monopole (that we have already discussed in our book on several occasions) can be chosen in the gauge (3.41). As expected, the radial projection of this potential $\propto xA_x + yA_y$ vanishes.

Let us find the spectrum of the Hamiltonian $H = -\not{\mathcal{D}}^2/2$ describing the angular motion of a charged particle in the field of the magnetic monopole. $S^2$ is a Kähler manifold, and the Dirac problem is equivalent to the Dolbeault one.

It is convenient, instead of decomposing $S^2$ into two hemispheres and imposing the matching condition (2.1) at the equator, to choose as one of the charts the *whole* sphere except the north pole. The second chart would in this case be a small neighbourhood of the north pole. For the first chart, we choose the complex stereographic coordinates

$$z = \tan\frac{\theta}{2}e^{-i\phi} = \frac{x+iy}{\sqrt{2}}, \quad \bar{z} = \tan\frac{\theta}{2}e^{i\phi} = \frac{x-iy}{\sqrt{2}}, \tag{14.26}$$

where $\theta$ is the polar angle on $S^2$ counted from the south to the north, and write the metric as in Eq. (2.6):

$$ds^2 = \frac{2dzd\bar{z}}{(1+\bar{z}z)^2}. \tag{14.27}$$

This metric is singular at $z = \infty$ (the north pole), but the singularity is integrable. In particular, the area of such sphere is

$$\mathcal{A} = \int \frac{dzd\bar{z}}{(1+\bar{z}z)^2} = 2\pi, \tag{14.28}$$

corresponding to the radius $R = \sqrt{2}/2$.

Consider the SQM system with the action

$$S = \int d\bar{\theta} d\theta dt \left[ \frac{\bar{D}\bar{Z}DZ}{4(1+\bar{Z}Z)^2} + \mathcal{W}(Z,\bar{Z}) \right]. \tag{14.29}$$

According to Theorem 14.5, this problem is isomorphic to the Dirac problem on $S^2$ with the gauge field (14.19). A particularly clever choice for $\mathcal{W}(Z,\bar{Z})$ is

$$\mathcal{W}(Z,\bar{Z}) = -\frac{q}{2} \ln(1+\bar{Z}Z) \tag{14.30}$$

giving

$$\boxed{A_M = \left( -\frac{iq\bar{z}}{2(1+\bar{z}z)}, \frac{iqz}{2(1+\bar{z}z)} \right).} \tag{14.31}$$

In this case, the form

$$F = dA = -2i\,\partial\bar{\partial}\mathcal{W}\, dz \wedge d\bar{z} = \frac{iq\, dz \wedge d\bar{z}}{(1+\bar{z}z)^2} \tag{14.32}$$

is proportional to the area 2-form $ih\, dz \wedge d\bar{z} = \sqrt{g}\, dx \wedge dy$ with the coefficient $q$, which is nothing but the magnetic field density $B$ — one and the same at all points. Its integral over $S^2$ is

$$\int_{S^2} F = q\mathcal{A} = 2\pi q. \tag{14.33}$$

Recalling Theorem 14.6, we derive that $q$ should be integer.

We can arrive at the same conclusion if we express the gauge potential (14.19) in the Cartesian coordinates $(x,y)$. One can be easily convinced that the circulation $\oint \boldsymbol{A} \cdot d\boldsymbol{x}$ over a small contour embracing the north pole is[5] $\pm 2\pi B$. This can be handled with the matching condition (14.23) between the two charts, specified above, with a *unique* gauge transition function $\omega = e^{i\chi}$.

In the physical literature, one can meet another interpretation. Physicists (starting from Dirac) did not cover the sphere by two different charts but considered a three-dimensional problem where the gauge field was singular at a ray going from the origin to infinity. Nonzero circulation $\oint \boldsymbol{A} \cdot d\boldsymbol{x}$ along an infinitesimal contour around the ray may be interpreted as an

[5]The sign depends on the direction of the circulation, which we are too lazy to fix.

infinitesimally narrow flux line (alias, the *Dirac string*) carrying the magnetic flux, an integer multiple of $2\pi$, which exactly compensates the flux due to the monopole magnetic field $\propto \boldsymbol{r}/r^3$. The point is that the Dirac string with such a quantized flux *is* not physically observable. One can draw it through any point on the sphere by choosing an appropriate gauge.

The problem (14.29) with $\mathcal{W}$ specified in (14.30) is so simple that the spectrum of its Hamiltonian can be explicitly found.[6] The quantum supercharges given by the sum of (9.12) and (9.14) read

$$\boxed{\begin{aligned} \hat{Q} &= -\, i(1+\bar{z}z)\psi \left[\partial - \frac{\bar{z}(1-q)}{2(1+\bar{z}z)}\right], \\ \hat{\bar{Q}} &= -\, i(1+\bar{z}z)\hat{\bar{\psi}} \left[\bar{\partial} - \frac{z(1+q)}{2(1+\bar{z}z)}\right]. \end{aligned}} \tag{14.34}$$

They act upon the two-component wave functions[7]

$$\Psi^{\text{cov}} = \Psi_{F=0}(\bar{z}, z) + \psi\Psi_{F=1}(\bar{z}, z) \tag{14.35}$$

normalized with the covariant measure

$$d\mu = \sqrt{g}\, d\bar{z}dz = \frac{d\bar{z}dz}{(1+\bar{z}z)^2}. \tag{14.36}$$

The operators $\hat{Q}$ and $\hat{\bar{Q}}$ are Hermitially conjugate to one another with the measure (14.36). The Hamiltonian is $\hat{H} = \{\hat{Q}, \hat{\bar{Q}}\}$. In the sector $F = 0$, it reads

$$\hat{H}_{F=0} = -(1+\bar{z}z)^2\partial\bar{\partial} - \gamma(1+z\bar{z})(\bar{z}\bar{\partial} - z\partial) + \gamma^2\bar{z}z - \gamma, \tag{14.37}$$

where

$$\gamma = \frac{q-1}{2}. \tag{14.38}$$

The expression for $\hat{H}_{F=1}$ is similar; one only has to interchange $z$ and $\bar{z}$ and inverse the sign of $q$.

[6]The Schrödinger problem for a scalar charged particle in the monopole field was solved by Igor Tamm [147]. He wrote this paper when he was visiting Cambridge in 1930 and learned about Dirac's discovery directly from Dirac. The effects of spin were taken care of by Banderet [148]. The supersymmetric problem that we are interested in was first solved in Ref. [149].

[7]Here, $F$ is not the field density form but the fermion charge — the eigenvalue of the operator $\psi\hat{\bar{\psi}} = \psi\frac{\partial}{\partial\psi}$.

Let us first find the vacuum zero-energy states. They should be annihilated by both $\hat{Q}$ and $\hat{\bar{Q}}$. Let $q$ be nonzero and positive. Then, the vacuum states must have zero fermion charge. Indeed, if $F = 1$, $\hat{Q}\Psi$ vanishes identically, while the condition $\hat{\bar{Q}}\Psi = 0$ gives the equation

$$\left[\bar{\partial} - \frac{z(1+q)}{2(1+\bar{z}z)}\right] \Psi^{(0)}_{F=1}(\bar{z}, z) = 0, \tag{14.39}$$

which does not have normalizable solutions.

The states in the sector $F = 0$ are all annihilated by $\hat{\bar{Q}}$. The action of $\hat{Q}$ is nontrivial and leads to the equation

$$\left[\partial + \frac{\gamma\bar{z}}{1+\bar{z}z}\right] \Psi^{(0)}_{F=0}(\bar{z}, z) = 0. \tag{14.40}$$

And this equation *has* exactly $q$ independent normalizable solutions:

$$\Psi^{(0)}_{F=0} = (1+\bar{z}z)^{-\gamma}, \quad \bar{z}(1+\bar{z}z)^{-\gamma}, \ldots, \bar{z}^{q-1}(1+\bar{z}z)^{-\gamma} \tag{14.41}$$

(we do not bother about the normalization coefficients). The Witten index of this system is equal to $q$.

Consider now a nonzero negative integer $q$. All the reasoning is repeated with the only difference that in this case there are no zero-energy states in the sector $F = 0$, but there are $|q|$ independent vacuum states in the sector $F = 1$. The Witten index is still equal to $q$.

Finally, if $q = 0$, there are no zero-energy states.[8]

The wave functions (14.41) (and also the wave functions of the excited states to be discussed a bit later) are defined in the chart covering the main body of $S^2$ and not including the north pole. In fact, for some of the functions in (14.41), one can include the point $z = \infty$ in the domain of their definition. If $q = 1$, there is only one such function, $\Psi = const$. If $q > 1$, the functions $(1+\bar{z}z)^{-\gamma}, \ldots, \bar{z}^{q-2}(1+\bar{z}z)^{-\gamma}$ vanish at $z = \infty$ and are hence well defined there. But the function $\bar{z}^{q-1}(1+\bar{z}z)^{-\gamma}$ does not vanish and does not have a definite value at $z = \infty$. This can be remedied if we recall that we are dealing here with the fiber bundle construction. The north pole belongs to the second chart. The wave functions in the different charts may be different, but they have to be related by a gauge transformation, $\Psi' = e^{i\chi}\Psi$, with the same $\chi(x)$ that enters the transformation law (14.23)

[8]Incidentally, this also follows from the so-called *Lichnerowicz formula*.

for the gauge potential. Then, the covariant derivatives $(\partial_M + iA_M)\Psi$ in the two charts are related by the same phase factor $e^{i\chi}$ as $\Psi$ is.

One can be convinced that in our case, the wave functions in the two charts are related as

$$\Psi_{\text{north pole}}(\bar{z}, z) = \left(\frac{z}{\bar{z}}\right)^{\gamma} \Psi_{\text{main body}}(\bar{z}, z), \tag{14.42}$$

and the left-hand side is regular at $z = \infty$ for all the functions in the spectrum.

The full spectrum was found in Ref. [149]. Consider first the sector $F = 0$. The Hamiltonian (14.37) commutes with the angular momentum operator $\hat{m} = \bar{z}\bar{\partial} - z\partial$ and the eigenstates are characterized by the principal quantum number $n = 0, 1, \ldots$ and the integer angular momentum eigenvalues:

$$\begin{aligned} m &= -n, \ldots, n + 2\gamma \quad \text{if} \quad q \geq 1, \\ m &= -n + 2\gamma, \ldots, n \quad \text{if} \quad q \leq 1. \end{aligned} \tag{14.43}$$

The energies do not depend on $m$[9]:

$$\begin{aligned} E_n^{(q)\, F=0} &= n(n+q) \quad \text{if} \quad q \geq 1, \\ E_n^{(q)\, F=0} &= n(n+2-q) + 1 - q \quad \text{if} \quad q \leq 1 \end{aligned} \tag{14.44}$$

or else

$$\boxed{E_L^{(q)\, F=0} = L(L+q)} \tag{14.45}$$

with integer $L \geq \max(0, 1-q)$.

As an illustration, we represent in Fig. 14.1 the spectrum of the Hamiltonian (14.37) for $q = 3$.

Note the relation

$$E_n^{2-q} = E_n^q + q - 1, \tag{14.46}$$

which follows naturally from (14.37) (perform there the inversion $\gamma \to -\gamma$, $z \leftrightarrow \bar{z}$). The eigenfunctions (called *monopole harmonics* [146, 147])

[9]This degeneracy is a corollary of spherical symmetry — cf. the familiar degeneracy of the states with a given $l$ and different $m$ for a scalar spherically symmetric potential.

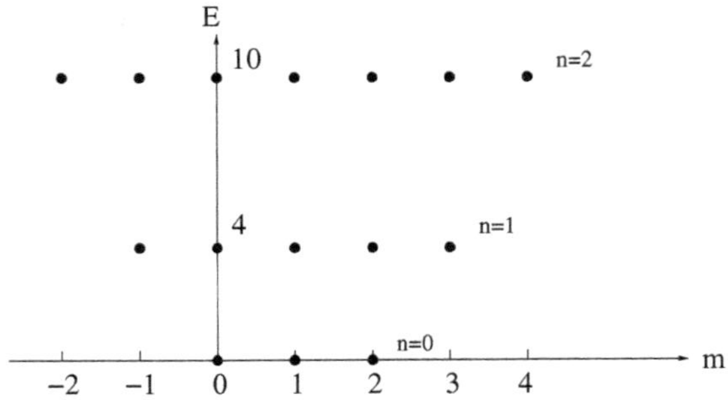


**Fig. 14.1.** The spectrum of the Hamiltonian for the magnetic charge $q = 3$ in the sector $F = 0$. It involves 3 zero-energy ground states.

have the form[10]

$$
\begin{aligned}
\Psi_{mn}^{(q\geq 1)\,F=0} &= e^{im\phi}(1-t)^{m/2}(1+t)^{\gamma-m/2}P_n^{m,2\gamma-m}(t),\\
\Psi_{mn}^{(q\leq 1)\,F=0} &= e^{im\phi}(1-t)^{-m/2}(1+t)^{m/2-\gamma}P_n^{-m,m-2\gamma}(t), \qquad (14.48)
\end{aligned}
$$

where

$$t = \cos\theta = \frac{1-\bar{z}z}{1+\bar{z}z}, \qquad (14.49)$$

$e^{i\phi} = \sqrt{\bar{z}/z}$, and $P_n^{\alpha,\beta}(t)$ are the Jacobi polynomials[11]:

$$P_n^{\alpha,\beta}(t) = \frac{\Gamma(\alpha+n+1)}{n!\Gamma(\alpha+\beta+n+1)}\sum_{k=0}^{n}\binom{n}{k}\frac{\Gamma(\alpha+\beta+n+k+1)}{\Gamma(\alpha+k+1)}\left(\frac{t-1}{2}\right)^k. \qquad (14.51)$$

[10]In fact, the first line in (14.48) describes also the states with $q < 1$, as one can be convinced using the identities [146]

$$
\begin{aligned}
P_{n+\alpha}^{-\alpha,\beta}(t) &= 2^{-\alpha}(t-1)^{\alpha}\frac{n!(n+\alpha+\beta)!}{(n+\alpha)!(n+\beta)!}P_n^{\alpha,\beta}(t),\\
P_{n+\beta}^{\alpha,-\beta}(t) &= 2^{-\beta}(t+1)^{\beta}\frac{n!(n+\alpha+\beta)!}{(n+\alpha)!(n+\beta)!}P_n^{\alpha,\beta}(t), \qquad (14.47)
\end{aligned}
$$

which are valid for integer $\alpha, \beta$. We have preferred to give two formulas, for $q \geq 1$ and for $q < 1$, to ensure that the proper range of $n$ is $n = 0, 1, \ldots$ in both cases.

[11]We quote the explicit expressions for first three polynomials:

$$
\begin{aligned}
P_0(\alpha,\beta) &= 1, \quad P_1(\alpha,\beta) = \alpha+1+\frac{\alpha+\beta+2}{2}(t-1),\\
P_2(\alpha,\beta) &= \frac{(\alpha+1)(\alpha+2)}{2}+\frac{(\alpha+2)(\alpha+\beta+3)}{2}(t-1)\\
&\quad+\frac{(\alpha+\beta+3)(\alpha+\beta+4)}{8}(t-1)^2. \qquad (14.50)
\end{aligned}
$$

The eigenfunctions and the eigenvalues in the sector $F = 1$ are given by the same expressions with the change $q \to -q$, $m \to -m, z \leftrightarrow \bar{z}$ (the eigenfunctions involve, of course, the extra Grassmann factor $\psi$). When $E \neq 0$, they can be obtained from the eigenfunctions (14.48) by the action of the supercharge $\hat{Q}$. Inversely, the eigenfunctions of the excited states in the sector $F = 0$ can be obtained from those in the sector $F = 1$ by the action of $\hat{\bar{Q}}$.

If one chooses $q = 1$ in the sector $F = 0$ (or $q = -1$ in the sector $F = 1$), the expressions are simplified. In this case, $\gamma = 0$ and the Hamiltonian reduces to the ordinary Laplacian. The eigenfunctions represent then the ordinary spherical harmonics. The spectrum (14.44) reduces to $E = n(n+1)$ as it should.

### 14.2.1 *Fractional magnetic charge*

For a fiber bundle on $S^2$, the magnetic charge must be an integer. The same is true for an ordinary Schrödinger problem with the Pauli Hamiltonian (14.10) including the field of a magnetic monopole: if $q$ is not integer, the angular wave function becomes singular at one point at least. This is so even if we allow for the wave functions not to coincide in the different charts but be related by a gauge transformation with a uniquely defined $\omega(x) = e^{i\chi(x)}$. Using a still different physical language, if $q$ is not integer, the Dirac string associated with the singularity of $A_M(x)$ becomes observable!

This looks weird, and that is why fractional charges are not usually considered, but we want to emphasize the following: all this *does* not mean yet that a spectral problem cannot be defined in this case. Indeed, we have seen in Section 9.3 that though $S^4$ is not a complex manifold, one can well define the Dolbeaux complex on $S^4\backslash\{\cdot\}$, and the corresponding spectral problem is supersymmetric. Is the same true for the Dirac problem with fractional magnetic charge on $S^2$? Can one define it on $S^2\backslash\{\cdot\}$ and keep the Hamiltonian supersymmetric? The answers to these two questions are different [150]:

**Theorem 14.7.** *1. A meaningful spectral problem can be defined even if the magnetic charge is fractional. 2. This problem is not supersymmetric.*

**Proof.** To define a spectral problem, we have to define the Hilbert space. As a first try, we do the same as we did in Section 9.3 for $S^4$ and include in the Hilbert space all square-integrable functions with the measure (14.36). With this condition, consider the spectrum of the Hamiltonian (14.37), where we do not assume $q$ to be integer any more. We obtain then *two* families of formal solutions to the Schrödinger equation:

$$\Psi_{nm} = \bar{z}^m (1+t)^\gamma P_n^{m,2\gamma-m}(t)$$
$$\text{with} \quad E_n = n(n+1+2\gamma) \tag{14.52}$$

and

$$\tilde{\Psi}_{nm} = \bar{z}^m (1+t)^{m-\gamma} P_n^{m,m-2\gamma}(t)$$
$$\text{with} \quad E_n = (n+m+1)(n+m-2\gamma). \tag{14.53}$$

The integers $n, m$ take the values $n = 0, 1, \ldots; m = 0, \pm 1, \ldots$, but we have to pick up only the normalizable solutions.

For an integer $q$, the normalizable wave functions in (14.53) are the same as in (14.52) [coinciding also with the functions (14.48)]. For example, three zero energy states for $q = 3$ represented in Fig. (14.1),

$$\Psi_{1,2,3} = \frac{1}{1+\bar{z}z}, \quad \frac{\bar{z}}{1+\bar{z}z}, \quad \frac{\bar{z}^2}{1+\bar{z}z}, \tag{14.54}$$

are described by (14.52), if choosing $n = 0,\ m = 0, 1, 2$, and also by (14.53), if choosing there $n = 2, m = 0$; $n = m = 1$; $n = 0, m = 2$ [the identities (14.47) are at work!].

But for a fractional charge, we meet a trouble. The functions (14.52) and (14.53) do not coincide anymore but it is not the only problem and not the most serious one. Take e.g. $q = \frac{1}{2}$, giving $\gamma = -\frac{1}{4}$, and choose $n = m = 0$. Equations (14.52) and (14.53) give two different eigenfunctions:

$$|1\rangle = (1+t)^{-1/4} = (1+\bar{z}z)^{1/4} \quad \text{with} \quad E = 0,$$
$$|2\rangle = (1+t)^{1/4} = \frac{1}{(1+\bar{z}z)^{1/4}} \quad \text{with} \quad E = \frac{1}{2}. \tag{14.55}$$

The function $|1\rangle$ is singular at the north pole $t = -1$, but it is square-integrable. The energies of the states (14.55) are different, but these states are not mutually orthogonal. And this means that the Hamiltonian (14.37) is not Hermitian in this case! One can also see it if one compares the matrix elements $\langle 1|\hat{H}|2\rangle$ and $\langle 2|\hat{H}|1\rangle$. They differ by a total derivative which does not vanish:

$$\begin{aligned}
&\langle 1|\hat{H}|2\rangle - \langle 2|\hat{H}|1\rangle \\
&\sim \int dz d\bar{z} \left[(1+\bar{z}z)^{-1/4}\bar{\partial}\partial(1+\bar{z}z)^{1/4} - (1+\bar{z}z)^{1/4}\bar{\partial}\partial(1+\bar{z}z)^{-1/4}\right] \\
&= \frac{1}{4}\int dz d\bar{z} \left[\bar{\partial}\left(\frac{\bar{z}}{1+\bar{z}z}\right) + \partial\left(\frac{z}{1+\bar{z}z}\right)\right] = \frac{1}{2}\int \frac{d\bar{z}dz}{(1+\bar{z}z)^2} \neq 0.
\end{aligned} \tag{14.56}$$

If one looks at the main reason of this mismatch, it is due to the fact that we included in the Hilbert space the functions that are singular on $S^2$. This suggests a remedy: let us *redefine* the Hilbert space in our problem and include there only the wave functions that are regular on $S^2$. Then, the state $|1\rangle$ is not admissible. We have no problem with hermiticity in the reduced Hilbert space and the spectral problem is well defined there. It is well defined in the sector $F = 0$ and also in the sector $F = 1$.

However, this spectral problem is not supersymmetric. To see this, consider the result of the action of the supercharge $\hat{Q}$ in (14.34) on the regular function $|2\rangle$. We obtain

$$\hat{Q}|2\rangle \sim \psi(1+\bar{z}z)\left[\partial - \frac{\bar{z}}{4(1+\bar{z}z)}\right]\frac{1}{(1+\bar{z}z)^{1/4}} \sim \frac{\psi\bar{z}}{(1+\bar{z}z)^{1/4}}. \tag{14.57}$$

*This* function is not regular as $z = \infty$. In other words, the state $|2\rangle$ is admissible and benign as far as the Hamiltonian spectral problem is concerned, but it does not belong to the *domain* of $\hat{Q}$: the action of $\hat{Q}$ brings us outside of the Hilbert space where the Hamiltonian is defined. As a result, certain states do not have superpartners and supersymmetry is lost. □

A similar phenomenon (the lost of apparent supersymmetry due to the fact that the superpartners of some states lie outside the Hilbert space and thus do not belong to the spectrum of the Hamiltonian) is known to show up for some other systems. The simplest example [151] is probably Witten's supersymmetric Hamiltonian (5.28):

$$\hat{H} = \frac{\hat{P}^2 + [W'(x)]^2}{2} + \frac{1}{2}W''(x)(\psi\hat{\bar{\psi}} - \hat{\bar{\psi}}\psi). \tag{14.58}$$

with the superpotential $W(x) = \omega x^2/2 - \ln x$, considered on the half-line $x > 0$. In the bosonic sector $F = 0$, the Hamiltonian is

$$\hat{H}_B = \frac{\hat{P}^2}{2} + \frac{\omega^2 x^2}{2} - \frac{3\omega}{2}. \tag{14.59}$$

The ground state $\Psi_0 \propto \exp\{-\omega^2 x^2/2\}$ has the negative energy $E = -\omega$, which is obviously inconsistent with supersymmetry. The reason for the trouble is that the function $\Psi_0$ does not belong to the domain of the supercharge $\hat{Q} \propto \hat{P} - iW'$: the function $\hat{Q}\Psi_0$ behaves as $\propto 1/x$ near $x = 0$ and is not square-integrable. To make the spectrum supersymmetric, we should restrict in this case the Hilbert space and consider only the functions that vanish at the origin, $\Psi(0) = 0$. With this restriction, the ground state $\tilde{\Psi}_0(x) = x\Psi_0(x)$ has zero energy.

On the other hand, for a magnetic field with noninteger flux, there is no way to make the Pauli Hamiltonian supersymmetric. We have just shown that the Hilbert space of regular functions on $S^2$ does not constitute the domain of $\hat{Q}$, but this is also true for any other restricted or enhanced Hilbert space. We address the reader to Ref. [150] for further details.

## 14.3 Gauge fields on $\mathbb{CP}^n$

Another name for 2-sphere is $\mathbb{CP}^1$. But one can also consider $\mathbb{CP}^{n>1}$ and study the gauge fields and the spectra of the corresponding SQM models there. This is our next task.

### 14.3.1 *Line bundles*

In the previous section, we showed that the model defined by Eqs. (14.29) and (14.30) maintains its apparent supersymmetry under quantization iff the topological charge $q$ is integer. The same concerns the cases when the sphere is "crumpled" and/or magnetic field is not homogeneous. Indeed, we know from Theorem 9.2 that, for Kähler manifolds (and $\mathbb{CP}^1$ is Kähler), the Dirac complex with the gauge field (14.19) maps to the twisted Dolbeault complex with a "shifted" gauge field

$$\tilde{A}_M = \left[ i\partial_m \left( \mathcal{W} - \frac{1}{4} \ln \det h \right), \; -i\partial_{\bar{m}} \left( \mathcal{W} - \frac{1}{4} \ln \det h \right) \right]. \tag{14.60}$$

For a complex manifold, a "pure" Dolbeault Hamiltonian $H^{\text{Dolb}} = \{\partial, \partial^\dagger\}$ is well defined and Hermitian. The same concerns the twisted Dolbeault Hamiltonian with the gauge field (14.60) provided that $\tilde{A}_M$ represents a fiber bundle and one can safely travel between the charts. Then, according to Theorem 14.6, the flux $\int_{S^2} \tilde{F}$ is quantized.

It is worth to emphasize that the line bundle for which Theorem 14.6 was proven is $\tilde{A}_M$ rather than $A_M = (i\partial_m \mathcal{W}, -i\partial_{\bar{m}} \mathcal{W})$: we start with the pure Dolbeault complex, which is well defined, then twist it by adding $\tilde{A}_M$ and require a benign behavior of the latter during the transitions from one chart to another. For $\mathbb{CP}^1$ (in contrast to $\mathbb{CP}^{2k}$ — see the following), it does not matter, however: the fluxes of the fields $F$ and $\tilde{F}$ are related as

$$\int_{S^2} F = \int_{S^2} \tilde{F} + 2\pi. \tag{14.61}$$

If $\int \tilde{F}$ is quantized to $2\pi q$, this is also true for $\int F$.

Let us now go to $\mathbb{CP}^{n>1}$. Each such manifold can be glued up from $n+1$ disks of complex dimension $n$ (see p. 45). In each chart, the metric can be chosen in the Fubini–Study form (2.80):

$$h_{m\bar{p}} = \frac{1}{1+\bar{z}^k z^k}\left(\delta_{m\bar{p}} - \frac{z^p \bar{z}^{\bar{m}}}{1+\bar{z}^k z^k}\right), \quad k = 1, \ldots, n. \tag{14.62}$$

Introduce now the gauge field with the prepotential

$$\mathcal{W} = \frac{q}{2(n+1)} \ln \det h. \tag{14.63}$$

We show that the parameter $q$ is not arbitrary but quantized. In this section, we prove it by pure geometric reasoning, and in the following chapter, we derive the quantization of $q$ using the supersymmetric methods, following the philosophy of our book.

**Theorem 14.8.** *For the field* (14.60) *with* $\mathcal{W}$ *given by* (14.63) *to represent a benign line bundle on* $\mathbb{CP}^n$ *so that the twisted Dolbeault complex is well defined,* $q$ *should be integer if* $n$ *is odd and half-integer if* $n$ *is even.*

**Proof.** Equation (14.62) implies that

$$\det h = \frac{1}{(1+\bar{z}^k z^k)^{n+1}}. \tag{14.64}$$

Then, Eq. (14.60) gives

$$\tilde{A}_m = i\left(\frac{n+1}{4} - \frac{q}{2}\right)\frac{\bar{z}^m}{1+\bar{z}^k z^k} \tag{14.65}$$

and the complex-conjugated expression for $\tilde{A}_{\bar{m}}$. This field is defined in the chart $\mathcal{C}_0$ including the origin $z^k = 0$ but not including the points of $\mathbb{CP}^n$ where one of $z^k$ hits infinity. To describe the whole manifold, one needs $n$ extra charts with the coordinates $\mathcal{C}_1 : (u^1 = 1/z^1, u^{2,\ldots,n} = z^{2,\ldots,n}/z_1)$, $\mathcal{C}_2 : (v^2 = 1/z^2, v^{1,3,\ldots,n} = z^{1,3,\ldots,n}/z^2)$, etc. We want to prolongate $\tilde{A}_M$ to these charts by performing a gauge transformation (14.23) in the overlap regions in such a way that the transformed field would not be singular.

When $z^1 \to \infty$, the gauge field (14.65) behaves as

$$\tilde{A}_1(z) \sim \frac{i\alpha}{z^1}, \quad \tilde{A}_2(z) \sim \frac{i\alpha\bar{z}^2}{\bar{z}^1 z^1}, \quad \tilde{A}_3(z) \sim \frac{i\alpha\bar{z}^3}{\bar{z}^1 z^1}, \ldots, \tag{14.66}$$

where $\alpha = (n+1)/4 - q/2$.

At first sight, it is not singular at $z^1 \to \infty$ and an extra gauge transformation is not needed, but one should recall that $\tilde{A}_M$ is a covariant vector, whose holomorphic component transforms under a variable change as

$$\tilde{A}_m(u) = \frac{\partial z^n}{\partial u^m} \tilde{A}_n(z). \tag{14.67}$$

We then see that $\tilde{A}_{2,3,\ldots}(u^k)$ are regular at $u^1 = 0$, but $\tilde{A}_1(u^k) \sim -i\alpha/u^1$ is singular and a gauge transformation (14.23) *is* required. We should perform it in the overlap region $\mathcal{C}_0 \cap \mathcal{C}_1$. The topology of the latter is $S^1 \times \mathbb{C}^{n-1}$. The function $\omega(x) = e^{i\chi(x)}$ defines a mapping $S^1 \times \mathbb{C}^{n-1} \to U(1)$, which has the same distinct topological classes as the mapping $S^1 \to U(1)$ that we had for $\mathbb{CP}^1$. Only certain discrete values of $\alpha$ are admissible. To find them, recall our convention:

$$u^1 = \frac{x + iy}{\sqrt{2}}, \quad \tilde{A}_1 = \frac{\tilde{A}_x - i\tilde{A}_y}{\sqrt{2}}. \tag{14.68}$$

Then, the singular part of $\tilde{A}_1 \sim -i\alpha/u^1$ is

$$\tilde{A}_x \sim -\frac{2\alpha y}{x^2 + y^2}, \quad \tilde{A}_y \sim \frac{2\alpha x}{x^2 + y^2}. \tag{14.69}$$

Formally, one can get rid of it applying the gauge transformation (14.23) with $\chi(x, y) = 2\alpha \arctan(y/x)$. To assure for $e^{i\chi(x,y)}$ to be uniquely defined, $2\alpha$ should be integer. In other words,

$$\boxed{s = q - \frac{n+1}{2} = \text{integer},} \tag{14.70}$$

from which the statement of the theorem follows. □

**Remark:** A corollary of this theorem is the fact mentioned above: $\mathbb{CP}^n$ with even $n$ do not admit spin structure; the "purely gravitational" Dirac operator (14.6) with $A_M = 0$ (implying $q = 0$) *is* not well defined in this case — there is a topological obstacle for that. For odd $n$, the value $q = 0$ is admissible.

### 14.3.2 *Vacuum states*

Similarly to what was the case for $\mathbb{CP}^1$, one can also completely solve the spectral problem and determine the spectra and the eigenstates for the ordinary and supersymmetric Hamiltonians for $\mathbb{CP}^{n>1}$ endowed with the metric (14.62) and the gauge field (14.65). We start our discussion with the vacuum states in the SQM model. As was explained in Chapter 6 and will be discussed further in the last chapter, their number is given by the Witten index and represents a topological invariant.

The vacuum states represent normalized solutions to the equations

$$\hat{Q}^{\rm cov}\Phi_0 = \hat{\bar{Q}}^{\rm cov}\Phi_0 = 0, \tag{14.71}$$

where the covariant supercharges were defined in Eqs. (9.12) and (9.14):

$$\begin{aligned} \hat{Q}^{\rm cov} &= -i\psi^k \left[\frac{\partial}{\partial z^k} + \frac{1}{2}\left(q - \frac{n+1}{2}\right)\frac{\bar{z}^k}{1+\bar{z}z} + \psi^a\hat{\bar{\psi}}^{\bar{b}}\omega_{a\bar{b},k}\right], \\ \hat{\bar{Q}}^{\rm cov} &= -i\hat{\bar{\psi}}^{\bar{k}} \left[\frac{\partial}{\partial \bar{z}^k} - \frac{1}{2}\left(q + \frac{n+1}{2}\right)\frac{z^k}{1+\bar{z}z} + \psi^a\hat{\bar{\psi}}^{\bar{b}}\omega_{a\bar{b},\bar{k}}\right] \end{aligned} \tag{14.72}$$

with the shorthand $\bar{z}z \equiv \bar{z}^l z^l$. By "normalized" we mean normalized with the covariant measure:

$$\int \prod_k dz^k d\bar{z}^k \prod_a d\psi^a d\bar{\psi}^a \times \frac{1}{(1+\bar{z}z)^{n+1}} e^{-\psi^b\bar{\psi}^b}\, \bar{\Phi}_0(z^k, \bar{z}^k, \bar{\psi}^a)\Phi_0(z^k, \bar{z}^k, \psi^a) = 1. \tag{14.73}$$

An analysis shows that the solutions to (14.71) may exist only in the sector of zero fermion charge or of fermion charge $n$. Let first $q$ be positive. In that case, the normalized solutions to (14.71) may exist only in the sector with zero fermion charge. Indeed, for the states of fermion charge $n$, the equation $\hat{Q}^{\rm cov}\Phi_0 = 0$ is satisfied automatically, while the equation $\hat{\bar{Q}}^{\rm cov}\Phi_0 = 0$ gives

$$\partial_{\bar{k}}\Phi_0 = c\frac{z^k}{1+\bar{z}z}\Phi_0$$

with $c > 0$. This equation has no normalized solutions.

Consider the states of zero fermion charge. In this case, the equation $\hat{\bar{Q}}^{\rm cov}\Phi_0 = 0$ is satisfied automatically, while the equation $\hat{Q}^{\rm cov}\Phi_0 = 0$ implies

$$\partial_k\Phi_0 = -\frac{s\bar{z}^k}{2(1+\bar{z}z)}\Phi_0 \tag{14.74}$$

with $s$ defined in (14.70). The normalized solutions to (14.74) exist only for nonnegative $s$. They read

$$\Phi_0 = (1+\bar{z}z)^{-s/2}P(\bar{z}), \tag{14.75}$$

where $P(\bar{z})$ is a polynomial of $\bar{z}^k$ of the degree not higher than $s$. The dimension of the vector space of such polynomials is equal to

$$\boxed{I_W^{q>0} = \binom{n+s}{n} = \binom{q+\frac{n-1}{2}}{n}}, \tag{14.76}$$

and this is the Witten index of our model [152,153]. It is not zero if $q \geq \frac{n+1}{2}$.

For negative $q$, the vacuum states are present only in the sector of fermion charge $F = n$. In this case, the equation $\hat{Q}^{\text{cov}}\Phi_0 = 0$ is satisfied automatically and the equation $\hat{\bar{Q}}^{\text{cov}}\Phi_0 = 0$ has normalized solutions provided $q \leq -\frac{n+1}{2}$. Their number is

$$I_W^{q<0} = \binom{\frac{n-1}{2} - q}{n}. \tag{14.77}$$

For odd $n$, the vacuum states in the sector $F = n$ are fermionic, and the index acquires the factor $(-1)^n$.

If $|q| < \frac{n+1}{2}$, the index is zero, which signalizes spontaneous breaking of supersymmetry.[12]

### 14.3.3 *Excited states*

The Witten index has topological nature. The number of vacuum states is thus robust, not depending on a particular form of the metric. Their wave functions satisfy the first-order differential equations like in (14.74), which are much simpler than the Schrödinger equation and can be solved also for "crumpled" 2-sphere and for "crumpled" $\mathbb{CP}^n$. But if the sphere is "round" being given by the metric (14.27), one can also solve the *full* spectral problem and determine the energies and wave functions of the excited states. The same concerns $\mathbb{CP}^{n>1}$. If the metric is "round" [has $SU(n+1)$ isometry] and is given by the Fubini–Study expression (14.62), the full spectral problem can be solved.

In nonsupersymmetric case, this was first done by Kuwabara [156] by mathematical methods: he considered the bundle $S^{2n+1} \to \mathbb{CP}^n$ and picked up the spherical harmonics on $S^{2n+1}$ that "survive" the projection. For supersymmetric Hamiltonian, it was done in Ref. [157]. The results coincide, as they should, with the results of Ref. [158] who studied the spectrum

[12]Note in passing that the index (14.76) is closely related to the Witten index in the three-dimensional supersymmetric Yang–Mills–Chern–Simons field theory [154]. See Ref. [155] or Chapter 10 of Ref. [75] for the detailed discussion.

of the Dirac operator. As we repeatedly mentioned above, the SQM problem, which boils down to the Dolbeault problem, and the Dirac problem are equivalent for Kähler manifolds.

#### 14.3.3.1 *Bosonic model*

We start with discussing the simplest problem of a point particle moving along $\mathbb{CP}^n$ in the absence of external fields. It is convenient to express the metric of $\mathbb{CP}^n$ in the form (2.79) depending on $n+1$ homogeneous coordinates $(w^0, \ldots, w^n) \equiv (\lambda w^0, \ldots, \lambda w^n)$:

$$ds^2 = 2\frac{\bar{w}^\alpha w^\alpha \, d\bar{w}^\beta dw^\beta - \bar{w}^\alpha w^\beta \, d\bar{w}^\beta dw^\alpha}{(\bar{w}^\alpha w^\alpha)^2}. \tag{14.78}$$

The $SU(n+1)$ isometry of this metric is explicitly seen — that is why the homogeneous description is so convenient.

The Lagrangian of the corresponding sigma model reads[13]

$$L = \frac{\dot{w}^\alpha \Pi_{\alpha\bar{\beta}} \dot{\bar{w}}^\beta}{\mathcal{X}}, \tag{14.79}$$

where $\mathcal{X} = w^\alpha \bar{w}^\alpha$ and $\Pi_{\alpha\bar{\beta}} = \delta_{\alpha\bar{\beta}} - \bar{w}^\alpha w^\beta/\mathcal{X}$ is the projector on the directions orthogonal to vector $w^\alpha$:

$$w^\alpha \Pi_{\alpha\bar{\beta}} = \bar{w}^\beta \Pi_{\alpha\bar{\beta}} = 0. \tag{14.80}$$

The Lagrangian (14.79) is invariant under the gauge transformations

$$w^\alpha \to \lambda(t) w^\alpha, \quad \lambda(t) \in \mathbb{C}^*. \tag{14.81}$$

The canonical momenta are

$$\pi_\alpha = \frac{\Pi_{\alpha\bar{\beta}} \dot{\bar{w}}^\beta}{\mathcal{X}}, \quad \bar{\pi}_\alpha = \frac{\Pi_{\beta\bar{\alpha}} \dot{w}^\beta}{\mathcal{X}}. \tag{14.82}$$

As a consequence of the gauge symmetry (14.81), they obey the constraints

$$w^\alpha \pi_\alpha = \bar{w}^\alpha \bar{\pi}_\alpha = 0, \tag{14.83}$$

in agreement with (14.80). The canonical classical Hamiltonian is

$$H = \mathcal{X} \pi_\alpha \bar{\pi}_\alpha. \tag{14.84}$$

To quantize the theory, we define the operators $\hat{\pi}_\alpha = -i\partial/\partial w^\alpha$, $\hat{\bar{\pi}}_\alpha = -i\partial/\partial \bar{w}^\alpha$. To write the quantum version of (14.84), we should prescribe a

[13]We are following Ref. [157] with a slightly changed convention: our $n$ is $n-1$ of Ref. [157]!

particular way of ordering of the operators in $\hat{H}$. We make the following choice:

$$\boxed{\hat{H} = -\mathcal{X}\,\frac{\partial^2}{\partial w^\alpha \partial \bar{w}^\alpha} \stackrel{\text{def}}{=} -\mathcal{X}\triangle,} \tag{14.85}$$

while the quantum constraints are chosen as

$$w^\alpha \frac{\partial}{\partial w^\alpha}\Psi = \bar{w}^\alpha \frac{\partial}{\partial \bar{w}^\alpha}\Psi = 0. \tag{14.86}$$

We are going to solve the spectral problem $\hat{H}\Psi = E\Psi$ in the Hilbert space involving the constraints (14.86) and equipped with the norm[14]

$$\langle\Psi|\Psi\rangle = \int \prod_{\alpha=0}^{n} dw^\alpha d\bar{w}^\alpha \exp\{-\mathcal{X}\}\,|\Psi|^2. \tag{14.89}$$

The eigenfunctions of the Hamiltonian (14.85) are easy to find. They are

$$\Psi(w,\bar{w}) = \frac{1}{\mathcal{X}^L} B_{\alpha_1\ldots\alpha_L|\bar{\beta}_1\ldots\bar{\beta}_L} w^{\alpha_1}\cdots w^{\alpha_L}\,\bar{w}^{\beta_1}\cdots\bar{w}^{\beta_L}, \tag{14.90}$$

where $L$ is a nonnegative integer and $B$ is a *traceless* tensor, $B_{\gamma\ldots\alpha_L|\bar{\gamma}\ldots\bar{\beta}_L} = 0$.

The ground state of $\hat{H}$ is $\Psi = const.$ The corresponding eigenvalue is $E = 0$. At the next level ($L = 1$), we have the functions

$$\Psi = \frac{B_{\alpha|\bar{\beta}}\, w^\alpha \bar{w}^\beta}{\mathcal{X}} \tag{14.91}$$

with $B_{\alpha|\bar{\alpha}} = 0$. There are $n^2+2n = (n+1)^2-1$ such independent functions. All of them have the energy $E = n+1$. These $(n+1)^2-1$ degenerate states

[14]One could take any rapidly falling function of $\mathcal{X}$ as the measure, but $e^{-\mathcal{X}}$ is the most natural choice. If we redefine the problem in terms of inhomogeneous coordinates $z^{j=1,\ldots,n} = w^j/w^0$, we arrive at the familiar from the previous chapters quantum problem with the unconstraint Hamiltonian

$$\hat{H} = -h^{\bar{k}j}\partial_j\bar{\partial}_{\bar{k}} = -(1+z^l\bar{z}^l)(\bar{z}^k z^j + \delta^{\bar{k}j})\,\partial_j\bar{\partial}_{\bar{k}} \tag{14.87}$$

and the norm

$$\langle\Psi|\Psi\rangle = \int \prod_{j=1}^{n} dz^j d\bar{z}^j\ \det(h)\,|\Psi|^2. \tag{14.88}$$

belong to the adjoint representation of $SU(n+1)$. The full spectrum is

$$\boxed{E(n,L) = L(L+n).} \tag{14.92}$$

The states at the level $L$ belong to the $SU(n+1)$ multiplet with the Dynkin labels

$$a_{i=1,\ldots,n} = (L, \underbrace{0, \ldots, 0,}_{n-2} L). \tag{14.93}$$

Such multiplet includes

$$\#(n,L) = \binom{L+n-1}{n-1}^2 \left(1 + \frac{2L}{n}\right) \tag{14.94}$$

degenerate states. For example, for $\mathbb{CP}^2$, $\#(2,L) = (L+1)^3$. We have a singlet, an octet, a 27-plet, etc.

We include now gauge field. We choose it in the form

$$A_M = \left(-\frac{im\bar{z}^j}{2(1+\bar{z}z)}, \ \frac{imz^j}{2(1+\bar{z}z)}\right). \tag{14.95}$$

Such a field represents a natural generalization of the Dirac monopole field (14.31). One can call this object *CP monopole.* The integer $m$ plays the role of magnetic charge.[15]

In the homogeneous coordinates, the vector potentials reads

$$A_\alpha = -i\frac{m\bar{w}^\alpha}{2\mathcal{X}}, \quad \bar{A}_\alpha = i\frac{mw^\alpha}{2\mathcal{X}}. \tag{14.96}$$

The Lagrangian (14.79) acquires an extra contribution:

$$L_m = L - \dot{w}^\alpha A_\alpha - \dot{\bar{w}}^\alpha \bar{A}_\alpha. \tag{14.97}$$

The canonical Hamiltonian is

$$H_m = \mathcal{X}(\pi_\alpha + A_\alpha)(\bar{\pi}_\alpha + \bar{A}_\alpha). \tag{14.98}$$

For its quantum counterpart, we choose a natural expression

$$\begin{aligned} \hat{H}_m &= -\frac{\mathcal{X}}{2}\{(\partial_\alpha + iA_\alpha), (\bar{\partial}_\alpha + i\bar{A}_\alpha)\} \\ &= -\frac{\mathcal{X}}{2}\left\{\left(\frac{\partial}{\partial w^\alpha} + \frac{m\bar{w}^\alpha}{2\mathcal{X}}\right), \left(\frac{\partial}{\partial \bar{w}^\alpha} - \frac{mw^\alpha}{2\mathcal{X}}\right)\right\} \end{aligned} \tag{14.99}$$

[15] We use here the notation $m$ rather than $q$ not to confuse it with $q$ that enters the superpotential (14.30) and which is half-integer for even $n$. We discuss the supersymmetric $\mathbb{CP}^n$ problem a bit later.

— the operator known as *Bochner Laplacian*. The constraints are modified as follows:

$$w^\alpha \frac{\partial}{\partial w^\alpha}\Psi = -\frac{m}{2}\Psi, \quad \bar{w}^\alpha \frac{\partial}{\partial \bar{w}^\alpha}\Psi = \frac{m}{2}\Psi. \tag{14.100}$$

Bearing in mind these constraints, the Hamiltonian (14.99) acquires a simple form

$$H_m = -\mathcal{X}\triangle - \frac{m^2}{4} \tag{14.101}$$

with $\triangle = \partial_\alpha \bar{\partial}_\alpha$. As was the case for the problem (14.85)–(14.86), the quantum problem (14.99)–(14.100) possesses $SU(n+1)$ symmetry. The eigenfunctions have the form

$$\Psi(w,\bar{w}) = \frac{1}{\mathcal{X}^{\frac{L+L'}{2}}} B_{\alpha_1\ldots\alpha_L|\bar{\beta}_1\ldots\bar{\beta}_{L'}} w^{\alpha_1}\cdots w^{\alpha_L}\,\bar{w}^{\beta_1}\cdots\bar{w}^{\beta_{L'}}, \tag{14.102}$$

where $B$ is a traceless tensor and $L$ and $L'$ are nonnegative integers with $L' - L = m$.

The eigenfunctions form $SU(n+1)$ multiplets that are characterized by the sets of Dynkin labels

$$(L, \underbrace{0, \ldots, 0,}_{n-2} L'). \tag{14.103}$$

($n \geq 2$). The dimensions of these representations are

$$\#(n,L,L') = \binom{L+n-1}{n-1}\binom{L'+n-1}{n-1}\left(1 + \frac{L+L'}{n}\right). \tag{14.104}$$

The energies are

$$\boxed{E(n,L,L') = LL' + n\frac{L+L'}{2}.} \tag{14.105}$$

#### 14.3.3.2 *Supersymmetric model*

The superfield action of the model in interest is

$$S = \int d\bar{\theta}d\theta dt \left[\frac{1}{4}h_{m\bar{k}}(Z,\bar{Z})\bar{D}\bar{Z}^{\bar{k}}DZ^m + \mathcal{W}(Z,\bar{Z})\right] \tag{14.106}$$

with the metric (14.62) and the prepotential (14.63). We go down to components and express it in homogenious coordinates including, on top of $w^\alpha$, their Grassmann superpartners $\chi^\alpha$. The quantum supercharges and the Hamiltonian were derived in this terms in Ref. [157]. They have the form

$$\hat{Q} = -i\chi^\alpha \frac{\partial}{\partial w^\alpha}, \quad \hat{\bar{Q}} = -i\mathcal{X}\frac{\partial^2}{\partial\chi^\beta \partial\bar{w}^\beta}, \tag{14.107}$$

$$\hat{H} = \{\hat{Q}, \hat{\bar{Q}}\} = -\mathcal{X}\triangle - \bar{w}^\alpha\chi^\alpha \frac{\partial^2}{\partial\chi^\beta\partial\bar{w}^\beta}. \tag{14.108}$$

The admissible wave functions $\Psi(w^\alpha, \bar{w}^\alpha; \chi^\alpha)$ satisfy two bosonic and one Grassmann constraints:

$$i\hat{C}_1\Psi = \left(w^\alpha \frac{\partial}{\partial w^\alpha} + \chi^\alpha \frac{\partial}{\partial \chi^\alpha}\right)\Psi = 0,$$

$$i\hat{C}_2\Psi = \bar{w}^\alpha \frac{\partial}{\partial \bar{w}^\alpha}\Psi = s\Psi, \quad i\hat{C}_3\Psi = w^\alpha\frac{\partial}{\partial\chi^\alpha}\Psi = 0 \tag{14.109}$$

with $s = q - (n+1)/2$. They are compatible with supersymmetry due to

$$[\hat{Q}, \hat{C}_1] = [\hat{Q}, \hat{C}_2] = 0, \quad \{\hat{Q}, \hat{C}_3\} = -i\hat{C}_1,$$
$$[\hat{\bar{Q}}, \hat{C}_1] = [\hat{\bar{Q}}, \hat{C}_2] = \{\hat{\bar{Q}}, \hat{C}_3\} = 0. \tag{14.110}$$

Here, $s$ must be integer, as is dictated by Theorem 14.8. It plays the role of $m$ of the bosonic model, to which the supersymmetric problem in the sector $F = 0$ is isomorphic. Note, however, the absence of the shift $-s^2/4$ in the Hamiltonian (14.108) and an asymmetric form of the constraints (14.109).

The supercharges (14.107) and the Hamiltonian (14.108) act on the Hilbert space with the inner product

$$\langle\Psi_1|\Psi_2\rangle = \int \bar{\Psi}_1\Psi_2\,\mathcal{X}^{n+1}\exp\left\{-\mathcal{X} - \frac{\chi^\gamma\bar{\chi}^\gamma}{\mathcal{X}}\right\}\prod_{\alpha=0}^{n} dw^\alpha d\bar{w}^\alpha d\chi^\alpha d\bar{\chi}^\alpha. \tag{14.111}$$

Then, the supercharges $\hat{Q}$ and $\hat{\bar{Q}}$ can be shown to be mutually conjugated and the Hamiltonian is Hermitian. The wave functions in the sectors with different fermion numbers have the form

$$\Psi^{F=0} = \Psi^{(0)}(w, \bar{w}), \quad \Psi^{F=1} = \Psi^{(1)}_\alpha(w, \bar{w})\chi^\alpha, \quad \text{etc.} \tag{14.112}$$

The maximal fermion charge is $F = n$ rather than $F = n+1$ because the function $\Psi \propto \prod_{\alpha=0}^{n}\chi^\alpha$ is not annihilated by the constraint $\hat{C}_3$ and is not admissible. This conforms to the standard description of the model with inhomogeneous variables where the number of fermion variables $\psi^{j=1,\ldots,n}$ is less by one.

The supersymmetric spectral problem is somewhat more complicated than the bosonic one due to the presence of the sectors with different fermion charges. We not go in all the details here referring the reader to Ref. [157].

The problem simplifies for the sectors with the minimal ($F = 0$) and maximal ($F = n$) fermion charges, where it is slightly more convenient to use inhomogeneous variables $z^j$ and $\psi^j$ and where the wave functions have effectively only one component: $\Psi^{F=0} = A(z^j, \bar{z}^j)$ and $\Psi^{F=n} = B(z^j, \bar{z}^j)\prod_{j=1}^{n}\psi^j$. The spectrum is[16]

$$E^{F=0}(n, q, L) = L(L + s + n), \quad L \geq \max(0, -s). \tag{14.113}$$

The spectrum in the sector $F = n$ is the same as in the sector $F = 0$ with the opposite sign of $q$.

In the intermediate sectors $0 < F < n$, it is more convenient to use homogeneous variables and focus on the *lower* components $\Psi_-$ of the supersymmetric doublets annihilated by $\hat{\bar{Q}}$. One can show that the functions $\Psi_-$ have the form

$$\begin{aligned}&\Psi_-(w, \bar{w}, \chi)\\&\quad = \frac{1}{\mathcal{X}^M} B_{\alpha_1\ldots\alpha_L|\bar{\beta}_1\ldots\bar{\beta}_{L'}|\gamma_1\ldots\gamma_F} w^{\alpha_1}\cdots w^{\alpha_L}\bar{w}^{\beta_1}\cdots\bar{w}^{\beta_{L'}}\chi^{\gamma_1}\cdots\chi^{\gamma_F},\end{aligned} \tag{14.114}$$

where $M = L + F$, $L' = M + s$ and the tensor $B$ satisfies the $\alpha\beta$ traceless condition:

$$B_{\delta\ldots\alpha_L|\bar{\delta}\ldots\bar{\beta}_{L'}|\gamma_1\ldots\gamma_F} = 0. \tag{14.115}$$

The energies are

$$\boxed{E_-(n, L, L', F) = (L + F)(L' - F + n),} \tag{14.116}$$

where $F = 0, \ldots, n-1$, $L' = L + F + s \geq 0$. When $F = 0$, the integer $L$ may also be positive or zero, but for nonzero $F$, $L$ is necessarily positive —

[16] For $n = 1$, it coincides with Eq. (14.45), as it should. For any $n$, it coincides with the spectrum (14.105),

$$E = L(L + m) + n\left(L + \frac{m}{2}\right),$$

of the bosonic model, with $m$ to be substituted by $s$ and an overall shift by $-ns/2$ added (this shift comes from slightly different conventions in the formulation of the quantum problems).

this is a corollary of the fermionic constraint $w^\alpha \partial\Psi/\partial\chi^\alpha = 0$. It cannot be fulfilled if the function (14.114) has no $w$ factors.[17]

The wave functions of the upper components may be found by acting on (14.114) with $\hat{Q}$.

The eigenstates of the Hamiltonian belong to degenerate $SU(n+1)$ multiplets. Nontrivialities start from $\mathbb{CP}^3$ for the multiplets whose lower components belong to the sector $F = 1$ and upper components to the sector $F = 2$ — such states cannot be "reached" from the sectors with the maximal or minimal fermion charge. The Dynkin labels of all such multiplets are

$$a_{i=1,\ldots,n} = (L-1, \underbrace{0, \ldots, 0}_{F-1}, 1, \underbrace{0, \ldots, 0}_{n-2-F}, L'), \qquad (14.117)$$

where $n \geq 3$ and $F$ $(1 \leq F \leq n-2)$ refers to the fermion charge of the lower component.

Dimensions of the representations are expressed via the Dynkin labels by the Weyl formula (see e.g. Eq. (5.5) in the review [159]) or the hook formula if one uses the equivalent description in terms of Young diagrams. The expressions thus derived are, however, rather bulky (see Eq. (51) in Ref. [158]).

## 14.4 Non-Abelian Fields □ Instantons on $S^4$

### 14.4.1 *Non-Abelian bundles*

In Chapters 5, 6, 9 and 13 and up to now in this chapter, we have only discussed Abelian gauge fields, the line bundles in the mathematical language. But the general construction of the principal fiber bundles outlined in Section 1.3 applies also to non-Abelian groups. The connection $A_M(x)$ belongs then to the Lie algebra of any simple or semi-simple gauge group $\mathcal{G}$. The connections in the different charts $\mathcal{C}_j$ are related as in (1.46) with the group-valued transition functions $\omega(x)$ that are uniquely defined in the overlap regions $\mathcal{C}_j \cap \mathcal{C}_k$. An interesting mathematics (and physics) emerges when the mappings $\mathcal{C}_j \cap \mathcal{C}_k \to \mathcal{G}$ are not trivial. For simple non-Abelian $\mathcal{G}$, this happens when the dimension of the manifold is $D = 4$ or more. If $D = 4$ (the most interesting case for physical applications), the overlap regions can have the topology of $S^3$. Topologically nontrivial mappings correspond then to nontrivial elements of $\pi_3(\mathcal{G})$.

[17] And it can if such factors are present. The simplest example is $\Psi = w^1\chi^2 - w^2\chi^1$.

Note that there may also be another source for nontrivial topology. The elements of the center of $G$ do not act faithfully on the gauge fields belonging to the adjoint representation. This means that the relevant gauge group is $\mathcal{G}/Z_{\mathcal{G}}$ rather than just $\mathcal{G}$. The fundamental group of $\mathcal{G}/Z_{\mathcal{G}}$ may be nontrivial. For example, for the unitary groups, $\pi_1[SU(N)/Z_N] = Z_N$. If the fundamental group of the four-dimensional manifold we are interested in is also nontrivial, this can bring about the configurations with *fractional* Pontryagin number (14.122). These configurations were first discovered by 't Hooft and are called *torons* [160]. Nontrivial $\pi_1[SU(N)/Z_N]$ may also bring about nontrivial non-Abelian bundles on two-dimensional manifolds with interesting physical consequencies for two-dimensional non-Abelian gauge theories including fermions in the adjoint group representation [161].

But these complications are beyond the scope of our book. Here, we consider only the simplest non-Abelian bundle that has the gauge group $\mathcal{G} = SU(2)$ and is defined on $S^4$. $S^4$ is simply connected, and there are no torons. $S^4$ can be covered by two charts — the northern and the southern hemispheres. The overlap region has the topology of $S^3$ and only $\pi_3(\mathcal{G})$ is relevant.

Topologically distinct bundles are characterized by an integer number — an element of $\pi_3[SU(2)] = \pi_3(S^3) = \mathbb{Z}$. Let us first derive an algebraic expression for this integer, an analog of the expression

$$q = -\frac{i}{2\pi}\int_{S^1} \omega^{-1} d\omega, \quad \omega \in U(1) \tag{14.118}$$

for the mapping $S^1 \to U(1)$ [see (14.24)].

**Theorem 14.9.** *The topological class of the mapping $S^3 \to SU(2)$ is given by the expression*

$$\boxed{q = \frac{1}{24\pi^2}\int_{S^3} \mathrm{Tr}\left\{\omega^{-1}d\omega \wedge \omega^{-1}d\omega \wedge \omega^{-1}d\omega\right\},} \qquad \omega \in SU(2). \tag{14.119}$$

**Proof.** Consider first the similar integral over the proper $SU(2)$ group:

$$q_{SU(2)} = \frac{1}{24\pi^2}\int_{SU(2)} \mathrm{Tr}\left\{\omega^{-1}d\omega \wedge \omega^{-1}d\omega \wedge \omega^{-1}d\omega\right\}. \tag{14.120}$$

The integrand in (14.120) is a 3-form $d\mu$ that is invariant under a group multiplication:

$$d\mu(\omega\alpha) = d\mu(\alpha\omega) = d\mu(\omega), \tag{14.121}$$

where $\alpha$ is an arbitrary constant $SU(2)$ matrix. Up to a coefficient, there is unique 3-form enjoying this property — the *invariant volume form.* The coefficient $1/24\pi^2$ has been chosen in such a way that the integral (14.120) is equal to 1.

The latter statement can be checked in an explicit calculation. As the integral does not depend on the parameterization, one may use any convenient one. One can e.g. choose $\omega(\boldsymbol{\xi}) = \xi_4 - i\sigma_j\xi_j$ with the condition $\xi_4^2 + \boldsymbol{\xi}^2 = 1$. (It is convenient to integrate only over a half of $S^3$ by posing $\xi_4 = \sqrt{1-\boldsymbol{\xi}^2}$ and multiply the result by 2.)

Going back to the integral (14.119), we make there a local variable change from the coordinates $\mathbf{x}$ parameterizing $S^3$ to the coordinates $\boldsymbol{\xi}$ parameterizing $SU(2)$. The integrand acquires the Jacobian factor $|\partial\mathbf{x}/\partial\boldsymbol{\xi}|$. Then, the integral acquires an extra integer factor indicating the number of times that the sphere $S^3$ covers the group. This factor can also be negative (for example, if $\mathbf{x} = -\xi$) or zero. □

Now, we can prove the following important theorem:

**Theorem 14.10.** *For a principal fiber bundle on $S^4$ with the $SU(2)$ gauge group, the integral*

$$\boxed{q = \frac{1}{8\pi^2}\int_{S^4} \mathrm{Tr}\{\mathcal{F}\wedge\mathcal{F}\},} \tag{14.122}$$

*where*

$$\mathcal{F} = dA + iA\wedge A \stackrel{\text{def}}{=} \frac{1}{2}F_{\mu\nu}dx^\mu \wedge dx^\nu$$

*is a field density 2-form representing a Hermitian $2\times 2$ matrix [see (1.50)], is an integer.*

The invariant (14.122) is called the *Pontryagin number*, the integral of the *second Chern class* $\mathrm{Tr}\{\mathcal{F}\wedge\mathcal{F}\}/8\pi^2$ (and the 2-form $\mathcal{F}/2\pi$ proportional to the magnetic field density, the integral of which over a compact 2-manifold is an integer, is called the *first Chern class*).

**Proof.** Note the identity[18]

$$\mathcal{F}\wedge\mathcal{F} = dK, \quad \text{where} \quad K = A\wedge dA + \frac{2i}{3}A\wedge A\wedge A. \tag{14.123}$$

[18]The term $\sim A\wedge A\wedge A\wedge A$ in $\mathcal{F}\wedge\mathcal{F}$ vanishes due to the property $t^{[a}t^b t^c t^{d]} = 0$. It is obvious for $SU(2)$, which has only three generators, and for an arbitrary compact group, this follows from the identity

$$f^{abe}f^{cde} + f^{ace}f^{dbe} + f^{ade}f^{bce} = 0.$$

By Stokes' theorem, the integral (14.122) is proportional to the integral $\int \mathrm{Tr}\{K\}$ over the border. $S^4$ has no borders, and integral seems to be zero. We have to recall, however, that, to define a fiber bundle, the manifold should be represented as a union of charts. In our case, it is sufficient to divide $S^4$ in two charts — the northern and the southern hemispheres. They overlap at the equator $S^3$. The gauge fields in the two charts are different, being related in the overlap region by a gauge transformation

$$A' = (\omega A + i d\omega)\omega^{-1}. \tag{14.124}$$

Then, the integral (14.122) over the whole $S^4$ is equal to the difference of two integrals (*Chern–Simons* invariants)

$$q = \frac{1}{8\pi^2}\int_{S^3} \mathrm{Tr}\{K'\} - \frac{1}{8\pi^2}\int_{S^3} \mathrm{Tr}\{K\}. \tag{14.125}$$

A short calculation gives

$$q = \frac{1}{8\pi^2}\int_{S^3} \mathrm{Tr}\left[-id(d\omega \wedge A\omega^{-1}) - \frac{1}{3} d\omega \cdot \omega^{-1} \wedge d\omega \cdot \omega^{-1} \wedge d\omega \cdot \omega^{-1}\right]. \tag{14.126}$$

The first term is the integral of a total derivative, which vanishes (we are now on the border of one of the charts, where $A$ is uniquely defined), and the second term coincides, up to a sign, with the invariant. (14.119). It is integer. □

### 14.4.2 *Self-dual fields □ Semi-dymanical variables*

An especially interesting class of topologically nontrivial gauge fields on $S^4$ are self-dual or anti-self-dual fields satisfying the condition

$$F_{\mu\nu} = \pm\tilde{F}_{\mu\nu} = \pm\frac{1}{2}E_{\mu\nu\alpha\beta}F^{\alpha\beta} = \pm\frac{1}{2}\sqrt{g}\,\varepsilon_{\mu\nu\alpha\beta}F^{\alpha\beta}. \tag{14.127}$$

They are interesting both from the physical and mathematical viewpoint. The physical interest come from the following noteworthy fact:

**Theorem 14.11.** *(Anti-)self-dual fields realize the minimum of the action functional*[19]

$$S = \frac{1}{2}\int \mathrm{Tr}\{F_{\mu\nu}F^{\mu\nu}\}\sqrt{g}\, d^4x \tag{14.128}$$

*in a given topological class.*

---

[19] In Chapter 4, we discussed only the action functionals like (4.15) that appear in classical mechanics. The functional (14.128) describes the field-theory action of a pure

**Proof.** Consider the functional

$$A = \int \mathrm{Tr}\{(F_{\mu\nu} - \tilde{F}_{\mu\nu})(F^{\mu\nu} - \tilde{F}^{\mu\nu})\}\sqrt{g}\, d^4x. \tag{14.129}$$

$A$ is positive definite and hence

$$S \geq \frac{1}{2}\int \mathrm{Tr}\{F_{\mu\nu}\tilde{F}^{\mu\nu}\}\sqrt{g}\, d^4x. \tag{14.130}$$

The RHS of this inequality is nothing but the Pontryagin number $q$ in (14.122) multiplied by $8\pi^2$. Obviously, when $F_{\mu\nu} = \tilde{F}_{\mu\nu}$, $q$ is positive, the inequality becomes an equality, and the minimum of the action is realized.

If $q < 0$, one should consider the functional

$$B = \int \mathrm{Tr}\{(F_{\mu\nu} + \tilde{F}_{\mu\nu})(F^{\mu\nu} + \tilde{F}^{\mu\nu})\}\sqrt{g}\, d^4x \tag{14.131}$$

and derive $S \geq 8\pi^2|q|$. The minimum is realized for anti-self-dual fields $F = -\tilde{F}$. □

If a field configuration realizes the minimum of the action, it represents a solution to the classical equations of motion. And such solutions are quite obviously relevant for physics.[20]

The mathematical interest of self-dual confugurations is that the corresponding SQM models enjoy the extended $\mathcal{N} = 4$ supersymmetry. Indeed, if we remove one point from $S^4$, we obtain a conformally flat four-dimensional manifold, which is HKT (see Section 9.3), and we know from Theorem 13.11 that adding a self-dual gauge field does not destroy the $\mathcal{N} = 4$ supersymmetry of the HKT sigma model.

An attentive reader can protest at this point: Theorem 13.11 was proven only for *Abelian* fields, and we are now interested in the non-Abelian case! However, a generalization of this theorem to the non-Abelian case exists. In particular, one can make the following observation [145]:

---

four-dimensional non-Abelian gauge field. And such fields and their actions are indispensible to describe the nature — the dynamics of strong and electroweak interactions. The factor 1/2 in (14.128) (which is, of course, irrelevant for the theorem) is a common physical convention and was included not to confuse a reader-physicist. (To make contact with the ordinary physical notation, we are also using here Greek letters for the world indices rather than the capital Latin ones.) And a reader-mathematician may simply ignore this footnote and this theorem.

[20] These are, however, *Euclidean* solutions, not the solutions in the physical Minkowski spacetime we live in. But practitioners in high energy physics know well that such solutions are also useful for theoretical evaluation of certain observable physical effects.

**Theorem 14.12.** *Consider a conformally flat four-dimensional manifold with the metric*[21]

$$ds^2 = \frac{dx^\mu dx^\mu}{f^2(x)}. \tag{14.132}$$

*Consider the matrix quantum supercharges*

$$\begin{aligned}
\sqrt{2}\,\hat{Q}^a &= \psi_j(\sigma_\mu^\dagger)^{ja}[(\hat{P}_\mu + A_\mu)f + i\partial_\mu f\,\psi_k\hat{\bar{\psi}}^k],\\
\sqrt{2}\,\hat{\bar{Q}}_a &= [f(\hat{P}_\mu + A_\mu) - i\partial_\mu f\,\psi_k\hat{\bar{\psi}}^k\,](\sigma_\mu)_{aj}\hat{\bar{\psi}}^j,
\end{aligned} \tag{14.133}$$

*where $\hat{P}_\mu = -i\partial_\mu$, $A_\mu$ are Hermitian matrices belonging to any semisimple Lie algebra $\mathfrak{g}$, $(\sigma_\mu)_{aj} = (\boldsymbol{\sigma}, i)_{aj}$ and $(\sigma_\mu^\dagger)^{ja} = (\boldsymbol{\sigma}, -i)^{ja}$, as in Eq. (3.11). The statement is that these supercharges satisfy the extended $\mathcal{N} = 4$ supersymmetry algebra iff the field density $F_{\mu\nu} = \partial_\mu A_\nu - \partial_\nu A_\mu + i[A_\mu, A_\nu]$ is self-dual.*

A similar $\mathcal{N} = 4$ supersymmetric system including an anti-self-dual $F_{\mu\nu}$ also exists. The supercharges of such system differ from (14.133) by the interchange of $\sigma_\mu$ and $\sigma_\mu^\dagger$. This is equivalent to the interchange of two spinor representations of $SO(4)$.

**Proof.** This observation can be verified by an explicit calculation,[22] but it represents in fact a particular case of a more general Theorem 14.13 to be proven at the end of this chapter. □

The following remarks are in order:

- The Hamiltonian corresponding to the supercharges (14.133) is[23]

$$\begin{aligned}
\hat{H} &= \frac{1}{2}f(\hat{P}_\mu + A_\mu)^2 f - \frac{i}{4}f^2 F_{\mu\nu}\,\psi\sigma_\mu^\dagger\sigma_\nu\hat{\bar{\psi}}\\
&\quad - if\partial_\mu f(\hat{P}_\nu + A_\nu)\,\psi\sigma_{[\mu}^\dagger\sigma_{\nu]}\hat{\bar{\psi}} + f\partial^2 f\left[\psi_j\hat{\bar{\psi}}^j - \frac{1}{2}(\psi_j\hat{\bar{\psi}}^j)^2\right].
\end{aligned} \tag{14.135}$$

[21]Following bad habits acquired in Section 9.3, where $S^4$ was also discussed, we do not distinguish here between the covariant and contravariant indices.

[22]To do so, use the identities (3.10) and the definitions (3.5) and (3.6). The Fierz identity

$$(\sigma_{[\mu}^\dagger)^{ja}(\sigma_{\nu]})_{bk} = \frac{1}{2}(\sigma_{[\mu}^\dagger\sigma_{\nu]})^j{}_k\,\delta_b^a + \frac{1}{2}(\sigma_{[\nu}\sigma_{\mu]}^\dagger)_b{}^a\,\delta_k^j \tag{14.134}$$

is also helpful.

[23]Attention: the factor 1/2 in the first term of the second line in Eq. (20) of Ref. [145] was redundant. Note also the different conventions for the sign of $A_\mu$.

- If $f = 1 + x_\mu^2/2$, the metric (14.132) coincides up to a factor with (9.70) and describes $S^4$. If the gauge field is absent, the supercharges (14.133) are equivalent to the supercharges (9.85), (9.86).
- When $A_\mu = 0$, the Hamiltonian (14.135) coincides up to a factor with the square of a *torsionful* Dirac operator (14.22).
- If $A_\mu = 0$ and $f(x^\mu)$ depends only of three out of four coordinates, one can perform a Hamiltonian reduction in the spirit of Chapter 12 to obtain the Hamiltonian (10.52) of the symplectic sigma model.

When the gauge field is Abelian, the Lagrangian of the model can be derived from the superspace formalism of Chapter 10 or by Legendre transformation of the classical Hamiltonian. The field-dependent part of such Lagrangian reads

$$L_A = -\dot{x}^\mu A_\mu + \frac{i}{4} f^2 F_{\mu\nu}\, \psi \sigma_\mu^\dagger \sigma_\nu \bar{\psi}. \tag{14.136}$$

An interesting question is how to derive the Lagrangian in the non-Abelian case. The conventional Legendre transformation does not work: the Hamiltonian (14.135) belongs to $\mathfrak{g}$ and has a matrix nature, while the action and the Lagrangian must be scalars. Still, the Lagrangian of the model can be written by including the so-called *semi-dynamical* variables.

The main idea is to introduce extra variables $\varphi_\alpha$ and the complex conjugated $\bar{\varphi}^\alpha$ belonging to the fundamental and antifundamental representations of the gauge group and to replace the Hermitian generators $t^p \in \mathfrak{g}$ in (14.135) ($A_\mu \equiv A_\mu^p t^p$) by the structure

$$T^p = \varphi_\alpha (t^p)^\alpha{}_\beta \bar{\varphi}^\beta. \tag{14.137}$$

The variables $\varphi_\alpha$ can have the ordinary or the Grassmann nature. Grassmann variables are somewhat more popular in this context; the idea comes back to the classical paper [55], where a triplet of Grassmann variables was introduced to describe spin dynamics. But the same idea can be realized with ordinary bosonic variables [162]. In the context of $\mathcal{N} = 4$ supersymmetric quantum mechanics, such extra bosonic variables were discussed in Ref. [163].

Thus, we introduce the ordinary complex variables $\varphi_\alpha$ and write instead of (14.136) the following expression [164]:

$$\boxed{\begin{aligned} L_A^{\text{non-Ab}} &= -i\bar{\varphi}^\alpha \dot{\varphi}_\alpha + B(\varphi_\alpha \bar{\varphi}^\alpha - 1) \\ &\quad - \dot{x}^\mu A_\mu^p T^p + \frac{i}{4} f^2 F_{\mu\nu}^p T^p\, \psi \sigma_\mu^\dagger \sigma_\nu \bar{\psi} \end{aligned}} \tag{14.138}$$

with $T^p$ defined in (14.137).

The field $B$ is a nondynamical auxiliary variable, alias a Lagrange multiplier. The variation over $B$ gives the constraint

$$\varphi_\alpha \bar{\varphi}^\alpha = 1. \tag{14.139}$$

The variables $\varphi_\alpha$ enter the Lagrangian with a time derivative, but the Lagrangian involves only the linear term in the generalized velocities $\dot{\varphi}_\alpha$, not the quadratic term, as is common for the conventional bosonic dynamical variables like $x^\mu$. That is why we called this kind of variable "semi-dynamical".[24] The variables $\varphi_\alpha$ can be interpreted as generalized holomorphic coordinates. Then, $-i\bar{\varphi}^\alpha$ are their canonical momenta. Under quantization, $\bar{\varphi}^\alpha \to \partial/\partial\varphi_\alpha$. The classical constraint (14.139) gives the following differential constraint to be imposed on the wave functions:

$$\varphi_\alpha \frac{\partial}{\partial \varphi_\alpha} \Phi(\varphi_\alpha, x^\mu, \psi_j) = \Phi(\varphi_\alpha, x^\mu, \psi_j). \tag{14.141}$$

The solution to (14.141) is simple:

$$\Phi(\varphi^\alpha, x^\mu, \psi_j) = A^\alpha(x^\mu, \psi_j)\, \varphi_\alpha. \tag{14.142}$$

This nicely maps to the matrix description where the wave functions depend only on $x^\mu$ and $\psi^j$, but are not ordinary scalars, but carry an extra group index, belonging to the fundamental representation of the gauge group.

The quantum version of the structure (14.137) is the operator

$$\hat{T}^p = \varphi_\alpha (t^p)^\alpha{}_\beta \frac{\partial}{\partial \varphi_\beta}. \tag{14.143}$$

It is easy to see that $\hat{T}^p$ satisfy the same commutation relations as the generators $t^p$. And the operator $\hat{T}^p \hat{T}^p$ boils down [bearing in mind (14.142)] to the multiplication by the Casimir invariant in the fundamental representation $c_F = t^p t^p$.

The full Lagrangian, which involves the term (14.138) and the other terms not including the gauge fields and semi-dynamical variables, is invariant under $\mathcal{N} = 4$ supersymmetry transformations. Its explicit form can be found in Ref. [164]. For the gauge group $SU(2)$, the $\mathcal{N} = 4$ invariance of the action can also be derived using the harmonic superspace formalism by introducing semi-dynamical harmonic superfields. Again, we address an interested reader to Ref. [164].

---

[24] I could not resist mentioning here that similar semi-dynamical variables appear in certain popular field theory models studied by physicists. I am talking about three-dimensional *Chern–Simons* theory with the action

$$S \propto \varepsilon_{ijk} \int d^3x \, \mathrm{Tr}\{A_i \partial_j A_k\}. \tag{14.140}$$

### 14.4.3 *Explicit solutions*

*Yang monopole*

In Section 14.2, we discussed at length the topologically nontrivial line bundle on $S^2$ with constant field density — the Dirac monopole field. A direct non-Abelian analog of this field confuguration is the *Yang monopole* [165].

The Yang monopole is a self-dual configuration with the same value of the invariant $\mathrm{Tr}\{F_{\mu\nu}F^{\mu\nu}\}$ (the same action density) at all the points of $S^4$. The metric of $S^4$ of radius $R$ is expressed in the stereographic coordinates as

$$ds^2 = \frac{16R^4}{(x^2+4R^2)^2}(dx^\mu)^2. \tag{14.144}$$

In a certain conveniently chosen gauge, the Yang solution has a simple form:

$$\boxed{A^p_\mu = \frac{2\eta^p_{\mu\nu}x^\nu}{x^2+4R^2}, \quad F^p_{\mu\nu} = \tilde{F}^p_{\mu\nu} = -\frac{16R^2\eta^p_{\mu\nu}}{(x^2+4R^2)^2},} \tag{14.145}$$

where $\eta^p_{\mu\nu}$ is the 't Hooft symbol (3.6). Then,

$$\frac{1}{2}\mathrm{Tr}\{F_{\mu\nu}F^{\mu\nu}\} = \frac{3}{R^4} \tag{14.146}$$

and the action (14.128) is

$$S = \frac{3}{R^4}V_{S^4} = \frac{3}{R^4}\frac{8\pi^2R^4}{3} = 8\pi^2, \tag{14.147}$$

as it should be for a self-dual configuration with unit topological charge.

*Instantons*

There are also self-dual topologically nontrivial configurations where $\mathrm{Tr}\{F_{\mu\nu}F^{\mu\nu}\}$ is not constant. The configurations where the characteristic support of the action integral (14.128) is much smaller than the radius $R$ of the sphere represent a particular physical interest. In this case, we are actually dealing with the self-dual configurations on $\mathbb{R}^4$ whose field strength vanishes at infinity. These configurations were discovered in Ref. [166] and are called *instantons*.[25] They look similar to (14.145),

$$\boxed{A^p_M = \frac{2\eta^p_{\mu\nu}x^\nu}{x^2+\rho^2}, \quad F^p_{\mu\nu} = \tilde{F}^p_{\mu\nu} = -\frac{4\rho^2\eta^p_{\mu\nu}}{(x^2+\rho^2)^2},} \tag{14.148}$$

[25]This name may look strange, but to explain its etymology, one would have to plunge deeply in physics of non-Abelian fields, which is not our intention here.

but the parameter $\rho$ has now the meaning of the instanton size, which is small compared to $R$. When $x^2 \gg \rho^2$, the field density falls down $\propto 1/x^4$, while the gauge potential tends to a pure gauge configuration

$$\hat{A}_\mu = i(\partial_\mu \omega)\omega^{-1} \quad \text{with} \quad \omega = \frac{x^0 - ix^j\sigma_j}{\sqrt{x^2}}. \tag{14.149}$$

The configuration (14.148) carries unit topological charge. It is also easy to write the anti-self-dual anti-instanton configuration with topological charge $-1$:

$$A^p_\mu = \frac{2\bar{\eta}^p_{\mu\nu}x^\nu}{x^2 + \rho^2}. \tag{14.150}$$

Generic (anti-)self-dual configurations of an arbitrary topological charge $q$ have more complicated form. They were constructed in Ref. [167] (see also Ref. [168]).

## 14.5 KLW theorem and its HKT generalization

We proved in Theorem 13.11 that a generic HKT geometry admits the inclusion of an Abelian gauge field with the tensor $F_{MN}$ (or rather its tangent space projection) belonging to $sp(n)$ subalgebra of the $so(4n)$ algebra so that $\mathcal{N} = 4$ supersymmetry of the model is left intact. In Theorem 14.12, we have shown that this result can be generalized to the non-Abelian case but only for a limited class of Obata-flat four-dimensional HKT manifolds with a conformally flat metric. A natural question is whether a non-Abelian generalization exists also for the generic HKT geometry. The answer to this question is positive.

**Theorem 14.13.** *The extended $\mathcal{N} = 4$ supersymmetry of an HKT SQM model is left intact also in the presence of a non-Abelian gauge field iff $e^M_A e^N_B F_{MN}$ belongs to the subalgebra $sp(n)$ of $so(4n)$ (and to the gauge algebra $\mathfrak{g}$, of course).*

In the *hyperkähler* case, a direct operator proof of this theorem was given by Kirschberg, Länge and Wipf in Ref. [153]. Another variant of such proof [which is similar to the proof of Theorem 11.1 asserting that the $(\mathbf{1}, \mathbf{2}, \mathbf{1})$ SQM models living on hyperkähler manifolds enjoy $\mathcal{N} = 8$ supersymmetry] and its generalization for the HKT manifolds was suggested in Ref. [106]. However, in the HKT case, this method works well only for four-dimensional manifolds. In higher dimensions, it meets certain difficulties.

Fortunately, we can prove this theorem is a rather simple way, capitalizing on the result of Theorem 13.11, which was proven in the preceding chapter.

**Proof.** In the absence of the gauge field, the classical and also quantum HKT supercharges were found to have the form (10.82). It is not difficult to include the field there: one only has to substitute $\hat{P}_M \to \hat{P}_M + A_M$. The commutators of the supercharges involve many different terms and most of them have exactly the same form in the non-Abelian case as in the Abelian one. The only specifics of the non-Abelian system is the presence of the terms involving the commutators $[A_M, A_N]$. These commutator terms come from the structures

$$[\hat{P}_M + A_M, \hat{P}_N + A_N] = -i(\partial_M A_N - \partial_N A_M) + [A_M, A_N] = -iF_{MN}; \tag{14.151}$$

they are added to the derivatives $\sim \partial_M A_N$ to form the non-Abelian field densities. If $F_{MN}$ commutes with all three complex structures [so that $F_{AB}$ belongs to $sp(n)$], the $\mathcal{N} = 4$ superalgebra holds by the same token as it holds (as we know from Theorem 13.11) in the Abelian case. The Hamiltonian of the system is obtained from the Hamiltonian of the system without the gauge field by substituting $\hat{P}_M \to \hat{P}_M + A_M$ and adding the extra term [cf. the term $\propto \partial\partial\mathcal{W}$ in Eq. (9.18)]:

$$\Delta_F \hat{H} = -\frac{i}{2} F_{MN} \hat{\Psi}^M \hat{\Psi}^N. \tag{14.152}$$

□

To give the reader more feeling how it actually works, compare the field-dependent terms in the squares of the Hermitian supercharges $\hat{\mathcal{Q}}_0$ and $\hat{\mathcal{Q}}_1$ in (10.82) modified by the presence of the field. We present the latter in the form

$$\begin{aligned} \hat{\mathcal{Q}}_0 &= \frac{1}{\sqrt{2}} \hat{\Psi}^M (\hat{P}_M + A_M) + \text{irrelevant bits}, \\ \hat{\mathcal{Q}}_1 &= \frac{1}{\sqrt{2}} \hat{\Psi}^N I_N{}^R (\hat{P}_R + A_R) + \text{irrelevant bits}. \end{aligned} \tag{14.153}$$

The field-dependent part in $(\hat{\mathcal{Q}}_1)^2$ is

$$\frac{1}{2} \hat{\Psi}^M \hat{\Psi}^N I_M{}^S I_N{}^R [\hat{P}_S + A_S, \hat{P}_R + A_R] = -\frac{i}{2} \hat{\Psi}^M \hat{\Psi}^N I_M{}^S I_N{}^R F_{SR}, \tag{14.154}$$

where we neglected the gauge-non-invariant terms involving the derivatives of the complex structure: they have the same form as in the Abelian case to be eventually cancelled out. Using the fact that $F$ and $I$ commute, we can bring (14.154) in the form (14.152) — the structure, which also shows up in $\hat{\mathcal{Q}}_0^2$. We conclude that $\hat{\mathcal{Q}}_0^2 = \hat{\mathcal{Q}}_1^2$.

In the anticommutator $\{\hat{\mathcal{Q}}_0, \hat{\mathcal{Q}}_1\}$, we obtain the structure $\sim \hat{\Psi}^M \hat{\Psi}^N I_{[N}{}^R F_{RM]}$, which vanishes due to skew symmetry of $I_{NR}$ and $F_{RM}$ and the property $[F, I] = 0$. The anticommutator $\{\hat{\mathcal{Q}}_1, \hat{\mathcal{Q}}_2\}$ gives the structure $\sim IFJ + JFI$, which vanishes due to $IJ = -JI = K$ and $[F, I] = [F, J] = [F, K] = 0$.

An immediate corollary of this theorem is the fact that, in the instanton background, the eigenstates with $\lambda \neq 0$ in the spectrum of the Dirac operator $\not{\mathcal{D}}$ are double degenerate, and hence the spectrum of the excited states in the Hamiltonian $-\not{\mathcal{D}}^2/2$ is four-fold degenerate, in accord with the old observation made in Ref. [169].

# Chapter 15

# Atiyah–Singer Theorem

## 15.1 Generalities

The Atiyah–Singer theorem is a very general statement concerning the properties of certain elliptic differential operators. Some special cases of this statement (e.g. the Riemann–Roch theorem) were known to mathematicians since the 19th century, but the general statement was first conjectured by Gelfand in 1960 [170]. It was announced as a theorem by Atiyah and Singer in 1963 [171], and the complete proof was published by them in a series of papers in 1968 and 1971 [66].

We are going to show in this chapter that the formulation and the proof of this theorem have in fact a very transparent and concise supersymmetric interpretation. But let us explain first what this theorem is about. To please our reader-mathematician, we use a moderately fancy mathematical language — the same as in the original papers, but a reader-physicist need not worry — we immediately give a clarifying example.

Let $X$ be a compact closed manifold.[1] Let $E$ and $F$ be two *vector bundles* over $X$, let $\Gamma(E)$ and $\Gamma(F)$ be the spaces of smooth sections of these bundles and let $\hat{D}$ be an elliptic differential operator acting on the sections in $\Gamma(E)$ in such a way that the results of this action belong to $\Gamma(F)$ and the action of $\hat{D}$ on any section in $\Gamma(F)$ gives zero. Correspondingly, the conjugate operator $\hat{D}^\dagger$ acts on the sections in $\Gamma(F)$ to produce the sections in $\Gamma(E)$ and annihilates the latter. Obviously, both $\hat{D}$ and $\hat{D}^\dagger$ are nilpotent.

[1] A generalization to manifolds with boundaries also exists [172].

**Definition 15.1.** The *index* of $\hat{D}$ is the difference between the number of the independent solutions of the equation $\hat{D}f = 0$ and the number of the independent solutions of the equation $\hat{D}^\dagger g = 0$ with $f \in \Gamma(E)$ and $g \in \Gamma(F)$.

In mathematical notation,

$$I = \dim \ker \hat{D} - \dim \ker \hat{D}^\dagger. \tag{15.1}$$

The AS theorem says that *this index is determined by the topology of the manifold and of its structures.* What particular topology and what particular structures depend on $\hat{D}$.

The promised important example of $\hat{D}$ is the even-dimensional Dirac operator $\not{\mathcal{D}}$ acting on the bispinors. To be more precise, $\hat{D}$ is the operator $\not{\mathcal{D}}(1+\gamma^{D+1})$ that acts nontrivially on the right-handed bispinors $\Psi_R$ satisfying $\gamma^{D+1}\Psi_R = \Psi_R$ (see Definition 14.3) to produce the left-handed spinors, which it annihilates. The conjugate operator $\hat{D}^\dagger = -\not{\mathcal{D}}(1-\gamma^{D+1})$ acts on the left-handed bispinors to produce the right-handed ones. Thus, $\hat{D}$ and $\hat{D}^\dagger$ act in complementary directions, as is required by the conditions of the theorem.

The Atiyah–Singer index coincides in this case with

$$\mathrm{ind}(\not{\mathcal{D}}) = n_+^{(0)} - n_-^{(0)}, \tag{15.2}$$

where $n_+^{(0)}$ is the number of zero modes of $\not{\mathcal{D}}$ with positive chirality and $n_-^{(0)}$ is the number of zero modes of $\not{\mathcal{D}}$ with negative chirality.

Without any doubt, our reader has already recognized in

$$\hat{D} = \not{\mathcal{D}}(1+\gamma^{D+1}), \quad \hat{D}^\dagger = -\not{\mathcal{D}}(1-\gamma^{D+1}) \tag{15.3}$$

the supercharges (14.8): $\hat{D}$ is proportional to[2] $\hat{Q}^\dagger$ and $\hat{D}^\dagger$ to $\hat{Q}$. Being nilpotent, these operators satisfy the algebra (5.1), with the anticommutator $\{\hat{D}, \hat{D}^\dagger\}$ playing the role of the Hamiltonian. And then the index (15.1) or (15.2) is, up to the sign flip, none other than the *Witten index* (6.55)!

[2]The reason for the different dagger attributions here can be traced back to Benjamin Franklin who defined the sign of the electric charge two and a half centuries ago in such a way that the charge of the electron (of the existence of which Franklin had no idea) turned out to be negative. As a result, the ground states of the Landau Hamiltonian for the electron in a magnetic field with a positive projection on the third axis have *negative* chiralities. And it was convenient for us when we were giving the supersymmetric interpretation to the Landau problem in Chapter 5, to ascribe to these ground states a *zero* fermion charge and call $\hat{Q}$ the operator that increases the fermion charge by 1.

In other words, the Atiyah–Singer theorem is in fact a statement about the ground (vacuum) states of the supersymmetric Hamiltonian known to us from Chapter 6. We noted there that the Witten index stays invariant under smooth deformations of the parameters of the model. But this is tantamout to saying that the index has topological nature. Similar to Mr. Jourdain back in 1670 and like Dr. Landau in 1930, Prof. Atiyah and Prof. Singer spoke in 1963 supersymmetric prose without realizing this!

To make the last statement absolutely clear, we gave in Table 15.1 a small dictionary of correspondences.

A nontrivial question is how to *evaluate* the index. In Chapter 6, we described a general method to do so which capitalizes on the fact that all the excited states of a supersymmetric Hamiltonian are doubly degenerate, and hence one can replace the index (6.55) by the supersymmetric partition function (6.51). The latter can be presented as (6.50), where the Euclidean evolution operator $\mathcal{K}(q_j, \psi_\alpha; q_j, \bar{\psi}_\alpha; -i\beta)$ is expressed as the path integral (6.48). The sum (6.51) does not depend on $\beta$ and it is convenient to assume $\beta$ to be small, in which case the path integral can be evaluated! Following Refs. [73], we did so by substituting for the complicated integral (6.48) its small $\beta$ approximation (6.47). After that, (6.50) was reduced to the finite-dimensional integral (6.56), which can be evaluated for all SQM systems we are familiar with.

In Chapter 6, we calculated the Witten index for the simplest supersymmetric systems which described one-dimensional motion (Witten's quantum mechanics) and two-dimensional motion in a magnetic field. And now we are interested in more complicated systems studied in the previous chapters and related to geometry of manifolds. How to evaluate Witten indices there?

**Table 15.1.** Supersymmetric interpretation of the AS theorem: the dictionary of correspondences. The bosonic or fermionic attribution of the sections $\Gamma(E)$ and $\Gamma(F)$ depends on convention.

| **AS theorem** | **Supersymmetry** |
|---|---|
| $\Gamma(E)$ | Bosonic (or fermionic) states |
| $\Gamma(F)$ | Fermionic (or bosonic) states |
| $\hat{D}, \hat{D}^\dagger$ | Supercharges |
| $\{\hat{D}, \hat{D}^\dagger\}$ | Hamiltonian |
| AS index | Witten index |

An example of such evaluation was already presented in Chapter 8. We calculated there the Witten index for the supersymmetric system (8.6) describing the de Rham complex and showed that the index is given by the integral (8.41) and coincides with the Euler characteristic of the manifold.

However, there are problems where the calculation of the index is not so easy. We discuss them in the rest of this chapter.

## 15.2 Hirzebruch signature

Like the Euler characteristic, the Hirzebruch signature is a topological invariant associated with the de Rham complex.

Consider the de Rham complex on a manifold of dimension $D = 4k$. Consider the Hodge duality operator (1.26). According to (1.27), the square of this operator is sometimes $+1$ and sometimes $-1$, which is not convenient for our purposes.

However, in a special case $D = 4k$, one can define the *signature operator*, the composition

$$\boxed{\hat{\tau} = (-1)^{p(p+1)/2} \circ \star,} \tag{15.4}$$

which actually means

$$\hat{\tau}\alpha_p = (-1)^{(4k-p)(4k-p+1)/2} \star \alpha_p = (-1)^{p(p-1)/2} \star \alpha_p \tag{15.5}$$

and which enjoys remarkable properties:

**Theorem 15.1.** *The operator $\hat{\tau}$ is Hermitian, the square $\hat{\tau}\hat{\tau}$ is the unity operator and it anticommutes with the operator $\hat{R} = d + d^\dagger$.*

**Proof.** We prove it in the simplest case $D = 4$. The generalization of the proof to higher $k$ is rather straightforward:

- To prove hermiticity, we use the property (1.27), which boils down for even $D$ to

$$\star \star \alpha_p = (-1)^p \alpha_p. \tag{15.6}$$

The definition (15.5) gives

$$\tau\alpha_p = \{\star\alpha_0, \star\alpha_1, - \star \alpha_2, - \star \alpha_3, \star\alpha_4\}. \tag{15.7}$$

That means

$$\tau^\dagger(\star\alpha_p) = \{\alpha_0, \alpha_1, -\alpha_2, -\alpha_3, \alpha_4\}. \tag{15.8}$$

On the other hand, applying the relations (15.7) for the forms

$$\beta_p = \star\alpha_p = (\star\alpha_4, \star\alpha_3, \star\alpha_2, \star\alpha_1, \star\alpha_4),$$

we derive

$$\begin{aligned}\tau(\star\alpha_p) &= \{\star\star\,\alpha_0, -\star\star\alpha_1, -\star\star\alpha_2, \star\star\,\alpha_3, \star\star\,\alpha_4\} \\ &= \{\alpha_0, \alpha_1, -\alpha_2, -\alpha_3, \alpha_4\}.\end{aligned} \tag{15.9}$$

The actions (15.8) and (15.9) coincide, from which the hermiticity follows. The property $\hat{\tau}\hat{\tau} = \mathbb{1}$ immediately follows from (15.6), (15.7) and (15.9):

- To prove that

$$\hat{\tau}\hat{R} + \hat{R}\hat{\tau} = 0, \tag{15.10}$$

use the definition (1.35), which boils down to $d^\dagger = -\star d\star$. Using (15.7) and (15.6), we can derive $\hat{\tau}d + d^\dagger\hat{\tau} = 0$ and $\hat{\tau}d^\dagger + d\hat{\tau} = 0$, from which (15.10) follows. □

It follows from $\hat{\tau}\hat{\tau} = \mathbb{1}$ that the eigenvalues of $\hat{\tau}$ are either $+1$ or $-1$. Thus, the whole set of $p$-forms can be divided in this case into two subsets: the subset $S_+$ of the forms with positive signature for which $\hat{\tau}\alpha = \alpha$ and the subset $S_-$ of the forms with negative signature. The property (15.10) implies that the action of the operator $\hat{R}$ on a positive-signature form gives a form with negative signature and vice versa. It also implies that $\hat{\tau}$ commutes with the Laplace–de Rham operator $\triangle = -\hat{R}^2 = -dd^\dagger - d^\dagger d$.

In other words, $\hat{R}$ plays in this setting the same role as the Dirac operator (multiplied by $i$) plays for the Dirac complex, while $\hat{\tau}$ plays the role of $\gamma^{D+1}$. In the full analogy with (15.3), the operators $\hat{R}(1 \pm \hat{\tau})$ are nilpotent, Hermitially conjugate to one another and play the role of the supercharges.

Note that these Hirzebruch supercharges do not coincide at all with the de Rham supercharges $d$ and $d^\dagger$. However, their anticommutator gives *the same* Hamiltonian as in the de Rham complex. Note also that the extended $\mathcal{N} = 4$ supersymmetry does not hold here: the anticommutators $\{\hat{R}(1 \pm \tau), d\}$ and $\{\hat{R}(1 \pm \tau), d^\dagger\}$ do not vanish.

**Definition 15.2.** The Hirzebruch signature of a $4k$-dimensional manifold is defined as

$$\boxed{sign = b_+ - b_-,} \tag{15.11}$$

where $b_\pm$ are the numbers of the closed but not exact forms, belonging to $S_+$ and $S_-$.

**Remarks:**

- The terminology is a little awkward here, but we hope that the reader will not confuse two different notions: the signature of a form (the eigenvalue of $\hat{\tau}$) and the Hirzebruch signature of a manifold.
- Do not mix up the Hirzebruch signature with the Euler characteristic (8.36), which also can be presented in the form (15.11), but the numbers $b_+$ and $b_-$ have in that case a completely different meaning being the sums of the Betti numbers $b_p$ with even and odd $p$.
- In fact, only the forms of degree $p = D/2 = 2k$ contribute to (15.11). Indeed, the duality operator transforms $p$-forms to $(D-p)$-forms. The same is true for $\hat{\tau}$. Bearing in mind this and the property $\hat{\tau}\hat{\tau} = \mathbb{1}$, we derive that, for any form $\alpha_{p\neq D/2}$, the form $\alpha_p + \hat{\tau}\alpha_p$ belongs to $S_+$, while the form $\alpha_p - \hat{\tau}\alpha_p$ belongs to $S_-$. The first form contributes to $b_+$, while the second one to $b_-$. The difference (15.11) is not sensitive for that.
- The Hirzebruch signature is a particular case of the Atiyah–Singer index and is a topological invariant.

On the supersymmetric side, we are dealing with the system with the Hamiltonian (8.31) defined on a manifold of dimension $D = 4k$.

**Theorem 15.2.** *The operator* (15.4) *has in this language a nice representation*

$$\hat{\tau} = \prod_{A=1}^{4k} \left( \frac{\partial}{\partial \psi_A} - \psi_A \right). \tag{15.12}$$

**Proof.** The operator (15.12) can be represented as a product of $k$ blocks:

$$\hat{\tau} = \prod_{A=1}^{4} \left( \frac{\partial}{\partial \psi_A} - \psi_A \right) \times \prod_{A=5}^{8} \left( \frac{\partial}{\partial \psi_A} - \psi_A \right) \times \cdots . \tag{15.13}$$

Each block commutes with "foreign" fermion variables (for example, the first block commutes with $\psi^{A=5,\ldots,4k}$). We leave it for the reader to check, using the definitions (1.26) and (1.53), that the action of, for example, the first block on the variables $\psi^{A=1,2,3,4}$ and their different products is the same as is prescribed by (15.5). □

The Hirzebruch signature is

$$sign = \mathrm{Tr}^* \left\{ \hat{\tau} e^{-\beta \hat{H}} \right\}, \tag{15.14}$$

where $\mathrm{Tr}^*$ stands for the operator trace in Hilbert space.

In Chapter 6, we discussed the ordinary partition function $Z = \text{Tr}^*\{e^{-\beta\hat{H}}\}$ and the supersymmetric partition function $\tilde{Z} = \text{Tr}^*\left\{(-1)^F e^{-\beta\hat{H}}\right\}$. We showed there that both $Z$ and $\tilde{Z}$ admit a path integral representation. The same is true for the expression (15.14). The corresponding path integral can be calculated. We do not do so here and only quote the result.

**Theorem 15.3.** *The signature is given by the integral*

$$\boxed{\textit{sign} = \int \det{}^{1/2}\left[\frac{\frac{\mathcal{R}}{2\pi}}{\tan\frac{\mathcal{R}}{2\pi}}\right],} \tag{15.15}$$

*where $\mathcal{R}$ is the matrix 2-form associated with the Riemann curvature tensor:*

$$\mathcal{R}^{AB} = \frac{1}{2}R^{AB}{}_{MN}\, dx^M \wedge dx^N. \tag{15.16}$$

**Proof.** It consists in a not-so-simple calculation of the appropriate path integral. The reader can find it in the original papers [67]. □

The integrand in (15.15) should be understood as a Taylor series,

$$\det{}^{1/2}\left[\frac{\frac{\mathcal{R}}{2\pi}}{\tan\frac{\mathcal{R}}{2\pi}}\right] = 1 - \frac{1}{24\pi^2}\text{Tr}\{\mathcal{R}\wedge\mathcal{R}\} + \text{higher-order terms}, \tag{15.17}$$

where only the term $\sim \overbrace{\mathcal{R}\wedge\ldots\wedge\mathcal{R}}^{D/2}$ proportional to the volume form is relevant. For four-dimensional manifolds,

$$\textit{sign} = -\frac{1}{24\pi^2}\int \text{Tr}\{\mathcal{R}\wedge\mathcal{R}\} = \frac{1}{24\pi^2}\int R^{AB}\wedge R^{AB}. \tag{15.18}$$

The simplest nontrivial example is $\mathbb{CP}^2$. This manifold is four-dimensional. The nonzero Betti numbers are $b_0 = b_2 = b_4 = 1$. The closed but not exact 2-form $\alpha_2$ happens to have the eigenvalue $+1$ with respect to $\hat{\tau}$ and the Hirzebruch signature is equal to 1. The integral (15.18) also gives this value.

## 15.3 Dirac index

### 15.3.1 *Kähler manifolds*

Consider now the index of the Dirac operator defined as

$$\boxed{I_D = \text{Tr}^*\left\{\gamma^{D+1}e^{\beta\not{D}^2/2}\right\}.} \tag{15.19}$$

We start our discussion with a more simple case when geometry is Kähler and the gauge field is Abelian. As we know from Theorem 14.4, the Dirac complex is equivalent in this case to the twisted Dolbeault complex, which is described by the SQM model (9.1). Let us try to calculate the Witten index in this model using the recipe (6.56). To this end, we need to know the classical Hamiltonian of the model. It was written in (9.18) with $\mathcal{P}_m$ defined in (9.10). In the Kähler case, the 4-fermion term in the Hamiltonian diappears and the Bismut spin connection (9.11) coincides with the ordinary Levi-Civita one. The Hamiltonian acquires the form

$$H_{\mathcal{W}}^{\rm Kahl} = h^{\bar{k}j}\left(p_j - i\omega_{a\bar{b},j}\,\psi^a\psi^{\bar{b}} + i\partial_j\mathcal{W}\right)\left(\bar{p}_{\bar{k}} - i\omega_{c\bar{d},\bar{k}}\,\psi^c\psi^{\bar{d}} - i\partial_{\bar{k}}\mathcal{W}\right)$$
$$- 2\,\partial_j\partial_{\bar{k}}\mathcal{W}\,\psi^j\bar{\psi}^{\bar{k}}. \tag{15.20}$$

The corresponding Lagrangian is

$$L_{\mathcal{W}}^{\rm Kahl} = h_{j\bar{k}}\dot{z}^j\dot{\bar{z}}^{\bar{k}} + \frac{i}{2}(\psi^a\dot{\bar{\psi}}^{\bar{a}} - \dot{\psi}^a\bar{\psi}^{\bar{a}}) + i(\dot{\bar{z}}^{\bar{k}}\omega_{a\bar{b},\bar{k}} + \dot{z}^j\omega_{a\bar{b},j})\psi^a\bar{\psi}^{\bar{b}}$$
$$+ i(\dot{\bar{z}}^{\bar{k}}\partial_{\bar{k}}\mathcal{W} - \dot{z}^j\partial_j\mathcal{W}) + 2\,\partial_j\partial_{\bar{k}}\mathcal{W}\,\psi^j\bar{\psi}^{\bar{k}}. \tag{15.21}$$

Let us do the calculation for $\mathbb{CP}^n$. Choose $\mathcal{W}$ in the form (14.63),

$$\mathcal{W} = \frac{q}{2(n+1)}\ln\det h = -\frac{q}{2}\ln(1+\bar{z}z), \tag{15.22}$$

(we use the shorthand $\bar{z}z \equiv \bar{z}^k z^k$), substitute it in (15.20) and do the integral

$$I_W = \int \prod_{k=1}^{n} \frac{dP_k d\bar{P}_k dz^k d\bar{z}^k}{(2\pi)^2} \prod_{a=1}^{n} d\psi^a d\bar{\psi}^a \, \exp\{-\beta H_{\mathcal{W}}^{Kahl}\}. \tag{15.23}$$

We derive

$$I_W = \left(\frac{q}{2\pi}\right)^n \int \frac{1}{(1+\bar{z}z)^{n+1}} \prod_k dz^k d\bar{z}^k = \frac{q^n}{n!}. \tag{15.24}$$

We have found in Chapter 14 that, for the gauge field to represent a benign line bundle, $q$ should be integer for odd $n$ and half-integer for even $n$ (see Theorem 14.8). For $n = 1$, when we are dealing with $\mathbb{CP}^1 \equiv S^2$, the expression (15.24) boils down to $I_W = q$, which is integer. But for all $n \geq 2$, we obtain a fractional result for the index, which makes no sense. This means that our calculation was wrong. At some point, we should have made an error.

We did, indeed, but it was not a trivial error. The point is that, no matter how small $\beta$ is, we are not allowed *in this particular case* to replace the functional integral (6.48) for the Euclidean evolution operator by its infinitesimal expression and assume that the variables $z^k(\tau)$, $\psi^a(\tau)$, etc. in the functional integral (6.52) do not depend on $\tau$.[3] Instead, we should do the following:

- Impose the periodic boundary conditions
$$z^k(\beta) = z^k(0), \quad \psi^a(\beta) = \psi^a(0), \quad \bar\psi^a(\beta) = \bar\psi^a(0). \tag{15.25}$$
- Expand all the variables in Fourier series,
$$z^k(\tau) = z^k_{(0)} + \sum_{m\neq 0} z^k_{(m)} e^{2\pi i m\tau/\beta}, \quad \text{etc.} \tag{15.26}$$
- Substitute this into (6.52).
- Do the integral over the nonzero Fourier modes in the Gaussian approximation. We show later that the Gaussian approximation is sufficient here — one need not take into account the quartic terms and terms of still higher power. Using the physical slang, it is not sufficient in this case to calculate the functional integral in the *tree approximation*, but the *one-loop approximation* is sufficient and gives the correct answer.
- Do the integral over zero Fourier modes with taking into account in the integrand the factor coming from the Gaussian integral calculated on the previous step.

The actual calculation of the path integral is not so simple. In the original papers devoted to the supersymmetric proof of the Atiyah–Singer theorem [67], the details of this calculation were not given. Some more details (but not all the details) were displayed in Refs. [173]. In the following section, we present this calculation in the most explicit terms. But before coming to grips with the technicalities, we present the *result* and discuss it.

The Witten index of the model (9.1) for a Kähler manifold reads

$$\boxed{I_W = \int e^{\mathcal{F}/2\pi} \det{}^{1/2}\left[\frac{\frac{\mathcal{R}}{4\pi}}{\sin\frac{\mathcal{R}}{4\pi}}\right]}, \tag{15.27}$$

where $\mathcal{F}$ is the field strength 2-form,

$$\mathcal{F} = \frac{1}{2} F_{MN}\, dx^M \wedge dx^N = -2i\, \partial_j \partial_{\bar k} W\, dz^j \wedge d\bar z^k, \tag{15.28}$$

[3] Hopefully, there will be no confusion between the Euclidean time $\tau$ and the signature operator $\hat\tau$. Alas, both Latin and Greek alphabets have a finite length.

and $\mathcal{R}$ was defined in (15.16). The symbol $e^{\mathcal{F}/2\pi}$ is deciphered as

$$e^{\mathcal{F}/2\pi} = 1 + \frac{\mathcal{F}}{2\pi} + \frac{\mathcal{F} \wedge \mathcal{F}}{8\pi^2} + \cdots \tag{15.29}$$

and the determinant factor as

$$\det{}^{1/2}\left[\frac{\frac{\mathcal{R}}{4\pi}}{\sin\frac{\mathcal{R}}{4\pi}}\right] = 1 + \frac{1}{192\pi^2}\int \mathrm{Tr}\{\mathcal{R} \wedge \mathcal{R}\} + \cdots . \tag{15.30}$$

The factor $e^{\mathcal{F}/2\pi}$ in the integrand is the tree contribution coming from the integral (15.23), while the determinant factor comes from nonzero Fourier modes. Only those terms of the expansion of $e^{\mathcal{F}/2\pi}$ and $\mathcal{R}/[4\pi \sin(\mathcal{R}/4\pi)]$ that are combined to give a volume form in the integrand ($2n$-form in the case of $\mathbb{CP}^n$) contribute.

Consider first the simplest case of $\mathbb{CP}^1$. In this case, the determinant factor is absent; there is only one term in the expansion (15.29) and the Witten index is simply equal to the magnetic flux $q$. This agrees with the results of Section 14.2.

For $\mathbb{CP}^2$, the index represents a sum of two terms:

$$I_W(\mathbb{CP}^2) = \frac{1}{8\pi^2}\int \mathcal{F} \wedge \mathcal{F} + \frac{1}{192\pi^2}\int \mathrm{Tr}\{\mathcal{R} \wedge \mathcal{R}\}. \tag{15.31}$$

In the first term in the RHS of (15.31), we recognize the Abelian version of the Pontryagin number (14.122) and the second term coincides with the Hirzebruch signature of $\mathbb{CP}^2$ given in Eq. (15.18) and taken with the coefficient $-1/8$. For higher $n$, the index is a sum of many different topological invariants.

The estimate (15.24) for the index would be reproduced if one ignores this curvature-dependent determinant factor. We include now this factor. Bearing in mind that *sign*$(\mathbb{CP}^2) = 1$, we derive from (15.31) and (15.18)

$$I_W(\mathbb{CP}^2) = \frac{q^2}{2} - \frac{1}{8}. \tag{15.32}$$

This expression is integer for half-integer $q$:

$$I_W(q = \pm 1/2) = 0, \quad I_W(q = \pm 3/2) = 1, \quad I_W(q = \pm 5/2) = 3, \tag{15.33}$$

etc. This coincides with (14.76), (14.77) for $n = 2$.

The final result (14.76), (14.77) for the Witten index in $\mathbb{CP}^n$ can in principle be derived from (15.27) in the same way as for $\mathbb{CP}^2$ by summing up all the relevant topological invariants. But it is complicated, and we even do not know whether anybody did so. A much simpler way is the direct counting of the number of independent ground states outlined in the preceding chapter.

An important remark is in order. Equation (15.27) is the result for the *Witten index* of the model (9.1) with the Hamiltonian (9.18). This index does not *exactly* coincide with the Dirac index defined in (15.19) — $I_W$ and $I_D$ may differ by a sign (see the footnote on p. 368). Indeed, we have seen that the ground state in the Landau problem with positive $B_z$ is bosonic so that the Witten index is positive, but in the matrix language, this ground state has negative chirality so that the Dirac index is negative.

This mismatch is not so surprising: the signs of both indices depend on convention. For the Witten index, the sign depends on what state is assumed to be bosonic and what state fermionic; it can be reversed by changing the convention $\psi \leftrightarrow \bar{\psi}$. For the Dirac index, it depends on the definition of what is left and what is right. We see a little later that the Dirac index (15.19) of an even-dimensional manifold (Kähler or not) is given by *almost* the same formula as (15.27) but with a negative sign before $\mathcal{F}$ in the exponent:

$$\boxed{I_D = \int e^{-\mathcal{F}/2\pi} \det{}^{1/2}\left[\frac{\frac{\mathcal{R}}{4\pi}}{\sin\frac{\mathcal{R}}{4\pi}}\right].} \tag{15.34}$$

In other words, $I_D = I_W$ for $4k$-dimensional manifolds and $I_D = -I_W$ for $4k+2$-dimensional manifolds.

### 15.3.2 *Non-Kähler manifolds*

Consider now a complex but not necessarily Kähler manifold. As we know from the previous chapter, one can define in this case two different complexes: (i) the Dolbeault complex, which is described by the SQM model (9.1) and which, according to Theorem 14.5, maps to a deformed torsionful Dirac operator, and (ii) the standard Dirac complex.

**Theorem 15.4.** *Torsions do not affect the value of the index.*

**Proof.** This simply follows from the fact that the Dirac complex in an even-dimensional manifold admitting spin structure enjoys supersymmetry associated with the pair of Hermitian supercharges proportional to $i\not{\mathcal{D}}$ and $\not{\mathcal{D}}\gamma^{D+1}$, irrespectively of whether the torsions are present or not. This supersymmetry holds under an arbitrary continuous deformation of the tensor $C_{MNL}$. Thus, the torsions can be unwound all the way to zero so that the index is not changed.[4] □

[4]See Ref. [174] for more detailed discussion. Note the essential difference with the gauge field. We have seen in Section 14.2 that, to keep supersymmetry, $A_M$ should be a

How to calculate the non-Kähler Dolbeault index? The first option that comes to mind is to study the generic SQM model (9.1) describing the Dolbeault complex and try to calculate the index there. However, this program meets serious difficulties due to the presence of the 4-fermion term in the generic Dolbeault Hamiltonian (9.9). We have seen above that the constant field approximation, giving the representation (6.56) for the index, does not work for the Dolbeault complex and the calculation of the path integral over nonzero modes in the Gaussian approximation is necessary. In the Kähler case, one does not need to go beyond the Gaussian approximation; this is proven in the following section devoted to explicit path integral calculations. And it is *there* that we show that it does not hold for the Hamiltonian (9.9), for which all the loops must be taken into account. Obviously, we are not able to do so.

We choose therefore another option: use the results of Theorem 15.4 and calculate the path integral for the index of the Dirac operator deprived of its torsion part. In this case, the Gaussian approximation turns out to be sufficient, as we show later. To this end, we consider instead of (9.1) the torsionless $\mathcal{N} = 1$ supersymmetric model[5] of Section 9.2:

$$L = i \int d\Theta \left( \frac{1}{2} g_{MN} \dot{\mathcal{X}}^N - A_M(\mathcal{X}) \right) \mathcal{D}\mathcal{X}^M = \frac{1}{2}\, g_{MN}(x)\, \dot{x}^M \dot{x}^N$$

$$+ \frac{i}{2} \Psi^A (\dot{\Psi}^A + \omega_{AB,M} \dot{x}^M \Psi^B) - A_M \dot{x}^M + \frac{i}{2} F_{AB} \Psi^A \Psi^B. \qquad (15.35)$$

In the generic non-Kähler case, it has only one classical conserved real supercharge

$$\mathcal{Q}_0 = e_C^M \Psi^C \left( P_M - \frac{i}{2} \omega_{AB,M} \Psi^A \Psi^B + A_M \right). \qquad (15.36)$$

The quantum counterpart of $i\mathcal{Q}_0$ maps to the standard Dirac operator. For a complex manifold, we may express everything in complex notation: $M = \{m, \bar{m}\}$, $A = \{a, \bar{a}\}$. Now, if geometry is not Kähler, the components $\omega_{ab,M}$ and $\omega_{\bar{a}\bar{b},M}$ are different from zero. As a result, the Dirac operator $\hat{\mathcal{Q}}_0$ and the Hamiltonian $\hat{H} = \hat{\mathcal{Q}}_0^2$ do not commute with the operator of the fermion charge $\hat{F} = \psi^a \hat{\bar{\psi}}^{\bar{a}}$ as they did in the Kähler case. They

connection of a regular fiber bundle belonging to a distinct topological class. It *cannot* be continuously deformed in an arbitrary way without losing supersymmetry.

[5]The component Lagrangian (15.35) is the sum of (8.4) expressed in flat fermionic variables and the integrand in (9.64).

commute, however, with the operator $(-1)^{\hat{F}}$. And this means that the mapping of the bispinor Hilbert space to the Hilbert space of the holomorphic wave functions $\Phi(x^M, \psi^a)$ in the spirit of Theorem 14.2, such that $\gamma^{D+1}$ is mapped to $(-1)^{\hat{F}}$, can still be defined. Then, the second Hermitian quantum supercharge can be defined as $\hat{\not{\mathcal{D}}}\gamma^{D+1}/\sqrt{2}$, and the Dirac index (15.19) is still mapped, up to a possible sign flip, to the Witten index $\text{Tr}^* \left\{(-1)^{\hat{F}} \exp\{-\beta \hat{H}\}\right\}$.

The latter may be evaluated as the functional integral of $\exp\left\{-\int_0^\beta L_E(\tau)\right\}$ with the Euclidean version of the Lagrangian (15.35) and the periodic boundary conditions for both the bosonic and fermionic fields being imposed. We will see in the following section that the result of this evaluation does not depend on whether the manifold is Kähler or not and coincides, if the conventions are properly adjusted, with the expression (15.34).

In fact, neither the Lagrangian (15.35) nor the result (15.34) depend on the assumption that the manifold is complex. We only need to assure the mapping $\gamma^{D+1} \to (-1)^{\hat{F}}$ (this is necessary to express the index as a path integral where fermion fields satisfy periodic boundary conditions). To this end, we have to define the Hilbert space of the wave functions $\Phi(x^M, \psi^a)$ depending on the holomorphic fermion variables. But complex Grassmann variables in the tangent space can be introduced for *all* even-dimensional manifolds: locally, any such manifold is complex. The only condition that we need to derive (15.34) is that the manifold admits spin structure so that the notion of the Dirac index makes sense.

## 15.4 Functional integral calculation

The Witten index of any supersymmetric system is given by the integral (6.50). So far, we calculated it in the approximation where the evolution kernel $\mathcal{K}$ was replaced by its infinitesimal form (6.47). As was mentioned above, this approximation does not work for the problem in interest and we have to evaluate the functional integral (6.48) for the kernel at small but finite $\beta$.

In our case, it is convenient first to do the Gaussian integral over $\prod_\tau dP_M(\tau)$ in the Hamiltonian path integral for our system [a generalization of (6.19) involving also the integration over Grassmann variables] and represent the path integral in the Lagrangian form[6]:

$$I_W = C \int \prod_\tau \sqrt{g[x^M(\tau)]} \prod_M dx^M(\tau) \prod_A d\Psi^A(\tau) \times \exp\left\{ - \int_0^\beta L_E(\tau) d\tau \right\} \tag{15.37}$$

with[7]

$$\boxed{L_E = \frac{1}{2} g_{MN}(x) \dot{x}^M \dot{x}^N + \frac{1}{2} \Psi^A (\dot{\Psi}^A + \omega_{AB,M} \dot{x}^M \Psi^B) + i A_M \dot{x}^M - \frac{i}{2} F_{AB} \Psi^A \Psi^B.} \tag{15.38}$$

Let first space be flat. In this case, the Cecotti–Girardello method works and we can reduce the path integral (15.37) for the index to the ordinary integral:

$$I_W^{\rm flat} = \left( \frac{i}{2\pi\beta} \right)^{D/2} \int d^D x \prod_C d\Psi^C \exp\left\{ \frac{i\beta}{2} F_{AB}(x) \Psi^A \Psi^B \right\}, \tag{15.39}$$

where the factor $(2\pi\beta)^{-D/2}$ comes from the integration over momenta in the original Hamiltonian path integral in the approximation (6.56) [that is how we fix the coefficient $C$ in (15.37)] and the factor $i^{D/2}$ appears due to the relation

$$\prod_c^{D/2} d\psi^c d\bar{\psi}^c = i^{D/2} \prod_C^D d\Psi^C \tag{15.40}$$

between the measure $\prod_C d\Psi^C$ and the canonical fermion measure expressed via the holomorphic variables $\psi^c, \bar{\psi}^c$ that we used in Eq. (6.56) and in Eq. (15.23).[8] Only the term of order $D/2$ in the expansion of the exponential

---

[6]To be *quite* meticulous, we had to write first a finite-number-of-points approximation of the Hamiltonian path integral, integrate over the momenta there to obtain a finite-number-of-points approximation of (15.37) and send the number of points to infinity at the end of the day. We do not do so here, addressing an interested reader to Ref. [9] where this calculation was accurately done in the Kähler case.

[7]It is worthwhile to note that, in contrast to the Minkowski Lagrangian (15.35), the Lagrangian (15.38) is not real. It need not be. On the other hand, the Euclidean path integral for the index is, of course, real.

[8]We remind our standard convention (7.32): $\psi^1 = (\Psi^1 + i\Psi^2)/\sqrt{2}$, $\psi^2 = (\Psi^3 + i\Psi^4)/\sqrt{2}$, etc., which dictates $d\psi^1 = (d\Psi^1 - i d\Psi^2)/\sqrt{2}$, etc.

in (15.39) works. Using

$$\int \left( \prod_{C=1}^{D} d\Psi^C \right) \Psi^{A_1} \cdots \Psi^{A_D} = (-1)^{D/2} \varepsilon_{A_1 \ldots A_D}, \qquad (15.41)$$

we derive

$$\boxed{I_W^{\text{flat}} = \frac{1}{(4\pi)^{D/2}(D/2)!} \varepsilon_{A_1 \ldots A_D} \int d^D x \, F_{A_1 A_2} \cdots F_{A_{D-1} A_D}.} \qquad (15.42)$$

The integral (15.42) corresponds to the integral $\int e^{\mathcal{F}/2\pi}$ in the language of forms.[9] The result (15.27) for the Witten index is thus reproduced in the case when the manifold is flat and the determinant factor in the integrand is absent.

We can as well reproduce in this way the result (15.34) for the canonically defined Dirac index, but to this end, we have to modify somewhat our conventions. Namely, we now suppose that

$$\psi^1 = (\Psi^1 - i\Psi^2)/\sqrt{2}, \text{etc.} \quad \text{and} \quad d\psi^1 = (d\Psi^1 + i d\Psi^2)/\sqrt{2}, \text{etc.} \qquad (15.43)$$

This gives the factor $(-i)^{D/2}$ rather than $i^{D/2}$ in (15.40), which brings about the factor $(-1)^{D/2}$ in (15.42), in agreement with (15.34).

For an arbitrary curved manifold, the determinant factor is nontrivial. As was mentioned above, to calculate it, we must go beyond the recipe (6.56), expand the fields in Fourier series and evaluate the path integral (15.37) in the Gaussian approximation. Substitute the expansion

$$x^M(\tau) = x^M_{(0)} + \sum_m{}' x^M_{(m)} e^{i\Omega_m \tau}, \quad x^M_{(-m)} = \overline{x^M_{(m)}},$$

$$\Psi^A(\tau) = \Psi^A_{(0)} + \sum_m{}' \psi^A_{(m)} e^{i\Omega_m \tau}, \quad \Psi^A_{(-m)} = \overline{\Psi^A_{(m)}}, \qquad (15.44)$$

with $\sum'_m \equiv \sum_{m \neq 0}$ and $\Omega_m = 2\pi m/\beta$, into (15.37). To calculate the functional integral in the Gaussian approximation (we justify the validity of this approximation later), we keep only quadratic in $x^M_{(m)}, \Psi^A_{(m)}$ terms and do the $\tau$ integral. The quadratic part of the relevant terms in the

[9]Do not forget about the factor 1/2 in the definition (15.28).

Lagrangian [we can now disregard the second line in (15.38) involving gauge fields: the latter have already been successfully handled] gives

$$\int_0^\beta L_E^{(2)} d\tau = \frac{\beta}{2} \sum_m{}' \Omega_m^2 \, g_{MN}^{(0)} \, x_{(-m)}^M x_{(m)}^N + \frac{i\beta}{2} \sum_m{}' \Omega_m \Psi_{(-m)}^A \Psi_{(m)}^A$$
$$+ \frac{i\beta}{4} \Psi_{(0)}^A \Psi_{(0)}^B \left( \partial_M \omega_{AB,N}^{(0)} - \partial_N \omega_{AB,M}^{(0)} \right) \sum_m{}' \Omega_m x_{(-m)}^M x_{(m)}^N$$
$$- i\beta \, \Psi_{(0)}^A \, \omega_{AB,M}^{(0)} \sum_m{}' \Omega_m x_{(-m)}^M \psi_{(m)}^B. \tag{15.45}$$

Now, we diagonalize the sum in (15.45) by the substitution

$$\Psi_{(m)}^A \Rightarrow \Psi_{(m)}^A - \omega_{AB,R}^{(0)} \Psi_{(0)}^B x_{(m)}^R. \tag{15.46}$$

The variable change (15.46) brings (15.45) in the simple form

$$\int_0^\beta L_E^{(2)} d\tau = \sum_m{}' A_{mAB} \, \Psi_{(-m)}^A \Psi_{(m)}^B + \sum_m{}' D_{mMN} \, x_{(-m)}^M x_{(m)}^N, \tag{15.47}$$

where

$$A_{mAB} = \frac{i\beta}{2} \Omega_m \delta_{AB},$$
$$D_{mMN} = \frac{\beta}{2} \left( \Omega_m^2 g_{MN}^{(0)} + \frac{i}{2} \Omega_m R_{MN,AB}^{(0)} \Psi_{(0)}^A \Psi_{(0)}^B \right) \tag{15.48}$$

and $R_{MN,AB}$ is the Riemann tensor (1.63). Note that the matrix of the partial derivatives corresponding to the variable change (15.46) is triangle and has a unit superdeterminant.

The functional integral over the nonzero modes is then given by a product of a large (in the continuous limit, infinite) number of finite-dimensional determinants, which can be symbolically written as[10]

$$\text{grav. factor} \propto \prod_m{}' \det{}^{1/2}(A_m) \cdot \det{}^{-1/2}(D_m). \tag{15.49}$$

Fixing the coefficient by the requirement for the determinant factor to reduce to unity in the flat case and the requirement of general covariance,

[10] *Grav.* means *gravitational.* For physicists, nonzero curvature is associated with general relativity and gravitation. But this association is not necessary, of course.

we deduce

$$\text{grav. factor} = \sqrt{g^{(0)}} \prod_m{}' \det{}^{-1/2} \left( \delta_{AB} + \frac{i\beta}{2\pi m} R_{AB} \right), \qquad (15.50)$$

where $R_{AB} = R^{(0)}_{AB,CD} \Psi^C_{(0)} \Psi^D_{(0)}/2$.

The infinite $m$-product in (15.50) can be done by writing the determinant as the product of its eigenvalues. All the eigenvalues of the real skew-symmetric matrix $iR_{AB}$ are imaginary. They split into $D/2$ pairs: $\pm i\lambda_1, \ldots, \pm i\lambda_{D/2}$. By orthogonal rotations, we can bring $iR_{AB}$ to the canonical form:

$$iR_{AB} = \text{diag}\left[ \begin{pmatrix} 0 & \lambda_1 \\ -\lambda_1 & 0 \end{pmatrix}, \begin{pmatrix} 0 & \lambda_2 \\ -\lambda_2 & 0 \end{pmatrix}, \ldots \right]. \qquad (15.51)$$

Each block associated with a given $\lambda_\alpha$ brings about the following factor in the determinant in Eq. (15.50):

$$\prod_m{}' \det{}^{-1/2}(\cdots)_\alpha = \prod_{m=1}^{\infty} \frac{(2\pi m)^2}{(2\pi m)^2 + (\beta\lambda_\alpha)^2} = \frac{\beta\lambda_\alpha/2}{\sinh(\beta\lambda_\alpha/2)}. \qquad (15.52)$$

We now use the following elementary mathematical fact:

*Let $Q$ be a real skew-symmetric matrix brought to the canonical form (15.51) and let $f(Q)$ be an even function of $Q$. Then,*[11]

$$\det{}^{1/2}[f(Q)] = \prod_\alpha |f(i\lambda_\alpha)|. \qquad (15.53)$$

In our case,

$$f(x) = \frac{\beta x/2}{\sin(\beta x/2)}$$

and $Q = iR$. Reading the relation (15.53) from the right to the left, we finally derive

$$\boxed{\text{g.f.} = \sqrt{g^{(0)}} \det{}^{1/2} \left( \frac{i\beta R/2}{\sin(i\beta R/2)} \right) = \sqrt{g^{(0)}} \det{}^{1/2} \left( \frac{\beta R/2}{\sinh(\beta R/2)} \right).} \qquad (15.54)$$

To obtain the Witten index, the factor (15.54) should be inserted in the integrand in (15.39), multiplied by $\exp\{i\beta F_{AB} \Psi^A_{(0)} \Psi^B_{(0)}/2\}$, by the factor $(i/2\pi\beta)^{D/2}$ to be integrated over the zero modes $\prod_M dx^M_{(0)} \prod_C d\Psi^C_{(0)}$

[11] Try it for

$$Q = \begin{pmatrix} 0 & \lambda \\ -\lambda & 0 \end{pmatrix}, \quad f(Q) = \mathbb{1} + Q^2.$$

afterwards. The resulting integrand represents a function $H$ of the bilinear fermion structure $N^{AB} = \Psi^A_{(0)}\Psi^B_{(0)}$. This function should be expanded into a series, where only the term $\propto N^{D/2}$ contributes in the integral. One can then incorporate the factor $i/(2\pi\beta)$ in the argument of $H$. We derive

$$I_W = \int d^D x \sqrt{g} \prod_C d\Psi^C \exp\left\{-\frac{F_{AB}\Psi^A\Psi^B}{4\pi}\right\} \det{}^{1/2}\left(\frac{iR/4\pi}{\sinh(iR/4\pi)}\right).$$

After integrating over the fermions using (15.41), we finally derive the result (15.27).

To derive the result (15.34) for the Dirac index, one should do the same, with the factor $(i/2\pi\beta)^{D/2}$ being replaced by $(-i/2\pi\beta)^{D/2}$. This change reverses the sign of $\mathcal{F}$ but does not affect the graviational determinant.

It is now clear why we had to insert in this particular case the 1-loop gravitational factor in the tree-level integral (15.23) for the index. Formally, the factor $\prod'_m$ in (15.50) tends to 1 for small $\beta$ and, naively, the corrections involving $\beta$ and its higher powers can be neglected. We see, however, that each factor of $\beta$ in the expansion is multiplied by a bi-fermion structure, as is also the case for the expansion of the integrand in (15.39). For the fermion integral not to vanish, we have to pick up the terms $\sim (\beta\Psi^2)^{D/2}$ coming from the expansions of both the factor $\exp\{i\beta F_{AB}\,\Psi^A_{(0)}\Psi^B_{(0)}/2\}$ and of the 1-loop factor (15.54) — they come on equal footing.

On the other hand, if one tries to evaluate the path integral more precisely, taking into account cubic and other higher-order terms in $x^M_{(m)}$ and $\Psi^A_{(m)}$ in the mode expansion for $\int_0^\beta L_E d\tau$, the corresponding corrections would involve more powers of $\beta$ than the powers of $\Psi^2$ and will be suppressed in the small $\beta$ limit. This justifies neglecting two-loop and higher-loop effects in the functional integral.

Now, we can also explain why we abandonned the idea to calculate the index for the Dolbeault Hamiltonian (9.9) involving a four-fermion term. The latter would give large contributions $\sim (\beta\psi^4)^k \sim \beta^{-k}$ to the integral (6.56). They even seem to diverge in the limit $\beta \to 0$. Of course, this does not actually happen — for example, for a manifold of real dimension 4, we obtain the contribution

$$\sim \int d^4x\, \varepsilon^{\mu\nu\rho\sigma}\, \partial_\mu C_{\nu\rho\sigma},$$

which is an integral of a total derivative and vanishes. But for higher dimensions, the large terms involving the expansion of $e^{-\beta H}$ may multiply some (normally, small) structures coming from higher loops and the vanishing of all these contributions is not at all obvious.

### 15.4.1 *Non-Abelian Dirac index*

The Euclidean supersymmetric Lagrangian (15.38) involves an Abelian gauge field, and the calculation above was performed for an Abelian gauge field. But the Dirac index can also be defined and calculated for a non-Abelian gauge background. To write the corresponding supersymmetric Lagrangian, one has to introduce on top of $x^M$ and $\Psi^A$ also extra bosonic or fermionic semi-dynamical variables in the fundamental representation of the gauge group (see the discussion on p. 361). But if one is only interested in the index, it is more convenient to work in the Hamiltonian approach.

To this end, we define the supersymmetric problem (14.8)–(14.10) where the Dirac operator has the form (14.5) but with a non-Abelian matrix-valued gauge field $A_M^{\text{non-Ab}} = A_M^a t^a$.

For simplicity, we consider now only flat geometry — the "gravitational" (nontrivial curvature) factor in the integrand of the index integral does not depend on whether the extra gauge field is Abelian or non-Abelian and was calculated before.

As was also the case for Abelian $A_M$, the pure gauge index can be calculated in the tree approximation. In the flat non-Abelian case, the integral (6.62) for the Witten index acquires the form

$$\begin{aligned}
I^{\text{flat}}_{\text{non-Ab}} &= \int \prod_M \frac{dP_M dx^M}{2\pi} \operatorname{Tr}\left\{(-1)^{D/2}\gamma^{D+1} e^{-\beta \hat{H}}\right\} \\
&= (-1)^{D/2} \frac{1}{(2\pi\beta)^{D/2}} \int d^D x \operatorname{Tr}\left\{\gamma^{D+1} e^{i\beta F_{AB}\gamma^A\gamma^B/4}\right\} \\
&= \frac{1}{(4\pi)^{D/2}(D/2)!}\,\varepsilon_{A_1\ldots A_D} \int d^D x \operatorname{Tr}\{F_{A_1A_2}\cdots F_{A_{D-1}A_D}\}
\end{aligned} \tag{15.55}$$

— an obvious generalization of (15.42). The Dirac index is given by the same expression with the extra factor $(-1)^{D/2}$.

The index depends on the representation to which $A_M$ belong. For a localized $SU(2)$ gauge field in $\mathbb{R}^4$ and if $A_M$ belongs to the fundamental representation, the index coincides with the Pontryagin number (14.122). In particular, for the instantons (14.148), this index is equal to unity. This means that an instanton supports one fundamental fermion (a quark) zero mode and that this zero mode is right-handed [17].

# Bibliography


[1] A.V. Smilga, *Differential Geometry through Supersymmetric Glasses*, World Scientific, 2020.

[2] E. Witten, *Supersymmetry and Morse theory*, J. Diff. Geom. **17** (1982) 661.

[3] Yu.A. Golfand and E.P. Likhtman, *Extension of the algebra of Poincare group generators by bispinor generators*, JETP Lett. **13** (1971) 323.

[4] L.D. Landau, *Diamagnetismus der Metalle*, Z. Phys. **64** (1930) 629.

[5] T. Eguchi, P.B. Gilkey and A.J. Hanson, *Gravitation, gauge theories and differential geometry*, Phys. Repts. **66** (1980) 213.

[6] P. Candelas, *Lectures on complex manifolds*, in Trieste 1987, Proceedings, *Superstrings 87*, pp. 1–88.

[7] N.E. Mavromatos, *A note on the Atiyah-Singer index theorem for manifolds with totally antisymmetric H torsion*, J. Phys. **A21** (1988) 2279.

[8] J.-M. Bismut, *A local index theorem for non Kähler manifolds*, Math. Ann. **284** (1989) 681.

[9] E.A. Ivanov and A.V. Smilga, *Dirac operator on complex manifolds and supersymmetric quantum mechanics*, Int. J. Mod. Phys. **A27** (2012) 1230024, arXiv:1012.2069 [hep-th].

[10] A. Newlander and L. Nirenberg, *Complex analytic coordinates in almost complex manifolds*, Ann. of Math. (2) **65** (1957) 391. See also L. Nirenberg, *Lectures on Linear Partial Differential Equations*, Amer. Math. Soc., 1973.

[11] A. Clebsch, *Über die simultane Integration linearer partieller Differentialgleichungen*, J. Reine Angew. Math. **65** (1866) 257–268; G. Frobenius, *Über das Pfaffsche Problem*, J. Reine Agnew. Math. **82** (1877) 230–315.

[12] A.V. Smilga, *Comments on the Newlander-Nirenberg theorem*, in Varna 2019, Proceedings, Lie Theory and its Application in Physics, arXiv:1902.08549 [math-ph].

[13] A. Opfermann and G. Papadopoulos, *Homogeneous HKT and QKT Manifolds*, arXiv:math-ph/9807026.

[14] C.M. Hull, *The Geometry of Supersymmetric Quantum Mechanics*, arXiv:hep-th/9910028.

[15] H. Hopf, *Zur Topologie der komplexen Mannigfaltigkeiten*, in *Studies and Essays Presented to R. Courant on his 60th Birthday, January 8, 1948*, Interscience Publishers, Inc., New York, p. 167.

[16] E. Calabi, *Métriques kähleriennes et fibrés holomorphes*, Ann. Ecol. Norm. Sup. **12** (1979) 269.

[17] G. 't Hooft, *Computation of the quantum effects due to a four-dimensional pseudoparticle*, Phys. Rev. **D14** (1976) 3432.

[18] V. Fock, *Proper time in classical and quantum mechanics*, Sov. Phys. **12** (1937) 404;
J. Schwinger, *On gauge invariance and vacuum polarization*, Phys. Rev. **82** (1951) 664.

[19] A.H. Taub, *Empty space-times admitting a three parameter group of motion*, Ann. Math., Second Series **53** (1951) 472;
E. Newman, L. Tamburino and T. Unti, *Empty-space generalization of the Schwarzschild metric*, J. Math. Phys. **4** (1963) 915.

[20] T. Eguchi and A.J. Hanson, *Asymptotically flat selfdual solutions to euclidean gravity*, Phys. Lett. **B74** (1978) 249.

[21] Ö. Kelekçi, *On Kähler structures of Taub-NUT and Kerr spaces*, Int. J. Geom. Meth. Mod. Phys. **20** (2023) 2350027, arXiv:2209.08823 [math-DG].

[22] S.W. Hawking, *Gravitational instantons*, Phys. Lett. **60A** (1977) 81.

[23] P.A.M. Dirac, *Quantised singularities in the electromagnetic field*, Proc. Roy. Soc. **A133** (1931), 821, p. 60.

[24] E.B. Bogomolny, *Stability of classical solutions*, Sov. J. Nucl. Phys. **24** (1976) 449;
M.K. Prasad and C.M. Sommerfeld, *An exact classical solution for the 't Hooft monopole and Julia-Zee dyon*, Phys. Rev. Lett. **35** (1975) 760.

[25] G.W. Gibbons and N.S. Manton, *Classical and Quantum Dynamics of BPS Monopoles*, Nucl. Phys. **B274** (1986) 183.

[26] M.F. Atiyah and N.J. Hitchin, *Low-Energy Scattering of Nonabelian Monopoles*, Phys. Lett. **A107** (1985) 21.

[27] G.W. Gibbons and N.S. Manton, *The Moduli space metric for well separated BPS monopoles*, Phys. Lett. **B356** (1995) 32, arXiv:hep-th/9506052;
G. Chalmers and A. Hanany, *Three-dimensional gauge theories and monopoles*, Nucl. Phys. **B489** (1997) 223, arXiv:hep-th/9608105.

[28] G. 't Hooft, *Magnetic monopoles in unified gauge theories*, Nucl. Phys. **B79** (1974) 276;
A.M. Polyakov, *Particle spectrum in quantum field theory*, JETP Letters **20** (1974) 194.

[29] N.J. Hitchin, A. Karlhede, U. Lindstrom and M. Roček, *Hyperkahler Metrics and Supersymmetry*, Commun. Math. Phys. **108** (1987) 535.

[30] G.W. Gibbons, P. Rychenkova and R. Goto, *Hyperkahler quotient construction of BPS monopole moduli spaces*, Commun. Math. Phys. **186** (1997) 581, arXiv:hep-th/9608085.

[31] K.G. Selivanov and A.V. Smilga, *Effective Lagrangian for* $3dN = 4$ *SYM theories for any gauge group and monopole moduli spaces*, JHEP **12** (2003) 027, arXiv:hep-th/0301230.

[32] A. Hanany and E. Witten, *Type IIB superstrings, BPS monopoles, and three-dimensional gauge dynamics*, Nucl. Phys. **B492** (1997) 152, arXiv:hep-th/9611230;

A. Hanany, B. Kol and A. Rajaraman, *Orientifold points in M theory*, JHEP **10** (1999) 027, arXiv:hep-th/9909028.

[33] A. V. Smilga, *Vacuum structure in quantum gravity*, Nucl. Phys. **B234** (1984) 402.

[34] P.S. Howe and G. Papadopoulos, *Twistor spaces for hyperkähler manifolds with torsion*, Phys. Lett. **B379** (1996) 80, arXiv:hep-th/9602108.

[35] G.W. Gibbons, G. Papadopoulos and K.S. Stelle, *HKT and OKT geometries on soliton black hole moduli spaces*, Nucl. Phys. **B508** (1997) 623, arXiv:hep-th/9706207.

[36] G. Grantcharov and Y.S. Poon, *Geometry of hyperkähler connections with torsion*, Commun. Math. Phys. **213** (2000) 19, arXiv:math/9908015.

[37] M. Verbitsky, *Hyperkähler manifolds with torsion, supersymmetry and Hodge theory*, Asian J. Math. **6** (2002) 679, arXiv:math/0112215.

[38] A. Fino and G. Grantcharov, *Properties of the manifolds with skew-symmetric torsion and special holonomy*, Adv. Math. **189** (2004) 439, arXiv:math.DG/0302358.

[39] M. Obata, *Affine connections on manifolds with almost complex, quaternion or Hermitian structure*, Japan J. Math. **26** (1956) 43.

[40] A. Soldatenkov, *Holonomy of the Obata connection on $SU(3)$*, Int. Math. Res. Not. (2012) 3483, arXiv:1104.2085[math.DG].

[41] K. Yano and M. Ako, *Integrability conditions for almost quaternion structures*, Hokkaido Math. J. **1** (1972) 63.

[42] F. Delduc and G. Valent, *New geometry from heterotic supersymmetry*, Class. Quant. Grav. **10** (1993) 1201.

[43] P. Spindel, A. Sevrin, W. Troost and A. Van Proeyen, *Extended supersymmetric $\sigma$ models on group manifolds. 1. The complex structures*, Nucl. Phys. **B308** (1988) 662.

[44] D. Joyce, *Compact hypercomplex and quaternionic manifolds*, J. Diff. Geom. **35** (1992) 743.

[45] H. Samelson, *A class of complex-analytic manifolds*, Portugal. Math. **12** (1953) 129.

[46] A.V. Smilga, *Group manifolds and homogeneous spaces with HKT geometry: the role of automorphisms*, arXiv:2002.06356[math-ph].

[47] F. Delduc and E. Ivanov, *$N = 4$ supersymmetric $d = 1$ sigma models on group manifolds*, Nucl. Phys. **B949** (2019) 114806, arXiv:1907.09518 [hep-th].

[48] L.D. Landau and E.M. Lifshitz, *Mechanics*, Pergamon Press, 1969.

[49] E. Noether, *Invariante Variationsprobleme*, Nachr. Ges. Wiss. Göttingen, Math-Phys. Klasse, 1918, p. 235.

[50] H. Goldstein, *Classical Mechanics*, Addison-Wesley, 1980;
V.I. Arnold, *Mathematical Methods of Classical Mechanics*, Springer, 1978.

[51] H.J. Grönewold, *On the principles of elementary quantum mechanics*, Physica **12** (1946) 405;
J.E. Moyal, *Quantum mechanics as a statistical theory*, Proc. Cambr. Phil. Soc. **45** (1949) 99.

[52] A.V. Smilga, *How to quantize supersymmetric theories*, Nucl. Phys. **B292** (1987) 363.

[53] F.A. Berezin, *The Method of Second Quantization*, Academic Press, NY, 1966.

[54] R. Casalbuoni, *The classical mechanics for Bose-Fermi systems*, Nuovo Cim. **A33** (1976) 389.

[55] F.A. Berezin and M.S. Marinov, *Particle spin dynamics as the Grassmann variant of classical mechanics*, Ann. Phys. **104** 336 (1977).

[56] M. Fierz, *Über die relativistische Theorie kräftefreier Teilchen mit beliebigen Spin*, Helv. Phys. Acta **12** (1939) 3; W. Pauli, *The connection between spin and statistics*, Phys. Rev. **58** (1940) 716.

[57] There is a vast literature on this subject. We refer the reader to the recently published book [C.M. Bender *et al.*, *PT Symmetry in Quantum and Classical Physics*, World Scientific, 2019].

[58] D. Robert and A.V. Smilga, *Supersymmetry vs. ghosts*, J. Math. Phys. **49** (2008) 042104, arXiv:math-ph/0611023.

[59] H. Nicolai, *Supersymmetry and spin systems*, J. Phys. **A9** (1976) 1497.

[60] A.V. Smilga, *Digestible Quantum Field Theory*, Springer, 2018.

[61] E. Witten, *Dynamical breaking of supersymmetry*, Nucl.Phys. **B188** (1981) 513.

[62] S. Bellucci and A. Nersessian, *(Super)oscillator on $CP^N$ and constant magnetic field*, Phys. Rev. **D67** (2003) 065013, arXiv:hep-th/0211070;
A.V. Smilga, *Weak supersymmetry*, Phys.Lett. **B585** (2004) 173, arXiv:hep-th/0311023.

[63] A.V. Smilga, *SUSY anomaly in quantum mechanical systems*, Phys. Lett. **B199** (1987) 516.

[64] J.W. van Holten and R.H. Rietdijk, *Symmetries and motions in manifolds*, J. Geom. Phys. **11** (1993) 559, arXiv:hep-th/9205074.

[65] V. De Alfaro, S. Fubini, G. Furlan and M. Roncadelli, *Operator ordering and supersymmetry*, Nucl. Phys. **B296** (1988) 402.

[66] M.F. Atiyah and I.M. Singer, *The index of elliptic operators*, Annals Math. **87** (1968) 484, 546; **93** (1971) 119, 139.

[67] L. Alvarez-Gaumé, *Supersymmetry and the Atiyah-Singer index theorem*, Commun. Math. Phys. **90** (1983) 161;
D. Friedan and O. Windey, *Supersymmetric derivation of the Atiyah-Singer index and the chiral anomaly*, Nucl. Phys. **B235** (1984) 395.

[68] R.P. Feynman and A.R. Hibbs, *Quantum Mechanics and Path Integrals*, McGraw-Hill, 1965.

[69] F.A. Berezin, *Feynman path integrals in a phase space*, Sov. Phys. Usp. **23** (1981) 763.

[70] S.I. Blinnikov and N.V. Nikitin, *Approximations to path integrals and spectra of quantum systems*, arXiv:physics/0309060.

[71] H. Cartan, *La transgression dans un groupe de Lie et dans un fibré principal*, Centre belge de recherches mathématiques, Masson et Cie, Paris, 1951, p. 57.

[72] E. Witten, *Constraints on supersymmetry breaking*, Nucl. Phys. **B202** (1982) 253.

[73] S. Cecotti and L. Girardello, *Functional measure, topology and dynamical supersymmetry breaking*, Phys. Lett. **B110** (1982) 39;
L. Girardello, C. Imbimbo and S. Mukhi, *On constant configurations and the evaluation of the Witten index*, Phys. Lett. **B132** (1983) 69.

[74] P. Yi, *Witten index and threshold bound states of D-branes*, Nucl. Phys. **B505** (1997) 307, arXiv:hep-th/9704098;
S. Sethi and M. Stern, *D-brane bound states redux*, Commun. Math. Phys. **194** (1998) 675, arXiv:hep-th/9705046;
M. Porrati and A. Rozenberg, *Bound states at threshold in supersymmetric quantum mechanics*, Nucl. Phys. **B515** (1998) 184, arXiv:hep-th/9708119;
V.G. Kac and A.V. Smilga, *Normalized vacuum states in $N = 4$ supersymmetric Yang–Mills quantum mechanics with any gauge group*, Nucl. Phys. **B571** (2000) 515, arXiv:hep-th/9908096.

[75] A.V. Smilga, *Witten Index*, World Scientific, 2025.

[76] A. Salam and J. Strathdee, *Supergauge transformations*, Nucl. Phys. **B76** (1974) 477.

[77] L. Brink, S. Deser, B. Zumino, P. Di Vecchia and P.S. Howe, *Local supersymmetry for spinning particles*, Phys. Lett. **B64** (1976) 435.

[78] A. Pashnev and F. Toppan, *On the classification of N extended supersymmetric quantum systems*, J. Math. Phys. **42** (2001) 5257, arXiv:hep-th/0010135.

[79] J. Wess and J. Bagger, *Supersymmetry and Supergravity*, Princeton University Press, 1992.

[80] J. Wess and B. Zumino, *Supergauge transformations in four dimensions*, Nucl. Phys. **B70** (1974) 39.

[81] S.J. Gates, Jr., C.M. Hull and M. Roček, *Twisted multiplets and new supersymmetric nonlinear sigma models*, Nucl. Phys. **B248** (1984) 157.

[82] S.A. Fedoruk and A.V. Smilga, *Bi-HKT and bi-Kähler supersymmetric sigma models*, J. Math. Phys. **57** (2016) 042103, arXiv:1512.07923 [hep-th].

[83] A.V. Smilga, *Low-dimensional sisters of Seiberg-Witten effective theory*, in M. Shifman, A. Vainshtein and J. Wheater (Eds.), *From Fields to Strings: Circumnavigating Theoretical Physics*, World Scientific, 2005, vol. 1, 523, arXiv:hep-th/0403294.

[84] A.V. Smilga, *Perturbative corrections to effective zero-mode Hamiltonian in supersymmetric QED*, Nucl. Phys. **B291** (1987) 241.

[85] E.A. Ivanov and A.V. Smilga, *Supersymmetric gauge quantum mechanics: superfield description*, Phys. Lett. **B257** (1991) 79.

[86] V.P. Berezovoj and A.I. Pashnev, *Three-dimensional $\mathcal{N} = 4$ extended supersymmetrical quantum mechanics*, Class. Quant. Grav. **8** (1991) 2141.

[87] S.J. Gates, M.T. Grisaru, M. Roček and W. Siegel, *Superspace or One Thousand and One Lessons in Supersymmetry*, Front. Phys. **58** (1983) 1, arXiv:hep-th/0108200.

[88] V. Nabokov, *Transparent Things*, McGraw-Hill, 1972.
[89] R.A. Coles and G. Papadopoulos, *The geometry of the one-dimensional supersymmetric non-linear sigma models*, Class. Quant. Grav. **7** (1990) 427.
[90] A.S. Galperin, E.A. Ivanov, V.I. Ogievetsky and E.S. Sokatchev, *Harmonic Superspace*, Cambridge Univ. Press, 2001.
[91] E.A. Ivanov and O. Lechtenfeld, $\mathcal{N}=4$ *supersymmetric mechanics in harmonic superspace*, JHEP **09** (2003) 073, arXiv:hep-th/0307111.
[92] F. Delduc and E. Ivanov, $\mathcal{N}=4$ *mechanics of general* $(\mathbf{4},\ \mathbf{4},\ \mathbf{0})$ *multiplets*, Nucl. Phys. **B855** (2012) 815, arXiv:1104.1429 [hep-th].
[93] D. Z. Freedman and P.K. Townsend, *Antisymmetric tensor gauge theories and nonlinear sigma models*, Nucl. Phys. **B177** (1981) 282.
[94] C.M. Hull, G. Papadopoulos and P.K. Townsend, *Potentials for* $(p,0)$ *and* $(1,1)$ *supersymmetric sigma models with torsion*, Phys. Lett. **B316** (1993) 291, arXiv:hep-th/9307013.
[95] A.C. Davis, A.J. Macfarlane, P. Popat and J.W. van Holten, *The quantum mechanics of the supersymmetric nonlinear sigma model*, J. Phys. **A17** (1984) 2945.
[96] M. Claudson and M.B. Halpern, *Supersymmetric ground state wave functions*, Nucl. Phys. **B250** (1985) 689.
[97] S.A. Fedoruk, E.A. Ivanov and A.V. Smilga, *Real and complex supersymmetric* $d=1$ *sigma models with torsion*, Int. J. Mod. Phys. **A27** (2012) 1250146, arXiv:1204.4105 [hep-th].
[98] E.A. Ivanov and A.V. Smilga, *Quasicomplex* $\mathcal{N}=2, d=1$ *supersymmetric sigma models*, SIGMA **9** (2013) 069, arXiv:1302.2902 [hep-th].
[99] S.S. Chern, *On the curvatura integra in a Riemannian manifold*, Ann. Math. **46** (1945) 674; *Complex Manifolds without Potential Theory*, Springer-Verlag, NY, 1979.
[100] H. Braden, *Supersymmetry with torsion*, Phys. Lett. **B163** (1985) 171.
[101] T. Kimura, *Index theorems of torsional geometries*, JHEP **08** (2007) 048, arXiv:0704.2111 [hep-th].
[102] L.I. Nicolaescu, *Notes on Seiberg-Witten Theory*, AMS, Providence, 2000.
[103] D.B. Ray and I.M. Singer, *Analytic torsion for complex manifolds*, Ann. Math. **98** (1973) 154;
J.-M. Bismut, H. Gillet and C. Soulé, *Analytic torsion and holomorphic determinant bundles*, Commun. Math. Phys. **115** (1988) 49;
V. Mathai, S. Wu, *Analytic torsion of* $Z_2$*-graded elliptic complexes*, Contemp. Math. **546** (2011) 199, arXiv:1001.3212[math.DG].
[104] S.A. Fedoruk, E.A. Ivanov and A.V. Smilga, *Generic HKT geometries in the harmonic superspace approach*, J. Math. Phys. **59**, 083501 (2018), arXiv: 1802.09675 [hep-th].
[105] H. Braden, $\sigma$ *models with torsion*, Ann. Phys. (N.Y.) **171** (1986) 433.
[106] A.V. Smilga, *Supercharges in the hyperkähler with torsion supersymmetric sigma models*, J. Math. Phys. **53**, 122105 (2012), arXiv:1209.0539 [math-ph].
[107] A.V. Smilga, *Dolbeault complex on* $S^4$ *and* $S^6$ *through supersymmetric glasses*, SIGMA **7** (2011) 105, arXiv:1005.3935[math-ph].

[108] L. Alvarez-Gaumé and D.Z. Freedman, *Geometrical structure and ultraviolet finiteness of the supersymmetric $\sigma$-model*, Commun. Math. Phys. **80** (1981) 443.

[109] S.A. Fedoruk and A.V. Smilga, *Comments on HKT supersymmetric sigma models and their Hamiltonian reduction*, J. Phys. **A48** (2015) 215401, arXiv:1408.1538 [hep-th].

[110] A.V. Smilga, *Taming the zoo of supersymmetric quantum mechanical models*, JHEP **05** (2013) 119, arXiv:1301.7438.

[111] B. Zumino, *Supersymmetry and Kähler manifolds*, Phys. Lett. **B87** (1979) 203.

[112] E.A. Ivanov and A.V. Smilga, *Symplectic sigma models in superspace*, Nucl. Phys. **B694** (2004) 473, arXiv:hep-th/0402041.

[113] A.V. Smilga, *Born-Oppenheimer corrections to the effective zero mode Hamiltonian in SYM theory*, JHEP **04** (2002) 054, arXiv:hep-th/0201048.

[114] A. Maloney, M. Spradlin, and A. Strominger, *Superconformal multiblack hole moduli spaces in four dimensions*, JHEP **04** (2002) 003, arXiv:hep-th/9911001.

[115] M. de Crombrugghe and V. Rittenberg, *Supersymmetric quantum mechanics*, Ann. Phys. **151** (1983) 99.

[116] A.V. Smilga, *Structure of vacuum in chiral supersymmetric quantum electrodynamics*, Sov. Phys. JETP, **64** (1986) 8.

[117] B.Yu. Blok and A.V. Smilga, *Effective zero-mode Hamiltonian in supersymmetric chiral non-Abelian gauge theories*, Nucl. Phys. **B287** (1987) 589;
A.V. Smilga, *Cartan monopoles*, Phys. Rev. **D110** (2024) 045026, arXiv: 2406.06042 [hep-th].

[118] See e.g. [H.B. Lawson and M.L. Michelson, *Spin Geometry*, Princeton University Press, Princeton, 1989].

[119] D.-E. Diaconescu and N. Seiberg, *The Coulomb branch of (4,4) supersymmetric field theories in two dimensions*, JHEP **07** (1997) 001, arXiv:hep-th/9707158;
A.V. Smilga, *Effective Lagrangians for* $(0+1)$ *and* $(1+1)$ *dimensionally reduced versions of* $D = 4$, $N = 2$ *SYM theory*, Nucl. Phys. **B652** (2003) 93, arXiv:hep-th/0209187.

[120] U. Lindström and M. Roček, *Scalar tensor duality and* $\mathcal{N} = 1, 2$ *non-linear* $\sigma$ *models*, Nucl. Phys. **B222** (1983) 285.

[121] J.F. Plebanski, *Some solutions of complex Einstein equations*, J. Math. Phys. **16** (1975) 2395.

[122] M. Roček, *Private communication.*

[123] G.W. Gibbons and S.W. Hawking, *Gravitational multi-instantons*, Phys. Lett. **B78**, 430 (1978).

[124] G. Gaeta and M.A. Rodriguez, *Hyperkähler structure of the Taub-NUT metric*, J. Nonlin. Math. Phys. **19** (2012) 1250014.

[125] T. Bridgeland and I.A.B. Strachan, *Complex hyperkähler structures defined by Donaldson-Thomas invariants*, Lett. Math. Phys. **111** (2021) 54, arXiv:2006.13059 [math.AG].

[126] M. Dunajski and T. Moy, *Heavenly metrics, hyper-Lagrangians and Joyce structures*, J. Lond. Math. Soc. **110** (2024) 5 arXiv:2402.14352 [math.DG].

[127] A.V. Smilga, *A few comments on (hyper)kähler geometry*, arXiv:2511.11786 [math.DG].

[128] E. Ivanov, O. Lechtenfeld and A. Sutulin, *Hierarchy of $\mathcal{N} = 8$ mechanics models*, Nucl. Phys. **B790** (2008) 493, arXiv:0705.3064 [hep-th].

[129] S.A. Fedoruk, E.A. Ivanov and A.V. Smilga, *$\mathcal{N} = 4$ mechanics with diverse $(\mathbf{4}, \mathbf{4}, \mathbf{0})$ multiplets: Explicit examples of HKT, CKT and OKT geometries*, J. Math. Phys. **55**, 052302 (2014), arXiv:1309.7253 [hep-th].

[130] S. Bellucci, E. Ivanov, S. Krivonos and O. Lechtenfeld, *ABC of $\mathcal{N} = 8$, $d = 1$ supermultiplets*, Nucl. Phys. **B699** (2004) 226, arXiv:hep-th/0406015.

[131] D.-E. Diaconescu and R. Entin, *A non-renormalization theorem for the $d = 1, \mathcal{N} = 8$ vector multiplet*, Phys.Rev. **D56** (1997) 8045, arXiv:hep-th/97060059.

[132] P.A.M. Dirac, *The Principles of Quantum Mechanics*, Oxford University Press, 1930.

[133] G.M. Asatrian and G.K. Savvidy, *Configuration manifold of Yang-Mills classical mechanics*, Phys. Lett. **A99** (1983) 290;
Yu.A. Simonov, *QCD Hamiltonian in the polar representation*, Sov. J. Nucl. Phys. **41** (1985) 835.

[134] S.J. Gates, Jr. and L. Rana, *Ultramultiplets: A new representation of rigid 2d $\mathcal{N} = 8$ supersymmetry*, Phys. Lett. **B342** (1995) 132, arXiv:hep-th/9410150;
S. Bellucci, S. Krivonos, A. Marrani and E. Orazi, *'Root' action for $\mathcal{N} = 4$ supersymmetric mechanics theories*, Phys. Rev. **D73** (2006) 025011, arXiv:hep-th/0511249;
F. Delduc and E. Ivanov, *Gauging $\mathcal{N} = 4$ supersymmetric mechanics*, Nucl. Phys. **B753** (2006) 211, arXiv:hep-th/0605211.

[135] N. Kozyrev, S. Krivonos, O. Lechtenfeld, A. Nersessian and A. Sutulin, *$\mathcal{N} = 4$ supersymmetric mechanics on curved spaces*, Phys. Rev. **D97** (2018) 085015, arXiv:1711.08734 [hep-th].

[136] J. Marsden and A. Weinstein, *Reduction of symplectic manifolds with symmetry*, Repts. Math. Phys. **5** (1974) 121.

[137] E. Witten, *Branes, instantons and Taub-NUT spaces*, JHEP **06** (2009) 067, arXiv:0902.0948 [hep-th].

[138] A.S. Galperin, E.A. Ivanov, V.I. Ogievetsky and E.S. Sokatchev, *Hyper-Kähler metrics and harmonic superspace*, Commun. Math. Phys. **103** (1986) 515.

[139] A.S. Galperin, E.A. Ivanov, V.I. Ogievetsky and P.K. Townsend, *Eguchi-Hanson type metrics from harmonic superspace*, Class. Quant. Grav. **3**, (1986) 625.

[140] A.V. Smilga, *An 8-dimensional Taub-NUT-like hyperkähler metric in harmonic superspace formalism*, J. Math. Phys., **61** (2020) 112301, arXiv:2006.11782 [hep-th].

[141] A.S. Galperin, E.A. Ivanov, V.I. Ogievetsky and E.S. Sokatchev, *Gauge field geometry from complex and harmonic analyticities. II Hyper-Kähler case.*, Ann. Phys. **185** (1988) 22.

[142] C.G. Callan, J.A. Harvey, A. Strominger, *World sheet approach to heterotic instantons and solitons*, Nucl. Phys. **B359** (1991) 611.

[143] G. Bonneau, G. Valent, *Local heterotic geometry in holomorphic coordinates*, Class. Quant. Grav. **11** (1994) 1133, arXiv:hep-th/9401003.

[144] G. Papadopolous, *Elliptic monopoles and (4,0) supersymmetric sigma models with torsion*, Phys. Lett. **B356** (1995) 249, arXiv:hep-th/9505119.

[145] M.A. Konyushikhin and A.V. Smilga, *Self-duality and supersymmetry*, Phys. Lett. **B689** (2010) 95, arXiv:0910.5162 [hep-th].

[146] T.T. Wu and C.N. Yang, *Dirac monopole without strings: monopole harmonics*, Nucl. Phys. **B107** (1976) 365.

[147] I. Tamm, *Die veralldeneinerten Kugelfunktionen und die Wellenfunktionen eines Elektrons im Felde eines Magnetpoles*, Z. Phys. **71** (1931) 141.

[148] P.P. Banderet, *Zur Theorie singulärer Magnetpole*, Helv. Phys. Acta **19** (1946) 503.

[149] S. Kim and C. Lee, *Supersymmetry-based approach to quantum particle dynamics on a curved surface with nonzero magnetic field*, Ann. Phys. **296** (2002) 390, arXiv:hep-th/0112120.

[150] A.V. Smilga, *Non-integer flux: why it does not work*, J. Math. Phys. **53** (2012) 042103, arXiv:1104.3986 [math-ph].

[151] M.A. Shifman, A.V. Smilga and A.I. Vainshtein, *On the Hilbert space of supersymmetric quantum systems*, Nucl. Phys. **B299** (1988) 79.

[152] E. Ivanov, L. Mezincescu and P.K. Townsend, *Fuzzy* $CP^* * (n|m)$ *as a quantum superspace*, in Salamanca 2003, Symmetries in gravity and field theory, p. 385, arXiv:hep-th/0311159.

[153] A. Kirschberg, J.D. Länge and A. Wipf, *Extended supersymmetries and the Dirac operator*, Ann. Phys. **315** (2005) 467, arXiv:hep-th/0401134.

[154] E. Witten, *Supersymmetric index of three-dimensional gauge theory* in: [Shifman, M.A. (Ed.) *The Many Faces of the Superworld*, World Scientific, 2000], p. 156 arXiv:hep-th/9903005.

[155] A.V. Smilga, *Vacuum structure in 3D supersymmetric gauge theories*, Phys. Uspekhi **57** (2014) 155, arXiv:1312.1804 [hep-th].

[156] R. Kuwabara, *Spectrum of the Schrödinger operator on a line bundle over complex projective spaces*, Tohoku Math. J. **40** (1988) 199.

[157] D. Bykov and A.V. Smilga, *Monopole harmonics on* $\mathbb{CP}^{n-1}$, SciPost Phys. **15** (2023) 195, arXiv:2302.11691 [hep-th].

[158] B.P. Dolan, I. Huet, S. Murray and D. O'Connor, *A universal Dirac operator and noncommutative spin bundles over fuzzy complex projective spaces*, JHEP **03** (2008) 029, arXiv:0711.1347 [hep-th].

[159] R. Slansky, *Group theory for unified model building*, Phys. Repts. **79** (1981) 1.

[160] G. 't Hooft, *Some twisted self-dual solutions for the Yang-Mills equations on a hypertorus*, Commun. Math. Phys. **81** (1981) 267.

[161] A.V. Smilga, *Instantons and fermion condensate in adjoint two-dimensional QCD*, Phys. Rev. **D49** (1994) 6836, arXiv:hep-th/9402066; D.J. Gross, I.R. Klebanov, A.V. Matytsin and A.V. Smilga, *Screening versus confinement in 1+1 dimensions*, Nucl. Phys. **B461** (1996) 109, arXiv:hep-th/9511104.

[162] A.P. Balachandran, S. Borchardt and A. Stern, *Lagrangian and Hamiltonian descriptions of Yang-Mills particles*, Phys. Rev. **D17** (1978) 3247; A.P. Balachandran, G. Marmo, B.-S. Skagerstam and A. Stern, *Gauge theories and fiber bundles*, Lecture Notes in Physics **188** 1, Springer, 1983, arXiv:1702.08910 [quant-ph].

[163] S. Fedoruk, E. Ivanov and O. Lechtenfeld, *Supersymmetric Calogero model by gauging*, Phys. Rev. **D79** (2009) 105015 arXiv:0812.4276 [hep-th]; S. Bellucci, S. Krivonos and A. Sutulin, *Three-dimensional $\mathcal{N} = 4$ supersymmetric mechanics with Wu-Yang monopole*, Phys. Rev. **D81** (2010) 105026, arXiv:0911.3257 [hep-th].

[164] E.A. Ivanov, M.A. Konyushikhin and A.V. Smilga, *SQM with non-Abelian self-dual fields: harmonic superspace description*, JHEP **05** (2010) 033, arXiv:0912.3289 [hep-th].

[165] C.N. Yang, *Generalization of Dirac's monopole to $SU(2)$ gauge fields*, J. Math. Phys. **19** (1978) 320.

[166] A.A. Belavin, A.M. Polyakov, A.S. Schwartz and Yu.S. Tyupkin, *Pseudoparticle solutions of the Yang-Mills equations*, Phys. Lett. **B59** (1975) 85.

[167] M.F. Atiyah, N.J. Hitchin, V.G. Drinfeld and Yu.I. Manin, *Construction of instantons*, Phys.Lett. **A65** (1978) 185.

[168] E. Corrigan, D.B. Fairlie, S. Templeton and P. Goddard, *A Green's function for the general selfdual gauge field*, Nucl.Phys. **B140** (1978) 31.

[169] R. Jackiw and C. Rebbi, *Conformal properties of a Yang-Mills pseudoparticle*, Phys. Rev. **D14** (1976) 517.

[170] I.M. Gelfand, *On elliptic equations*, Russ. Math. Surv. **15** (1960) 113.

[171] M.F. Atiyah and I.M. Singer, *The index of elliptic operators on compact manifolds*, Bull. Amer. Math. Soc. **69** (1963) 422.

[172] M.F. Atiyah, V.K. Patodi and I.M. Singer, *Spectral asymmetry and Riemannian geometry*, Math. Proc. Camb. Phil. Soc. **77** (1975) 43; **78** (1976) 405; **79** (1976) 71.

[173] P. Windey, *Supersymmetric quantum mechanics and the Atiyah-Singer index theorem*, Acta Phys.Polon. **B15** (1984) 435;
A. Mostafazadeh, *Supersymmetry and the Atiyah-Singer index theorem. 1: Peierls brackets, Green's functions, and a supersymmetric proof of the index theorem*, J. Math. Phys. **35** (1994) 1095, arXiv:hep-th/9309059.

[174] A.V. Smilga, *Supersymmetric proof of the Hirzebruch-Riemann-Roch theorem for non-Kähler manifolds*, SIGMA **8** (2012) 003, arXiv:1109.2867 [math-ph].

# Index

www.ingramcontent.com/pod-product-compliance
Lightning Source LLC
Chambersburg PA
CBHW071136040426
41971CB00022B/25
* 9 7 8 9 8 1 9 8 1 7 6 6 5 *